Delivery and controlled release of bioactives in foods and nutraceuticals

Related titles:

Understanding and controlling the microstructure of complex foods
(ISBN 978-1-84569-151-6)
Awareness of the interrelationship between the nanoscopic, microscopic and macroscopic features of foods and their bulk physicochemical properties, sensory attributes and healthfulness is increasing. It is now widely recognised that creation of novel foods or improvement of existing products largely depends on a better understanding of this complex interrelationship. This important collection presents current understanding of significant aspects of food structure and methods for its control. The main structural elements in foods, their interactions and newer analytical techniques for structure assessment are reviewed. Subsequent chapters then discuss the use of structural design principles in product development and structural issues in particular foods, including ice cream and chocolate.

Phytochemical functional foods
(ISBN 978-1-85573-672-6)
Phytochemicals are non-nutritive components that provide plants with colour, flavour and toxicity to pests. There is now a growing body of research which also suggests they may also help to reduce the risk of chronic diseases such as cancer, osteoporosis and heart disease. Edited by two leading experts, this collection provides an authoritative review of the range of phytochemicals. The first part of the book considers individual groups of phytochemicals such as phenolic compounds and their health benefits. Later parts of the book discuss how functional benefits are tested, and ways of producing phytochemical functional products.

Starch in food: Structure, function and applications
(ISBN 978-1-85573-731-0)
Starch is both a major component of plant foods and an important ingredient for the food industry. *Starch in food* reviews starch structure and functionality and the growing range of starch ingredients used to improve the nutritional and sensory quality of food. Part I illustrates how plant starch can be analysed and modified, with chapters on plant starch synthesis, starch bioengineering and starch-acting enzymes. Part II examines the sources of starch, from wheat and potato to rice, corn and tropical sources. The third part of the book looks at starch as an ingredient and how it is used in the food industry. There are chapters on modified starches and the stability of frozen foods, starch–lipid interactions and starch-based microencapsulation. Part IV covers starch as a functional food, including the impact of starch on physical and mental performance, detecting nutritional starch fractions and analysing starch digestion.

Details of these books and a complete list of Woodhead titles can be obtained by:

- visiting our web site at www.woodheadpublishing.com
- contacting Customer Services (email: sales@woodhead-publishing.com; fax: +44 (0) 1223 893694; tel.: +44 (0) 1223 891358 ext. 130; address: Woodhead Publishing Limited, Abington Hall, Abington, Cambridge CB21 6AH, England)

Delivery and controlled release of bioactives in foods and nutraceuticals

Edited by
Nissim Garti

CRC Press
Boca Raton Boston New York Washington, DC

WOODHEAD PUBLISHING LIMITED
Cambridge England

Published by Woodhead Publishing Limited, Abington Hall, Abington,
Cambridge CB21 6AH, England
www.woodheadpublishing.com

Published in North America by CRC Press LLC, 6000 Broken Sound Parkway, NW,
Suite 300, Boca Raton, FL 33487, USA

First published 2008, Woodhead Publishing Limited and CRC Press LLC

British Library Cataloguing in Publication Data
A catalogue record for this book is available from the British Library.

Library of Congress Cataloging in Publication Data
A catalog record for this book is available from the Library of Congress.

Woodhead Publishing Limited ISBN 978-1-84569-145-5 (book)
Woodhead Publishing Limited ISBN 978-1-84569-421-0 (e-book)
CRC Press ISBN 978-1-4200-7436-9
CRC Press order number WP7436

The publishers' policy is to use permanent paper from mills that operate a sustainable forestry policy, and which has been manufactured from pulp which is processed using acid-free and elementary chlorine-free practices. Furthermore, the publishers ensure that the text paper and cover board used have met acceptable environmental accreditation standards.

Project managed by Macfarlane Production Services, Dunstable, Bedfordshire, England
(e-mail: macfarl@aol.com)
Typeset by Godiva Publishing Services Ltd, Coventry, West Midlands, England
Printed by TJ International Limited, Padstow, Cornwall, England

Contents

Contributor contact details xi

Preface xv

Part I The effectiveness of controlled release and delivery systems

1 Assessing the bioavailability of nutraceuticals 3
R. M. Faulks and S. Southon, Institute of Food Research, UK
1.1 Introduction 3
1.2 Measuring absorption, metabolism and tissue targeting 5
1.3 Study design and interpretation 13
1.4 Other considerations 19
1.5 Health response 21
1.6 Implications of controlled absorption for product development 22
1.7 Future trends 24
1.8 Sources of further information and advice 24
1.9 References 25

2 Structure of the gastrointestinal mucus layer and implications for controlled release and delivery of functional food ingredients . 26
G. Lafitte, Centre for Chemistry and Chemical Engineering, Sweden
2.1 Introduction 26
2.2 Understanding the structure of the gastrointestinal mucus layer 27
2.3 Implications of the mucin molecules and the mucus layer in controlled release and delivery of functional food ingredients 36

2.4 Conclusion 47
2.5 Acknowledgements 47
2.6 References 47

3 Testing the effectiveness of nutrient delivery systems 53
E. Acosta, University of Toronto, Canada
3.1 Introduction: a holistic view of delivery systems 53
3.2 Delivery systems: formulation, processing and post-processing 54
3.3 *In vivo* tests 70
3.4 *In vitro* tests 77
3.5 Future trends 95
3.6 Sources of further information and advice 96
3.7 References 97

4 Lyotropic liquid crystals as delivery vehicles for food ingredients 107
J. Barauskas, Institute of Biochemistry, Lithuania and T. Nylander, Lund University, Sweden
4.1 Introduction 107
4.2 Lyotropic lipid liquid crystals 109
4.3 Lamellar phase colloids 111
4.4 Nonlamellar liquid crystalline lipid-based nanoparticles 112
4.5 Future trends 125
4.6 Acknowledgements 126
4.7 References 126

Part II Materials and techniques for controlled release and delivery of nutrients

5 Structured lipids as delivery systems 135
E. M. Hernandez, OmegaPure Technology and Innovation Center, USA
5.1 Introduction 135
5.2 Synthesis of structured lipids 138
5.3 Nutrition, absorption, transport and metabolism of structured lipids 140
5.4 Foods applications of structured lipids 141
5.5 Structured lipids in drug delivery 143
5.6 References 146

6 Micro- and nano-emulsions for delivery of functional food ingredients 149
N. Garti and I. Yuli-Amar, The Hebrew University of Jerusalem, Israel
6.1 Introduction 149
6.2 Microemulsions 152

6.3 Loaded microemulsions 162
6.4 Bioavailability of nutraceuticals 172
6.5 Microemulsions as microreactors 174
6.6 Conclusions 179
6.7 References 179

7 Emulsion droplet interfacial engineering to deliver bioactive lipids into functional foods 184
L. A. Shaw, H. Faraji, T. Aoki, D. Djordjevic, D. McClements and E. A. Decker, University of Massachusetts, USA
7.1 Introduction 184
7.2 Dietary ω-3 fatty acids 185
7.3 Dietary ω-3 fatty acids and health 186
7.4 Potential for ω-3 fatty acids in functional foods 188
7.5 Mechanisms of lipid oxidation 190
7.6 Lipid oxidation in emulsions 193
7.7 Interfacial engineering 194
7.8 Electrostatic layering in oil-in-water emulsions 196
7.9 Microencapsulation of oil-in-water emulsion systems 198
7.10 Emulsion droplet engineering to stabilize ω-3 fatty acids 200
7.11 Conclusions 202
7.12 References 203

8 Lipid self-assembled particles for the delivery of nutraceuticals 207
O. Ramon and D. Danino, Technion – Israel Institute of Technology, Israel
8.1 Introduction 207
8.2 Structure and properties of lipids 208
8.3 Self-assembly and microstructure of lipid systems 209
8.4 Liposomes 211
8.5 Cubic and hexagonal mesophases and their dispersed nanoparticles 217
8.6 Future trends 224
8.7 References 225

9 Complexes and conjugates of biopolymers for delivery of bioactive ingredients via food 234
Y. D. Livney, Technion – Israel Institute of Technology, Israel
9.1 Introduction 234
9.2 Complexes and coacervates 236
9.3 Conjugates – covalently bonded biopolymers 243
9.4 Complexes of a protein with a polysaccharide-protein conjugate 245
9.5 Future trends 246

9.6 Sources of further information and advice 246
9.7 References 246

10 Food-protein-derived materials and their use as carriers and delivery systems for active food components 251
M. Subirade and L. Chen, Laval University, Canada
10.1 Introduction 251
10.2 Food protein hydrogels and related networks as nutraceutical carriers 252
10.3 Food protein micro- and nanoparticles 258
10.4 Future trends 270
10.5 Sources of further information and advice 271
10.6 Acknowledgements 271
10.7 References and further reading 271

11 Starch as an encapsulation material to control digestion rate in the delivery of active food components 279
E. Shimoni, Technion – Israel Institute of Technology, Israel
11.1 Introduction 279
11.2 Uses of starch to control delivery in microencapsulation 280
11.3 Starch as nanoencapsulation material 285
11.4 Conclusion 288
11.5 References 289

Part III Delivery and controlled release of particular nutraceuticals

12 Encapsulation and controlled release of antioxidants and vitamins 297
C. M. Sabliov and C. E. Astete, Louisiana State University Agricultural Center, USA
12.1 Introduction 297
12.2 Antioxidants and vitamins in protecting human health 298
12.3 Advantages of encapsulation over traditional delivery methods 304
12.4 Top-down techniques used for encapsulation of antioxidants and vitamins in polymeric nanoparticles 309
12.5 Characterization methods 313
12.6 Controlled release of antioxidants and vitamins 313
12.7 Conclusions 317
12.8 Future trends 318
12.9 References 319

13 Encapsulation and controlled release of folic acid 331
K. Kailasapathy, University of Western Sydney, Australia
13.1 Introduction 331
13.2 Encapsulation technologies for folates 334
13.3 Controlled release of folate in the gastrointestinal system ... 337
13.4 Applications in functional foods 338
13.5 Future trends 340
13.6 References 341

14 Encapsulation of probiotics 344
C. P. Champagne, Agriculture and Agri-Food Canada, Canada and K. Kailasapathy, University of Western Sydney, Australia
14.1 Introduction 344
14.2 Encapsulation technologies for probiotics 346
14.3 Applications in nutraceuticals 354
14.4 Applications in functional foods 358
14.5 Future trends 363
14.6 Sources of further information and advice 364
14.7 References 364

15 Encapsulation of fish oils 370
S.-J. Lee, Massey University, New Zealand and D. Y. Ying, HortResearch, New Zealand
15.1 Introduction 370
15.2 Encapsulation technologies used to deliver fish oils 377
15.3 Using encapsulated fish oils in food products 389
15.4 Suppliers of microencapsulated fish oil powder 393
15.5 Future trends 394
15.6 Sources of further information and advice 395
15.7 References 395

16 Encapsulation approaches for proteins 404
A. Millqvist-Fureby, YKI Institute for Surface Chemistry, Sweden
16.1 Introduction 404
16.2 Solid formulations for proteins 405
16.3 Surface formation in drying 408
16.4 Coating in spray drying 411
16.5 Polyelectrolyte microcapsules for protein encapsulation 416
16.6 Summary and future trends 421
16.7 References 422

Part IV Regulatory issues and future trends

17 Regulatory aspects of nutrient delivery systems 429
E. Acosta, University of Toronto, Canada
17.1 Introduction 429
17.2 Selection of ingredients 430
17.3 Stability of the product 439
17.4 Bioavailability of the active ingredient 443
17.5 Future trends 447
17.6 Sources of further information and advice 448
17.7 References 448

18 The future of controlled release and delivery technologies 450
N. Garti, The Hebrew University of Jerusalem, Israel
18.1 Introduction 450
18.2 The food additives 454
18.3 Bioavailability and performance – the need for new vehicles . 456
18.4 Bioavailability 458
18.5 Nanoscience, nanovehicles, and controlled release 459
18.6 Specific targets 459
18.7 The future of controlled release systems 460
18.8 Encapsulation with organogels and lyotropic liquid crystals . 462
18.9 Reactivity at interfaces 463
18.10 Conclusions 463
18.11 References 463

Index 465

Contributor contact details

(* = main contact)

Editor

Professor Nissim Garti
Casali Institute of Applied Chemistry
Givat Ram Campus
The Hebrew University of Jerusalem
Jerusalem 91904
Israel

E-mail: garti@vms.huji.ac.il

Chapter 1

Richard M. Faulks* and Susan Southon
Institute of Food Research
Norwich Research Park
Colney
Norwich NR4 7UA
UK

E-mail: richard.faulks@bbsrc.ac.uk

Chapter 2

Geraldine Lafitte
Centre for Chemistry and Chemical Engineering
Physical Chemistry I
Lund University
PO Box 124
SE-221 00 Lund
Sweden

E-mail: geraldine.lafitte@bimgi.com

Chapters 3 and 17

Edgar Acosta
Department of Chemical Engineering and Applied Chemistry
University of Toronto
Toronto, ON
M5S 3E5
Canada

E-mail: acosta@chem-eng.utoronto.ca

Chapter 4

Justas Barauskas
Department of Bioanalysis
Institute of Biochemistry
Mokslininku 12
LT-08662 Vilnius
Lithuania

E-mail: Justas.Barauskas@bchi.lt

Tommy Nylander*
Centre for Chemistry and Chemical Engineering
Physical Chemistry 1
Lund University
PO Box 124
SE-221 00 Lund
Sweden

E-mail: Tommy.Nylander@fkem1.lu.se

Chapter 5

Ernesto M. Hernandez
OmegaPure Technology and Innovation Center
6961 Brookhollow W Drive, Ste. 190
Houston, TX 77040
USA

E-mail: ehernandez@OmegaProteinInc.com

Chapter 6

Nissim Garti* and Idit Yuli-Amar
Casali Institute of Applied Chemistry
Givat Ram Campus
The Hebrew University of Jerusalem
Jerusalem 91904
Israel

E-mail: garti@vms.huji.ac.il

Chapter 7

Lauren A. Shaw, H. Faraji, T. Aoki, D. Djordjevic, D. J. McClements and Eric A. Decker*
Department of Food Science
Chenoweth Laboratory
University of Massachusetts
Amherst, MA 01003
USA

E-mail: edecker@foodsci.umass.edu

Chapter 8

Dr Dganit Danino and Ory Ramon
Department of Biotechnology and Food Engineering
Technion – Israel Institute of Technology
Haifa 32000
Israel

E-mail: dganitd@tx.technion.ac.il; pramon@tx.technion.ac.il

Chapter 9

Yoav D. Livney
Department of Biotechnology and Food Engineering
Technion – Israel Institute of Technology
Haifa 32000
Israel

E-mail: livney@technion.ac.il

Chapter 10

Muriel Subirade* and Lingyun Chen
Université Laval/Laval University
Chaire de recherche du Canada sur les protéines, les bio-systèmes et les aliments fonctionnels/Canadian Research Chair in Proteins, Bio-system and Functional Foods
Institut de recherche sur les nutraceutiques et les aliments fonctionnels (INAF/Stela)/ Nutraceuticals and Functional Foods Institute
Pavillon Paul Comtois, local 1415
QC
G1K 7P4
Canada

E-mail: muriel.subirade@aln.ulaval.ca

Chapter 11

Eyal Shimoni
Faculty of Biotechnology and Food Engineering
Technion – Israel Institute of Technology
Haifa 32000
Israel

E-mail: eshimoni@techunix.technion.ac.il

Chapter 12

Cristina M. Sabliov* and C.E. Astete
Biological and Agricultural Engineering Department
Louisiana State University Agricultural Center
USA

E-mail: csabliov@bae.lsu.edu

Chapter 13

Kasipathy Kailasapathy
Probiotics and Encapsulated Functional Foods Research Unit
Centre for Plant and Food Science
School of Natural Sciences
University of Western Sydney
Hawkesbury Campus
Locked Bag 1797
Penrith South DC
NSW 1797
Australia

E-mail: k.kailasapathy@uws.edu.au

Chapter 14

Claude P. Champagne*
Bioproducts and Bioprocesses
Agriculture and Agri-Food Canada
Food R & D Centre
3600 Casavant
St Hyacinthe, QC
J2S 8E3
Canada

E-mail: ChampagneC@agr.gc.ca

Kasipathy Kailasapathy
Probiotics and Encapsulated Functional Foods Unit
Centre for Plant and Food Science
School of Natural Sciences
University of Western Sydney
Hawkesbury Campus
Locked Bag 1797
Penrith South DC
NSW 1797
Australia

E-mail: k.kailasapathy@uws.edu.au

Chapter 15

S.J. Lee
Institute of Food, Nutrition and Human Health
Massey University at Albany campus
Private Bag 102 904
North Shore Mail Centre
Auckland
New Zealand

E-mail: S.J.Lee@massey.ac.nz

D.Y. Ying
HortResearch
Private Bag 3123
Hamilton
New Zealand

E-mail: danyang.ying@hortresearch.co.nz

Chapter 16

Anna Millqvist-Fureby
Pharmaceuticals and Food Section
YKI Institute for Surface Chemistry
Box 5607
SE-114 86 Stockholm
Sweden

E-mail: anna.fureby@surfchem.kth.se

Chapter 18

Nissim Garti
Casali Institute of Applied Chemistry
Givat Ram Campus
The Hebrew University of Jerusalem
Jerusalem 91904
Israel

E-mail: garti@vms.huji.ac.il

Preface

Food is an essential part of life. Without food humans and/or animals could not survive. However, this basic need is always only one face of the term 'food'. In the modern world we live in, food is nutrition and delight, or, as was already stated by others, food is CHIEF, which stands for Convenience, Health, Indulgence, Excitement and Freshness. We scientists, engineers and technologists have to meet the challenges to provide the basis for the production of delightful products and to deliver all the CHIEF requirements.

The food that we intake must meet all the above specifications and criteria and, therefore, we have to meet the performance challenges in any of the stages of food production. These include:

- selection of raw materials and understanding their properties and their limitations;
- ingredient preparation – learning to manipulate and work out methods as well as find the tools with which to integrate the ingredients into a specific mixture or composition;
- processing – the engineering techniques, mixing, evaporation, boiling, freezing, etc., must be done to perfection;
- packaging – this is a very critical stage in food preparation, manufacture and storage.

However, in recent years, consumers have been much more concerned with the eating stage (taste, feel, odour, etc.) and with the effect of food on our body. The concern simply for nutrition, calories and building blocks is being replaced by the concern for 'quality of life' and 'nutrition for improved health'. This concern also extends, if possible, to food supplements as pharmaceutical replacers to treat disease. As a result, food producers, health authorities and consumers all focus on technology-driven food products.

The slogan is changing very rapidly from 'nutrition systems for a healthy tomorrow' to 'healthier food for today's health'. This change in the concepts of food leads directly to the so-called 'technology-driven foods' rather than 'concept-driven foods'. This means that scientists are looking not just into 'high stability foods' – pre-prepared food, flavoured food, energy food and sport food – but rather into food with 'encapsulated nutraceuticals', food with 'controlled release of bioactives', or food with supplements that can be transported through the gut–blood barrier for improved bioavailability and bioefficacy.

Based on these new concepts, many of the leading food producers are changing gear and are now claiming to have new technologies, in many cases nano-technologies, for better health-life. Manufacturers are very much interested in new technologies that can help them make such claims, so that they can show their consumers that they care about their health.

Therefore, the old concepts of foods low on sugar, carbohydrates, salt, fat, and cholesterol, are changing to foods with high fibre content and vitamins, enriched with nutraceuticals and, more specifically, food reinforced with delivery technologies for improved efficacy of the food ingredients.

The aim of this book is to contribute to the understanding of the terms: nutraceuticals, food additives, new delivery methods (food goes nano), and their correlation to our health. The book is organized in four parts.

In Part I, we bring some physical aspects of the effectiveness of controlled release and delivery systems. Chapter 1 by Faulks and Southon gives a general scientific assessment of the major controversial issues, and deals with the idea that nutraceuticals must be bioavailable to enhance their activity. Otherwise, the consumer pays for nutraceuticals with very limited effect. Delivery systems are essential to turn food additives into nutraceuticals. In Chapter 2, Lafitte deals with three issues: the gut–blood barrier, the gastrointestinal membrane and the mucus layer that is coating the membrane, and the implications for controlled release and delivery of functional foods.

The following chapters in Part I describe specific delivery technologies, carriers and vehicles. In Chapter 3, Acosta reviews a large number of delivery systems and analyses each of them (scope and limitations), including engineering aspects for treating the final products. In Chapter 4, Barauskas and Nylander discuss the use of lyotropic liquid crystals as delivery vehicles. These new exciting vehicles are gaining the interest of pharmacology and food science.

In Part II in Chapter 5, the view of Hernandez on structured lipids is an interesting way to explore lipid structures and to learn how to make delivery vehicles out of them. In Chapter 6, Garti and Yuli-Amar discuss progress in utilizing new U-type of microemulsions that are based on food-grade emulsifiers as dilutable vehicles for delivery of nutraceuticals. In Chapter 7, Shaw, Faraji, Aoki, Djordjevic, McClements and Decker deal with recent attempts at making emulsion loaded with nutraceuticals. In Chapter 8, Ramon and Danino concentrate on lyotropic liquid crystals with special focus on cryo-TEM techniques.

In Part II, we concentrate mainly on biopolymers and other materials as building blocks for making new delivery vehicles. The chapters also deal with analytical techniques to evaluate efficacy of solubilization of the bioactives. In Chapter 9, Livney discusses new developments in utilizing milk-based proteins and complexes as delivery systems. In Chapter 10, Subirade and Chen explore other food proteins as carriers for nutraceuticals. In Chapter 11, Shimoni uses starch (amylose) as encapsulation material to control digestion rate in the delivery of active foods.

In Part III, we discuss examples of particular compounds that have been encapsulated/entrapped within new carriers and vehicles such as vitamins (Chapter 12 by Sabliov and Asete), folic acid (Chapter 13 by Kailasapathy), probiotics (Chapter 14 by Champagne and Kailasapathy), fish oil (Chapter 15 by Lee and Ying) and proteins (Chapter 16 by Millqvist-Fureby).

In Part IV, we present two chapters that assess the delivery systems from a regulatory point of view (Chapter 17 by Acosta) and a personal view on the future of controlled release and delivery technology (Chapter 18 by Garti).

We hope that this book, which should be of interest to food scientists, food technologists, food engineers, nutritionists and food colloid experts, will broaden the reader's horizons, while emphasizing what has already been achieved and what still needs to be done in order to make our future food more delightful and healthy.

Part I

The effectiveness of controlled release and delivery systems

1

Assessing the bioavailability of nutraceuticals

R. M. Faulks and S. Southon, Institute of Food Research, UK

1.1 Introduction

A targeted delivery system, natural or engineered, for the efficient release of an active compound from a food, nutraceutical or pharmaceutical matrix, can provide a means of optimising both extent and rate of absorption of the compound into the body. Delivery systems may also have considerable benefit by protecting essential nutrients and other 'bioactive' food components during food processing and digestion. For example, it may prevent the loss of the component, or adverse interactions with inhibitors of absorption.

However, optimum absorption should not be confused with maximum absorption, as so often happens in the bioavailability field. The concept of 'if a little does you good, more will be better' has pervaded the nutraceutical market for many years. Even with the essential nutrients, getting 'more' into the body at a 'faster' rate does not necessarily equate to 'improved' function and 'better' health, in some instances quite the reverse is the case (Omenn *et al.*, 1996). Different rates of delivery of the same load may have a profound effect on subsequent health outcome. For example, there are proven health benefits from 'slow release' carbohydrate foods; they do not overstimulate the secretion of insulin, cause unnecessary large excursions in blood glucose, or undesirable glycosylation of proteins. By analogy, the slower delivery of other food components may maximise health benefits by not overloading transport systems, or causing undesirable excursions in plasma and tissue concentration. Furthermore, the fact that some portion of nutrients escape absorption in the small intestine and are 'lost' to the colon should not automatically be interpreted negatively, since they may contribute positively to colon health through the maintenance of an appropriate microflora and the production of beneficial products of large bowel fermentation.

Achieving optimum (amount, rate and site) rather than maximum absorption via controlled or targeted delivery is particularly pertinent to the focus of this book and highlights the importance of understanding the bioavailability concept and measuring its components intelligently. The following sections examine the bioavailability concept and approaches to measuring its component parts.

1.1.1 The bioavailability concept

Each year there are approximately 200+ research publications with 'bioavailability' in the title. It is obvious from these publications that 'bioavailability' means different things to different people. Its most common usage by far is to describe absorption, or 'being in a form' that can be absorbed. For drugs and the correction of nutritional deficiencies in plants and animals, the endpoints are closely monitored and evaluated. Where there is a putative benefit to human health, the endpoints of greatest interest and utility, namely the metabolic consequences and impact on wellbeing, are unfortunately seldom monitored and evaluated adequately. Although changes in plasma concentration, nutrient retention/excretion etc., induced by interventions, may be simply measured, the failure to evaluate holistic health benefit in 'healthy volunteers' seriously undermines any subsequent dietary recommendations.

The term 'bioavailability' has arisen (1) from the nutritional concept that some nutrients in food are only partially 'available', e.g. lysine, niacin and iron, and (2) in the field of pharmacology where it is a useful working concept of the 'rate and extent to which a drug reaches its site of action'.

Clearly bioavailability has a number of components, embracing absorption, distribution, metabolism, excretion (ADME), with subsequent biochemical and physiological effects. It is therefore a concept and as such has no numerical value or units, it cannot be increased or diminished, only changed. Nevertheless, as a concept, it has value in providing a framework for studying links between the administration of a compound, the metabolic and physiological consequences and impact on health.

The pharmacological concept is concerned with site-specific drug delivery in order to treat a disorder, where effective and timely treatment requires the drug to reach a site at a concentration for sufficient duration to elicit the desired effect. Within this concept consideration is given not only to absorption but to concentration and persistence in the body, metabolism, and whether the compound (or its effective metabolites) can actually reach the target organ or site and elicit a clearly defined and characterised response. This involves considerations of, for example, toxicity, hydrophobicity, solubility, molecular size and shape, affinity for target, membrane permeability, carriers and transport mechanisms. Some of these properties can be manipulated by the construction of analogues, or derivatives, to achieve the desired outcome, while minimising toxicity and loss of active component. Thus it is possible to make an injected drug more 'bioavailable' by improving the affinity of the target, or constructing a derivative drug with greater affinity for the target. This amounts to manipulating the way the drug is partitioned into the various body pools. The selection of drug candidates and their modification for delivery and efficacy is a very sophisticated process; in contrast, the foods we eat contain many components about which we know almost nothing and which do not lend themselves to modification as a way of manipulating 'bioavailability'.

As with drugs, the concept of nutrient bioavailability should embrace all these key issues. Bioavailability should not be limited (as so often happens) to measurement of nutrient release from the food (or ingredient) matrix and subsequent absorption. However, a drug is generally a known synthetic construct for a specific purpose. Prior to bioavailability trials, the drug will not be present (or will be present at a very low concentration) in the tissues of human volunteers, giving a clear baseline. When the drug is administered, it and its metabolites can be seen clearly to appear and disappear from the plasma pool or other assayed tissue. This makes absorption and metabolic studies much simpler to perform and interpret. Additionally, a drug is normally consumed in tiny quantities and is less likely to affect satiety, consumption patterns and lifestyle, whereas foods need to be supplementary to (or replace items of) habitual diet. A drug is usually incorporated into a simple matrix with predictable 'release' characteristics. Unlike food components, a drug is not a compound to which the body is normally exposed from the cradle to the grave. It is, however, possible to treat food components like drugs and to feed the extracted component and/or package it in simple or complex characterised delivery systems. This does not really simplify measurement in a tissue already containing the same component and its metabolites that have been derived from the diet. Furthermore, the extrapolation from populations who derive health benefits from eating a particular diet or food to inferring those benefits derive from a single component should be critically examined. Studying and predicting the absorption, metabolic fate and functional effects of food components is hugely more challenging.

In summary, understanding the concept of bioavailability, how its component parts are best measured, limitations to interpreting the results of experimentation, the pros and cons of modifying the extent and rate of compound release at specific sites in the intestine and definition of 'optimum', is crucial to ethical product development and good nutrition.

1.2 Measuring absorption, metabolism and tissue targeting

Measuring the quantity of nutrients, or non-nutrient components, consumed in foods and nutraceutical products may have little predictive value with respect to their interaction with the human organism for a variety of reasons.

- The 'native' compound present in the consumed product may not be the chemical form to which human tissues are exposed after digestion and post-absorptive metabolism.

- Diet matrix may significantly affect the extent and rate at which compounds are liberated and become available for absorption.
- A proportion of some compounds released into the lumen of the intestine may need a carrier to be potentially absorbable.
- Only a proportion of a released compound may be absorbed.
- The absorbed compound may not appear systemically because it is metabolised by the enterocyte or re-excreted back into the gut lumen.
- Non-absorbed compounds may still exert physical, physiological and/or biochemical effects on the intestine itself, with subsequent consequences for whole body metabolism.
- The genetic heterogeneity in human populations may result in individual differences in the site, extent and/or rate of absorption, metabolism and tissue targeting, which in turn will influence functional response.

Exploration of the mechanisms of action of food (and nutraceutical) components on human health thus requires an understanding of the factors that constrain their release from the matrix in which they are contained, the rate and extent of conversion or transfer to absorbable species, the extent, rate and site of absorption and their health impact in genetically diverse individuals.

Measuring how much of a compound (drug, nutrient, toxin or carcinogen) reaches its site of action, irrespective of its route of administration, requires sampling the tissue/cell site and making the measurements. A more pragmatic and less interventionist approach is to administer a range of doses to groups of similar cases, or volunteers, to see which dose is 'effective' and to relate this to some more easily performed measurement, for example, concentration over time in whole blood, or a blood fraction such as plasma, erythrocytes or a specified population of white blood cells. This surrogate approach is defined by Schumann *et al.* (1997) as 'that fraction of an oral dose (parent compound or active metabolite) that reaches the systemic circulation'. From the drug or nutraceutical perspective, the assessment of 'effective' is based on some clearly defined and usually short-term objective, for example, suppressing an infection, relieving pain or altering blood chemistry. In the case of foods the time scale may be short, for example, correcting a vitamin deficiency, but is more often than not long term. Will the intervention reduce the risk of cancer or cardiovascular disease in 10 or 50 years time, and does the intervention have to be continued over this period? Clearly there is a hierarchy of effectiveness from a 100 per cent short-term 'cure' to a lifelong reduction of risk and this needs to be considered during the design of any study.

Schumann's definition (1997), however, precludes injection as a route of administration. The injection route is precluded because it bypasses complications of intestinal metabolism and absorption and any effects of time. An injected compound is considered to be 100 per cent 'bioavailable', despite the fact that this is erroneous in terms of the drug definition because it does not take into consideration how much gets to the target. Injections or infusions are therefore more correctly 100 per cent absorbed although these routes exclude orally administered compounds that are absorbed into, and are metabolised by, the enterocyte, or which enter the enterocyte and are subsequently re-excreted into the gut lumen. In both cases, the authors consider that such compounds are 'bioavailable'.

This distinction between routes is crucial because many 'bioavailability' studies use tracer–tracee methods where the absorption kinetics of an oral dose is quantified in the plasma pool by using a simultaneous injected (real or virtual) labelled tracer, i.e. absorption is quantified by reference to the 100 per cent 'absorbed' injected dose. This approach, however, must only be used for compounds that are not affected by the delivery route, for example, the approach would not be suitable for hydrophobic compounds packaged in chylomicrons or for folates. The plasma area under the curve (AUC) of an injected hydrophilic compound is likely to be much larger than an equivalent oral dose, since the oral dose is absorbed via the portal vein and may be more or less metabolised (cleared) on first pass of the liver. However, clearance kinetics of the compound of interest, once equilibrated throughout the body will be the same in both cases provided metabolism, or clearance mechanisms, are not overloaded.

Where the orally administered compound is strongly hydrophobic, and follows lipid absorption, it will be packaged into lipoproteins and enter via the thoracic duct. This is very different to an injected or infused dose because the injected compound may not be packaged in the same way as that which has passed through the enterocyte and will therefore not be picked up by the lipoprotein receptors. The injected material may therefore have different kinetics in the plasma pool, in which case the percentage 'comparative' absorption measurement for the oral dose is of unknown value.

Measuring the absorption and metabolism of isolated food components *in vivo* may be valid if used to address mechanisms and for the construction of models but the results may bear little resemblance to the case of foods where:

1. most of the compound in the food is not the same as that absorbed and/or appearing in the plasma;
2. an absorbed compound may be metabolised by the enterocyte, or re-excreted back into the gut lumen and never seen in the plasma pool;
3. rate and limit of production of bioaccessible forms of substances (a term used by the authors to define the amount of a compound released from the matrix during digestion and/or fermentation and thus available for absorption) will be constrained by buccal, gastric and small intestinal processes, diet composition and food matrix interactions; and
4. non-bioaccessible components may become bioaccessible (in one form or another) through the action of the large bowel biota.

Any attempt to take into account or model the effects of all the variables that may affect the absorption of a potential product of digestion are bound to fail because there are too many unknown factors. As an example, West and Castenmiller's (1998) 'SLAMENGHI' mnemonic for carotenoids (S = species, L = molecular linkage, A = amount consumed in a meal, M = matrix containing the carotenoid, E = effectors of absorption and bioconversion, N = nutrient status of the host, G = genetic factors, H = host related factors and I = the mathematical interactions) may be a useful way of remembering some of the variables (as applied in this case to carotenoids) but affords no insight into how they can be effectively combined to construct a universal predictive model of 'bioavailability' and 'bioefficacy' of β-carotene.

Where absorption of nutrients or other compounds from foods is concerned, the pragmatic approach is to treat the gastrointestinal tract as a 'black box' in the sense that it is not necessary to know how systems work, only their effect. With input measures (food, food form, composition, etc.) and output measures (absorbed components or surrogates, health measurements) it is possible to train neural networks to accurately predict outcomes (a model of reality) from a selected range of inputs, without knowledge of the algorithms by which the prediction is made. However, if informed rather than intuitive manipulation is required, for example, 'What should I do to this food to increase the health benefit?', Bayesian statistical analytical methods, using our knowledge and beliefs, can be applied to identify the main effectors. The main effectors can then become new, or revised, inputs for the neural network to refine the predictive power of the model (model closer to reality).

However, once the nutrient has become systemic, understanding its distribution, metabolism and excretion requires compartmental modelling (Jacquez, 1999) to understand how the component affects, for example, gene expression, mRNA, protein synthesis; and ultimately, some integral of functionality that describes the health outcome.

Compartmental models are used to describe the flux of nutrients and other compounds *in vivo* into and out of body compartments (Fig. 1.1). The use of compartmental models requires either:

- perturbation of the steady state within physiological limits
- or use of tracers (stable/radio isotope)
- and serial sampling and analysis of as many accessible body pools as possible.

Compartmental models can be constructed from a linguistic description of the process. For example, the following statement can be easily converted to a linear compartmental mathematical model. 'Most of the nutrient is absorbed from the gut into the blood. It exchanges with liver stores and is eventually cleared into the urine.' The above description indicates 'pools', process, sequence and time and the equations are derived from a consideration of mass balance and assumption that the nutrient is each pool is well mixed.

In this case, that fraction of the component that is excreted in the faeces is unlikely to be 'bioavailable' but that which is excreted in the urine must have been absorbed and passed through the body and hence would be considered 'bioavailable'. The simplest 'compartment' is the whole body but any number of compartments (at any scale) can be used, for example, plasma, muscle, adipose tissue, bone, liver, faeces, urine. As a general rule, the number of compartments should be kept to the minimum that can be biologically justified. Data can easily

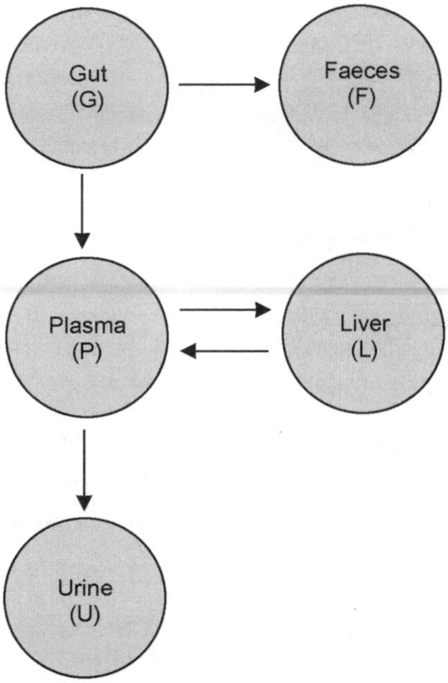

Rates of change in any compartment can be described as follows:

$$\frac{dG}{dt} = -K_{fg} * G - K_{pg} * G$$

$$\frac{dF}{dt} = K_{fg} * G$$

$$\frac{dP}{dt} = K_{pg} * G + K_{pl} * L - K_{lp} * P - K_{up} * P$$

$$\frac{dL}{dt} = K_{lp} * P - K_{pl} * L$$

$$\frac{dU}{dt} = K_{up} * P$$

Fig. 1.1 Arrows represent the transfer of nutrient from one compartment (*i*) to another (*j*) and are associated with the rate constant K*ji*. Circles represent well-mixed pools.

be overinterpreted by including more compartments (variables) and although this improves the data fit it also increases the error in the parameters of interest, i.e., the rate constants. A good data fit obtained by adding additional compartments must be justified by sound physiological or biochemical argument (see Fig. 1.2) (Jacquez, 1999).

Here, the substance enters into a dynamic pool, or pools, the concentration of which varies as fluxes (mass per unit time, indicated by arrows) in and out change with metabolic demand. The size of a 'pool' in such models is the distribution volume or mass. Pool volume or mass can be measured, found in tables of human body data, or predicted from anthropometric data.

If steady state is assumed (the volunteer is not growing, sick or perturbed by intervention) any compound should not be described as 'used', 'utilised', 'stored' or 'accumulated' but simply moving into the body from the diet, distributed dynamically between various compartments and excreted unchanged, or as downstream metabolites. After dosing, the ratios of the amount of material in pools does not remain constant because there are always delays in pool mixing during this dynamic period. They should, however, return to normal after an acute dose. If pool ratios are found to have changed after a chronic regimen then bioavailability has changed. However, there are cases where the rate controlling mechanism is not the creation of absorbable species or absorption but is an inherent feature of post-absorptive metabolism. In this case the body

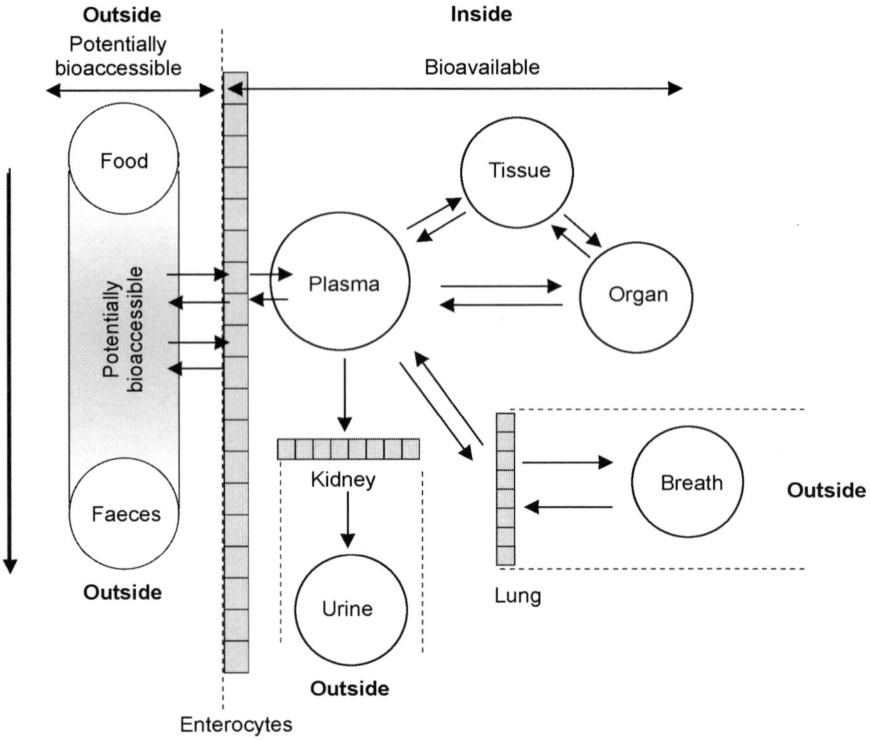

Fig. 1.2 A more generalised compartmental model.

will freely and rapidly absorb as much as it is given, pools will become saturated and the body will 'dump' 'unused' material into excretory pathways. This can be considered in the same way as exceeding the physiological dose leading to nonlinearity. In both these cases excess dose 'spills over' into the disposal system (provided it too is not overloaded). In the view of the authors, overloads and non-metabolisable organic compounds should still be considered bioavailable if they require the expenditure of energy for their excretion.

For organic compounds that can be metabolised, the amount metabolised in a pool is simply the difference between influx and efflux over time, hence the importance of accurately establishing the rate constants. It is not normally possible to identify a downstream metabolite as being specific to a particular pool other than perhaps the liver, which has the specific capacity to detoxify otherwise intractable compounds. Thus it will not be possible to measure the carbon dioxide from a particular pool (other than the whole body) but it is possible to identify specific downstream metabolites produced by the liver and excreted in the urine.

Where the food or nutraceutical contains isomers of the same compound, the isomer profile should be checked at all points to ensure that statements like 'isomer [A] is selectively absorbed' is in fact true. Isomer [A], for example, may persist in the plasma longest at a measurable concentration, while the other isomers disappear quickly. In addition, isomer [A] may be formed in the lumen or enterocyte and therefore is the predominant form transferred to the serosal side. Finally, compounds do not 'accumulate' in any particular pool, even if there are receptors. The pool is simply a thermodynamically favourable environment – the system is dynamic.

Unfortunately compartmental modelling is often misunderstood or abused. The most common errors are the use of incomplete area under the curve (AUC), the use of injectable standards or calibrants where this route is inappropriate and the assumption that changes in plasma concentration are due only to the compound being absorbed from the test food/diet and not a normal physiological response to the meal.

Figure 1.3 illustrates the problem with using partial AUC. The intravenous (IV) dose starts to clear from the time of injection, but the oral dose is delayed by the stomach and constrained by the rate of gastric emptying and absorption. The two curves are therefore temporally displaced and analysis of AUC before return to baseline (for example, point A on the curve) will remove different proportions of the curves. The area under the incomplete curves cannot therefore be compared. Even when two different vehicles are given orally to deliver the same dose, incomplete curves cannot be compared because the delivery vehicles may have different effects on gastric processing, emptying and subsequent absorption kinetics.

The third point, ensuring that the plasma response is only derived from the oral dose, is essential since it has now been demonstrated that oral folate containing meals stimulate the release of endogenous folate into the plasma (Wright *et al.*, 2003, 2005). It is also recognised that the sight, smell or anticipation of foods can trigger responses so where placebos or blanks are used they should be chosen with care.

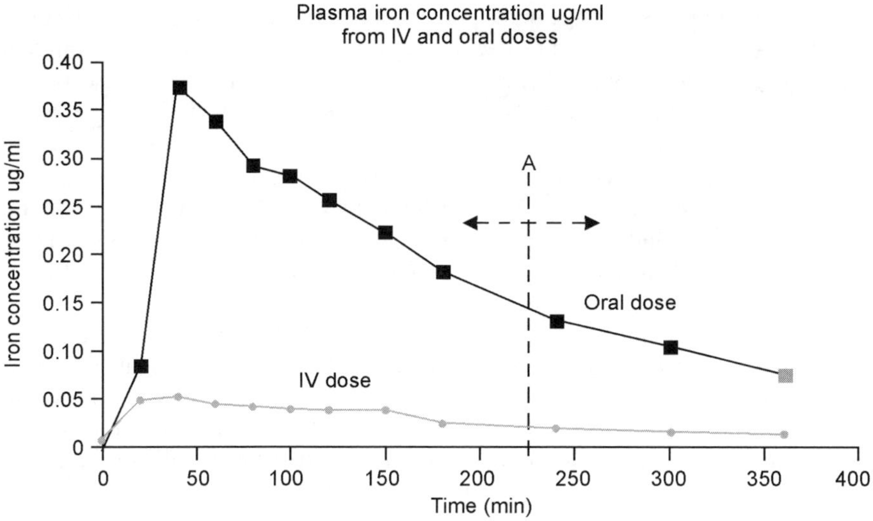

Fig. 1.3 Illustration of the error of using partial AUC.

Finally, a question that is rarely asked is 'are the changes induced by the test meal/compound significantly different from day-to-day variation within an individual?'. In most cases it is not possible to answer this question because only a single baseline measurement has been made. The authors recommend that researchers do not rely on a single baseline measurement but take sufficient measurements to establish a threshold where the intervention becomes significant.

An essential feature of the compartmental modelling concept is that the body attempts to distribute nutrients throughout the body (via the plasma pool) to various organs and tissues and the equilibrium between pools is linear and constant in the steady state. This condition is termed the 'homeostatic set point' by the authors. This state is an integration of genetic make-up and exposure to environmental factors and includes body mass, body composition, physiological and biochemical conditions that an individual attains at steady state. The homeostatic set point does not imply that this is the 'healthiest' condition the body can attain given the conditions.

Increasing the amount of a compound entering a pool does not necessarily change 'bioavailability' because the compound will partition proportionally to all the pools according to distribution volumes/masses and pool affinities and will be subjected to the same processes. The equilibrium between pools changes only if the homeostatic set point moves, for example, growing, gaining or losing weight, changes in habitual energy expenditure, episodes of illness, ageing, environment and lifestyle. A logical conclusion from this equilibrium model is that 'dose response' in terms of changes in pool concentration is predictable unless the dose saturates, or overwhelms, normal physiological processes and the model becomes nonlinear.

More commonly, the term 'dose response' is used to describe how efficacious a dose is at eliciting the desired physiological/health/pharmacological outcome rather than simply a change in a pool concentration. An organism responds holistically (not just as a result of achieving a particular pool concentration) and this response is usually hormetic (exhibits hormesis), that is, response ranges from the no observable effect level (NOEL), through the most effective dose to overt toxicity. Thus, although pool concentrations may be linearly related to dose injected or absorbed, the response of the organism is seldom that simple. 'Dose response' can be either a relationship between dose and amount in a pool, or between dose and organism response and these are not the same. Dose response should therefore always be carefully defined and accessible to precise measurement.

As a cautionary note, before embarking on food, diet or nutraceutical interventions to increase the intake or absorption of 'beneficial' compounds, it is as well to know where the subject(s) is/are on the hormetic dose response curve. In the curve illustrated in Fig. 1.4, only those subjects with sub-optimal exposure will 'gain' from an increase in exposure, the rest will 'lose'. Where dissimilar individual outcomes are observed, the individuals should be stratified and the reasons for differences identified. In this way, the target groups who will, or will not, benefit can be identified.

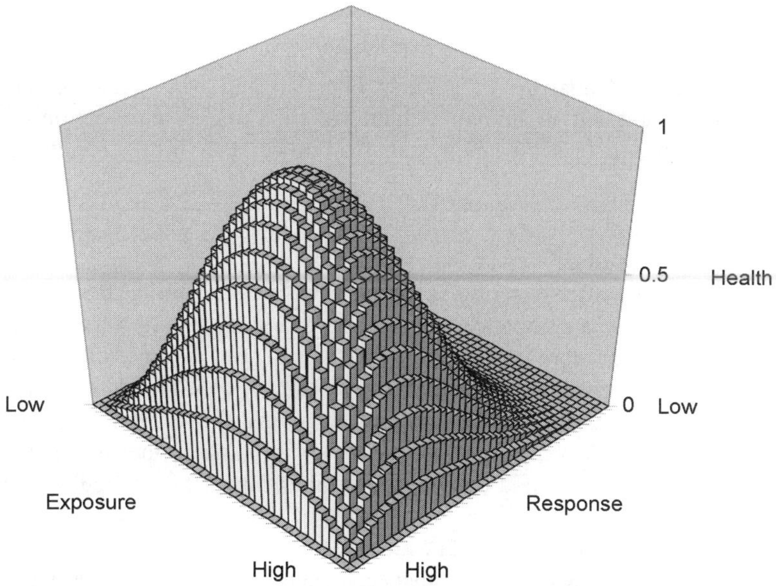

Fig. 1.4 'Hormetic' exposure, response and health 3D plot.

Within physiological limits, any absorbed substance is 100 per cent available to participate in biochemical processes but its distribution to the different 'pools' and ultimate fate(s) is an inherent feature of the individual's metabolism. This in turn is a requirement of maintaining the homeostatic set point of that individual. Absorption is easily defined but 'bioavailability' also includes consideration of how efficacious the absorbed dose is at eliciting the desired effect(s), just as in the drug model.

The following section provides some recommendations for best practice in the design and interpretation of nutrient absorption and metabolism studies, using specific nutrients to illustrate approaches for hydrophobic and hydrophilic compounds.

1.3 Study design and interpretation

1.3.1 Ethics

All studies involving human volunteers must have received a 'favourable ethical opinion' from a recognised ethics committee that has jurisdiction in the country or state where that work will be carried out. Ethics committees are helpful in providing advice to applicants, as well as passing judgement on the value of the science and its impact on the volunteers. Furthermore, well thought out ethics application forms ask a variety of relevant questions which may not have been considered by the researcher but which are vital if the work is to be carried out safely to a publishable and acceptable international standard.

1.3.2 What sort of study?

At the outset there are a number of decisions to be made. These decisions are based on a critical analysis of what is intended and why. Can the intended studies be carried out in human volunteers, time scale, intervention effects, compliance, chronic or acute study, statistical power and will the results provide a clear answer?

1.3.3 Chronic studies

Chronic studies are often described as measuring 'bioavailability' but, in fact, can only be used to give a comparative indication of absorption. In such studies it is normal to use two or more large, matched groups of 'free-living' volunteers, each of which is given a different long-term intervention with the object of determining which intervention is most efficacious in achieving the desired outcome. The regimen normally takes the form of supplementation to the normal diet over a period of weeks with isolated compounds (tablets/capsules), or foods rich in the compound of interest. Concentration changes in the blood plasma or serum are tracked over time until a new plateau is reached and confirmed. The relative absorption from the 'test' food is then calculated against a 'standard food' or isolated compound. With this design, a new plasma concentration plateau must be reached, the same compound from different sources can be compared, and relative absorption can be determined. With chronic dosing it is essential that the data are not compromised by seasonal fluctuations in diet or differences in the timing of supplement/food ingestion, or taking of blood samples.

It is not usual to use a crossover design with a washout period, because of the long-term nature of the intervention, although this can be done so that each volunteer is their own control. If a crossover design is used, it is essential that the pool concentration of the compound of interest returns to the 'habitual' concentration before embarking on the second arm. The end point may be the mean (group) change in plasma concentration, or this may be coupled with other assessments, for example, changes in body weight, lifestyle, cognition, ability to carry out tasks, incidence of infectious or non-infectious diseases, progression or reversal of disease states, severity of symptoms.

Chronic studies are useful in a more holistic sense, since the time scale allows the body to adapt and respond. This then permits nutrigenomic, transcriptomic and proteomic studies to be carried out to provide a fundamental understanding of the effect of the intervention on the whole body and test hypotheses of the functionality of foods and nutraceuticals. Care should be taken to ensure that the intervention has not changed the rate of compound clearance or pool ratios. If there is evidence for either of these effects then it can be concluded that the 'homeostatic set point' has moved and 'bioavailability' has changed.

1.3.4 Acute studies

With acute dosing, the objective is to measure either relative or absolute absorption by following the absorption and disposal of a single dose through monitoring appearance and disappearance of the compound in plasma. The experimental design can be a crossover, where each volunteer acts as their own control, or can be based on a single dose if mathematical modelling is to be applied to the data. Fractional absorption of an oral dose can also be assessed by comparison to an IV dose, using isotopic labelling to differentiate the two sources.

$$\text{Fractional absorption} = \frac{\ln 2}{t_{1/2}} \times \frac{\text{AUC oral} \times \text{plasma volume}}{\text{Oral dose}}$$

Absolute absorption can be measured from the peak plasma excursion, or by measuring the area under the complete curve using the clearance rate ($t_{1/2}$) from the IV dose, and it is assumed that unit absorption provokes unit excursion. Absolute absorption may also be measured against the AUC induced by an IV dose assuming there are no confounding factors. A seldom used alternative is to calculate what the virtual dose would be to simulate the plasma excursions found and express this as a percentage of the dose given (O'Neill and Thurnham, 1998). This requires that the clearance kinetics and pool size are known.

1.3.5 Mass balance studies

As the name implies mass balance studies measure intake and excretion over a period of time. They are most commonly acute studies, this being a more practical alternative to chronic studies which can be confounded by the intake of the compound of interest from dietary sources. They are non-invasive, compliance is good and intervention effects are minimal. Mass balance studies offer a relatively simple way of obtaining information on absorption at the whole body level. They are particularly useful for inorganic ions, organic compounds which are excreted unchanged, or which have a specific metabolite that can be easily detected, and if necessary isolated (for isotope measurements) in urine/ faeces, or breath. These simple whole body studies can be coupled with the sampling and analysis of body tissues.

The model assumes that the compound recovered from the faeces has not been absorbed and that the compound, or downstream metabolites recovered from the urine, represents material that has been absorbed. Point A (Fig. 1.5) represents a variant of the mass balance approach using ileostomy volunteers, where the unabsorbed material is intercepted at the end of the terminal ileum and therefore has not been subject to fermentation by the large bowel microflora. The ileostomy approach is useful for hydrophobic organic compounds that do not have identifiable urinary metabolites. Point B (Fig. 1.5) indicates body pool samples (plasma, biopsy) that may be necessary or provide additional information.

The disadvantage of these studies is that in many cases it is necessary to use isotopically labelled material and mass spectroscopy to enable discrimination between material arising from the test dose and that which arises from the diet or body. The mass balance between food and faecal excretion of unabsorbed material may have to be continued for several days to ensure that all stools

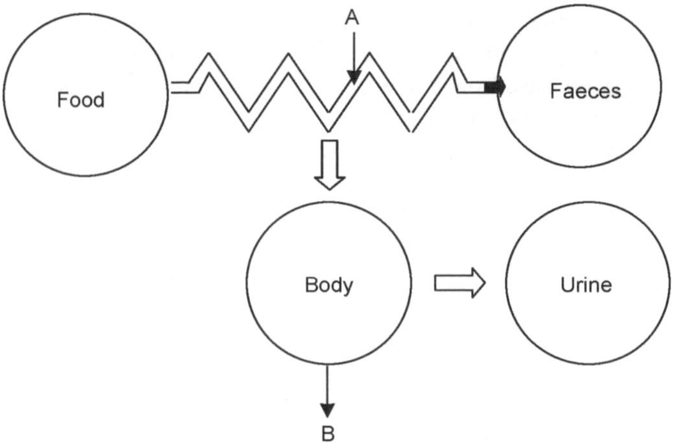

Fig. 1.5 Mass balance schematic.

containing unabsorbed material from the test diet are recovered and often timing is erratic depending on bowel habit. With the ileostomy model, gastrointestinal transit is normally complete in about 14 hours and complete effluent collections can be made at regular intervals and rapidly preserved. Absorbed material or downstream metabolites recovered from the urine can give a direct estimate of the amount of material absorbed. Where the half-life of the whole body pool into which the dose has been incorporated is weeks, months or even years, urine and/or faeces recovery will need to continue until a satisfactory decay curve is obtained and the equation established, so that the line can be extrapolated on to baseline for a whole AUC. This is often difficult to obtain because the dose is commonly only a small fraction of the body pool and even with mass spectroscopy of stable isotopes (mass ratio measurements) it may be difficult to obtain accurate data at the high dilutions found days after the test dose.

Because both faecal and urinary excretion can be measured at the same time, absorption measurement can be either (intake–faecal excretion/intake) or (urinary excretion/intake) so it is possible to cross-check the data and to test some of the assumptions, for example, loss during gastrointestinal transit is only due to absorption.

Mass balance studies provide a value for absorption, they do not measure 'bioavailability' and give no information on, for example, plasma concentrations or clearance rates (other than from whole body), which are essential in making the link between intake and health outcome.

1.3.6 Absorption, distribution, metabolism and excretion studies

For all studies a decision needs to be taken as to whether a controlled clinical study or a study of 'free-living' individuals is needed. A controlled study in a clinical environment means that the researcher can exercise control over every aspect of the study and this can be applied to all volunteers and repeated on subsequent occasions. The value of controlling the study closely is repeatability, avoiding confounding factors, ensuring compliance and medical assistance if required. A study of 'free-living' individuals will be more variable, so may require a larger number of volunteers to achieve statistical significance. Volunteer compliance may be a problem, timing of samples may be erratic and complete sample sets may not be available.

The next issue to address is whether the design is a supplementation study, where test material is given in addition to normal diet, or is a replacement study, where one (or more) foods are replaced by the test material. Supplementation studies are generally easier to conduct because, in theory, they are additional to habitual intakes, because the baseline remains constant. Any measured changes are therefore due to the intervention. Nevertheless some steps should be taken to ensure that the additional intake does not displace foods, or amounts of foods, habitually eaten. This is most easily detected by using intake diaries, before, during, and if necessary, after the study. Diaries will also pick up seasonal dietary changes in long-term studies. Replacement studies automatically preclude those volunteers who do not habitually eat the food. Replacing a food, or a range of foods, with a 'superfood' or non-food vehicle will have an ill-defined effect due to the loss of intake of compounds from the habitually consumed foods and may confound the study.

Studies can have different levels of complexity depending on requirements:

- absorption, distribution, metabolism and excretion (ADME) study
- definitive absorption study
- comparative absorption study.

The ADME study

This is the most complex, the most invasive and may well involve the use of isotopically labelled foods, orally administered isolates (isotopically labelled), intravenous dosing (isotopically labelled), or continuous infusions (isotopically labelled). It will also involve collecting serial blood and perhaps other tissue samples (biopsies), urine, faeces and breath (CO_2) over extended time periods during which the volunteer will need to eat and drink. These studies create a large number of different samples with diverse analytical requirements, for example, analysis of the compound of interest and downstream metabolites in different sample matrices, including breath CO_2. Because of the intensive nature of such studies they are best carried out on one or two highly motivated individuals in a closely controlled clinical environment. Such studies are extremely valuable in resolving the detail and allowing the construction of informed general absorption and metabolic models.

Definitive absorption studies

Definitive absorption studies are needed to measure exactly how much of a compound is absorbed from a food, nutraceutical or drug delivery system. Although it is possible to vary intake and dose to achieve a specific plasma pool concentration or outcome, it is desirable to know exactly how much active component has been absorbed and over what time scale. Without this information, mis-dosing is likely to occur if the vehicle, or route of delivery, is changed. Additionally, where the compound is a precursor to an active metabolite it is not possible to make an estimate of conversion efficiency. In order to establish definitive absorption it is necessary to know either the clearance rate from the sample pool, or to use a comparison to a 100 per cent absorbed standard. To make such a comparison care should be taken to avoid confounding factors which may be introduced with the 100 per cent absorbed comparator, for example, metabolism or isomerisation in the enterocyte, the *in vivo* carriers (if any), to understand the implications of entry route (portal vein, thoracic duct, IV, ip) and any time effects.

Where it is possible to establish clearance kinetics from the body (or a major body pool and the size of the pool) the absolute amount absorbed can be calculated. This is not always possible for compounds that are very rapidly cleared, or where clearance is disproportionally changed in response to the presence of the compound in the pool, unless the concentration dependent change can be established, for example, as can be done for glucose. To establish the clearance rate for compounds administered orally, sampling the pool over at least the time taken for the food to pass through the stomach and ileum is required to ensure that at least five data points are acquired during the clearance phase, uncomplicated by continuing absorption from the gut. The concentration-dependent clearance rate can then be applied to the whole absorption/clearance curve to establish the total amount absorbed. For such studies, sampling times and frequency should be set to capture an accurate concentration/time curve. Curves can be extrapolated to the point where the pool concentration returns to baseline. The more data available on the clearance-only part of the curve, the better the mathematical fit and more accurate the predicted absorption.

An alternative way of establishing the clearance rate is to load the pool of interest at a single time point, for example by IV injection. After a short period of mixing throughout the body, ca. 10 min, the change in concentration in the pool sampled is solely due to clearance. Such studies can be designed so that a single volunteer receives a food or nutraceutical preparation on one occasion and an IV on another occasion, or both can be accomplished at one session if one of the doses is isotopically labelled and can therefore be discriminated from the other and from the endogenous background. Care should be taken to ensure that the delivery routes do not have different clearance rates. Although absolute absorption studies are relatively simple, they are best carried out on small well-defined groups of volunteers in a carefully controlled environment.

Comparative absorption studies

Comparative absorption studies are perhaps the easiest to perform but they do not provide any information on the absolute amounts absorbed. The only information they provide is that one delivery system or food, when compared to another containing the same (or normalised) load of component, elicits a greater or smaller concentration excursion in the sampled pool. This type of study is often carried out to compare absorption from two or more foods in different physical states, for example, raw vs cooked, or cooked vs some processed product, or a standard product with more or less added compound of interest. It is implicit in such studies that the complete concentration-vs-time AUC is a measure of absorption because it will take account of any differences in time taken to ingest the food, gastric processing time, gastric emptying rate and differences in delivery profile. Occasionally it is possible to use the maximum plasma excursion concentration instead of the AUC provided that delivery profile is a constant.

Studies should be conducted within an individual so that they act as their own control, allowing a paired *t*-test which will increase the power of the statistical treatment. Both AUC and peak excursion can be normalised to take account of different loads, provided it is known that clearance rate is not unpredictably load sensitive.

1.4 Other considerations

Using an intravenous dose (assumed 100% 'absorbed') as a standard for comparison is not always valid for hydrophobic compounds. Such compounds, fed orally, may passively follow lipid absorption and are therefore transferred to the serosal side incorporated into chylomicrons synthesised in the enterocytes. This situation cannot usually be replicated and applied to human studies; thus an intravenous preparation would not behave in the same way as the chylomicron packaged material. Because of the rapid clearance of the chylomicrons (half-life 3–5 min) that carry the newly absorbed compound, it is not easy to provoke significant plasma response curves with a physiological dose. This has led to the belief that some hydrophobic compounds are poorly absorbed by some individuals. An added complication to using plasma is that hydrophobic compounds carried by chylomicrons are sequestered by the liver and re-exported/transported in other lipoproteins. Thus, distribution among lipoprotein carriers may change rapidly over the period of study.

There are two potential resolutions to this problem: (1) isolate chylomicrons (or triglyceride rich lipoprotein (TRL) fraction) or (2) continue the study over several days until the compound has cleared from the long-lived lipoprotein particles. The latter approach is impractical since it would mean maintaining volunteers on a diet free of the compound of interest for several days and during this time baseline concentration would start to fall. Analysis of the chylomicron (TRL) fraction is therefore preferable since appearance and clearance of the compound in the fraction is over within 12–16 h. This is a tolerable time for the volunteer to fast, or they can be given a meal free of the compound of interest at 5–6 h when the original test meal has completely cleared the stomach. This procedure has the advantage that, whereas a concentration change may not be seen against the endogenous background in plasma, it is more easily seen in the chylomicrons (TRL), which carry only a small fraction of the total amount in plasma.

As with hydrophilic compounds, a classical approach to predicting the relative absorption of hydrophobic nutrients, or other food-derived compounds, is to compare the plasma (or other blood fraction) concentration AUC response of a test dose of an isolated, well-absorbed compound (the 'reference' dose) to a test meal containing the compound of interest. Before using this method it is essential to establish that change in post-prandial plasma concentration is directly, or predictably, a response to the oral test dose. For example, recent research has demonstrated that this is not the case for folate and vitamin C, both of which are water soluble.

It is accepted that both synthetic folic acid and natural (or synthetic nature-identical) folates given orally would be bio-transformed in the intestinal enterocytes and passed to the hepatic portal vein only as 5-methyltetrahydrofolic acid (5-MTHF). The liver then removes a fraction of the 5-MTHF and the remainder enters the systemic blood system to create a systemic post-prandial 5-MTHF response. Since the proportion removed by the liver is unknown, the most common approach is to relate the plasma 5-MTHF AUC response to that of an equal reference dose of folic acid, and then calculate relative absorption. Recent publications, using isotopically labelled folate demonstrate, however, that a substantial portion of the total plasma 5-MTHF response to an oral dose is of endogenous origin, i.e. does not originate from the test dose. This endogenous response is large and unpredictable, and is not related to fasting baseline plasma or erythrocyte folate concentration. Thus it was concluded, that relative absorption calculated from all previous studies, which used unlabelled test doses (or foods) and an unlabelled reference dose, are invalid (Wright *et al.*, 2005).

The same phenomenon has been demonstrated for vitamin C, where at least 50 per cent of the plasma total post-prandial vitamin C response came from an endogenous source, also in an as yet unpredictable manner (Bluck *et al.*, 2005; Bates *et al.*, 2004). Once more, this invalidates previous studies conducted without the aid of 'labelled' material. Whether other food compounds/nutrients behave in a similar manner must be determined prior to the conduct of studies using 'unlabelled' compounds.

The intestine is a major organ, works in concert with the body via feedback mechanisms, and has the capacity not only to digest and absorb compounds but also to metabolise, temporarily store and regulate serosal delivery post absorption. There are therefore many potential mechanisms by which a plasma response cannot be reconciled to the absorbed dose. While the gut lumen is empty it may be logical to assume that the gut sequesters and temporarily stores nutrients from the plasma to maintain its high metabolic requirements over extended time periods, only to dump them back into circulation once nutrients start to be absorbed from a meal.

1.5 Health response

There are many epidemiological studies linking the prevalence of non-communicable diseases (NCD) to diet and lifestyle. There is strong evidence that stopping smoking reduces the risk of lung cancer and even some evidence that adopting a diet similar to a population with a low incidence of NCD reduces the risk of developing some of these diseases. Where the difficulty arises is the identification of a single compound or group of compounds found in 'good' diets and attaching the putative benefits to these compounds. In other words, it is the holistic response of the individual to all the external factors that determines the health outcome. Optimum health can therefore only be achieved if both internal and external factors are sufficiently flexible and adaptable to attain congruence throughout life.

Most 'bioavailability' (absorption, mass balance, retention, tracer–tracee) studies are performed with the volunteer at steady state, although 'snapshots' can be taken during periods where the homeostatic set point is moving, for example, during growth, provided the time window for the study is relatively short with respect to the rate of change. A growing child and an adult may both absorb the same amount of a nutrient, such as calcium, but the child is likely to retain more calcium for growth, while the adult uses it for recycling established stores. Thus, calcium distribution, metabolism, requirement and functional response will be different in each case. Absorption is not the same as bioavailability, and bioavailability does not equate with health.

With the ever increasing interest in diet–health relationships creating opportunity for new food products and spawning innumerable programmes for 'healthy eating', the questions that must be answered are:

1. Does changing the diet simply change the concentration of the body pools of compounds and thus alter health outcome?
2. Does changing the diet move the homeostatic set points of the body, and hence bioavailability, and thus alter health outcome?

The latter is more likely because a persistent change or larger or more frequent oscillations in the concentration of body pools of compounds will in itself move the set point (a change in body pool ratios). As a simple example, reducing energy intake in an individual at steady state will alter both the overall size of the pool (body mass) and pool ratios (lean : fat). Thus, eating a diet rich in fruits and vegetables may not confer health benefits simply through increased intake of substances found in these foods *per se* but by stimulating changes in homeostasis which cumulatively confer health benefits.

It is still a matter of speculation how large or persistent changes in intake and absorption have to be to provoke significant changes in gene expression or the proteome and how long these changes persist if the diet is returned to pre-intervention conditions. This raises the question, not only of amounts and delivery profile, but also of frequency and duration.

For nutritional studies, the term 'bioavailability' should therefore not be used to describe studies from which it is hoped to measure a value; nor should the term be used for comparative studies where absolute absorption is not known. Furthermore, it cannot be claimed that for two foods containing a similar quantity of the same compound that one food induces a greater or smaller perturbation in the measured pool, and thus the compound of interest from that food is more or less bioavailable. It is more or less absorbed which is a function of bioaccessibility. In addition, 'bioavailability' cannot be used to compare two different compounds, or even two different isomers of the same compound.

1.6 Implications of controlled absorption for product development

The upper digestive tract has a complex of control mechanisms which govern the entire process of ingestion and digestion. The main switches for these processes are oral stimuli, gastric stretch receptors, gut hormone responses to food in the GI tract and the systemic presence of absorbed nutrients.

It is commonly accepted that the half-life of solid foods in the stomach is about 60–90 min. and that most digestion and absorption occurs in the upper 30 per cent of the small intestine. The potential time window for digestion and absorption, mouth to terminal ileum, is around 6 h. Foods, nutrients and structures which are resistant to digestion in this 6 h time frame are predominantly lost to the large bowel. The examination of terminal ileal effluent from ileostomy volunteers routinely shows recognisable fruit and vegetable structures (leaf, skin, seeds and tissue fragments), nut fragments, cereal structures and mushrooms. These materials, and much of their potential nutrient content, survive because the cell walls are not digestible. Although most nutrients can be absorbed throughout the small intestine the ability to absorb them diminishes distally, thus glucose infused into the distal ileum is poorly absorbed and may trigger osmotic diarrhoea. It can therefore be envisioned that the presence of absorbable compounds at different points along the small intestine will have different effects on the control and feedback mechanisms that regulate digestion.

With the notable exception of carbohydrate foods where the 'Glycaemic Index' is seen to be important, putative health benefits have been related to overall amount, and not a combination of load and rate/site/limit of digestion/absorption. However, optimal rate of nutrient delivery, as well as optimal amounts, is also potentially important for maintenance of good health, and has long been an important concept in enteral and parenteral feeding in a clinical setting. Experimental studies in patient volunteers have demonstrated significant differences in physiological response associated with altered rates of delivery. Under 'normal' feeding conditions, it has been shown that even modest variations in the initial rate of duodenal glucose entry may have profound effects on subsequent glycaemic, insulin and incretin responses (O'Donovan *et al.*, 2004). Physiological response to different rates and sites of delivery of micro-constituents of food has been poorly researched but such investigations are crucial for the ethical design of controlled or site-specific release products. Modifying how and where nutrients (and other compounds) are released and absorbed can have both beneficial and adverse consequences.

Epidemiological studies provide a regular supply of data indicating associations between dietary habits, specific foods, or specific constituents of foods, and increased/decreased risk of a range of chronic diseases. Well-known examples include dietary fibre and colon cancer and lycopene and prostate cancer. Following disclosure of these epidemiological associations, dietary supplements, or foods containing more readily absorbed, or higher amounts of these putative beneficial compounds are promoted in the marketplace.

However, the bridge between observation and exploitation may not be supported by rigorous examination of the hypothesis linking a specific compound to altered disease risk and, crucially, there is often inadequate understanding of the effects of dose and rate of delivery of the putative beneficial compound. A recent example of where too little information has resulted in adverse response in potentially 'at risk' individuals relates to beta-carotene (Omenn *et al.*, 1996). The purported health-giving properties of plant polyphenols also represent a case in which enthusiastic marketing claims may far exceed the current scientific evidence. Even when good experimental evidence exists, results need to be interpreted with caution in relation to human health benefits at different time scales.

The analysis of nutrient risk–benefit curves, to define both the optimal intakes and safe limits of consumption of bioactive food components, is a focus of investigation using the 'omic' technologies (transcriptomics, proteomics, metabolomics), which address this issue from a nutritional systems biology perspective. By adopting a multi-parameter approach, it is envisaged that this will allow the quantification of subtle, physiologically relevant changes and establish methodologies for the assessment of effects of bioactive food components in the context of the diet, at relevant intake, and as they occur over long periods of time in the whole human body.

Currently, the approach is seeking to integrate the effects of different components of specific elementary cellular and physiological processes, for example, how do different food-derived components improve (or hamper) mitochondrial function. This 'holistic' approach distinguishes risk–benefit analysis from disease prevention-directed nutritional sciences, and from toxicology. Nutrigenomics approaches are crucial to this goal, since the technologies allow simultaneous analysis of the large numbers of parameters that need to be assessed. Clearly also, inter-individual variation, gender, age and lifestyle, must be considered, both in relation to personalised nutrition and to identify requirements of specific subgroups. Analysis of gene and subsequent metabolic response in humans requires quantitative measurement of site-specific nutrient delivery and emphasises the paramount importance of definitive measures of nutrient absorption and distribution.

This new approach to nutrition is now being explored and developed in many research laboratories around the world. One excellence source of information is the European Nutrigenomics Organisation.

1.7 Future trends

We are all concerned with the health of society and consequently the individual, and if nothing else, the cost of healthcare will be a major incentive to optimise our personal health. There is a debate to be had as to what constitutes personal health and the social and economic objectives of society. It is undoubtedly the case that there will be developments in 'personal nutrition' but these, at least initially, may only be targeted to, or adopted by, those whose genetic make up predisposes them to a specific disease state. The disease state must have a clear functional relationship between intake/absorption and long-term health outcome and could be the result of a missing or damaged gene, a suppressed or silenced gene, auto-immune deletion of specific cells or proteins, or production of a dysfunctional protein. The overcoming of a single 'error' in an individual whose health will deteriorate if there is no intervention is just the tip of a very big iceberg. At first sight, for a 'healthy' person, analysis of the health impact of a specific food or diet that takes into account all of the genes in the human genome, their interactions and all of the interactions possible with diverse lifestyles, age and sex, seems impossible. Such complex interactive systems are notoriously difficult to perturb, simply because there is a high degree of compensation arising from multiple copies of genes or other systems that can up- or downregulate. However, a close analysis of this complexity may reveal that there are only a few key drivers and the rest of the systems are basically slaves. Should this turn out to be the case, it greatly diminishes the number of probable variables that need to be manipulated to attain optimum health.

It is well documented that poor nutrition *in utero* (Barker, 1998) may have a profound effect on health, with respect to both communicable and non-communicable disease. The reader is directed to the output of a large, ongoing EC-supported project concerned with early nutrition programming (EARNEST, FOOD-CT-2005-007036). What is less well documented is that the nutritional status of parents and grandparents may also have heritable epigenetic effects (Pembrey, 2002; Whitelaw, 2006; Richards, 2006).

It is clear that further work needs to be undertaken to evaluate nutrient–gene interactions and the subsequent health effects induced by wide fluctuations of intake within individuals (e.g. seasonal reliance on specific foods, periods of starvation) which lead to cyclical changes in body mass and metabolic profile and which may have transient or heritable epigenetic effects.

The time has come to stop trying to pick 'magic bullets' out of foods, discard quick fix diets and to adopt flexible (age, sex, lifestyle) dietary regimens that are congruent with our 'optimum metabolic set point' and thus to minimise the risk of developing non transmissible diseases.

1.8 Sources of further information and advice

Handbook of Nutraceuticals and Functional Foods. Ed. Robert Wildeman, CRC Series in Modern Nutrition, CRC Press, 2001.

Web sites

The European Nutrigenomics Organisation linking genomics, nutrition and health research NuGO, www.nugo.org

EARNEST – Early Nutrition Programming Project. http://earnest.web.med.uni-muenchen.de.

1.9 References

BARKER D. *In utero* programming of chronic disease. *Clin.Sci.*, 1998; 95: 115–128.

BATES CJ, JONES KS, BLUCK LJ. Stable isotope-labelled vitamin C as a probe for vitamin C absorption by human subjects. *Br. J. Nutr.*, 2004; 91(5): 699–705.

BLUCK LJ, JONES KS, COWARD WA, BATES CJ. The anomalous absorption of labelled and unlabelled vitamin C in man. *Br. J. Nutr.*, 2005; 93(5): 627–632.

JACQUEZ JA. *Modeling with Compartments*, 3rd edn. BioMedware, Ann Arbor, MI, 1999.

O'DONOVAN DG, DORAN S, FEINLE-BISSET C, JONES KL, MEYER JH, WISHART JM, MORRIS HA, HOROWITZ M. Effect of variations in small intestinal glucose delivery, insulin and incretin hormones in healthy subjects and type2 diabetes. *Journal of Clinical Endocrinology & Metabolism*, 2004; 89(7): 3431–3435.

O'NEILL ME, THURNHAM DI. Intestinal absorption of beta-carotene, lycopene and lutein in men and women following a standard meal: response curves in the triacylglycerols-rich lipoprotein fraction. *Br. J. Nutr.*, 1998; 79: 149–159.

OMENN GS, GOODMAN GE, THORNQUIST MD, *et al.* Effects of a combination of beta-carotene and vitamin A on lung cancer and cardiovascular disease. *N. Engl. J. Med.*, 1996; 334: 1150–1155.

PEMBREY ME. Time to take epigenetic inheritance seriously. *European Journal of Human Genetics*, 2002; 10: 669–671.

RICHARDS EJ. Inherited epigenetic variation – revisiting soft inheritance. *Nature Reviews Genetics*, 2006; March 14. AOP. *www.nature.com/reviews/genetics*.

SCHUMANN K, CLASSEN HG, HAGES M, PRINZ-LANGENOHL R, PIETRZIK K, BIESALSKI HK. Bioavailability of oral vitamins, minerals and trace elements in perspective. *Arzneim-Forsch/Drug Res.*, 1997; 41: 369–380.

WEST CE, CASTENMILLER JJM. Quantification of the 'SLAMENGHI' factors for carotenoid bioavailability and bioconversion. *Int. J. Vitam. Nutr. Res.*, 1998; 68: 371–377.

WHITELAW E. Sins of the fathers, and their fathers. *European Journal of Human Genetics*, 2006; 14: 131–132.

WRIGHT AJA, FINGLAS PM, DAINTY JR, HART DJ, WOLFE CA, SOUTHON S, GREGORY JF. Single oral doses of 13C forms of pteroylmonoglutamic acid and 5-formyltetrahydrofolis acid elicit differences in short-term kinetics of labelled and unlabelled folates in plasma: potential problems in interpretation of folate bioavailability studies. *Br. J. Nutr.*, 2003; 90: 363–371.

WRIGHT AJA, FINGLAS PM, DAINTY JR, WOLFE CA, HART DJ, WRIGHT DM, GREGORY JF. Differential kinetic behaviour and distribution for pteroylmonoglutamic acid and reduced folates: a revised hypothesis of the primary site of PteGlu metabolism in humans. *J. Nutr.*, 2005; 135: 619–623.

2

Structure of the gastrointestinal mucus layer and implications for controlled release and delivery of functional food ingredients

G. Lafitte, Centre for Chemistry and Chemical Engineering, Sweden

2.1 Introduction

The rate and extent of absorption of drugs and nutrients depend upon the route of administration. In the case of controlled-release formulations, the oral administration is one of the most common routes (York, 2002). In the gastrointestinal tract (GIT), the efficiency of any formulations depends on the physiology at the administration site, on the composition and the thickness of the membrane barriers present at those sites. Among the various membrane barriers in the GIT, the first one that a drug and/or functional food encounters is the mucus layer. This viscoelastic translucent layer secreted throughout the GIT and covering the majority of the epithelium is a potential issue for drug and nutrient absorption, since it prevents direct adhesion to the epithelial cells and retards the transport of active molecules (Widdicombe, 1997; Thomson and Dietschy, 1977; Ashford, 2002; Allen and Flemstrom, 2005). The main constituents of the mucus, independently of its anatomical origin, are water, mineral salts, free proteins, mucins, lipids and DNA (Wikman-Larhed *et al.*, 1998; Khanvilkar *et al.*, 2001). In Table 2.1, the composition of pig gastric mucus is presented as an example. The composition may vary depending on its site of secretion, its physiological and mechanical role, as well as the presence of any disease. For example, patients suffering from cystic fibrosis show increased concentration of serum proteins and DNA in mucus secretions as well as some modifications of the composition of the mucin molecules (Bhat *et al.*, 1996).

Table 2.1 Composition of pig gastric mucus

Constituents	Composition (% of total weight)
Water	~ 95
Mucins	0.5–5
Lipids	Traces
Proteins	1
Mineral salts	0.5–1
DNA	Traces

The complex composition of the mucus layer renders it interesting in terms of physiology and physical chemistry. In the first part of this chapter, the physico-chemical properties of the mucus layer are outlined. Next, the discussion is focused on the composition and the properties of the mucin molecules, since they play an important role in the structure and the properties of the layer. Indeed, the mucus would not be a viscoelastic layer without the presence of these molecules.

In the second part of this chapter, the mechanism of the adhesion of delivery vehicles to the mucus layer is presented. The efficiency of a controlled-release formulation is subjected to the balance between the kinetics of its adhesion and the rate of erosion of the mucus. The amount of the mucus adherent to the epithelial surface is determined by a balance between the rate of secretion by the underlying epithelium and the rate of erosion by mechanical shear associated with the digestive processes (Allen and Pearson, 2000). It is then important to understand the mechanism of the adhesion of delivery vehicles in order to be able to improve the efficiency of controlled-release formulations. An improved adhesion can be obtained by including functional molecules (e.g. chitosan, lectin) to the formulation of delivery vehicles.

Food and/or drug delivery is completed when the active molecules reach the target cells. This means that molecules must reach the blood stream and consequently should not only adhere to the mucus layer but they must diffuse through it. The various components of the mucus layer can affect the diffusion of the controlled-release formulation through this membrane. In the last part of this chapter, the influence of some of these components (the mucin molecules and the lipids) is discussed.

2.2 Understanding the structure of the gastrointestinal mucus layer

2.2.1 Physico-chemical properties of the mucus layer

The mucus layer is able to act as a protective barrier of the epithelial cells in the GIT thanks to some advanced physico-chemical properties. Among others, three properties are characteristic and will be discussed: the viscoelasticity, hydrophobicity and the varying acidity within the layer.

Viscoelasticity
The most important and extensively studied property of the mucus layer is its viscoelasticity. *In vivo* and *in vitro* investigations of mucus gels from the entire GIT lead to the conclusion that two types of mucus, differing in terms of viscoelasticity, are secreted. Recently, a schematic figure of the thickness of the mucus layer with a distinction between these two types has been proposed (see Fig. 2.1; Atuma *et al.*, 2001).

Although both types of mucus have gel properties (i.e. shear storage modulus higher than shear loss modulus), they differ in terms of physiological properties because of some rheological differences. The 'loosely adherent gel' can be removed by suction and readily broken down by low applied shear forces (Allen and Flemstrom, 2005; Taylor *et al.*, 2004). Thus, *in vivo*, under shear forces due to the digestive process, the movement of solid matter through the gut is facilitated by the loose adherent gel, which has the ability to act as a lubricant (Atuma *et al.*, 2001; Taylor *et al.*, 2004; Allen and Flemstrom, 2005). A more firm mucus layer can be observed underneath this loose mucus. The firm layer is more resistant to applied shear and collapses only at shear forces two orders of magnitude higher than the loose gel (Taylor *et al.*, 2004). Such rheological properties allow formation of a firm protective barrier and a stable unstirred layer, which is resistant to shear.

Regarding the thickness of the layer of each type of mucus, the relative value varies with the regions of secretion within the gut. In the stomach, both layers have similar thicknesses, whereas in the small intestine, the firm mucus is much thinner than the loosely adherent one. This is in accordance with the role of the

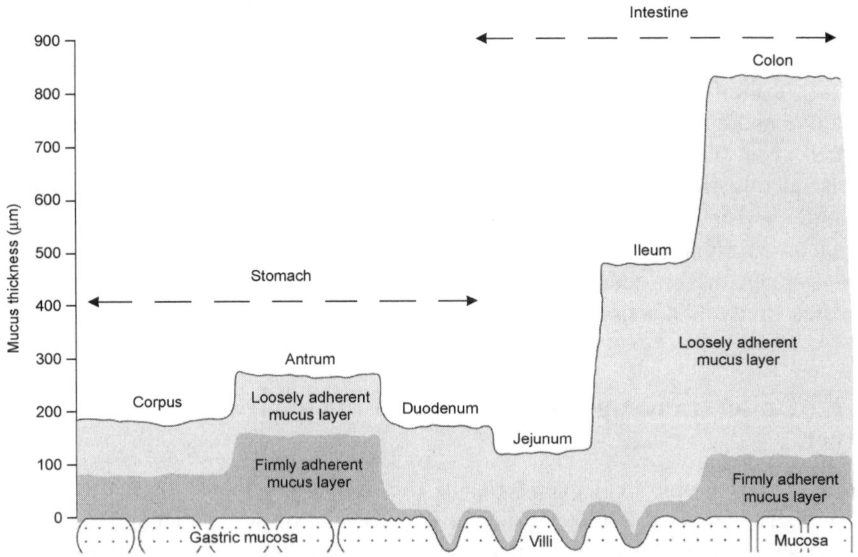

Fig. 2.1 Thickness of the entire mucus layer in the gastrointestinal tract, from Atuma *et al.* (2001).

mucus layer in the different parts of the gut. Indeed, since the mucus mainly acts as a lubricant in the small intestine, it is consistent to have a thick layer of loose mucus. On the other hand, in the stomach lubrication and protection are both important features. So, it is not surprising that both layers have similar thickness. It is still unclear why two different types of mucus can be formed from similar components. One reason may be that the dilution of the molecules within the mucus layer is gradually increased going from the epithelial cell to the luminal surface.

The viscosity of the mucus layer can be affected by external factors. For example, the extent of the viscoelastic properties depends on the nature of the mucin molecules and the water content. Small variations in the amount of water in the mucus result in significant changes in its rheological properties (Wolf *et al.*, 1977; Fleetwood *et al.*, 1986). Changes in the amount of ions, soluble proteins within the mucus modulate also the swelling of the mucus (Tam and Verdugo, 1981).

Hydrophobicity of the mucus layer and surface activity

In addition to the viscoelastic properties outlined above, the hydrophobic properties of the mucus also plays a role in the protection of the epithelium (Jimenezcastellanos *et al.*, 1993; Khanvilkar *et al.*, 2001). The hydrophobicity of the mucus has been the subject of many investigations in the past (Allen, 1981; Smith and Lamont, 1984; Roussel *et al.*, 1988; Bansil *et al.*, 1995; Lichtenberger, 1995; Shi and Caldwell, 2000; Atuma *et al.*, 2001; Taylor *et al.*, 2003; Allen and Flemstrom, 2005; Sanders *et al.*, 2000). All conclusions agree with the fact that lipids (20% of the organic material in the gastric mucus), glycero-glucolipids and ceramides including various oligosaccharide compositions are responsible for the hydrophobic character of the mucus layer. Even if the role of each type of lipids in the hydrophobicity is still unclear, evidence supports the idea that, in the case of the phospholipids, their established amphiphilic properties allow them to form a hydrophobic barrier (Lichtenberger, 1995).

In addition to their role in the global hydrophobicity of the mucus layer, phospholipids included in the mucus also act at the surface. At the lumen surface, strong surface activity (characterized by low wetting and spreading properties) is the result of the presence of the phospholipids. In this part of the layer, some phospholipid molecules may be localized such that their hydrophilic heads are oriented towards the mucus, with their hydrophobic tails dangling in the lumen (Sanders *et al.*, 2000). In the stomach, this hydrophobic layer is necessary for the protection of the epithelial cells since it offers a good barrier towards the protons present in the gastric fluid. In contrast, at the epithelial interface, the phospholipids are not involved in the strong cohesional properties of the mucus to the epithelial cells.

Varying pH within the mucosal layer

Among the molecules that can be harmful for the epithelial cells and that must be stopped by the mucus layer are acid molecules. These molecules are dangerous because they can damage the epithelial cells and even cause their death. The formation of a stable, unstirred layer at the mucosal surface that prevents immediate mixing of the secreted HCO_3^- with the excess acid in the lumen, ensures a neutral pH at the mucosal surface and generates a pH gradient across the mucus (Allen and Flemstrom, 2005; Corfield *et al.*, 2001). This phenomenon can occur because of the diffusion of bicarbonate ions from the epithelial cells to the lumen and the neutralization of the back-diffusing acid (Phillipson, 2003). The existence of a pH gradient in the gastric mucus during acid secretion is a physiological paradox, since acid molecules, secreted by the gastric glands, have to pass through the mucus layer to reach the gastric lumen. Hydrodynamic theory gives some clues to explain these phenomena. Indeed, hydrochloric acid secreted under pressure in the gastric mucus may reach the lumen thanks to the phenomenon of viscous fingering. This phenomenon is a process by which a complex interface is developed when a fluid of low viscosity is injected into a more viscous fluid; the flowing fluid may penetrate the stationary fluid but part of this stationary fluid remains behind the interface and forms 'fjords', named 'viscous fingers' (Bhaskar *et al.*, 1992). These 'viscous fingers', maintained as single channels, may ensure the diffusion of HCl from the mucosal layer to the lumen and simultaneously these channels prevent the backwards diffusion of HCl after emerging at the surface (Bansil *et al.*, 1995).

2.2.2 Introduction to the mucin molecules

It is worth noticing that the viscoelastic properties of the mucus detailed above are mainly governed by the mucin molecules (0.5–5% of total weight of mucus in the gastrointestinal tract). Mucin molecules are glycoproteins of high molecular weight ranging from hundred thousands to several million Daltons, containing hundreds of carbohydrate chains, attached to the peptide backbone by O-glycosidic linkages between *N*-acetylgalactosamine and a hydroxylated amino acid (serine or threonine). The mucin molecules can be divided into two families with respect to their ability to form viscoelastic gels as detailed below (see pages 34–35) (Filipe, 1979; Roussel *et al.*, 1988; Strous and Dekker, 1992). However, the generic structure and composition of the mucin molecules are the same for all types of mucins.

Chemical composition of the mucin molecules

An exact chemical composition of mucins cannot be obtained since it depends on the origin of the sample, the physiological conditions of the animals or the humans from whom the sample was taken. It is worth noting that to be able to study mucins, purification of an extract of the mucus must be performed. This purification involves many steps during which degradations can occur. This explains why the molecular weight of mucin molecules have been claimed to range from 0.5 to 20 MDa. It is now widely accepted that the mucin molecules of higher molecular weight are made up of multiples of a basic unit (Harding,

Fig. 2.2 Structure and composition of a basic unit of mucin, from Dekker *et al.* (2002).

1989). Recently, the architecture of a typical mucin molecule was presented and is reproduced in Fig. 2.2 (Dekker *et al.*, 2002). Mucin molecules are composed of the following:

- *A protein backbone* (Sharon, 1966; Ginsburg and Neufeld, 1969; Kornfeld and Kornfeld, 1976). The protein backbone, which represents about 10–30% of the total weight of the basic unit of the mucin molecules, consists of a single polypeptide chain. This chain is composed of a large number of repeating units that are rich in serine, threonine and proline. The presence of the two first amino acids is important, since they are linking the protein backbone to the carbohydrate side chains. As seen in Table 2.2, it is the hydroxyl group of these molecules, which enables such links. The large proportion of proline plays an important role in the conformation of the mucin molecules, as discussed below (see pages 33–34).
- *Carbohydrate side chains.* The carbohydrate side chains are composed mainly of five monosaccharides including *N*-acetylglucosamine, *N*-acetylgalactosamine, galactose, fucose and sialic acid (See Table 2.2). The side chains are composed of 2–20 monomers. They usually contribute to more than 50% of the molecular weight of the mucin molecule (Strous and Dekker, 1992; Dekker *et al.*, 2002). Although the carbohydrate side chains exhibit moderate branching, they are considerably heterogeneous in terms of length and complexity (Harding, 1989). As one can observe in Fig. 2.2, the carbohydrate side chains form a 'bottle-brush' around the protein backbone. It is worth noting that the carboxylic acid group on the sialic acid molecules is the major source of negative charges in the mucin molecules.
- *Cysteine-rich domains.* The cysteine-rich domains are an important part of the mucin molecules, and more precisely of the protein backbone. These regions have been shown to be involved in dimerization via disulfide bond formation, and subsequently polymerization of the dimers to form large mucin molecules. These domains are non-glycosylated regions also called 'naked regions'. In addition to being polymerization sites, these regions are also hydrophobic sites, which is another important feature for the mucin molecules. Their localization and their characteristics are not well defined.

Table 2.2 Molecular structure of the main constituents of each part of the mucin molecules

Constituents of mucin molecules	Molecular structure
Carbohydrate chains	
N-acetylglucosamine	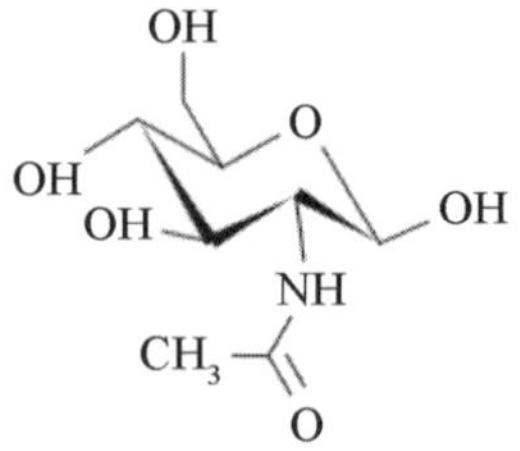
N-acetylgalactosamine	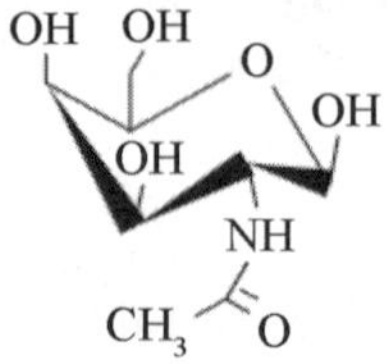
Galactose	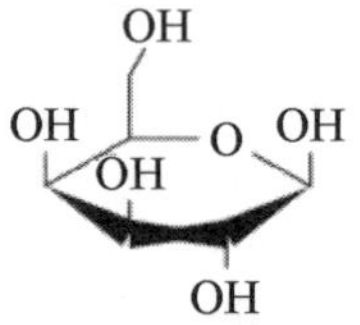
Fucose	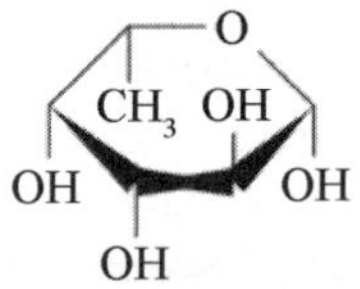
Sialic acid	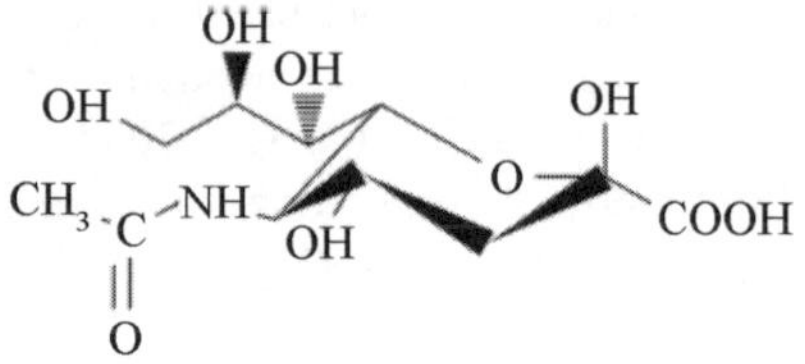

However, some investigations have indicated that in gallbladder mucin, the naked region containing hydrophobic domains is in proximity to an acidic or negatively charged environment (Smith and Lamont, 1984).

- *N- and C-terminal ends*. Few data can be found about the C- and N-terminal ends. This is most probably due to the absence of any specificity as compared to the others parts of the mucin molecules. The amino acid residues constituting these ends are similar to those found in other peptides and proteins (Strous and Dekker, 1992).

Table 2.2 *Continued*

Constituents of mucin molecules	Molecular structure
Protein backbone	
Serine	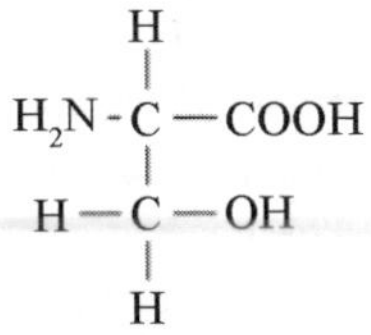
Threonine	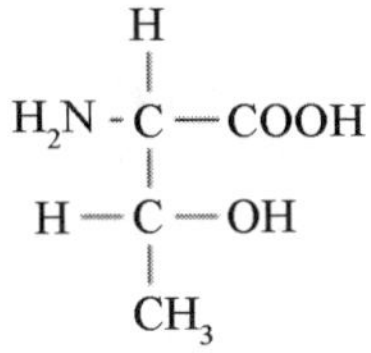
Proline	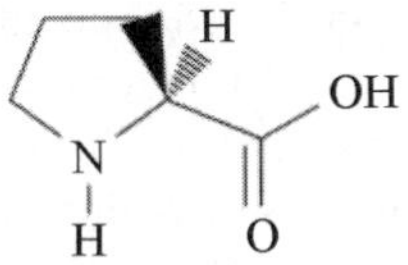
Naked regions	
Cysteine	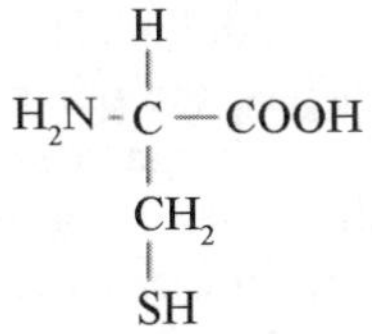

Conformation of single mucin molecules in solution

The exact conformation of the mucin molecules in solution has been difficult to determine because of the complexity of handling this biopolymer. Different models have been suggested after investigations based on imaging methods such as electron microscopy and atomic force microscopy. All studies seem to converge towards the same conclusion. The mucin molecules are suggested to be flexible, and to form filamentous structures with highly elongated rodlike or wormlike conformations (Slayter *et al.*, 1974; Roussel *et al.*, 1988; Bansil *et al.*, 1995). This conformation has been observed as early as 1974 for ovarian cyst mucin (Slayter *et al.*, 1974). In the case of pig gastric mucin molecules, which are similar to the human gastric ones, the length of an elongated mucin molecule ranges from 200 to 1500 nm (Fiebrig *et al.*, 1995; Round *et al.*, 2002; Hong *et al.*, 2005). Such a large range illustrates the heterogeneity of the mucin molecules, in terms of contour length and fibre length.

If one looks in more detail at the conformation of the backbone of the mucin molecules, it appears that although being a glycoprotein, no evidence of folded structures, such as α-helix and/or β-sheets is observed, except for the N- and C-terminal ends, which are similar to globular proteins. Thanks to the high proportion of proline and the trans configuration of this amino acid (see Table 2.2), the protein backbone has a remarkable structure, which is presented as a loosely coiled structure. The ease to adopt the trans configuration makes it ideal for forming bends and turns in the polypeptide backbone (Schimmel and Flory, 1968; Harding, 1989).

Physico-chemical properties of the mucin molecules

The complicated chemical composition of the mucin molecules has the advantage to develop interesting physico-chemical properties. First of all, this polymer is highly negatively charged due to the presence of residues of sialic acid at the extremity of the carbohydrate side chains. Indeed, the carboxylic acid group is an important source of negative charges. Functional groups on the protein core as well as on the side chains also contribute to the charge of the molecules, for example, glutamic acid and aspartic acid in the protein core and sulfate groups on the side chains. It is commonly mentioned that the mucin molecules are fully charged at pH above 2.6, which is the pKa value of sialic acid. However, since negative charges come from several functional groups with different pKa values, it seems more correct to consider that the global charge of the mucin molecules varies with the pH in a more or less continuous way for pH values between 1 and 7. Thus, the mucin molecules are fully charged at pH 7 and the amount of negative charges decreases by decreasing the pH until the neutralization point at about pH 1. From a physiological point of view, this means that the mucin molecules evolve from fully charged at the epithelial surface to neutralized or weakly charged at the lumen surface within the mucus layer in the stomach.

The second important feature from a physico-chemical point of view is the hydrophobicity of the mucin molecules. Although most hydrophobic groups belong to the protein core and more precisely to the naked regions, some functional groups in the carbohydrate side chains have similar properties. This is the case for fucose. The methyl group at the equatorial position confers a significant degree of hydrophobicity to that region of the molecule (Jimenezcastellanos *et al.*, 1993).

Finally, the rheological properties of mucin are the result of disulfide bridges between peptide backbones. Thus, the mucin molecules form a viscoelastic gel with specific physical properties, which allows the gel to flow (Strous and Dekker, 1992). The importance of the disulfide bridge in the viscoelasticity of the mucin gels is well illustrated with the two types of mucin molecules. The first type, named *secretory* mucin, is capable of forming a viscous gel due to its ability to form intermolecular disulfide bridges. The second type of mucin lacks this ability and a much looser gel can be formed (Strous and Dekker, 1992). However, the relative proportion of the two types of mucin molecules within the firmly and the loosely adherent mucus layers is still unclear.

Since the mucin molecules consist of hydrophobic patches, negative charges and disulfide bridges, it is not surprising that they have been compared to hydrophobically modified polyelectrolytes in order to explain the mechanism of the structural changes of cervical mucus (Harding, 1989) and of pig gastric mucin (Lafitte *et al.*, 2007a). The degree of flexibility of the macrostructure is also in agreement with the behaviour of polyelectrolytes, which have among other properties the ability to form a loose, highly solvated, and greatly expanded structure (Evans and Wennerström, 1999; Holmberg *et al.*, 2003).

2.2.3 Structure of the mucin molecules and their arrangement within the mucus layer: concentration, pH and temperature dependence

We have presented above the conformation of single mucin molecules when they are in solution. However, by increasing the concentration above 0.5 wt%, aggregation of molecules occurs and network starts to be formed (Cao *et al.*, 1999). The conformation of the mucin molecules is then affected by this aggregation, which depends on the concentration. By increasing the concentration of the mucin molecules in solution, different aggregates are formed.

First, at low concentration, i.e. in the range 0.5–2% of total weight of the sample, mucin molecules form a loose gel. The conformation of the mucin molecules is also dependent on the pH value of the solution. At neutral pH, the mucin molecules have been seen as fibre-like molecules, which form loose (Hong *et al.*, 2005), polydisperse and partially penetrating aggregates (Cao *et al.*, 1999). At pH 4, the conformation of the mucin molecules has been suggested to change drastically due to unfolding of the protein backbone (Lee *et al.*, 2005). Hydrophobic patches, trapped within the protein-fold at pH above 4, would then be able to interact and create intermolecular connection nodes. At low pH, the mucin molecules seem to form a heterogeneous network, with a wide distribution of pore sizes of the clusters, stabilized mainly by hydrophobic interactions (Celli *et al.*, 2005, Maleki *et al.*, 2007). Lowering the pH from 4 to 2 results in an increase of mean aggregate sizes as found by AFM (Hong *et al.*, 2005) and DLS (Maleki *et al.*, 2007).

If the concentration of the mucin molecules is increased even more up to the upper limit of physiological conditions, i.e. 5% of total weight of the mucus, a strong gel is formed. At such concentrations, the mucin molecules, extracted from pig gastric mucus, form a loose network of elongated chains at pH 7. In this type of network, the inter-molecular interactions are weak. By decreasing the pH to 4, the network gains in stability and the mucin molecules have a low motion (Maleki *et al.*, 2007). Finally, by decreasing the pH even more, down to 1, the mucin molecules collapse and large hydrophobic clusters are formed. The gel constituted by these clusters interspersed by large water domains is then much weaker than the gel at pH 7 (Lafitte *et al.*, 2007a).

The changes in the structure of the mucin network with pH generate modifications in the viscosity of the mucin gels (Bhaskar *et al.*, 1991; Bansil *et al.*, 1995; Celli *et al.*, 2005; Lafitte *et al.*, 2007a). Because of the heterogeneity of the mucin molecules, it is difficult to predict exactly the viscosity profile as a function of pH. However, in all studies the same trends were observed; i.e. at pH above 4, the mucin molecules form a strong gel with a tendency to be strong at pH close to 4. At pH below 4, the gels lose their viscoelastic properties and form a weak gel at pH around or below 2.

The physico-chemical properties of the mucin molecules have also been proposed to be salt-sensitive in several studies (Bhaskar *et al.*, 1991; Cao *et al.*, 1999; Lee *et al.*, 2005; Lafitte *et al.*, 2007a). For example, at pH 7, the aggregate size and the conformation of the mucin molecules appear to be modified in the presence of salt, whereas no changes were observed at pH 2–1. The addition of salt to a network formed without salt has a similar effect as a decrease of pH (Lafitte *et al.*, 2007a). This is due to the importance of electrostatic interactions in the equilibrium of the mucin network. Indeed, a decrease of pH as well as an increase of the ionic strength screens the electrostatic interactions, leading to an increase of the relative strength of the hydrophobic interactions.

Finally, contradictory results have been presented concerning the effect of the temperature on the structure of the mucin gels (Cao *et al.*, 1999; Bromberg and Barr, 2000; Lafitte *et al.*, 2007a). The different results may stem from the different origin of the mucin molecules. As far as pig gastric mucin is concerned, changes arising from a change of temperature are of minor importance. Indeed, increasing the temperature from 25 °C to 37 °C (Lafitte *et al.*, 2007a) or even to 55 °C (Cao *et al.*, 1999) in concentrated or diluted systems, respectively, was not responsible for major changes in the structure and arrangement of the mucin molecules.

2.3 Implications of the mucin molecules and the mucus layer in controlled release and delivery of functional food ingredients

An efficient release and delivery of functional food ingredients is subjected to two main phenomena. First, the delivery system has to adhere to the mucus layer. Then, the functional food ingredients included or not in a delivery formulation must diffuse through the mucus layer in order to reach the blood stream. Both phenomena are then influenced by mucin molecules and the other molecules forming the mucus.

2.3.1 Mucoadhesion: a complex mechanism capable of improving the oral delivery in the gastrointestinal tract

Definition and role of mucoadhesion in controlled release

In the GIT, the physiological activity is intense, mainly due to contractions from propulsive and mixing movements (Guyton and Hall, 2000). A controlled-release formulation, such as medicine or functional food, is then submitted to strong forces, which limit its efficiency. One way to improve the efficiency is to enhance the adhesion of the controlled-release formulation to the mucus layer. In the case of the adhesion of two materials where at least one is of a biological nature, we talk about bioadhesion. When the bioadhesion involves the attachment of a carrier to the mucus, the phenomenon is referred to as *mucoadhesion* (Park and Robinson, 1987; Gu *et al.*, 1988). A mucoadhesive controlled-release formulation offers advantages over more traditional dosage forms, including:

- increased residence time leading to enhanced adsorption
- improved intimacy of contact with various biological membranes
- improved bioavailability through the protection of bioactive molecules from physical and chemical degradation
- targeting and localization of a release device at a specific site.

Theories and factors affecting mucoadhesion

The theory behind mucoadhesion is quite complex and a complete understanding is not yet available, though certain elements of the process are clear such as the two main stages. First, the mucoadhesive polymers must spread over the mucus layer to initiate close contact and to increase the surface area of contact. This is the contact stage (Smart, 2005). Then, the adhesive joints must be strengthened and consolidated, leading to a prolonged adhesion (Lee *et al.*, 2000). This is the consolidation stage (Smart, 2005).

One single theory cannot explain such a complex process, and five alternative theories can be found in the literature (Jimenezcastellanos *et al.*, 1993; Peppas and Sahlin, 1996; Huang *et al.*, 2000; Lee *et al.*, 2000; Hägerström, 2003; Smart, 2005). They include electronic, adsorption, wetting, diffusion and fracture theories. Recently, a sixth theory called mechanical theory has been presented as a potential explanation for the phenomenon.

Electronic theory

If the mucoadhesive polymer and the mucus have different electronic characteristics, transfer of electrons across the interface occurs. This movement of coulombic charges sets up an electrical double layer and, subsequently, a system of attractive forces that maintains the mucoadhesive formulation and the surface of the mucus layer in intimate contact.

Adsorption theory

The adsorption theory relies on the interactions between a bioadhesive polymer and the molecules of the mucus via hydrogen bonds, Van Der Waals forces and electrostatic interactions. Hydrophobic interactions can also occur if the mucoadhesive polymer is partially hydrophobic. The adsorption theory also includes the chemisorption theory. The chemisorption is one specific type of adsorption since it involves strong covalent binding between the polymers and the mucus.

Wetting theory

The wetting theory, which applies mainly to liquid bioadhesive systems, is based on the difference in surface tension between the bioadhesive polymers and the mucus layer. This theory can then predict the adhesion of the polymer by calculating the spreading coefficient introduced by Young (Evans and Wennerström, 1999), which has been suggested as the driving force for the wetting process. Indeed, if the spreading coefficient is positive, spreading will occur and consequently an intimate contact and a strong adhesive bond are favoured. On the other hand, a negative value is characteristic of non-spreading and consequently no adhesion will occur.

Diffusion theory

The diffusion theory emphasizes that the adhesion is strengthened by the diffusion of polymer chains into the mucus layer and by the diffusion of strands of the mucin molecules into the bioadhesive polymer. This process is driven by concentration gradients and it is affected by the size of the molecules, their molecular weight and their mobility. The depth of interpenetration of the polymeric chains depends on the diffusion coefficient and the time of contact. In order to have strong adhesion, the interpenetration must be in the range of 0.2–0.5 μm. In general terms, the bond strength for a given polymer is obtained when the depth of penetration is approximately equal to the end-to-end distance of the polymer chains. Another important factor is the nature of the two diffusing species. The diffusion of a bioadhesive polymer is enhanced for polymer having similarities to the mucin molecules regarding the functional groups.

Fracture theory

The fracture theory can hardly be used to explain the mucoadhesion phenomenon. But it is useful to predict the strength of the bond formed between the two adherent layers. This theory relates the adhesive force to the force that is required to separate two surfaces. In this theory, one must assume that the detachment occurs at the interface. However, failure normally occurs at the weakest component, which is typically a cohesive failure within one of the adhering surfaces. This theory is considered to be appropriate for the calculation of fracture strength of adhesive bonds involving rigid mucoadhesive materials.

Mechanical theory

The mechanical theory deals with the enhancement of adhesion when the bioadhesive polymers are in contact with the mucus layer. Indeed, the roughness of the surface may favour the adhesive force because of an increased contact area.

None of these theories can give a precise description of the mechanisms occurring during adhesion, but they identify variables that are important to the bioadhesion process. They imply that different types of bonds may be formed during the mucoadhesion process. Two families of bonds can be discerned: chemical and mechanical bonds.

First, the chemical bonds include strong primary bonds such as covalent bonds and weaker secondary forces (i.e. ionic bonds, Van der Waals interactions and hydrogen bonds). These bonds are usually formed independently of the theory considered to explain the mucoadhesion phenomenon. Even if individually these forces are very weak, the strength of the resulting force at the interface is enhanced by the large number of interaction sites.

Second, the mechanical bonds can be thought of as physical connections between surfaces. Macroscopically, they involve the inclusion of one substance in the cracks or crevices of another. On a microscopic scale, they can involve physical entanglement of mucin strands with flexible polymer chains and/or interpenetration of mucin strands into a porous polymer substrate.

Factors influencing the strength of the mucoadhesion

The formation of chemical and mechanical bonds is subject to many parameters, which can influence their strength. These factors may be related either to the nature of the polymer or to the surrounding media. The main factors are presented below.

Molecular weight and crosslinking of the polymer (Park and Robinson, 1987; Blanco-Fuente *et al.*, 1996; Lee *et al.*, 2000; Woolfson *et al.*, 2002)

The adhesive strength increases by increasing the molecular weight of the mucoadhesive polymer. Good adhesion is only obtained within a range of molecular weights. In general, below a threshold of 100 000 Da, the bioadhesion is poor. On the other hand, the adhesion is not improved above a critical molecular weight. There is no general value for this critical value since it depends on the type of polymer in question.

In addition to the molecular weight, adhesion can also be affected by the degree of crosslinking. By increasing the crosslinking density, the polymer loses part of its flexibility and elasticity. Consequently, the interfacial regions for contact are less deep and subsequently provide a worse environment for entanglement between the adhesive polymer and the mucin molecules. For example, the profound changes in density and mobility of the chains of Carbopol® polymers, which occur when the polymer is subjected to crosslinking prevent a strong adhesion.

The spatial configuration of the polymer (Woolfson *et al.*, 2002)

The optimum adhesion is highly dependent on the spatial configuration of the molecules. In the case of a linear polymer such as polyethylene oxide, bioadhesion is improved by increasing the molecular weight, meaning that the optimum adhesion is obtained for a large polymer. An increased molecular weight implies a larger amount of functional groups and consequently a stronger binding to the mucus layer. In comparison, in the case of a nonlinear polymer such as dextran, an increase of the molecular weight also results in an increased mucoadhesion but to get similar adhesion to that of a linear polymer, a much larger molecular weight is necessary. The nonlinear polymer has a tendency to form a helicoidal structure, where many adhesive groups of the polymer are trapped inside the coil. These groups cannot participate actively in the process of adhesion and thus larger molecules must be used to get similar results as with linear polymers.

Concentration (Gurny *et al.*, 1984; Duchene *et al.*, 1988; Jimenezcastellanos *et al.*, 1993; Smart, 2005)
An optimum bioadhesion is obtained for an optimum concentration, which is not necessarily the highest. A concentration beyond the optimum concentration results in a drop of the bioadhesion because the polymer molecules start to form a coil and the number of chains available for interpenetration becomes limited.

Chain flexibility (Park and Robinson, 1987; Jimenezcastellanos *et al.*, 1993; Lee *et al.*, 2000; Smart, 2005)
Since bioadhesion may occur by interpenetration of polymer chains through the mucus layer, it is conceivable that the polymer chains must have a high flexibility. Park and Robinson (1987) showed that poly(acrylic) acid chains with a high flexibility may create significant depth of the interfacial region for contact and subsequently provide a good environment for entanglement between the adhesive polymer and the mucin molecules. However, if the polymer molecules are too flexible, the entanglement with the mucin molecules is hindered.

Hydration capability (Jimenezcastellanos *et al.*, 1993)
The adhesion properties of polymers also vary according to the degree of hydration. When the degree of hydration is high, the adhesive properties of a polymer are lost due to the formation of slippery, non-adhesive mucilage near the interface, which results in the absence of adhesion between the polymer and the mucus layer.

Functional groups (Huang *et al.*, 2000; Woolfson *et al.*, 2002)
As mentioned in the description of the diffusion theory, similarities between the mucoadhesive polymers and the mucus components favour mucoadhesion. Thus, polymers containing hydroxyl, carboxyl, and some amines and sulfates are potentially good bioadhesives. With such functional groups, hydrogen and dipole–dipole bonding are predominant.

Molecules improving the mucoadhesion of controlled-release formulations
The span of molecules tested for their mucoadhesion is broad. Nevertheless, polymers can be separated into two groups usually called generations. The ‘first generation’ of mucoadhesives comprises the hydrophilic macromolecules containing numerous hydrogen bond forming groups. Recently, a second generation has been investigated. It contains new enhanced materials, which are characterized by their availability to adhere at specific sites.

First generation of mucoadhesives

The first generation of mucoadhesive polymers presents significant formulation challenges, since they are only partly hydrophilic. In addition, they form high viscosity aqueous solutions already at low concentration, which are often pH sensitive. In a more functional type of classification, bioadhesive polymers can be grouped into 'water-soluble' molecules, which are typically linear, and 'water-insoluble' molecules, which form in most cases a swollen network by covalent or ionic bonds via a crosslinking agent. The list of polymers belonging to the first generation is quite long. The most common ones are presented in Table 2.3.

It is a general conclusion that the mucoadhesion of this type of polymer is non-specific and is highly dependent on the structure of the polymers and the pH

Table 2.3 Main bioadhesive polymers and some of their characteristics

Bioadhesives	Properties[a]	Characteristics
Polycarbophil (polyacrylic acid crosslinked with divinyl glycol)	• Mw 2.2×10^5 • Viscous colloid in cold water	• Swellable depending on pH, but insoluble in water • Entangle the polymer with mucus on the surface of the tissue • Hydrogen bonding between the non-ionized carboxylic acid and mucin
Carbopol/carbomer (carboxy polymethylene)	• Mw 1×10^6– 4×10^6 • pH 2.5–3.0	• Excellent thickening, emulsifying, suspending, gelling agent • Common component in bioadhesive dosage forms
Sodium carboxymethyl cellulose (CMC)	• Mw 9×10^4–7×10^5 • pH 6.5–8.5 • Soluble in water	• Emulsifying, gelling, binding agent • Good bioadhesive strength
Hydroxypropylmethyl cellulose (cellulose 2-hydroxypropyl-methylether)	• Mw 8.6×10^4 • Soluble in cold water	• Suspending, viscosity-increasing and film-forming agent • Adhesive ointment ingredient
Alginate	• pH 7.2 • Soluble in water	• Stabilizer in emulsion, suspending agent
Chitosan	• Derivatives of chitin • Mw 3800–2×10^6 • Polycation at acidic pH • Soluble in dilute aqueous acidic solution ($pH < 6.5$)	• Non-toxic polymer • Strong electrostatic interaction with mucus

[a] Mw: molecular weight; pH measured at 1.0% aqueous solution

of the surroundings. As mentioned above, it appears that the higher the cross-linking density, the lower is the effectiveness of the mucoadhesion. Another parameter that has been under investigation in many studies is the role of pH on mucoadhesion. The reason is that most of the polymers studied are derivatives of polyacrylic acid. Guggi and coworkers showed that decreasing the pH below 5 within the polymer could increase the mucoadhesion up to 2–4 fold (Guggi *et al.*, 2004). This may result from the fact that the mucoadhesion happens only when the carboxyl groups are in their acid form. At pH higher than 6, charge repulsion dominates as carboxyl groups are in the basic form, which limits the adhesion. A favourable adhesion is obtained when about 80% of the carboxyl groups are protonated (Park and Robinson, 1987).

One typical example of the first generation of mucoadhesive polymers is chitosan (see Table 2.3). Chitosan is a natural linear biopolysaccharide obtained by alkaline deacetylation of chitin, which is one of the most abundant natural saccharides. These molecules are characterized by a charged NH_3^+ group and a hydrophobic group, which are two potential sites for interactions with the mucin molecules (Harding, 2006). The adhesion of chitosan to the mucin molecules was clearly demonstrated by transmission electron microscopy and atomic force microscopy (Fiebrig *et al.*, 1995, 1997; Deacon *et al.*, 2000). The adhesion results in the formation of large complexes of the order of 1 μm in size (see Fig. 2.3) (Harding, 2006). Under physiological conditions, the formation of these aggregates is due to electrostatic interactions and hydrogen bonding between the positively charged chitosan and the negatively charged mucosal surfaces (Deacon *et al.*, 2000; Dodou *et al.*, 2005). The mucoadhesion of chitosan follows the theories explained above. Chitosan swells and forms a gel-like layer in an aqueous environment, which is favourable for interpenetration of polymer and glycoprotein chains. Then, the electrostatic interactions favour a strong adhesion (Sinha *et al.*, 2004). Although chitosan has presented interesting characteristics and the potential to be a good enhancer of drug permeability, its action is limited. In the case of hydrophilic molecules such as atenolol and mannitol (Schipper *et al.*, 1996, 1999), the enhancement was limited by the propensity for the chitosan to bind to the luminal gastrointestinal mucin (Smart, 2005), and/or by the poor diffusion of the chitosan through the mucus. Another factor that

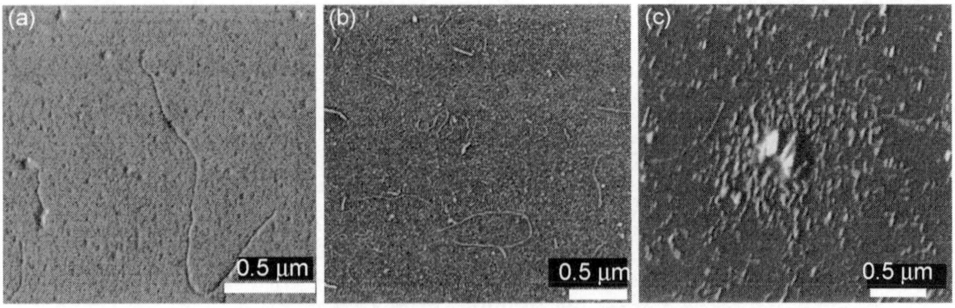

Fig. 2.3 Atomic force microscopy image of (a) mucin, (b) chitosan and (c) chitosan/mucin aggregates in 'phase' mode from Harding (2006).

must be considered in the formulation of drugs is the insolubility of chitosan at pH higher than 6.5, which restricts the mucoadhesion in the stomach and prevents it in the intestine (Dodou *et al.*, 2005). This is the reason why chitosan has recently been modified with thiolated groups (Kast and Bernkop-Schnurch, 2001; Bernkop-Schnurch *et al.*, 2003). These functional groups enhance the solubility of the chitosan and keep or even improve the mucoadhesion properties (Hornof *et al.*, 2003; Dodou *et al.*, 2005). Thus, the thiolated chitosan have the ability to create disulfide bridges between the thiolated groups and the mucin molecules, creating a better adhesion force.

Second generation of mucoadhesives

The second generation of mucoadhesive polymers is characterized by the ability to target the specific mucosal surfaces. One of the most common mucoadhesives of second generation is lectin. Lectins can be defined as proteins or glycoproteins of non-immununoglobulin nature capable of specific recognition and of reversible binding to carbohydrate moieties of complex glycoconjugates, without altering the covalent structure of any of the recognized glycosyl ligands (Irache *et al.*, 1994). Because of their specificity, different lectin molecules interact in different ways with mucin molecules. Table 2.4 presents a list of lectins with their specific sites of binding. The binding capacity of these mucoadhesive polymers depends on the accessibility of the specific sites of binding on the mucin backbone rather than the chemical composition of the mucin molecules. Actually, lectins have efficient mucoadhesive properties in various parts of the body, which indicates that the mucoadhesive properties are

Table 2.4 Main lectins found in drug and food delivery and their specificities

Lectin (*Latin* and common name)	Sugar specificity	Specific site
Lycopersicon esculentum Tomato (TL)	*N*-acetyl-D-glucosamine	Small intestine epithelial surface and strong binding to M cells
Triticum vulgaris Wheat germ (WGA)	*N*-acetyl-D-glucosamine/ Sialic acid	Epithelial cells in stomach, caecum, colon and absorptive enterocytes in small intestine
Galanthus nivalis Agglutinin (GNA)	Mannose	Epithelial cells in stomach, caecum and colon
Lotus tetragonolus Lotus/Asparagus pea(AL)	Fucose	
Ulex europaeus I Furze (UEA)	Fucose	Small intestine and very specific for M cell
Mycoplasma gallisepticum (ML)	Sialic acid	

independent of the type of mucin encountered (Bies *et al.*, 2004). Better mucoadhesion is obtained for lectins binding specifically to sialic acid of the mucin molecules. This is due to the terminal position of sialic acid on the oligosaccharide side chains on the mucin molecules as presented above (see page 31) and consequently to the facility of the lectin molecules to interact with these acid groups. On the other hand, if lectin molecules bind specifically to one amino acid of the protein backbone, one can expect a low mucoadhesion due to limited accessibility. Whatever the efficiency of the lectin-mucin binding, the phenomenon appears to be fully reversible and time-independent.

2.3.2 Transport through mucin gels and mucus layer

Food and/or drug oral uptake is subjected both to the mucoadhesion and to the diffusion of molecules through the mucosal layer before reaching their intended targets. Several investigations have shown that the presence of the mucus retards the transport of drug compounds to a greater extent than an equivalent unstirred aqueous layer (Khanvilkar *et al.*, 2001). Species such as ergot alkaloids (Nimmerfall and Rosenthaler, 1980), aminoglycosides (Niibuchi *et al.*, 1986), fatty acids (Smith *et al.*, 1986) and some antibodies (Kearney and Marriott, 1987) have a significantly retarded diffusion in the presence of the mucus. The retardation phenomenon is most probably a global result of all the components of the mucus. However, two of these components appear to be of great importance: the mucin molecules, which are the most important molecules in terms of gel properties, and the lipids, which greatly influence the diffusion of molecules through the mucus layer.

Transport through the mucin gels: viscosity and proportion of 'free' water

As mentioned above, the mucin molecule is one of the most important components of the mucosal layer in terms of physico-chemical properties. The mucin molecules form a viscoelastic gel with a complex structure. The diffusion of large molecules is then highly retarded by this gel since molecules must find their way through the porous material. In the case of COOH-polystyrene nanoparticles, the mean square displacement was primarily sub-diffusive, indicating that a majority of these particles are temporarily trapped within cages formed by the mucin molecules. Similar results have been obtained for the diffusion of several drugs through the mucus gel. The overall diffusional resistance of the gel layer was evaluated at 50–57% (Bhat *et al.*, 1995). The diffusional reduction caused by the mucin molecules is non-specific and similar mechanisms are probably involved for the diffusion of any diffusing molecules. In addition, the proportion of water molecules trapped within the mucus network and consequently within a mucin gel, which is also called 'free' water, has an important role in the diffusion of molecules. A decrease of the 'free' water results indeed in a decrease of the diffusion rate of a molecule (Bhat *et al.*, 1995).

Hydrophobic and electrostatic interactions between diffusing molecules and the mucin molecules

The mucin molecules are characterized by the presence of hydrophobic patches on the protein backbone and negative charges. One must consider that the existence of hydrophobic and electrostatic interactions between the diffusing molecules and the mucin molecules can decrease the diffusion rate. Various studies indicate that positively charged, low-molecular weight molecules (for example gentamicin, tobramycin (Niibuchi *et al.*, 1986) and some antibiotics (Niibuchi *et al.*, 1986; Bolister *et al.*, 1991)) bind electrostatically to the negatively charged mucin molecules. Even more simple species such as potassium ions have their diffusion retarded due to electrostatic interactions (Desai *et al.*, 1992). On the other hand, the ζ-potential of negatively charged nanoparticles does not play a significant role in the transport of nanospheres through the mucus, as has been demonstrated with the study of modified nanoparticles (Norris and Sinko, 1997).

A good example to illustrate the importance of the hydrophobic interaction between diffusing particles and the mucin molecules is the study of the transport properties of molecules of testosterone. These molecules have been investigated since the 1980s. These studies have been based on the fact that the diffusion of testosterone was reduced by more than four-fold as compared to two-fold for most other drugs, suggesting an unusually strong interaction between testosterone and mucus (Hughes, 1988). However, more recently, a study of diffusion of testosterone and other hydrophobic drugs in solution with mucin has shown that the effect of lipophilicity was general and not specific to testosterone (Wikman-Larhed *et al.*, 1997). In this study, it is shown that the diffusion of mannitol (a hydrophilic molecule) was not influenced by pig intestinal mucin concentrations, whereas a significant reduction in the diffusion coefficient of testosterone (a hydrophobic molecule) was observed which increased the concentrations of pig intestinal mucin. Similar results were obtained for drugs with a higher degree of hydrophobicity. The hydrophobic binding occurs at the non-glycosylated regions of the mucin macromolecules. The naked protein regions, depending on their amino acid sequence, may provide a sufficiently hydrophobic region for efficient drug–protein interactions.

The electrostatic interactions also have an important role in the retardation of the diffusion of molecules. However, the strength of these interactions is quite weak and they can be compensated easily by addition of other molecules. For example, a study of the diffusion of a model carrier for hydrophobic molecules formed by cationic and non-ionic surfactants in solution has shown that the global charge of the carrier is important. Indeed, in the case of a highly positive charged carrier, strong electrostatic interactions between the carrier and the mucin molecules result in the formation of a precipitate. *In vivo,* these strong interactions may result in the formation of solid-like aggregates at the lumen surface of the mucus layer and consequently may hinder the diffusion of the cationic molecules. Simultaneously, the formation of these aggregates may favour the shrinkage of the mucus network and enhances the diffusion of other components of the model carrier such as the non-ionic molecules (Lafitte *et al.*, 2007b,c). However, after addition of non-ionic surfactant, the global charge of the model carriers for hydrophobic molecules decreases, the strength of the interaction between the model carrier and the mucin molecules decreases and no more precipitates are formed (Lafitte *et al.*, 2005, 2007b). *In vivo,* the absence of interactions prevents any physical changes of the structure of the mucus gel and consequently prevents the enhancement of the diffusion of model carrier (Lafitte *et al.*, 2007b,c). This shows the important role of electrostatic interactions during the diffusion of molecules through the mucus layer. Another example is the binding of tetracycline molecules to the mucin molecules (Macadam, 1993).

Finally, other types of interactions can occur when molecules such as drugs and nutrients are diffusing through the mucus layer (Wikman-Larhed *et al.*, 1998). For example, in the case of diffusion of hydrocortisone, it is suggested that other factors, such as hydrogen-bonding capacity, contribute to the interactions between drugs and mucus, decreasing substantially the diffusion of hydrocortisone.

The diffusion of molecules through a mucin gel and consequently to the mucus layer is also affected by the molecular structure of the diffusing molecules. The reduction of the diffusion coefficients is then dependent on the molecular weight of the diffusing particles. For example, over a molecular weight range of 126 000–186 000 Da, a consistent increase of the retarding effect is seen with increasing molecular weight of the diffusing molecules (Desai *et al.*, 1992). The diffusion of large molecules may be inhibit because of their large size compared to the mesh size of the gel (Lafitte *et al.*, 2007a). In the case of microspheres, the diffusion is limited when the particles are larger than 0.5 μm (Norris and Sinko, 1997).

Effects of the other components of the mucus layer: importance of lipids

Although it is difficult to get an exact composition of the mucus, it is accepted that, whatever the origin of the sample, one of its main constituents are the lipids. This explains why many studies have focused on the role of lipids on the diffusion of molecules through the mucus layer. The effect is dependent on the nature of the lipids. Thus, in the case of diffusion of protons through the gastric mucus, most of the lipids appear to play a major role in the retardation of hydrogen ion diffusion through the mucus layer. This effect may be the result of hydrophobic character of the mucin molecules. On the other hand, the way in which phospholipids exert their effect on the retardation of proton diffusion is not readily apparent. This retardation effect may be due to the ability of phospholipids to stabilize the extended molecular structure of mucin molecules (Sarosiek *et al.*, 1984). Similar results have been found with the diffusion of testosterone. By studying different types of mucus (differentiated by the purification process, the reconstitution), Larhed and coworkers (Wikman-Larhed *et al.*, 1998) showed that a diffusing molecule sees its diffusion coefficient varying with the proportion of the lipids in the mucus extract. A similar decrease was reported by Karlsson and co-workers using a cell culture system (Karlsson *et al.*, 1993).

2.4 Conclusion

The mucus layer is a complex aqueous layer in the gastrointestinal tract. It plays an important role in the delivery process of food and drugs since it retards the diffusion of molecules. However, it has also the ability to favour the adhesion of certain polymer molecules and consequently to improve the delivery of the active substances from such polymer matrices. All these steps are subject to many external factors such as the pH of the lumen, the physiological activity of the stomach, etc. Consequently, it is of great importance to investigate further mucoadhesive polymers in order to improve the adhesion of the controlled-release formulations at such interface. Focus must also be put on the diffusion of the active substances through the mucus layer. Indeed, although this process has been under investigation for the past decade, a full understanding is still lacking, which hinders the improvement of controlled-release formulations.

2.5 Acknowledgements

I would like to thank Krister Thuresson (Camurus AB, Lund, Sweden) and Iseult Lynch (School of Chemistry and Chemical Biology, University College, Dublin, Ireland) for discussions about the content of this chapter.

2.6 References

ALLEN A. (1981), 'Structure and function of gastrointestinal mucus', in Press R. (ed.), *Physiology of the Gastrointestinal Tract*, New York, Johnson, L.R., 617.

ALLEN A. and FLEMSTROM G. (2005), 'Gastroduodenal mucus bicarbonate barrier: protection against acid and pepsin', *Am J Physiol Cell Physiol*, 288 (1), C1–19.

ALLEN A. and PEARSON J. P. (2000), 'The gastrointestinal adherent mucous gel barrier', in Corfield A. P. (ed.), *Methods in Molecular Biology: Glycoproteins: Methods and Protocols: The Mucins*, Totowa, Humana Press, 57–64.

ASHFORD M. (2002), 'The gastro-intestinal tract – physiology and drug absorption', in Aulton M. E. (ed.), *Pharmaceutics: The Science of Dosage Form Design*, Leicester, Harcourt Publishers Limited, 217–233.

ATUMA C., STRUGALA V., ALLEN A. and HOLM L. (2001), 'The adherent gastrointestinal mucus gel layer: thickness and physical state *in vivo*', *Am J Physiol Gastrointest Liver Physiol*, 280 (5), G922–929.

BANSIL R., STANLEY, E. and LAMONT, J.T. (1995), 'Mucin biophysics', *Annual Review of Physiology*, 57, 635–657.

BERNKOP-SCHNURCH A., KAST C. E. and GUGGI D. (2003), 'Permeation enhancing polymers in oral delivery of hydrophilic macromolecules: thiomer/GSH systems', *Journal of Controlled Release*, 93 (2), 95–103.

BHASKAR K. R., GONG D., BANSIL R., PAJEVIC S., HAMILTON J. A., TURNER B. S. and LAMONT J. T. (1991), 'Profound increase in viscosity and aggregation of pig gastric mucin at low pH', *American Journal of Physiology*, 261 (5), G827–G833.

BHASKAR K. R., GARIK P., TURNER B. S., BRADLEY J. D., BANSIL R., STANLEY H. E. and LAMONT

J. T. (1992), 'Viscous fingering of HCI through gastric mucin', *Nature*, 360 (6403), 458–461.

BHAT P. G., FLANAGAN, D. R. and DONOVAN, M. D. (1995), 'The limiting role of mucus in drug absorption: drug permeation through mucus solution', *International Journal of Pharmaceutics*, 126 (1–2), 179–187.

BHAT P. G., FLANAGAN D. R. and DONOVAN M. D. (1996), 'Drug diffusion through cystic fibrotic mucus: steady-state permeation, rheologic properties, and glycoprotein morphology', *Journal of Pharmaceutical Sciences*, 85 (6), 624–630.

BIES C., LEHR C. M. and WOODLEY J. F. (2004), 'Lectin-mediated drug targeting: history and applications', *Advanced Drug Delivery Reviews*, 56 (4), 425–435.

BLANCO-FUENTE H., ANGUIANO-IGEA S., OTERO-ESPINAR F. J. and BLANCO-MENDEZ J. (1996), '*In-vitro* bioadhesion of carbopol hydrogels', *International Journal of Pharmaceutics*, 142 (2), 169–174.

BOLISTER N., BASKER M., HODGES N. A. and MARRIOTT C. (1991), 'The diffusion of beta-lactam antibiotics through mixed gels of cystic fibrosis-derived mucin and pseudomonas-aeruginosa alginate', *Journal of Antimicrobial Chemotherapy*, 27 (3), 285–293.

BROMBERG L. E. and BARR D. P. (2000), 'Self-association of mucin', *Biomacromolecules*, 1 (3), 325–334.

CAO X. X., BANSIL R., BHASKAR K. R., TURNER B. S., LAMONT J. T., NIU N. and AFDHAL N. H. (1999), 'pH-dependent conformational change of gastric mucin leads to sol-gel transition', *Biophysical Journal*, 76 (3), 1250–1258.

CELLI J., GREGOR B., TURNER B., AFDHAL N. H., BANSIL R. and ERRAMILLI S. (2005), 'Visco-elastic properties and dynamics of porcine gastric mucin', *Biomacromolecules*, 6 (3), 1329–1333.

CORFIELD A. P., CARROLL D., MYERSCOUGH N. and PROBERT C. S. J. (2001), 'Mucins in the gastrointestinal tract in health and disease', *Frontiers in Bioscience*, 6, D1321–D1357.

DEACON M. P., McGURK S., ROBERTS C. J., WILLIAMS P. M., TENDLER S. J. B., DAVIES M. C., DAVIS S. S. and HARDING S. E. (2000), 'Atomic force microscopy of gastric mucin and chitosan mucoadhesive systems', *Biochemical Journal*, 348, 557–563.

DEKKER J., ROSSEN J. W. A., BULLER H. A. and EINERHAND A. W. C. (2002), 'The MUC family: an obituary', *Trends in Biochemical Sciences*, 27 (3), 126–131.

DESAI M. A., MUTLU M. and VADGAMA P. (1992), 'A study of macromolecular diffusion through native porcine mucus', *Experientia*, 48 (1), 22–26.

DODOU D., BREEDVELD P. and WIERINGA P. A. (2005), 'Mucoadhesives in the gastrointestinal tract: revisiting the literature for novel applications', *European Journal of Pharmaceutics and Biopharmaceutics*, 60 (1), 1–16.

DUCHENE D., TOUCHARD F. and PEPPAS N. A. (1988), 'Pharmaceutical and medical aspects of bioadhesive systems for drug administration', *Drug Development and Industrial Pharmacy*, 14 (2–3), 283–318.

EVANS D. F. and WENNERSTRÖM H. (1999), *The Colloidal Domain: Where Physics, Chemistry, Biology and Technology Meet*, New York, Wiley-VCH.

FIEBRIG I., HARDING S. E., ROWE A. J., HYMAN S. C. and DAVIS S. S. (1995), 'Transmission electron microscopy studies on pig gastric mucin and its interactions with chitosan', *Carbohydrate Polymers*, 28 (3), 239–244.

FIEBRIG I., VARUM K. M., HARDING S. E., DAVIS S. S. and STOKKE B. T. (1997), 'Colloidal gold and colloidal gold labelled wheat germ agglutinin as molecular probes for identification in mucin/chitosan complexes', *Carbohydrate Polymers*, 33 (2–3), 91–99.

FILIPE M. I. (1979), 'Mucins in the human gastrointestinal epithelium: a review', *Invest. Cell Pathol.*, 2 (3), 195–216.

FLEETWOOD L., LANDGREN, B. and ENEROTH P. (1986), 'Quantitation of human cervical mucin during consecutive days and hourly during one day at midcycle', *Gynecologic and Obstetric Investigation*, 22 (3), 145–152.

GINSBURG V. and NEUFELD E. F. (1969), 'Complex heterosaccharides of animals', *Annual Review of Biochemistry*, 38 371–388.

GU J. M., ROBINSON J. R. and LEUNG S. H. S. (1988), 'Binding of acrylic polymers to mucin epithelial surfaces – structure–property relationships', *CRC Critical Reviews in Therapeutic Drug Carrier Systems*, 5 (1), 21–67.

GUGGI D., MARSCHUTZ M. K. and BERNKOP-SCHNURCH A. (2004), 'Matrix tablets based on thiolated poly(acrylic acid): pH-dependent variation in disintegration and mucoadhesion', *International Journal of Pharmaceutics*, 274 (1–2), 97–105.

GURNY R., MEYER J.-M. and PEPPAS N. A. (1984), 'Bioadhesive intraoral release systems: design, testing and analysis', *Biomaterials*, 5 (6), 336–340.

GUYTON A. C. and HALL J. E. (2000), 'General principles of gastronitestinal function-motility, nervous control, and blood circulation', in Schmitt W. (ed.), *Textbook of Medical Physiology*, Philadelphia, Saunders, 718–727.

HäGERSTRÖM H. (2003) 'Polymer gels as pharmaceutical dosage forms: rheological performance and physicochemical interactions at the gel-mucus interface for formulations intended for mucosal drug delivery'. Department of Pharmacy, Uppsala, Uppsala Biomedical Center.

HARDING S. E. (1989), 'The macrostructure of mucus glycoproteins in solution', *Advances in Carbohydrate Chemistry and Biochemistry*, 47, 345–381.

HARDING S. E. (2006), 'Trends in mucoadhesive analysis', *Trends in Food Science & Technology*, 17 (5), 255–262.

HOLMBERG K., JÖNSSON B., KRONBERG B. and LINDMAN B. (2003), *Surfactants and Polymers in Aqueous Solution*, Chichester, John Wiley & Sons.

HONG Z., CHASAN B., BANSIL R., TURNER B. S., BHASKAR K. R. and AFDHAL N. H. (2005), 'Atomic force microscopy reveals aggregation of gastric mucin at low pH', *Biomacromolecules*, 6 (6), 3458–3466.

HORNOF M. D., KAST C. E. and BERNKOP-SCHNURCH A. (2003), '*In vitro* evaluation of the viscoelastic properties of chitosan-thioglycolic acid conjugates', *European Journal of Pharmaceutics and Biopharmaceutics*, 55 (2), 185–190.

HUANG Y. B., LEOBANDUNG W., FOSS A. and PEPPAS N. A. (2000), 'Molecular aspects of muco- and bioadhesion: tethered structures and site-specific surfaces', *Journal of Controlled Release*, 65 (1–2), 63–71.

HUGHES D. R. L. (1988), 'The influence of gastrointestinal mucus on drug absorption'. Brighton Polytechnic.

IRACHE J. M., DURRER C., DUCHENE D. and PONCHEL G. (1994), '*In-vitro* study of lectin–latex conjugates for specific bioadhesion', *Journal of Controlled Release*, 31 (2), 181–188.

JIMENEZCASTELLANOS M. R., ZIA H. and RHODES C. T. (1993), 'Mucoadhesive drug delivery systems', *Drug Development and Industrial Pharmacy*, 19 (1–2), 143–194.

KARLSSON J., WIKMAN A. and ARTURSSON P. (1993), 'The mucus layer as a barrier to drug absorption in monolayers of human intestinal epithelial Ht29-H goblet cells', *International Journal of Pharmaceutics*, 99 (2–3), 209–218.

KAST C. E. and BERNKOP-SCHNURCH A. (2001), 'Thiolated polymers – thiomers: development and *in vitro* evaluation of chitosan-thioglycolic acid conjugates', *Biomaterials*, 22 (17), 2345–2352.

KEARNEY P. and MARRIOTT C. (1987), 'The effects of mucus glycoproteins on the bioavailability of tetracycline. 3. Everted gut studies', *International Journal of Pharmaceutics*, 38 (1-3), 211–220.

KHANVILKAR K., DONOVAN M. D. and FLANAGAN D. R. (2001), 'Drug transfer through mucus', *Advanced Drug Delivery Reviews*, 48 (2–3), 173–193.

KORNFELD R. and KORNFELD S. (1976), 'Comparative aspects of glycoprotein structure', *Annual Review of Biochemistry*, 45, 217–237.

LAFITTE G., THURESSON K. and SÖDERMAN O. (2005), 'Mixtures of mucin and oppositely charged surfactant aggregates with varying charge density. Phase behavior, association, and dynamics', *Langmuir*, 21 (16), 7097–7104.

LAFITTE G., SÖDERMAN O., THURESSON K. and DAVIES J. (2007a), 'PFG-NMR diffusometry: a tool for investigating the structure and dynamics of noncommercial purified pig gastric mucin in a wide range of concentrations', *Biopolymers*, 86 (2), 165–175.

LAFITTE G., THURESSON K. and SÖDERMAN O. (2007b), 'Diffusion of nutrients molecules and model drug carriers through mucin layer investigated by magnetic resonance imaging with chemical shift resolution', *Journal of Pharmaceutical Sciences*, 96 (2), 258–263.

LAFITTE G., THURESSON K. and SÖDERMAN O. (2007c), 'Transport properties of cationic/nonionic mixed micelles through a mucin gel studied by MRI: effects of the charge of the mixed micelle, the chain length of the cationic surfactant, and the structure of the gel', submitted to *Journal of Pharmaceutical Sciences*.

LEE J. W., PARK J. H. and ROBINSON J. R. (2000), 'Bioadhesive-based dosage forms: the next generation', *Journal of Pharmaceutical Sciences*, 89 (7), 850–866.

LEE S., MULLER M., REZWAN K. and SPENCER N. D. (2005), 'Porcine gastric mucin (PGM) at the water/poly(dimethylsiloxane) (PDMS) interface: influence of pH and ionic strength on its conformation, adsorption, and aqueous lubrication properties', *Langmuir*, 21 (18), 8344–8353.

LICHTENBERGER L. M. (1995), 'The hydrophobic barrier properties of gastrointestinal mucus', *Annual Review of Physiology*, 57 (1), 565–583.

MACADAM A. (1993), 'The effect of gastrointestinal mucus on drug absorption', *Advanced Drug Delivery Reviews*, 11 (3), 201–220.

MALEKI A., LAFITTE G., KJONIKSEN A. L., THURESSON K. and NYSTROM B. (2007), 'Effect of pH on the association behaviour in aqueous solutions in pig gastric mucin', *Carbohydrate Research* (in press).

NIIBUCHI J. J., ARAMAKI Y. and TSUCHIYA S. (1986), 'Binding of antibiotics to rat intestinal mucin', *International Journal of Pharmaceutics*, 30 (2–3), 181–187.

NIMMERFALL F. and ROSENTHALER J. (1980), 'Significance of the goblet-cell mucin layer, the outermost luminal barrier to passage through the gut wall', *Biochemical and Biophysical Research Communications*, 94 (3), 960–966.

NORRIS D. A. and SINKO P. J. (1997), 'Effect of size, surface charge, and hydrophobicity on the translocation of polystyrene microspheres through gastrointestinal mucin', *Journal of Applied Polymer Science*, 63 (11), 1481–1492.

PARK H. and ROBINSON J. R. (1987), 'Mechanisms of mucoadhesion of poly(acrylic acid) hydrogels', *Pharmaceutical Research*, 4 (6), 457–464.

PEPPAS N. A. and SAHLIN J. J. (1996), 'Hydrogels as mucoadhesive and bioadhesive materials: A review', *Biomaterials*, 17 (16), 1553–1561.

PHILLIPSON M. (2003) 'Acid transport through gastric mucus: a study *in vivo* in rats and mice'. Department of Medical Cell Biology, Uppsala, Uppsala University.

ROUND A. N., BERRY M., MCMASTER T. J., STOLL S., GOWERS D., CORFIELD A. P. and MILES M. J.

(2002), 'Heterogeneity and persistence length in human ocular mucins', *Biophysical Journal*, 83 (3), 1661–1670.

ROUSSEL P., LAMBLIN G., LHERMITTE M., HOUDRET N., LAFITTE J., PERINI J., KLEIN A. and SCHARFMAN A. (1988), 'The complexity of mucin', *Biochimie*, 70 (11), 1471–1482.

SANDERS N. N., DE SMEDT S. C. and DEMEESTER J. (2000), 'The physical properties of biogels and their permeability for macromolecular drugs and colloidal drug carriers', *Journal of Pharmaceutical Sciences*, 89 (7), 835–849.

SAROSIEK J., SLOMIANY A., TAKAGI A. and SLOMIANY B. L. (1984), 'Hydrogen ion diffusion in dog gastric mucus glycoprotein: effect of associated lipids and covalently bound fatty acids', *Biochemical and Biophysical Research Communications*, 118 (2), 523–531.

SCHIMMEL P. R. and FLORY P. J. (1968), 'Conformational energies and configurational statistics of copolypeptides containing L-proline', *Journal of Molecular Biology*, 34 (1), 105–120.

SCHIPPER N. G. M., VARUM K. M. and ARTURSSON P. (1996), 'Chitosans as absorption enhancers for poorly absorbable drugs. 1. Influence of molecular weight and degree of acetylation on drug transport across human intestinal epithelial (Caco-2) cells', *Pharmaceutical Research*, 13 (11), 1686–1692.

SCHIPPER N. G. M., VARUM K. M., STENBERG P., OCKLIND G., LENNERNAS H. and ARTURSSON P. (1999), 'Chitosans as absorption enhancers of poorly absorbable drugs. 3. Influence of mucus on absorption enhancement', *European Journal of Pharmaceutical Sciences*, 8 (4), 335–343.

SHARON N. (1966), 'Polysaccharides', *Annual Review of Biochemistry*, 35 485–520.

SHI L. and CALDWELL K. D. (2000), 'Mucin adsorption to hydrophobic surfaces', *Journal of Colloid and Interface Science*, 224 (2), 372–381.

SINHA V. R., SINGLA A. K., WADHAWAN S., KAUSHIK R., KUMRIA R., BANSAL K. and DHAWAN S. (2004), 'Chitosan microspheres as a potential carrier for drugs', *International Journal of Pharmaceutics*, 274 (1–2), 1–33.

SLAYTER H. S., COOPER A. G. and BROWN M. C. (1974), 'Electron microscopy and physical parameters of human blood group i, A, B, and H antigens', *Biochemistry*, 13 (16), 3365–3371.

SMART J. D. (2005), 'The basics and underlying mechanisms of mucoadhesion', *Advanced Drug Delivery Reviews*, 57 (11), 1556–1568.

SMITH B. F. and LAMONT J. T. (1984), 'Hydrophobic binding-properties of bovine gallbladder mucin', *Journal of Biological Chemistry*, 259 (19), 2170–2177.

SMITH G. W., WIGGINS P. M., LEE S. P. and TASMANJONES C. (1986), 'Diffusion of butyrate through pig colonic mucus *in vitro*', *Clinical Science*, 70 (3), 271–276.

STROUS G. J. and DEKKER J. (1992), 'Mucin-type glycoproteins', *Critical Reviews in Biochemistry and Molecular Biology*, 27 (1–2), 57–92.

TAM P. Y. and VERDUGO P. (1981), 'Control of mucus hydration as a Donnan equilibrium process', *Nature*, 292 (5821), 340–342.

TAYLOR C., ALLEN A., DETTMAR P. W. and PEARSON J. P. (2003), 'The gel matrix of gastric mucus is maintained by a complex interplay of transient and nontransient associations', *Biomacromolecules*, 4 (4), 922–927.

TAYLOR C., ALLEN A., DETTMAR P. W. and PEARSON J. P. (2004), 'Two rheologically different gastric mucus secretions with different putative functions', *Biochimica et Biophysica Acta (BBA) – General Subjects*, 1674 (2), 131–138.

THOMSON A. B. R. and DIETSCHY J. M. (1977), 'Derivation of equations that describe effects of unstirred water layers on kinetic-parameters of active-transport processes in intestine', *Journal of Theoretical Biology*, 64 (2), 277–294.

WIDDICOMBE J. G. (1997), 'Airway liquid: a barrier to drug diffusion?' *European Respiratory Journal*, 10 (10), 2194–2197.

WIKMAN-LARHED A., ARTURSSON P., GRÅSJÖ J. and BJÖRK E. (1997), 'Diffusion of drugs in native and purified gastrointestinal mucus', *J. Pharm. Sci.*, 86 (6), 660–665.

WIKMAN-LARHED A., ARTURSSON P. and BJÖRK E. (1998), 'The influence of intestinal mucus components on the diffusion of drugs', *Pharm. Res.*, 15 (1), 66–71.

WOLF D. P., SOKOLOSKI J., KHAN M. A. and LITT M. (1977), 'Human cervical mucus. III. Isolation and characterization of rheologically active mucin', *Fertility and Sterility*, 28 (1), 53–58.

WOOLFSON A. D., MALCOLM R. K., McCARRON P. A. and JONES D. S. (2002), 'Bioadhesive drug delivery systems', in Dumitriu S. (ed.), *Polymeric Biomaterials*, New York, Marcel Dekker, 1063–1081.

YORK P. (2002), 'The design of dosage forms', in Aulton M. E. (ed.), *Pharmaceutics: The Science of Dosage Form Design*, Leicester, Harcourt Publishers Limited, 1–12.

3

Testing the effectiveness of nutrient delivery systems

E. Acosta, University of Toronto, Canada

3.1 Introduction: a holistic view of delivery systems

To formulate effective controlled-release systems, especially those intended for food applications, it is necessary to consider the different stages that the product goes through, and make sure that the formulation protects the integrity of the active ingredient throughout these stages, but releases the ingredient where and when required. The purpose of this chapter is to identify the key variables that determine the effectiveness of the delivery system at different stages of this journey, and the appropriate test methods to evaluate their ability to protect and release the active ingredient.

This chapter is divided into three parts. The first part of the chapter (Section 3.2) consists of a brief review of the different vehicles and encapsulation methodologies currently in use (or being potentially useful) in food products, and pertinent test methods used to validate the production process, and the stability of the product. The second part of the chapter (Section 3.3) presents a brief review of the test methods employed to evaluate the *in vivo* bioavailability of micronutrients and nutraceuticals. The last part of the chapter (Section 3.4) introduces the issues surrounding the validation of *in vitro* release test methods, and in particular the aspects surrounding the evaluation of controlled-release formulas. A final reflection on the current state of affairs and outlook on these evaluation methods is offered in Section 3.5.

3.2 Delivery systems: formulation, processing and post-processing

3.2.1 Wet (liquid) processing

Most of the processes currently utilized in active ingredient encapsulation for controlled-release purposes can be divided into three categories: wet processing, drying/coating and solid processing. Wet or liquid phase processing not only includes water-based systems but also systems that use organic solvents and hot melts. Controlled-release formulas are initially prepared in liquid suspensions where the formulator has various tools at his disposal to produce and control the desired morphology. Most of these tools have been evolved from food encapsulation technologies.[1,2] Among these tools, the formulator could use carbohydrates, surfactants, proteins and lipids to emulsify or suspend insoluble active ingredients in smaller aggregates (e.g. emulsions, microemulsions); could introduce ionic polymers to produce self-assembled monolayers to coat the aggregates (complex coacervation); could use cyclodextrin and other 'cage' trapping molecules in solution (inclusion complexation); conduct suspension-emulsion-microemulsion polymerization; or use simple coacervation with partially soluble proteins. In the systems that do not require high shear rates, typically a mixer is used. In addition to modifying the chemistry of the suspension, the formulator can also control the hydrodynamics of the system (e.g. colloid mills, wet grinding mills) to produce the desired morphology. New developments in microfluidics have also made possible the production of complicated geometries that could be used in more complex controlled-release strategies. Colloidal mills are used in the production of liposomes, and other lipid nanoparticles. In most cases liquid formulas undergo a process of drying or coating (explained in the next section), but they may also be left as stabilized liquid suspensions.[3,4] However, due to instability issues, drying/coating processes are commonly used to produce more stable solid forms.

There are also other mechanical methods to produce microencapsulated, controlled-release powder formulas such as spray drying, spray cooling, spray chilling, spinning disk atomization, freeze drying (lyophilization), fluidized bed coating, centrifugal suspension separation, and extrusion that will be discussed in the next section.

There follows a short description of the most common liquid phase processing technologies, their uses, and common efficiency parameters.

Emulsification

Emulsification methods are commonly used to produce encapsulations and controlled-release systems for oils (essential oils – flavors, omega-3 fatty acids, antioxidants) in aqueous suspensions that are later dried to fix the droplet into a more stable powder form. There are numerous review articles on emulsions and encapsulation.[1,2,5–8] Figure 3.1 illustrates the process of emulsification and spray drying to produce an encapsulated powder form. The most common form of emulsion encapsulation consists of dispersing the oil of interest (by mechanical

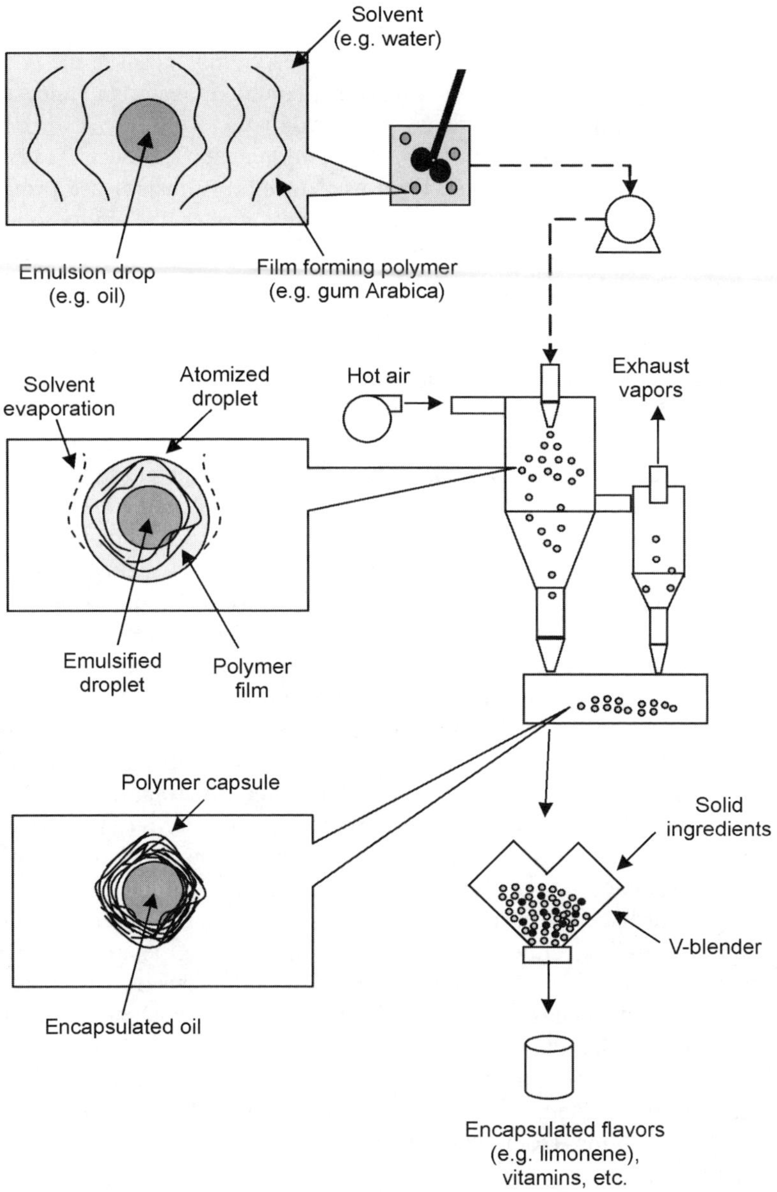

Fig. 3.1 Mechanism of oil encapsulation using a combination of emulsification and spray drying.

agitation) in an aqueous solution of a 'film forming' polymer (typically a carbohydrate), which, upon drying (typically carried out using spray drying), would produce a polymer matrix containing the entrapped oil.

Film forming (wall) materials include gum arabic, maltodextrin, modified starches, modified celluloses among others.[1,7–11] These conventional film forming polymers typically render the microencapsulated oil stable, but do not produce controlled-release properties in aqueous solutions because the polymer (especially in the case of gum arabic) is readily soluble in water and releases the ingredient upon dilution.[7] However, the introduction of more hydrophobic polymeric matrices, such as proteins, Maillard reaction products, and ionic carbohydrates (e.g. chitosan) have been used to modify the release profile of these systems.[3,9,12–14] In the case of protein-based matrices, the ability to induce gelation by heat treatment or crosslinking is the key to reduce the permeability of the polymer matrix which typically results in a more stable and gradual release of the active ingredient.[14] In the case of chitosan, the release of the active ingredient is triggered by the ionization of the polymer when exposed to the acidic environment found in the stomach.[12]

The aqueous solubility, viscosity and emulsification properties of the film forming polymer should be considered at the time of designing the encapsulation formula. Typically, it is desirable that concentrated solutions of the film forming polymer would produce relatively low viscosities (less than 1000 cP) to attain optimal atomization in the spray dryer. Rosenberg *et al.*[15] found that when using a mixed gum arabic–sodium alginate system to encapsulate ethyl caproate, a maximum retention was achieved when the viscosity of the polymer solution ranged between 150 and 200 cP. The emulsifying properties of the polymer are defined in terms of its ability to reduce the interfacial tension between the emulsified oil and water, and its ability to prevent emulsion coalescence either through steric hindrance or by inducing electrostatic repulsion.[6] Gum arabic, modified starch, whey protein and gelatin are good emulsifiers because of the presence of oil-liking (lipophilic) and water-liking (hydrophilic) moieties in their backbone, which allow them to adsorb at the oil–water interface, reducing the interfacial tension and hindering the coalescence of oil droplets.[7] Unmodified starches are poor emulsifiers because of the lack of lipophilic moieties in their backbone. Natural emulsifiers such as lecithin can be added to facilitate the emulsification process. In the latter case, lecithin (a natural surfactant) adsorbs at the oil–water interface, reducing the interfacial tension and forming lamellar phases on the surface of the droplets which contribute to the stabilization of the emulsion.

There are two important efficiency parameters related to emulsion-based microencapsulated systems: one is the encapsulation load (or payload), and the other is the retention (sometimes referred to as core retention) or microencapsulation efficiency.[7,10,14] The encapsulation load refers to the weight fraction of active ingredient in the encapsulated product (dry basis). The retention or microencapsulation efficiency refers to the fraction of the active ingredient (initially added to the emulsion) that is eventually encapsulated in the final (solid) product.

Typical encapsulation loads for emulsion systems range between 10 and 50%, although technically it is possible to encapsulate systems containing as much as 80% of the active ingredients. However, it has been found that as the load increases, the microencapsulation efficiency of volatile oils decreases.[10,15]

This is explained by the fact that as the load increases, the 'thickness' of the polymer matrix encapsulating the oil reduces, which produces both a higher rate of oil evaporation during drying and more defects on the surface of the particle, leading to poorer microencapsulation efficiency and stability.[15]

Another important factor to consider when preparing emulsion systems for the microencapsulation of semi-volatile and volatile oils is the total concentration of solids in the emulsion. It is a general practice to use high concentration of total solids to reduce the amount of water evaporated per unit mass of encapsulated product. Reducing the mass of water that needs to be evaporated reduces the required retention time in the dryer (to achieve a given moisture level in the final product) which in turn reduces the chances of oil volatilization and thermal degradation (especially in the case of unsaturated oils).[15] The correlation between emulsion droplet size and retention efficiency is typically confused by factors associated with encapsulation load (higher loads tend to produce larger particle sizes – see reference 10). The particle size distribution of the final product typically ranges between 0.3 and 10 μm. As the particle size reduces, surface area exposed per unit volume increases, which tends to accelerate the rate of drying, but at the same time also accelerates the rate of oil volatilization and the porosity of the encapsulation system.

In addition to regular emulsion systems, multiple emulsions have also been used in microencapsulation technologies.[3,5,14,16–18] Stepwise preparation is perhaps the most common method of preparation. Water-in-oil-in-water (W/O/W) double emulsion systems have been used to encapsulate proteins and probiotics in hydrophilic environments.[16,18] However, liposome systems are more common for those applications. Oil-in-water–in-oil (O/W/O) systems have been used to formulate controlled-release systems for essential oils and fish oils (polyunsaturated fatty acids).[3,4,17]

O/W/O formulations are produced by first emulsifying the oil to be encapsulated in an aqueous solution containing proteins or other 'curable' film forming polymers, followed by a second step where the first O/W emulsion is now emulsified in a second oil phase (such as vegetable oil) which is maintained at high temperature required to 'cure' the wall polymer. In the case of proteins, this 'curing' process consists of inducing the dehydration/gelation of proteins at high temperature. In the case of Maillard reaction products, 'curing' is produced by the crosslinking condensation reaction between proteins and carbohydrates used as wall materials. The 'solvent' or external oil phase is typically removed by filtration. These 'cured' walls are less permeable than the ones produced by simple emulsification/spray drying, which explains why these systems are more stable and typically produce larger encapsulation efficiency (more than 90%) and loads of up to 50%. By controlling the degree of curing, it is possible to adjust the release profile of the product, and avoid 'burst' release effect of conventional emulsion encapsulation.[3,14]

Nanoemulsions are yet another emulsion-based technology that could be used in controlled-release formulations (although it has not been used in food products thus far).[19] To produce these nanoemulsions, a precursor W/O (Type

II) microemulsion system is diluted in water, to promote a phase change to an O/W system, but in this phase transition the surfactants and/or lipids adsorbed at the oil–water interface go through a lamellar phase which forms a protective 'skin' around the oil droplets.

One efficiency parameter that has not been mentioned so far is cost. This is especially important in the case of food manufacturing due to the small margins in this industry. The simplicity and relative low cost of spray-dried emulsions make them the technology of choice for most applications. However, the higher cost of gum arabic (due to weather and political instability in sub-Saharan Africa) is prompting new developments of film forming polymers, especially in the area of protein-based coatings.

Coacervation

Coacervation is a more advanced version of emulsion encapsulation. In a way, the main difference with the emulsification approach is that the wall around the emulsion droplet is produced in liquid phase and does not require drying the solvent to produce the encapsulating wall.[1,2,7,20] There are two methods of coacervation, a simple polymer coating, and a complex polymer coating. The simple polymer coating requires the use of a polymer which is initially soluble, but can be rendered insoluble after changes in pH and/or temperature (this is especially true for proteins and ionic polysaccharides). Additionally, the polymer has to have a certain affinity for the emulsified phase in order to coat the surface of the droplet. The simple encapsulation begins as a regular emulsion with the polymer adsorbed at the interface between the emulsified phase and the solvent. After that point, the temperature or pH is changed in a way that the polymer becomes insoluble in the solvent and forms a separate phase (coacervate) that coats the emulsified phase.

The complex coacervation approach is illustrated in Fig. 3.2. This complex coacervation consists of mixing two polymers (e.g. one with net negative charge, and another with net positive charge) that form an insoluble complex on the surface of the emulsified droplet. For example, Weinbreck *et al.*[21] microencapsulated various essential oils using a mixture of whey protein (WP) and gum arabic (GA) as coating polymers. According to their procedure, they first produced an emulsion of the essential oil in water in the presence of WP. After that, a concentrated solution of gum arabic was added to the emulsion at a pH when both WP and GA are negatively charged. Then the pH was reduced to a value when WP has a net positive charge and GA has a net negative charge, inducing the complexation of these polymers and their coacervation (phase separation). This combination produced an insoluble protective coating that prevents the emulsion droplets from coalescing. Typically, gelatin is used instead of WP. Another optional step is the use of glutaraldehyde (when allowed by local regulations) as a crosslinking agent.

The encapsulated product is typically separated using filtration/centrifugation. This technology avoids the use of high temperature processes to dry the solvent (an advantage for heat labile active ingredients). However, some systems need to be

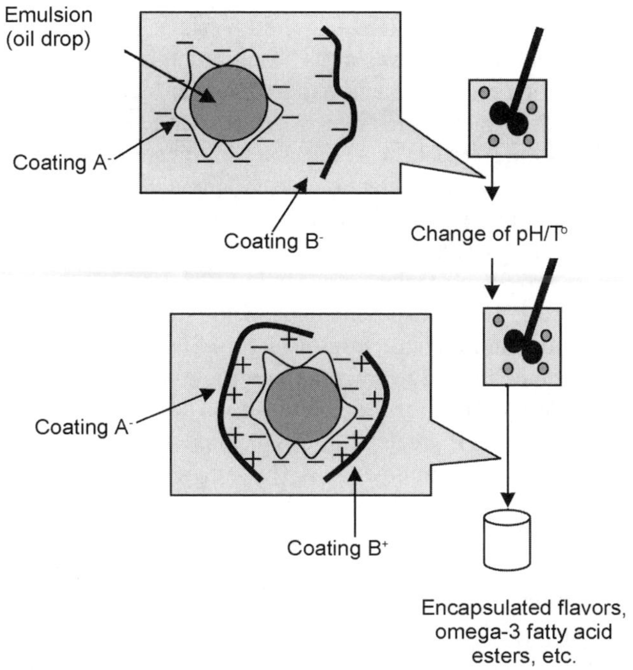

Fig. 3.2 Schematic of the complex coacervation method.

exposed to relatively high temperatures for a prolonged period of time to induce coacervation. In the case of complex coacervation, payloads as high as 90% can be achieved without sacrificing microencapsulation efficiency.[7,21] Unfortunately, due to the high processing cost and complexity of the process, coacervation technology is reserved for special applications such as the encapsulation of expensive flavoring ingredients or to mask strong odors or flavors of nutraceutical ingredients (in particular omega-3 fatty acids and its esters).[22]

If the product of the coacervation process is meant to be left as a liquid suspension, then emulsion stability studies have to be conducted to evaluate the potential of coalescence or phase separation. Weinbreck *et al.*[21] used Turbiscan® measurements to determine the rate of emulsion creaming as a measure of emulsion stability. However, there is no standard emulsion stability test for these formulas. Alternatively, ASTM accelerated emulsion stability tests for emulsions could be adapted for specific applications (ASTM D 1791 Standard Test Method for Accelerated Aging of Liquid Water-Emulsion Floor Polishes). The accompanying regulation chapter has additional details on the conditions and protocols used to evaluate the stability of liquid formulations.

Liposome encapsulation

Liposomes are produced by the association of amphiphilic compounds (mainly phospholipids) into bilayer structures. Using various techniques (solvent evaporation, sonication, microfluidization), this bilayer forms spherical core-shell structures (vesicles). In some cases, these vesicles are nested in onion-like structures called multilamellar vesicles. The use of liposomes in the encapsulation and controlled release of active ingredients in food products is still in its early stages, but has attracted the attention of academic and industrial research groups.[1,20,22] Liposomes are best suited for the encapsulation of hydrophilic compounds such as enzymes, water-soluble vitamins (vitamin C in particular), flavors, and food antimicrobials in liquid suspensions.[23] Among the different methods of liposome production, microfluidization with colloidal mills is perhaps the most promising technology for industrial scale production in food applications. Figure 3.3 presents a schematic of the liposome production method using colloidal mills, and the structure of large unilamellar vesicles (LUV).

It is widely accepted that large unilamellar vesicles (LUV, vesicles larger than 100 nm) are more appropriate for food applications due to the larger payload (up to 45% loading efficiency), greater stability against vesicle fusion, and ease of production.[1,24] However, nanoliposomes (less than 50 nm in diameter) have proven to be an effective method for simultaneous encapsulation of hydrophilic compounds (inside the vesicle) and hydrophobic antioxidants (typically α-tocopherol) solubilized in the hydrophobic part of the lipid bilayer.[25]

The encapsulation efficiency (fraction of active ingredient encapsulated) of liposome depends on the rigidity of the bilayer membrane. Laridi *et al.*[26] encapsulated nisin Z (a lantibiotic) in commercial liposome formulations finding that the encapsulation efficiency of 'soft' liposomes (having melting temperatures near 25 °C) was close to 10%, whereas a 'hard' liposome (melting temperature of 65 °C) had an encapsulation efficiency of 35%. The addition of fluidizing agents such as cholesterol reduced the encapsulation efficiency of these systems.

To determine the encapsulation efficiency of liposomes, Laridi *et al.*[26] used a two-step method. The first step consisted of separating the liposomes from the aqueous suspension using ultracentrifugation. The liposomes (now separated

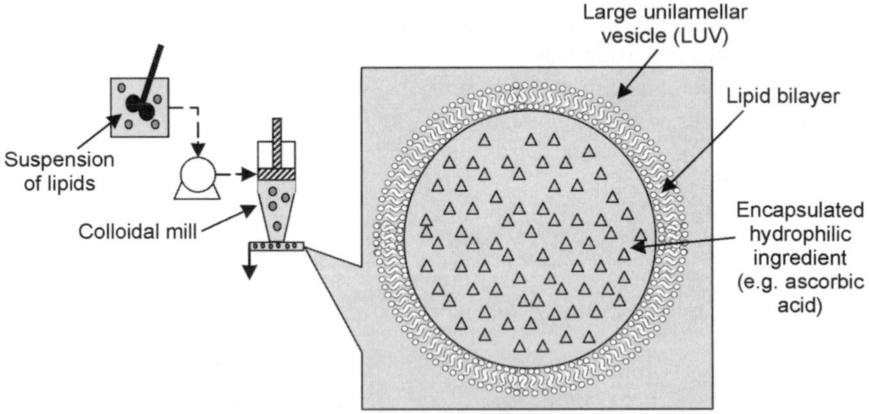

Fig. 3.3 Schematic of liposome (in the form of large unilamellar vesicles, LUV) production using colloidal mills.

from their original aqueous suspension) were then exposed to a detergent (Triton X-100) solution to 'dissolve' the liposomes and release the active ingredient which was quantified using liquid chromatography. The encapsulation efficiency is then calculated as the ratio between the mass of the active ingredient recovered from the liposomes and the mass of the active ingredient added into the initial suspension.[27] It has been proposed that separation methods like the one used by Laridi *et al.*[26] are too disruptive for liposomes and that they tend to underpredict the value of encapsulation efficiency.

The stability of the liposome needs to be considered when assessing the effectiveness of these formulas. Lasic considers that there are four factors that affect the stability of liposomes:[28,29] physical, chemical, colloidal, and biological. The physical stability depends on the natural curvature of the lipid mixture (equilibrium curvature of the liposome), and the rigidity of the bilayer. More rigid membranes (higher melting points) with curvatures (the reciprocal value of the liposome radius) close to their natural curvature would be more stable against perturbations such as temperature increase, shear, vibration, freeze/thawing cycles. Chemical stability refers to the ability of the liposome to maintain the level of encapsulation efficiency with changes in solution chemistry such as pH, electrolyte composition, liposome concentration, oxidizing agents, and presence of surface active compounds (e.g. surfactants, cholesterol, bile salts). Colloidal stability refers to the ability of the liposomes to maintain their size under various storage conditions. The natural curvature of most phospholipids used to formulate liposomes is near zero (flat bilayers); therefore, liposomes tend to fuse and grow. This explains why large liposomes (LUV) are more stable than small ones. Some formulators introduce ionic lipids to maintain a net charge on the surface of the liposome to produce an electrostatic barrier against liposome fusion. Steric barriers such as hydrocolloids (e.g. gum arabic, starch, cellulose) have also been used to prevent liposome fusion. Biological stability refers to the ability of the liposome to withstand the attack of biological agents, specially enzymes and macrophages.

There is not a standard stability test protocol for liposomes, so the formulator needs to consider the specific conditions and the life cycle of the formulation up to the point of release, and simulate these conditions. After the protocol has been carried out, the activity of the encapsulated ingredient, the morphology of the remaining liposomes, and the release profile are usually the main parameters used to evaluate the stability of the system.

Inclusion complexation

Inclusion complexation entails the use of β-cyclodextrin as a 'molecular cage' to encapsulate the active ingredient. β-Cyclodextrin are cyclic carbohydrates where their hydrophilic moieties oriented outwards and their lipophilic interior provide a suitable environment to solubilize organic compounds. Figure 3.4 presents a schematic of inclusion complexation and the structure of cyclodextrins. These formulas are used in the encapsulation of volatile organic molecules (essential oils and vitamins), and are especially useful to mask odors

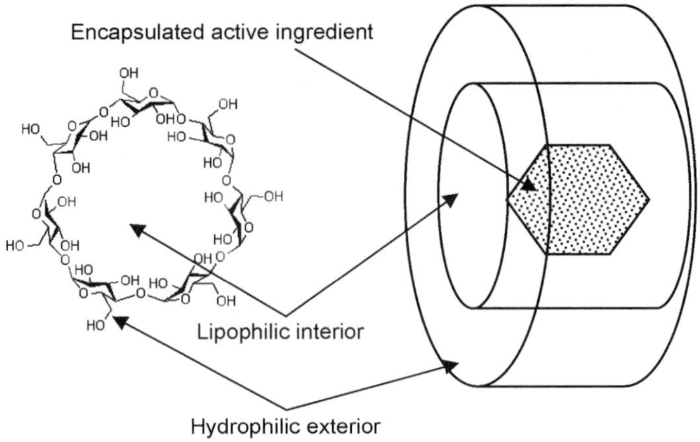

Fig. 3.4 Schematic of inclusion complexation using β-cyclodextin.

and flavors, and preserve aromas. The loading efficiency is typically low (about 11%), but the encapsulation efficiency is high and the encapsulated material is stable up to 200 °C. However, the elevated cost of β-cyclodextrin limit the application of these systems.[1,7,20] Dendrimers are, hypothetically, an alternative to β-cyclodextrin; however, their high cost and potential toxicity are still limiting factors for their application in food encapsulation technologies.

Solid lipid nanoparticles

Solid lipid nanoparticles are produced by melting a lipid matrix such as hydrogenated vegetable oils, and dissolving the hydrophobic active ingredient in this hot melt. After that, this hot melt is incorporated into an oil-in-water nanoemulsion (produced by high shear) or microemulsion system at high temperature. This hot-melt nanoemulsion or microemulsion is then spray congealed to produce submicron particles containing a solid solution of the active ingredient.[30,31] This solid solution typically produces a faster release of active ingredients that tend to form insoluble crystals. The solid solution maintains the active ingredient in a glassy state in which this ingredient is more active. When the release profile of cyclosporine formulated in solid lipid nanoparticles is compared to that of microemulsion systems, the latter produced a 'burst' release effect that is not observed with nanoparticle systems.[32]

The loading efficiency of these systems can vary between 5 and 50% depending on the drug and the lipid matrix. However, most studies have reported 20% loading efficiency.[31] The encapsulation efficiency for lipophilic ingredients has been reported to be has high as 99.6%.[33]

Figure 3.5 presents a schematic of the method of production of solid lipid nanoparticles using microfluidic devices (high shear). This technology has found its way into the formulation of cough syrups.[34] Recently, microemulsion-based solid nanoparticles have been formulated for the controlled release of tea polyphenols.[34,35]

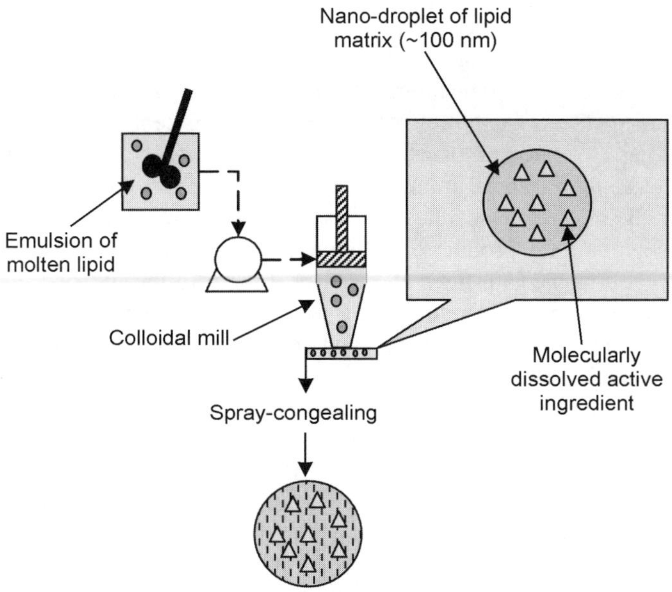

Fig. 3.5 Production of solid lipid nanoparticles using colloidal mills/microfluidic devices.

Microemulsions

Microemulsion systems are composed of oil and/or water nano-domains that coexist in thermodynamic equilibrium.[5,6] Perhaps the best known application of microemulsion in oral formulations is Neoral™ for the delivery of the hydrophilic peptide cyclosporine A, and HIV protease inhibitors Ritonavir and Saquinavir which are commercialized under the name of Norvir® and Invirase®.[36] In those systems, the microemulsion is encapsulated with an enteric coating to protect the integrity of the microemulsion during its passage through the stomach. Once the formulation reaches the small intestine, the encapsulated ingredients form a microemulsion system with or without the aid of bile salts (natural surfactants). These formulations are called self-microemulsifying drug delivery systems (SMEDDS). SMEDDS are used to increase the bioavailability of otherwise poorly adsorbed drugs by reducing the drop size of the emulsified system to less than 100 nm.[36,37]

SMEDDS are formulated as microemulsion precursors, typically with gel-like paste consistency, containing the surfactants, lipids, carrier oil (such as a medium chain triglyceride), and the active ingredient (hydrophobic drugs in most cases). SMEDDS are expected to spontaneously form microemulsion (oil-in-water microemulsions, in most cases) upon dilution in the stomach or the intestines. SMEDDS have been recently used to deliver Isoflavones extracted from Pueraria Lobata (a traditional Chinese medicinal herb), finding that this form of delivery increased the bioavailability (absorption) of this active ingredient by 2.5-fold when compared to a tablet formula.[38] The increased bioavailability obtained with SMEDDS is due to the increased solubilization of the active ingredients in micelles which can either be absorbed directly or increase the local concentration of the active ingredient next to the luminal surface. It has also been proposed that SMEDDS increase paracellular transport through the epithelial cells lining the intestines, increase the residence time of the drug in the intestine (bioadhesion), increase the permeability of the epithelial tissue, and activate the production of chylomicrons (lipoproteins that collect the lipids adsorbed in the intestines and transport them to the liver).[39]

The fluid nature of the surfactant film in microemulsion systems helps increase the rate of mass transfer and adsorption through membranes and epithelial tissues. However, the fluidity of the interface also facilitates phase transitions induced by changes in temperature, composition (e.g. dilution), electrolyte concentration, and pH (these variables influence the thermodynamic equilibrium of microemulsions). One alternative to prevent these phase transitions is to polymerize the core of the internal phase of the microemulsion. For example, enzymes have been successfully encapsulated in silica nanoparticles produced using oil-in-water microemulsions.[40,41] Another interesting alternative that has been explored recently to stabilize microemulsion aggregates is the use of complex coacervation. According to this approach, an oil-in-water microemulsion was coated with a coacervate phase formed after combining gum arabic and gelatin.[42] The encapsulation efficiency of this complex was 60%, and the size of the coated aggregates ranged between 30 and 100 nm.

Leser *et al.* have recently reviewed some of the applications of microemulsions on the encapsulation of antioxidants and flavors.[43] They point out that the thermodynamic stability of microemulsions and the ability of these systems to form spontaneously (minimum energy input) are the main drivers to pursue the use of microemulsions systems in food formulations. However, they summarize the main limitations of microemulsions as follows.

- There is a relatively small pool of potential surfactant candidates with GRAS status.
- High concentration of surfactant and cosurfactant are needed (for example, to encapsulate 10% of oil, typically 20% or more surfactant is required, i.e. loading efficiencies of 30–40%).
- It is not possible to attain a conventional bicontinuous microemulsion when the oil phase is a triglyceride compound (difficult formulations).
- High temperature is needed to produce microemulsion systems with triglycerides.

Similar observations have been made by Flanagan and Singh in a recent review on the use of microemulsions as delivery of bioactives in food.[44]

One way to get around these limitations is the use of GRAS oil cosolvents (e.g. limonene) and aqueous cosolvents (e.g. ethanol, glycerol, polyethylene glycol, etc.) to form 'dilutable' or U-type microemulsions introduced by the group of Garti and Aserin.[45] A detailed description of these systems is provided later in this book. These microemulsions are formulated in such a way that they can be diluted in water without undergoing phase transition. Dilution (dissolution) tests are important to guarantee the stability of microemulsions during food preparation and/or digestion. There are certain dissolution protocols that are discussed in Section 3.4.2. However, there is not a standard dilution test for microemulsions. The formulator should be very careful at the moment of selecting the composition of the diluting solution since changes in the chemistry of the aqueous solutions can bring about phase transitions that affect the ultimate fate of microemulsions. Porter and Charman have indicated that the failure of some microemulsion formulas is likely a result of not considering the chemistry involved in the digestive system.[37]

Another way to circumvent some of the limitations of microemulsions involves the use of linker-based microemulsions.[46] Using the linker approach, lecithin-based microemulsion systems have been produced using biocompatible ingredients with high oil solubilization capacities.[47] In linker-based formulations, lecithin is used as the main surfactant, sorbitol monoleate is used as lipophilic linker (to enhance the lecithin–oil interaction), and a hydrophilic linker such as C6-C8 polyglucosides or a partially saponified C7-C8 fatty acid enhance the lecithin–water interaction. In general, linker systems have high oil solubilization capacity; however, the formulation of single phase microemulsions with pure tryglicerides is still a challenge even for linker systems.

The best approach to formulate a single phase bicontinuous microemulsion with triglycerides involves the use of extended surfactants.[48] Unfortunately, extended surfactants do not have GRAS status.

Certainly there are challenges in microemulsion-based encapsulation but there are a number of success stories that continue to inspire scientists and engineers in this field.

3.2.2 Drying/coating

The spray drying process begins when a suspension containing the active ingredient, and the dissolved polymer matrix is pressurized and then atomized (after passing through a nozzle) to form a 'mist' in the drying chamber (illustrated in Fig. 3.1). A hot gas (air or possibly nitrogen if one or more ingredients tend to oxidize) is also blown into the drying chamber with the objective of evaporating the solvent.[1,2,7,8,15,20,49] Gum arabic and starches are commonly used as the polymer matrix because they tend to form spherical microparticles during the drying process.[50] The microcapsules produced using spray drying have been particularly useful for the controlled release of volatile flavors and aromas.[2,8,50,51]

There are various books on spray drying (in particular, see reference 52) and the discussion on the applications and limitations of the technique is beyond the scope of this chapter. However, it is necessary to briefly identify various parameters of the spray drying process that affect the effectiveness of the delivery system. The first thing one must realize is that spray drying is a kinetically controlled process, and that the properties of the final product (size, shape, porosity) depend on the interplay of simultaneous factors. The spray drying process starts with the formation of the atomized droplet, whose size depends on the surface tension and viscosity of the liquid, pressure drop across the nozzle, and the velocity of the spray. Smaller drop sizes typically lead to smaller particles and faster drying. As the droplet contacts the hot gas, the solvent starts to evaporate and the liophilic polymer (e.g. starch, gum arabic) begins to form a film or 'skin' on the surface of the droplet.[53] As the solvent evaporates, and the vapors permeate the polymer skin, a porous structure within the particle begins to form. At the same time, the liophilic polymer and the active ingredient concentrate, inducing a phase transition into various liquid crystal and solid phases. The mechanical properties of the resulting solid will control the strength of the solid (friability, plasticity), and in some cases, the swelling (dissolution kinetics).[49,52]

Unfortunately, the complexity of the spray drying process is such that, to a great extent, the properties of these systems are adjusted using an empirical (trial and error) approach rather than being set by design (although new trends reveal a move in the industry towards formulation by design). However, there are certain rules of thumb to consider. Particle size (typically ranging from 1 to 1000 μm) affects the appearance (smaller particles are less visible), the density (smaller particles are typically denser – less trapped gasses, and pack better), the wettability (smaller particles are more difficult to wet), and the flowability (smaller particles tend to pack better and agglomerate). Increasing the 'hydrophobicity' of the liophilic polymer tends to decrease the wettability and flowability of the particles. The porosity of the polymeric matrix also has an effect on the rate of swelling/dissolution, but more importantly, it plays a role on the stability of the formulation, since the porosity of the polymer also regulates the rate of oxygen and water uptake, and the rate of evaporation of the active ingredient.[8]

Particle size distribution is determined through static light scattering. Particle morphology (size and shape) can be determined using optical microscopy, and scanning electron microscopy. Particle density and porosity are typically measured using gas displacement pycnometry.[54] Particle flowability can be measured in different ways such as funneling time, and the drained and poured angle of repose.[55] The surface hydrophobicity can be assessed through electron spectroscopy for chemical analysis (ESCA) which quantifies the concentration of fats, proteins, carbohydrates on the surface of the particle.[55] The wettability or dispersability of the particles is measured as the time that it takes to redisperse the powder in water.[56]

Other spray techniques include spray chilling and spray cooling, which are used to encapsulate water-soluble active ingredients (e.g. minerals) in a hydrophobic shell. Typically, a hot melt of fatty compounds (e.g. stearin) is atomized and contacted with chilled/cold air. As the droplet begins to cool, the fatty compounds begin to solidify on the surface of the droplet forming a 'skin' similar to that produced by the film forming polymer in spray drying.

Fluidized bed operations have been used to introduce additional coatings and agglomeration of the primary powder,[57] or as alternative to spray drying/cooling processes.[1,2] Various additives can be added during bed coating: inorganic fillers (silica, carbonates, kaolin) are used to adjust the density of the particle; powdered pigments (e.g. titanium dioxide) are added to mask strong colors; anticaking agents (e.g. starches, celluloses, stereates) are also added to modify the mechanical strength, adhesiveness and wettability of the powder. Diosady *et al.*[58] used fluidized bed coating to encapsulate potassium iodide and potassium iodate with modified starches, gelatin, and sodium hexametaphosphate, finding the method suitable to protect the encapsulated mineral against moisture in extreme environment.

Another drying method that is gaining popularity is the use of centrifugal atomizers and extruders.[1,2,8,20] In centrifugal atomizers, the emulsion or suspension is impinged into a rotating wheel that disperses small liquid droplets into a drying chamber. These atomizers are especially useful for viscous liquids. Centrifugal extruders are based on the same principle, only that in these extruders the suspension of the active ingredient is co-injected with a coating fluid into a perforated rotating disk that produces droplets where the encapsulated ingredient is coated by a film of the coating fluid. From the formulation point of view, it is important to consider the surface tension of the suspension which impacts the Raleigh instabilities that lead to the formation of the droplets.[20] Both centrifugal atomization technologies are typically a part of spray drying or spray congealing processes.

3.2.3 Solid processing

After the powder formula (sometimes called the premix) is produced, one could blend it with another powder carrier such as sugar, salt, flour, powdered milk, cocoa, etc. One of the challenges of incorporating powdered premixes into a powdered carrier is to guarantee homogeneous blending to prevent particle segregation. Various test protocols have been proposed to determine the particle size after powder mixing and handling.[59,60] In the pharmaceutical industry, ultraviolet (UV) spectroscopy and near infrared (NIR) imaging or spectroscopy have been used to perform blend uniformity analysis (BUA).[61–63] For infrared imaging, heat lamps and an infrared camera (e.g. InSb focal array plane camera) are placed on the top of the blender (typically V blenders), and the gray level obtained in the image reflects the level of near infrared light absorbed. After analyzing the gray level for each pixel, the level of dispersion is measured by the relative standard deviation (RSD – the ratio of the standard deviation divided by the average value of the measurement) of the gray level distribution. More commonly NIR spectroscopy is used in modern operations.[62,63] To conduct NIR spectroscopic analysis, sapphire windows are placed in different parts of the blender, and a fiber optic-based NIR spectrophotometer is used to obtain the absorbance spectrum in the near infrared range (1000–2200 nm).[62] The absorbance spectrum represents a 'fingerprint' of the mixture. To apply this methodology it is necessary to identify a specific set of wavelengths that are most sensitive to changes in composition of the active ingredient.[64] Variations of the absorbance levels from the average value are reported in terms of RSD. The actual levels of active ingredient in the mixture and RSD of the distribution are important parameters for claim substantiation. It has been proposed that active ingredient concentration within ±10% of that indicated by the label and a maximum 5% RSD are target values for blending processes.[65] In early 2000, the US Food and Drug Administration (FDA) evaluated the possibility of introducing BUA standards. However, after comments received from various stakeholders, a decision was made to request BUA monitoring and reporting, but no standards were established.[66] Part of the discussions that led to this decision is that the release profile (measured through dissolution tests) supersedes any other processing variable. These tests will be covered in Chapter 17.

Extrusion processes are solid processing operations that can be used to integrate powdered encapsulated materials into formulations of various products such as cereals and pasta. However, within the context of encapsulation and controlled release, extrusion is used to mix the active ingredient (typically a flavor) with a molten encapsulating material (typically a polysaccharide).[1,2,20] The product that exits the die of the extruder is passed through a dehydrating solvent such as isopropyl alcohol where the product cools down and solidifies. The fibers produced in this way are cut into small pieces and dried. The encapsulated products have a shelf life of up to five years.[20] However, the loading efficiency is relatively low (close to 8%).

3.2.4 Post processing: quality control and stability tests

In the preceding sections we have considered the formulation and production technologies involved in controlled-release formulations. Along the way, we have considered the different efficiency parameters of these technologies (e.g. loading efficiency, encapsulation efficiency, and stability). However, the ultimate formulation effectiveness test is to obtain the release profile after simulating the life cycle of the formula leading to the point of release.

In this section we will consider the intermediate steps leading to the final evaluation of simulated stomach dissolution and intestine absorption.

Figure 3.6 illustrates various post production stages. While the conditions of storage in the first three stages of Fig. 3.6 (storage – transport – distribution) are different, it is convenient to simulate them (stability tests) as one, and consider the extreme conditions to which the product will be exposed.

The protocols used to evaluate the stability of liquid and solid products are specified by the good manufacturing practices (GMP) of each country. However, the International Conference on Harmonization (ICH, a multinational organization led by the United States, the European Union, and Japan) has established a set of standard conditions that have been adopted by numerous countries in drafting their GMP regulations.[67] The specific conditions and explanation of the regulations are explained in Chapter 17.

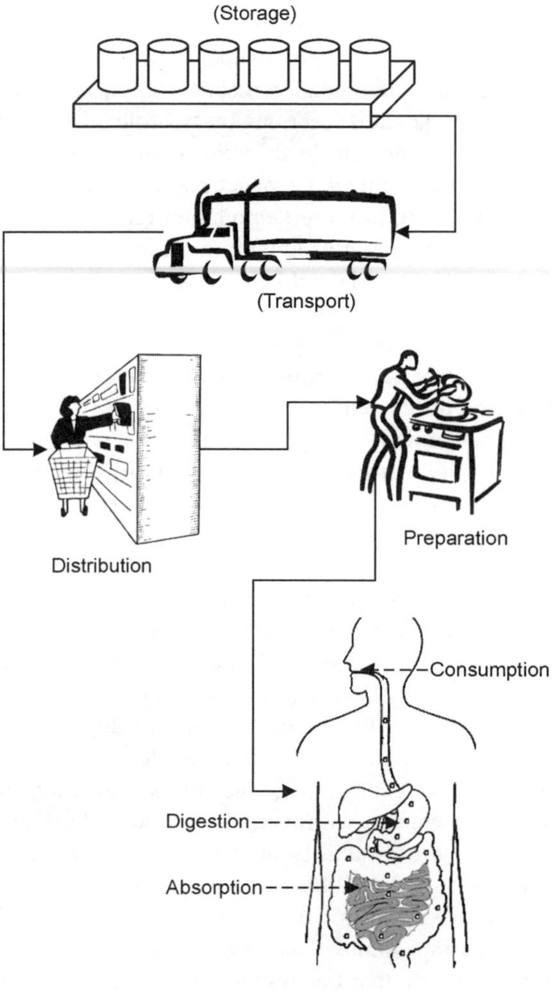

Fig. 3.6 Post-processing stages.

In summary, the ICH regulations indicate that time, temperature, and humidity are the main variables that affect the stability of formulations. A formulation is stable (within a given time) if the concentration of the active ingredient does not vary by more than 5% of the original concentration. Such variations in concentration may be due to the volatilization of the active ingredient or the carrier (in the case of liquid products), or degradation of the active ingredient (e.g. oxidation). In addition to chemical stability, biological stability (presence/growth of microbes), and physical stability (phase separation, coalescence of emulsions or liposomes) are also factors that need to be considered.

The ICH protocols define the conditions of temperature and humidity for different types of products (solids, liquids, non-refrigerated, refrigerated, products with permeable and impermeable packing). Some of the most extreme conditions (used for accelerated stability tests) require exposing the formulations to 40 °C and 75% R.H. In the case of temperature alone, an increase in temperature increases the rate of degradation reaction (an increase of 10 °C typically increases the rate of degradation reactions by twofold). Unfortunately, there is not much that the formulator can do to prevent this effect other than use the appropriate coatings to minimize the penetration of potential reactants (e.g. oxygen) that may participate in these degradation reactions.

The combination of temperature and humidity can bring about changes in the permeability of coatings, as explained below.

Effect of temperature and humidity on stability

As discussed above, the increase in temperature and/or humidity represents a threat to the stability of the encapsulated ingredient. Ubbink and Krüger[68] describe this effect in terms of the glass transition temperature of the encapsulating matrix and the activity (relative humidity) of water. When the temperature of the system is below the glass transition temperature (T_g) of the polymer (e.g. a modified starch), the molecules of the polymer are relatively immobile and water molecules can only diffuse (slowly) through the polymer matrix as single molecules. As the temperature of the system approaches T_g, the molecules of the polymer matrix become mobile and facilitate the formation of small clusters of water molecules that move through channels within the polymer. The formation of these clusters of water depends on the relative humidity of the system. As the relative humidity approaches 100% (water activity = 1), more water molecules tend to 'condensate' in these clusters. To complicate this situation even more, the glass transition temperature of hydrophilic polymers decreases with increasing relative humidity.[69,70]

One way to address the instability of high temperature and high humidity environment is the use of hydrophobic coatings (such as hydrogenated vegetable oils) produced by spray cooling or bed coating.[71] However, it is necessary to guarantee that the melting point of the hydrophobic coating would be higher than the highest temperature that the system will experience during storage and transport.

As a result of the increased permeability of the encapsulating material due to the high temperature and humidity, vapors of volatile flavors and aromas are released, leading to losses in active ingredient. It has also been reported that this increased polymer permeability, induced by humidity, leads to an increase rate of oxidation of encapsulated unsaturated oils.[70,72,73]

3.3 *In vivo* tests

In the preceding sections we have discussed some of the most common formulation and processing alternatives for controlled-release products applied to the food and nutraceutical industry. We have also discussed some of the tests used to evaluate the effectiveness of controlled-release formulas during their production, storage, and distribution. However, the ultimate effectiveness test for controlled-release formulas is to obtain its release profile using *in vivo* tests.

The test protocols for *in vivo* tests used in validation or claim substantiation are regulated by government agencies or professional organizations. The tests protocols prescribed by the US FDA are described in Chapter 17.

In this section of the chapter, the *in vivo* tests and test methods used for research purposes (with emphasis on mineral and hydrophobic nutraceuticals) are described, as well as the physiological parameters that affect the bioavailability of micronutrients and nutraceuticals.

3.3.1 *In vivo* test methods applied to micronutrients and nutraceuticals

Food products (including beverages and nutrition supplements) are not held to the same bioavailability standards as pharmaceutical products, unless there is a specific health-related claim on the label. However, the regulations regarding these products are changing fast. Many countries are starting to recognize that preventive medicine (health through better eating habits, and healthy life style) is a cost-effective alternative to curative medicine. As nutraceuticals gain recognition as health-related products, their bioavailability becomes a greater issue for regulators and manufacturers.[74] A sign of these changes is the initiative of the United States Pharmacopeia (USP) of introducing the program 'USP-verified dietary supplement' seal of approval. This seal guarantees that the product contains the nutrients indicated in the bottle, that the product does not contain harmful contaminants, that the product dissolves and releases the ingredients in the body, and that the product is manufactured following good manufacturing practices.

Another important fact to consider is that food scientists and nutritionists have been studying the bioavailability of micronutrients and nutraceuticals to address concerns regarding the deficiency of these components in our diets. Unfortunately, the interpretation of bioavailability studies in food and related products is often challenging due to the food – active ingredient interactions and the effects of food processing and food preparation techniques. In the next sections, selected *in vivo* studies for minerals and antioxidants will be discussed.

Bioavailability of hydrophilic ingredients (minerals)

Fairweather-Tait reviewed the different methods used to evaluate the bioavailability of micronutrient minerals, in particular iron, zinc, copper, manganese, and selenium.[75] There are three types of minerals, those that are poorly absorbed (<25%, such as Fe, Si, Mn, V, Ni, Cr), those that experience intermediate absorption (25%-75%, Zn, Cu, Se, Mo), and those that are highly absorbed (F, I, B, Co). The methods used to evaluate the bioavailability of minerals include metabolic balance, rate of repletion, plasma appearance, radioisotopes, and stable isotopes.

Metabolic balance consists of monitoring the mass of the mineral consumed, and the mass of the mineral excreted over an extended period of time (weeks). In conducting these experiments it is important to account for other sources of the mineral of interest in the diet. Unfortunately, without any specific tagging, it is not possible to discern the source of the mineral absorbed (from the diet or from the nutritional product). One way to avoid this limitation is the use of isotopes that could be easily identified. The metabolic balance method is best suited for minerals that have intermediate or high absorption, such that the uncertainties of the measurement are significantly lower than the calculated difference between the mass of the mineral ingested and the mass excreted.

Determining the rate of repletion is an alternative method to the metabolic balance for poorly absorbed minerals. The repletion method requires subjects with marked deficiency of the mineral of interest such that the tendency of the body to absorb the mineral is accentuated. The idea is to evaluate the metabolic absorption as a function of time until the organism reaches its homeostatic level (negligible absorption). This time of repletion can be used as an indication of the bioavailability, in particular if one compares the bioavailability from food or from nutritional products.[76]

Using plasma concentration (as suggested by FDA regulations) is suitable for minerals dosed in high concentrations where the rate of absorption is higher than the rate of elimination (in order to avoid the onset of flip-flop at the earlier stages of the test). This method may be appropriate for nutritional products containing minerals that are highly absorbed.[77] The plasma concentration of minerals is typically measured using atomic absorption.

One efficient method to track the mass balance (fraction excreted, metabolized and eliminated) of minerals ingested in a specific food product is through the use of radioisotopes. Radioisotopes can also be used in plasma concentration studies, whole body retention and imaging studies.[78] Radioisotopes were introduced shortly after World War II, when accelerated neutrons produced by nuclear reactors were used to irradiate compounds to render them radioactive. Iron (^{59}Fe) was the first mineral investigated using this technique.[79] The detection of radioactive isotope in bodily fluids is accomplished using scintillation counters that detect the emission of ionizing radiation (beta and gamma photons). In principle, a significant fraction of the mineral in the food product should be present in the radioactive form in order to measure its concentration and complete the mass balance – this is called the intrinsic labeling method. However, it has been found that due to isotope exchange, in some cases one only needs to introduce a radiotracer of the element (in the case of iron, this is accomplished using $^{59}FeCl_3$ or $^{59}FeSO_4$) that could serve as a marker – this is the extrinsic tagging method.[80,81] However, the problem with radioisotopes is the risk associated with the exposure to ionizing radiation. For the most part, the use of radioisotopes for nutritional research in humans is discouraged (and unacceptable for children and pregnant or lactating women) but it is still a useful tool for animal and *in vitro* studies.

Stable isotopes are a safer alternative to radioisotopes. These stable isotopes do not emit ionizing radiation and are naturally present in numerous food products. There are two problems associated with the use of stable isotopes. The first is to find a reasonably pure form of the isotope, and the second involves the detection of the isotope. In a recent review, Abrams and Wong explained that the development of new separation methods for stable isotopes and the rapid expansion of the field of mass spectrometry (ICP-MS, GC-MS, LC-MS) have made this alternative the method of choice for bioavailability studies.[82] Fairweather-Tait and Dainty[83] agree with the previous observation but advise considering the potential interferences with the fraction of the stable isotope already present in the body and in other food ingredients. To this end, it is necessary to carefully select the isotope to be considered (e.g. ^{57}Fe, ^{44}Ca, ^{64}Zn are popular choices for bioavailability studies) and the concentration of the isotope in the foodstuff (intrinsic) or mixed with the food ingredients (extrinsic tag). In addition to minimizing the interference of naturally occurring isotopes, it is necessary to use elaborated multi-compartment mass balance models to separate the changes in the isotope concentration due to the absorption of the isotopic tracer and that due to naturally occurring isotope. Commercial mathematical packages such as SAAM II (University of Washington) have been developed to solve the simultaneous differential equations that arise from such multi-compartmental models. In addition to these mass balance models, phenomenological models have been introduced to account for enhancing or inhibiting factors that affect the rate of mineral absorption.[84]

It is beyond the scope of this chapter to discuss the bioavailability studies for all the minerals used in food fortification or nutritional supplements. However, the bioavailability of iron and calcium will be discussed to illustrate the complexity of designing bioavailability studies and the interpretation of the data.

The first fact that one has to realize is that, unlike with pharmaceutical drugs, the body has evolved to metabolize minerals and micronutrients in order to maintain a homeostatic level of these minerals in bodily fluids and tissue. Therefore, the bioavailability of a mineral dose depends on the degree of mineral deficiency in the test subject (the higher the deficiency, the greater the bioavailability). The homeostatic level of the mineral of interest is not the same for all subjects. In the case of iron, for example, children and pregnant women have greater needs for iron than other groups of the population. Additionally, there are compounds such as ascorbic acid, citric acid, succinic acid and fructose that contribute to iron absorption. On the other hand, phytate, wheat bran, polyphenols, and other minerals hinder the absorption of iron.[75,85]

Iron salts such as ferrous sulfate or ferrous fumarate are found to be readily absorbed by the intestine, thus they are considered bioavailable species of the mineral, although their absorption is less than 30%.[85] Iron fortification premixes or iron supplement formulas are typically designed to release the bioavailable form of iron in simulated stomach dissolution conditions; however, this approach does not take into account the homeostatic level of iron or iron deficiency of the subject. In iron deficient groups, the simple dissolution method is a good indication of the bioavailability of iron.[58] Nutritional scientists point out that, unless there is a severe deficiency of iron, iron supplement formulas (even controlled-release formulas) are typically less bioavailable than iron naturally present in various food products.[77,85,86] The advantage of iron supplementation is that high doses of iron force the body to absorb the iron faster (although not efficiently) and achieve slightly elevated levels of iron in tissues.

Calcium bioavailability is also a matter of concern, especially for young children, and for pregnant, lactating and post-menopausal women. Various calcium salts such as calcium carbonate, calcium citrate, calcium lactate, calcium phosphate, calcium gluconate, and others have been approved for food fortification and supplementation.[85] Calcium absorption typically ranges from 20 to 30%, thus it can be classified as a poorly absorbed mineral. The bioavailability of calcium depends on the level of calcium deficiency, and on the presence of vitamin D which mediates the active transport of calcium through the small intestine.[85,87] Active transport is independent of the concentration gradient across the epithelial tissue, and it is mediated through specific transport channels present in epithelial cells of the small intestine. Passive transport (diffusion/permeation) is less important for the permeation of hydrophilic compounds and minerals.

Calcium fortification in dairy products, soy milk, and mineral water has been evaluated with intrinsic and extrinsic tagging methods using the stable isotopes ^{44}Ca and ^{43}Ca. The concentration of these ions in bodily fluids was measured using ICP-MS, and the mass balance of these ions in the different samples was adjusted using multiple compartment models.[88,89] The consensus of these and other studies is that the calcium in supplement formulas has the same bioavailability as calcium ingested through other common calcium sources such as dairy products. However, a higher calcium dosage in fortification products does lead to higher and faster intake, at a cost of lower bioavailability (fractional calcium absorption).[87] It has been suggested that multiple but smaller dosages throughout the day is a more efficient approach than a large single dosage once a day.[85] This is, in principle, what could be achieved with a controlled-release formula. However, if the encapsulated mineral does not remain in the intestine for the expected release time, then there is no point in pursuing controlled-release products for these applications. To address this concern, bioadhesive polymers and proteins have been introduced into controlled-release formulas with the objective of extending the retention time of the carrier in the small intestine.[90]

Bioavailability of hydrophobic ingredients (antioxidants)

Antioxidants such as lycopene, polyphenol and polyunsaturated fatty acids (PUFAs, including omega-3 fatty acids) have been introduced in various nutraceutical and food products as a way of trapping free radicals that initiate numerous degenerative processes in the body.

The methods of evaluating the bioavailability of organic products (hydrophobics or not) are similar to those employed for minerals. In most cases, however, the bioavailability of hydrophobic organic compounds is higher than that of highly soluble minerals (such as iron and calcium). Therefore the use of metabolic mass balance (ingested minus excreted) without isotopic tracers is a common practice in the field.[91–93] Nevertheless, there is increased interest in the use of stable isotope and mass spectrometry to track the metabolic products of these antioxidants.[94–97]

Unlike minerals, the bioavailability of antioxidants and their metabolic pathways is poorly understood. At this time, it is not possible to find a systematic study on the effect of controlled-release formulas on the bioavailability of antioxidants. It is possible that this lack of systematic studies may be a reflection of the fact that the low solubility of antioxidants in water is the limiting factor in increasing the absorption of these compounds.[22,98–100] Thus most of the studies concentrate on dissolution tests which will be reviewed in Section 3.4.1.

3.3.2 Factors affecting *in vivo* absorption

After a brief overview of the methods to evaluate the bioavailability of minerals and nutraceuticals, it is evident that the uptake of these compounds involves a combination of processes, most of them operating simultaneously. The schematic of Fig. 3.7 illustrates some of these processes.

After the food has been partially digested (mainly by mastication) in the oral cavity, the food goes through a dissolution process in the stomach in acidic conditions (typically simulated using a 0.1 N HCl solution) during a period of time that ranges from 1 to 3 hours. Various enzymes are also released at this point; however pepsin is typically used to simulate the enzymatic action in the

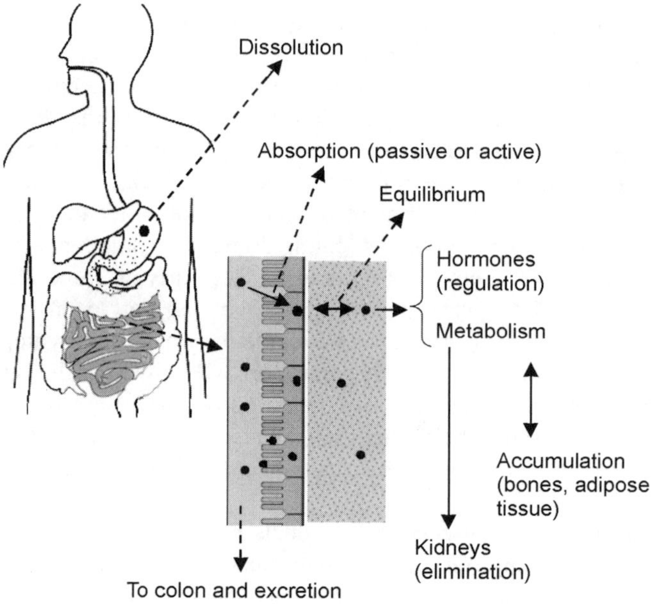

Fig. 3.7 Simplified (generic) mechanism of uptake and elimination of nutrients and nutaceuticals.

stomach. After the suspension leaves the stomach it makes its way to the duodenum where the gall bladder releases the bile salts (such as sodium glycocholate, sodium taurocholate, and lecithin) that help emulsify the fatty components in the suspension. Also in the duodenum the pancreas releases a bicarbonate solution containing a cocktail of enzymes (trypsin among others) that increases the pH of the solution to 6–7. After its short passage through the duodenum, the suspension enters the largest part of the small intestine (4–7 m) where it resides for about 3 to 5 hours before entering the large intestine.

The inner surface of the small intestine is covered with small 'finger-like' protuberances called villi. Each epithelial cell is covered with even smaller protuberances called microvilli (illustrated in Fig. 3.7) that help increase the effective area for nutrient absorption. The absorption of nutrients through the small intestine occurs through two main mechanisms: active and passive transport.

Active transport involves the uptake of the mineral through specific channels on the surface of the epithelial cells. The cells use their own energy to capture and absorb nutrients even when the concentration of the nutrient inside the cell is higher than the concentration outside the cell. As indicated in the preceding sections, this is the main mechanism by which cells capture and absorb highly soluble minerals like calcium and iron. This active uptake is controlled by hormones that regulate the concentration of minerals and other nutrients in the body (see Fig. 3.7). This explains why the bioavailability of minerals and nutrition additives actively absorbed is highly dependent on the level of its deficiency in the body. Increasing the concentration of the mineral or nutrient in the suspension (higher dose) is not an effective method of increasing its bioavailability if its uptake is controlled by an active transport mechanism.

Passive transport occurs by a simple diffusion across the epithelial tissue. In this case the rate and extent of uptake is a function of the difference in the activity of the mineral or nutrient across the epithelial tissue. The activity of a solute is calculated as the product of its concentration times its activity coefficient. The activity coefficient reflects the solubility of the solute in the solvent. Poorly soluble ingredients (e.g. hydrophobic compounds in water) have a large activity coefficient, thus inducing a large driving force for the nutrient (active ingredient) to permeate via a passive transport even when they are present in very dilute solutions. This explains why most hydrophobic compounds permeate through a passive transport while hydrophilic ones permeate via active transport. In the case of emulsified or microemulsified oils that are poorly soluble in water, their rate of uptake also depends on the droplet size, but it is considered that droplets with sizes less than 100 nm are readily absorbed by the intestine.[37,101–103]

The formulator of nutraceutical products has its best chances of modifying the bioavailability profile (using controlled-release formulas) if the active ingredients are primarily absorbed through a passive transport mechanism.

Both active and passive transport are highly dependent on the residence time of the active ingredient in the small intestine. The mucin (glycoprotein) coating of the small intestines helps increase the retention of nutrients and minerals, thus increasing the chances of absorption. Increasing the retention time using bioadhesives such as chitosan and lectins is being actively pursued in controlled-release formulas for pharmaceuticals.[90,104,105] There is preliminary evidence that the use of mucoadhesive lectins improves the retention and absorption of mineral and nutrients.[106,107] However, this area of research is still in its infancy.

Another important factor to consider in the evaluation of the bioavailability of minerals and nutraceuticals is the mechanisms of accumulation of these compounds in the body. As indicated in Fig. 3.7, bones and adipose tissue serve as reservoirs for minerals and hydrophobic molecules, respectively. During infancy and pregnancy there is an increased need for minerals and other nutrients (thus greater chances of deficiency) because a significant fraction of the mass absorbed is accumulated in the tissue. Obesity is also a factor on the absorption of hydrophobic compounds (such as antioxidants). When hydrophobic compounds such as vitamin A or vitamin D enter the plasma, the molecule tends to partition into the adipose tissue where these molecules are preferentially absorbed, thus reducing its concentration in plasma, and creating a deficiency of these compounds in obese subjects.[108–110] In the case of vitamin D deficiency (in obese individuals) this also interferes with the availability of calcium.

This brief review of the methods to evaluate the *in vivo* bioavailability of minerals and nutraceuticals, and the factors that affect these evaluations is by no means complete, but does reflect the level of complexity involved in the evaluation of controlled-release formulas. Because of the difficulties in controlling all the variables that affect bioavailability in human subjects, the costs and logistical issues around these evaluations, *in vitro* test methodologies have been developed over the years. The next section provides a brief review of these methods.

3.4 *In vitro* tests

In vivo tests are typically performed with the objective of demonstrating the effectiveness of the formula and obtaining the approval of regulatory agencies. During the early formulation stages, however, *in vitro* test methods are preferred because they do not require the complicated logistics and resources of *in vivo* tests, and because using *in vitro* tests it is possible to control the variables that affect the dissolution or absorption of the active ingredient. *In vitro* test methods can also be used as quality control tools.

When *in vitro* tests methods are used as a quality control tool or as a replacement for *in vivo* test methods, they are required to follow protocols specified by the regulations of government agencies. Chapter 17 summarizes the test protocols specified under the US FDA regulations. In this part of the chapter, we will explore specific examples of *in vitro* test methods applied to mineral and micronutrients, with a short review of the mathematical framework that relates these *in vitro* evaluations with *in vivo* studies.

Section 3.4.1 describes some of the common experimental protocols used for specific ingredients, and recipes used to simulate gastric and intestinal fluids. Section 3.4.2 presents the mass balance equations that determine the relation between dissolution profile, absorption and *in vivo* (plasma) response. Section 3.4.3 describes the absorption tests used in formulation and research, and the use of tissue and cell cultures to evaluate passive and active transport through the intestine. Finally, Section 3.4.4 describes different tests used to evaluate the bio-adhesive properties of various coatings used in controlled-release formulations.

3.4.1 Dissolution tests

There are two types of dissolution tests that have been used to evaluate the *in vitro* bioavailability of minerals and nutraceuticals: the simple dissolution test (following the USP/FDA guidelines for *in vivo in vitro* correlation or IVIVC), and the combination of dissolution and dialysis (used to simulate the intestine permeation barrier).

Roig *et al.*[111] compared the bioavailability of calcium from human milk, cow milk and infant formula, and compared these values with the fraction of calcium dissolved using dissolution and dialysis tests. For their dissolution tests, these researchers followed the procedure that Miller *et al.* used for *in vitro* iron bioavailability studies.[112,113] In short, this dissolution test consisted of two stages: the first stage involved a digestion at 37 °C with a solution containing 0.1 N HCl, and 0.5% w/v of pepsin for 2 h (gentle shaking), followed by a titration step with a solution containing 0.4% w/v pancreatin, 2.5% bile salts and 0.1M $NaHCO_3$ to a pH of 5. After this pH was attained, the solution was incubated for 2 h at 37 °C. In the case of simple dissolution test, the remaining suspension is centrifuged and a sample of the clear supernatant is taken to measure the concentration of the mineral (calcium in the case of Roig *et al.*[111]). The dialysis method involves the use of a dialysis membrane (10 000 MWCO for Roig *et al.*[111]) to separate the calcium ions from calcium in suspension or bound to macromolecules. In the original method of Miller *et al.*[112] a 7000 MWCO membrane was used. The calcium dialysis experiment is supposed to emulate fraction of calcium that is easily absorbed, thus the bioavailability predicted using this value tends to be lower than the bioavailability predicted by the total calcium dissolution. Roig *et al.*[111] did not find a satisfactory correlation between the *in vivo* bioavailability ranking of the different calcium sources and the ranking they obtained from the dialysis experiments, but found good correlation with the simple calcium dissolution experiments. Roig *et al.*[111] acknowledge that the literature presents conflicting data on which of the *in vitro* dissolution methods (if any) is more accurate to describe *in vivo* bioavailability. They indicate that variables such as fat and protein content in the food can alter the results significantly. It is perhaps not a surprise that this happens since calcium ions that are not bound to proteins are rapidly dissolved, but their permeability *in vivo* is also dictated by mechanisms of active transport that, as discussed before, are controlled by hormonal regulation processes.

The case of iron dissolution tests seems to be a different story. Miller *et al.*[112] showed that even when simulating complex meals, the bioavailability of iron correlated well with the results from the dialysis experiments. He cautions that pH is a variable that needs to be closely monitored (controlled if possible) because the activity of the enzymes is highly dependent on pH.

Judprasong *et al.*[114] introduced a new modification to Miller's digestion method. In short, these researchers constructed a continuous flow dialysis (CFD) apparatus where the simulated gastric fluid (HCL–pepsin solution) containing the digested food was injected in the internal side of the dialysis sac (at time zero). After 30 minutes a solution containing sodium bicarbonate, pancreatin, and bile salts was also injected in the internal side of the dialysis sac. Throughout the experiment, a bicarbonate solution is used as a receiver (permeate) solution. As the minerals dissolve from the dialysis sac and permeated through the membrane they were detected using an online inductively coupled plasma optical emission spectrometer (ICP-OES). This method has two advantages: first, that the data is not taken as one point in time (as is the case with the Miller method) but as a function of time which simulates the time dependence of plasma concentration experiments. The second advantage of this method is that it allows the determination of various ions simultaneously (Ca, Mg, P, Fe and Zn). The authors comment on the good qualitative agreement of their method with known values of absorptions for these minerals.

Reboul *et al.*[115] introduced an *in vitro* digestion (dissolution) method to predict the bioavailability of vitamin E and carotenoids from various dietary sources. The method is similar to the *in vitro* digestion method of Miller *et al.*,[112] with the difference that the stomach dissolution step is shorter (30 min. versus 2 h in the method of Miller *et al.*) and that the digestion takes place at a pH of 3 to 4. The formulation for the pancreatic + bile salt secretion is also similar to the one used in the Miller digestion method, only that the concentration of bile salts is larger (approximately 4% w/v for Reboul *et al.*[115] versus 2.5% for Roig *et al.*[111]) to promote the solubilization of these hydrophobic compounds (in particular lycopene). After the introduction of the simulated pancreatic juice and bile salts, the suspension was incubated for another 30 minutes at 37 °C. The suspension was then centrifuged at 20 000 rpm and 10 °C for 18 h. The hydrophobic compounds solubilized in bile salt micelles were then recovered from the supernatant, and their composition was investigated using reverse phase HPLC. The authors obtained a good correlation between the fraction of alpha and beta carotenes solubilized in micelle solutions *in vitro* and the fraction solubilized *in vivo*.[116] However, the fraction of lycopene (a highly hydrophobic carotenoid) solubilized was underpredicted (50% of the *in vivo* value), and the fraction of lutein (a polar carotenoid) was overpredicted by nearly fivefold. The authors conclude that the method is suitable for carrying out comparative studies of vitamin E and carotenoid bioavailability from different dietary sources, including nutraceutical (food supplement) formulas.

Bajpai *et al.*[117] also compared the release profile of vitamin B12 encapsulated in a hydrogel matrix using a traditional dissolution test and a flow through diffusion cell. Their concern was that the traditional dissolution test does not account for the fact that, as the suspension moves through the small intestine it loses water through the intestinal wall, which in turn concentrates the suspension, increasing the driving force towards the passive transport of the active ingredient. To test their hypothesis they used a 'dripping funnel' configuration whereby the hydrogel beads were exposed to a dripping stream of the simulated intestinal fluid suspension. Samples of the effluent were collected as a function of time and analyzed for their vitamin B12 content. The authors do observe an increase in the fraction of vitamin B12 released using their flow through diffusion cell, which does support their initial hypothesis. However, there was not enough information to form an opinion on the level of water that the intestinal fluid has to lose before this effect becomes substantial.

3.4.2 Mathematical models for *in vitro* dissolution and absorption

The schematic of Fig. 3.7 presents, in a simplified form, the processes that take place after the ingestion of a meal or a dose of a nutritional supplement. According to Fig. 3.7, dissolution is a necessary but not sufficient condition to guarantee the adsorption of the active ingredient. It is necessary to couple this process with the process of permeation through the intestinal wall to evaluate the absorption of the active ingredient. However, in order to couple absorption with *in vivo* plasma concentration, it is also necessary to take into account the processes of elimination and absorption into tissue (e.g. adipose tissue, bones, etc.). The purpose of this section is to give a short introduction to the mathematical models used to account for these processes.

Oh *et al.*[102] and Amidon *et al.*[118] studied the mass balance around the intestine. Figure 3.8 presents a schematic of the process of dissolution and absorption considered in their model. In summary, they used the following assumptions to establish the mass balance.

- The administered dose enters the intestine in the form of monodisperse spherical particles or droplets, and no aggregation or attrition occurs during intestinal transit.
- The intestine is a cylindrical tube of radius R and length L.
- There are no reactions (metabolism) in the intestine.
- The solubility of the active ingredient (C_S) is independent of particle size (r_Z) or changes in pH through the intestinal transit.
- The suspension of the particles moves in plug flow conditions (no dispersion, high Peclet numbers).
- The mass transfer coefficient (for the dissolution of the active ingredient from the particle) obeys the limiting Sherwood Number for viscous flow ($k_l = \mathcal{D}/r_Z$), where $\mathcal{D}$ is the diffusion coefficient of the active ingredient in the luminal solution.
- The concentration of the dissolved active ingredient in the lumen (C_Z) is several times greater than the concentration of the active ingredient in plasma

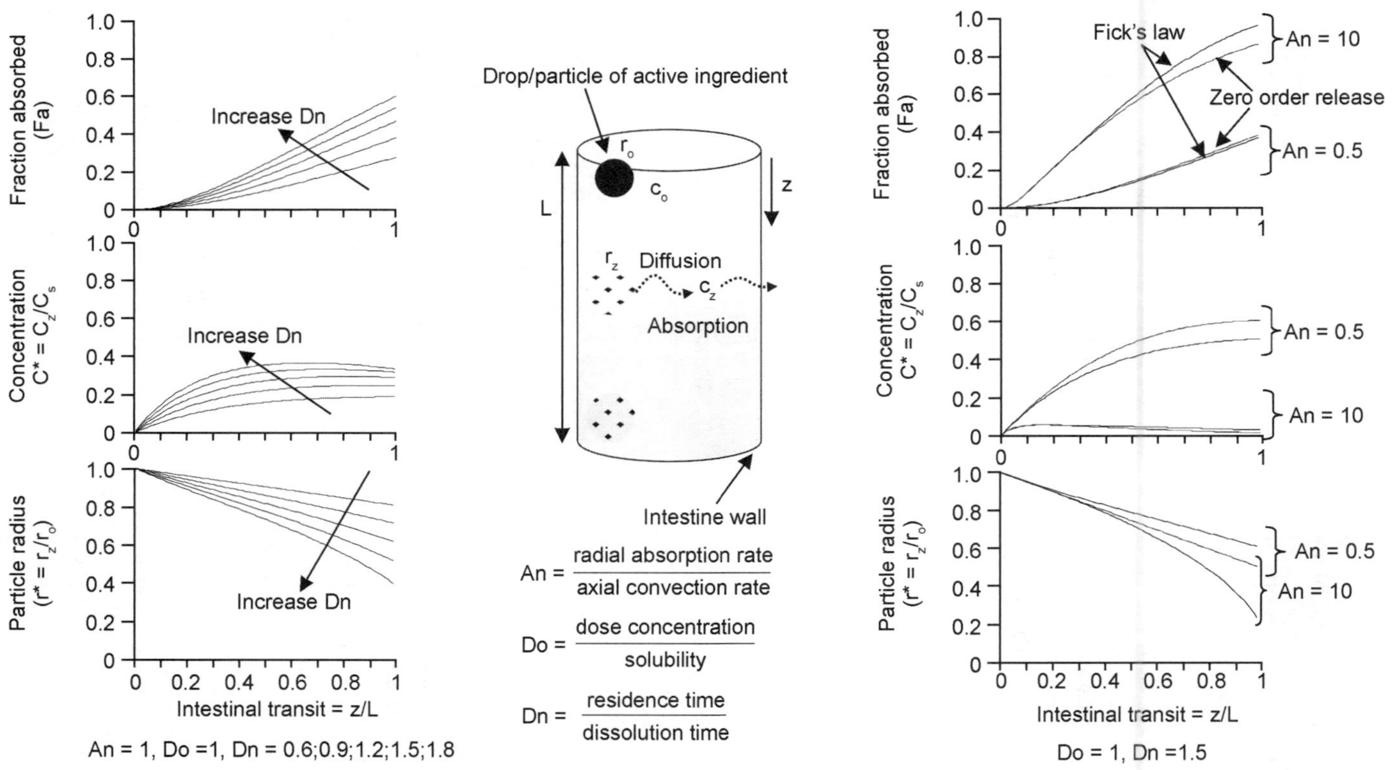

Fig. 3.8 Schematic of mass balance around the intestine according to the model of Oh *et al.*[102] Concentration profiles obtained with equations (3.1) and (3.3), and the dose/absorption parameters indicated above.

(C_P) such that $C_Z \gg C_P$, and the flux across the intestinal membrane is $P_{eff} * C_Z$, where P_{eff} is the effective permeability of the active ingredient through the intestinal membrane.

Using these assumptions, the mass balance around the particles yield:

$$\frac{dr^*}{dz^*} = -\frac{Dn}{3}\frac{(1 - C^*)}{r^*} \tag{3.1}$$

where r^* is a normalized form of the particle radius ($r^* = r_z/r_0$, r_0 is the initial particle size), z^* is a normalized form of the position or transit of the particle through the intestine ($z^* = z/L$), C^* is the normalized concentration of the active ingredient in solution ($C^* = C_Z/C_S$), and Dn is defined as the dissolution number (see Fig. 3.8), and is calculated using the following expression:

$$Dn = \frac{\mathcal{D}}{r_0}\frac{C_S(4\pi \times r_0^2)}{(4/3\pi \times r_0^3\rho)}\frac{\pi R^2 L}{Q} \tag{3.2}$$

where ρ is the density of the particle, and Q is the flow rate of the intestinal fluid (assumed constant).

The mass balance for the dissolved active ingredient in the intestine is:

$$\frac{dC^*}{dz^*} = Dn \times Do \times r^*(1 - C^*) - 2 * An * C^* \tag{3.3}$$

where Do is the 'dose number' (see Fig. 3.8) and is calculated as

$$Do = \frac{Mo}{Vo}\frac{1}{C_S} \tag{3.4}$$

where Mo is the mass of the active ingredient dosed, and Vo is the volume of liquid taken with the dose. The absorption number, An, is defined as:

$$An = \frac{P_{eff}\pi \times R \times L}{Q} \tag{3.5}$$

Finally, using the total mass balance of the active ingredient, the fraction absorbed (F) is given by:

$$F = 1 - r^{*3} - \frac{C^*}{Do} \tag{3.6}$$

The importance of these equations is that they help understand the effect of different formulation strategies on the effectiveness (absorption) of the formulated product. For example, the curves of fraction absorbed, dissolved concentration and particle radius on the left of Fig. 3.8 have been produced using different dissolution numbers (Dn). The dissolution number (Dn) is the parameter that reflects the effect of controlled-release formulas. For systems that increase the rate of release (such as microemulsions or solid lipid nanoparticles), Dn tends to increase (small r_0) which leads to larger fraction absorbed (F, bioavailability), larger dissolved concentration (C^*), and faster reduction in particle size (r^*).

If instead of an accelerated solubilization, a controlled-release formula seeks a slower but sustained release (lower *Dn*) then the absorption of the active ingredient may be significantly compromised (see top left graph in Fig. 3.8). One way to solve this problem is to increase the absorbance number (*An*) for the formulation. According to the top right graph in Fig. 3.8, by increasing *An* from 0.5 to 10 (for $Do = 1$, $Dn = 1.5$), the fraction absorbed (see curves labeled as Fick's law) is increased by nearly threefold. To increase this *An* value, the formulator can introduce various bioadhesives in the formula in order to increase the retention time in the intestine. Bioadhesives will be discussed later in this section.

The equations of Oh *et al.*[102] and Amidon *et al.*[118] assume that Fickean diffusion is applicable to evaluate the dissolution profile. However, numerous controlled-release methodologies approach a zero order release to provide a constant dose to the body. Equations (3.3) and (3.6) can be modified for zero order release, as follows:

$$\frac{dr^*}{dz^*} = -\frac{Dn^0}{3} \tag{3.7}$$

$$\frac{dC^*}{dz^*} = Dn^0 \times Do \times r^{*2} - 2 * An * C^* \tag{3.8}$$

where Dn^0 is a zero-order dissolution number

$$Dn^0 = 3\frac{J^0}{r_0 \rho}\frac{\pi R^2 L}{Q} \tag{3.9}$$

and J^0 is the constant flux achieved with the controlled-release formula.

Using this new definition, the dashed curves for the graphs located in the right-hand side of Fig. 3.8 were obtained. According to these graphs a zero order release (with similar *Dn* values) does not have a profound impact on the fraction absorbed. However, as with the first order (Fick's law) counterpart, increasing the absorbance number improves the absorption of the active ingredient.

The mathematical framework of Oh *et al.*[102] and Amidon *et al.*[118] discussed above is the basis for the Biopharmaceutical Classification System (BCS). According to this classification, there are four classes of active ingredients. Class I actives (high solubility, high permeability – high *An*, *Dn*) are typically highly absorbed as long as 85% of the dose is dissolved within 15 minutes. Class II actives have low solubility, but high permeability (which is the case for hydrophobic actives with low molecular weight), and the *in vitro* dissolution tests correlate well with the fraction absorbed. Class III systems have high solubility, but low permeability (e.g. hydrophilic peptides) in which case the dissolution tests do not correlate with the fraction absorbed, and permeation tests (described in the next section) are necessary in order to assess the bioavailability of the product. Class IV systems have low permeability and low solubility (e.g. large molecular weight hydrophobic actives).

The use of dissolution tests to evaluate the bioavaility of controlled-release formulas has advantages and disadvantages.[119,120] On the one hand, by introducing controlled (extended)-release formulas for products that fit the BCS class I category the correlation between *in vitro* dissolution and *in vivo* absorption is enhanced, and the controlled release is a valuable tool to modify not only the release profile, but also the absorption profile. However, for all other classes, the introduction of controlled-release formulations may further slow down the dissolution of the active ingredient and reduce the bioavailability of the active ingredient, as discussed above. Again, the use of nanoparticle delivery systems (smaller r_0, higher Dn) and/or bioadhesives (higher An) should be considered when introducing controlled-release formulations.

The model of Oh *et al.*[102] and Amidon *et al.*[118] could be coupled with the mass balance of the active ingredient in the plasma in order to produce a plasma concentration profile. In this case, the transport across the intestine should consider the plasma concentration (C_P), and the flux should be calculated as $P_{eff} * (C_z - C_P)$. Therefore equations (3.3) and (3.8) become:

$$\frac{dC^*}{dz^*} = Dn \times Do \times r^*(1 - C^*) - 2 * An * (C^* - C_P^*) \tag{3.10}$$

$$\frac{dC^*}{dz^*} = Dn^0 \times Do \times r^{*2} - 2 * An * (C^* - C_P^*) \tag{3.11}$$

where $C_P^* = C_p/C_S$.

To calculate the mass balance on the 'plasma compartment' illustrated in Fig. 3.9, it is assumed that only elimination takes place and that this elimination is first order with respect to the plasma concentration. In this case, the mass balance equation is:

$$\frac{dC_P^*}{dz^*} = An \times Vn \times (C^* - C_P^*) - \ln(2) \times En * C_P^* \tag{3.12}$$

where Vn = (volume of the dose)/(plasma volume), and En = (residence time in the intestine)/(elimination half life). Figure 3.9 presents the luminal (intestinal) and plasma concentration profile for Fickean (solid line, equations (3.10) and (3.12)) and zero-order release (dashed line, equations (3.11) and 3.12)) for fast ($En = 10$) and slow ($En = 0.01$) elimination.

According to the fraction absorbed curve in Fig. 3.9, it is confirmed that even for an active ingredient that is highly absorbed ($An = 10$), including the concentration of plasma in the mass balance equations, it is of little consequence in terms of fraction absorbed. The reason for this is that the volume of the dose is typically smaller than the volume of the plasma (about $3L$ in an adult). The plasma concentration graph (bottom left of Fig. 3.9) reflects the effect of elimination and controlled release. As expected, as the elimination rate (En) increases, the plasma concentration reduces. Additionally, introducing zero-order controlled release (dashed line) tends to modulate the plasma concentration, which is useful to avoid 'burst release' effects that may induce undesirable side effects.

The purpose of Fig. 3.9 is to illustrate, using simple mathematical equations, how the *in vivo* response (plasma concentration) is connected to *in vitro*

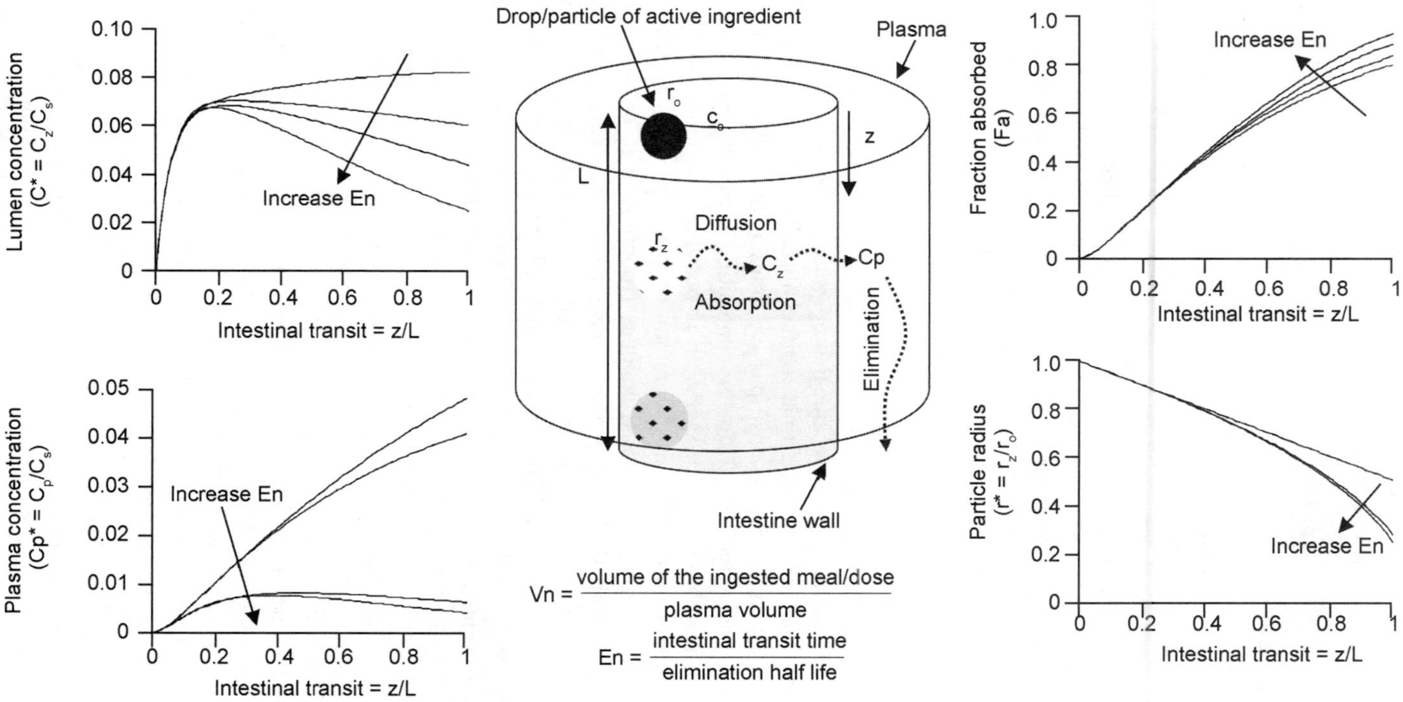

$$Vn = \frac{\text{volume of the ingested meal/dose}}{\text{plasma volume}}$$

$$En = \frac{\text{intestinal transit time}}{\text{elimination half life}}$$

An = 10, Do = 1, Dn = 1.5, Vn = 0.1, En = 0.01, 10

Fig. 3.9 Coupled intestine-plasma mass balance model.

dissolution tests (that help us determine *Do*, *Dn*) and permeability tests (that help us determine *An*). The latter aspect will be covered in the following section.

3.4.3 Permeability tests

Permeability (typically expressed in cm/s) is defined as the ratio between the flux of the active ingredient that diffuses through the intestinal wall (e.g in $\mu mol/cm^2 s$) and the difference in active ingredient concentration across this wall (e.g. in $\mu mol/cm^3$).

The challenge of designing *in vitro* experiments to evaluate the permeation of active ingredients is finding an appropriate membrane with physiological, mechanical, and surface chemistry properties similar to that of the small intestine. There are three types of permeability membranes that aim at simulating the transport of the active ingredient through the intestinal wall: artificial membranes, cell cultures and *ex vivo* tissue models.

Parallel artificial membrane permeability assay (PAMPA)

The concept of artificial membranes is to produce a synthetic mixture of lipids that emulate the surface and mechanical properties of the lamellar phase that constitute the wall of the intestinal epithelial cells. This lamellar phase represents the main barrier against passive transcellular (through the cell) diffusion of active ingredients. Thus, by conducting permeability assays using these artificial membranes (i.e. PAMPA) it is possible to evaluate the passive transcellular permeability (the most common transport mechanism) of active ingredients using high throughput techniques (robotics). The original idea of producing these synthetic lamellar phases dates back to the early 1960s when bilayer lipid membranes were produced by depositing a mixture of phospholipids and alkanes over small holes on a hydrophobic surface, and evaporating the alkane from the solution.[121] However, the fragile nature of these bilayer membranes prevented their use as a high throughput technique. It was not until 1998 when Kansy *et al.*[122] demonstrated that by coating a filter insert with a phospholipid film (also deposited using an alkane-phospholipid mixture) that researchers were able to produce synthetic lamellar phases suitable for permeability studies. One issue that needs to be considered is that, for highly permeable molecules, their mass transfer is controlled by their diffusion through the 'unstirred' water layer on the surface of the filter insert. For ionizable molecules, the contribution of membrane and unstirred water layer permeability can be deconvoluted by determining the permeability as a function of pH.[123]

The PAMPA methodology correlates relatively well with human bioavailability when the permeability of the active ingredient is less than 1×10^{-6} cm/s.[123] Active ingredients with permeabilities larger than 10×10^{-6} cm/s are considered highly permeable. PAMPA permeabilities lower than 1×10^{-6} cm/s are indicative that the active ingredient will not be absorbed (if the molecule obeys a passive transport mechanism). Another issue with PAMPA membranes is that high concentrations of phospholipids (e.g. more than

10% in the solvent phase) or a high fraction of charged lipids will induce the absorption of the active ingredient in the membrane, which is an undesirable effect if the target is to measure permeability. In order to minimize membrane absorption, 'sink' conditions are maintained in the receptor (or permeate) side of the assembly using 1% w/v sodium dodecyl sulfate.

PAMPA is typically used in the process of screening potential drug candidates. It is still far from being used as an alternative to IVIVC, but the methodology continues to gain recognition as an alternative to cell cultures.[121,123–125] At the time of this review it was not possible to identify any articles where PAMPA has been applied to evaluate the permeability of micronutrients or nutraceuticals. In addition, it was not possible to identify any articles using PAMPA to evaluate the permeation of an active ingredient (of any kind) delivered through a controlled-release mechanism. The reason for the lack of literature for these particular applications is likely due to the fact that PAMPA methods are relatively new and that most research efforts have been concentrated on developing drug screening methodologies. While there is no reason to believe that this methodology would not work for hydrophobic nutraceuticals such as carotenoids and polyunsaturated fatty acids, it is still necessary to consider the effect of vehicle (if it is a controlled-release formula) and its interaction with the PAMPA membrane. Absorption is not necessarily an undesirable side effect of using lipid membranes if the membrane composition is designed in a way that is able to reproduce the *in vivo* absorption.

Cell cultures

The first cell culture suitable for permeability studies was obtained using adenocarcinoma cell lines derived from human colonic epithelia, better known as Caco-2 cell cultures. Pinto *et al.*[126] reported the first fully functional cell culture with biophysical properties similar to that of the epithelial cells of the intestine in 1983, and proposed that these cell cultures could be used to evaluate the permeability of drugs through the intestine. In 1985, Vincent *et al.*[127] carried out one of the first Caco-2 studies to investigate the permeability of folic acid in the hope of correlating this value with the *in vivo* absorption of this nutrient. However, most researchers in the area recognize the study by Hidalgo *et al.*[128] in 1989 as the turning point for Caco-2 permeability studies. There are various reviews that describe the procedure to grow these cell cultures, the methodology to evaluate the permeability of active ingredients, example applications, and the limitations of the method.[125,129–132] In short, cells are cultivated on permeable supports using cells from standard sources such as the American Type Culture Collection (ATCC) or the European collection (ECACC), using cell densities of approximately 100 000 cells/cm^2. These cells are grown at 37 °C for a period of 3 weeks using standard culture media replaced every 2 days and under a 95% R.H. and 5% CO_2 atmosphere.[125] Artursson and Tavelin comment that deviation from the standard method (particularly in growing time) results in large deviations on the value of permeability, and they also advise against the use of automated (high throughput) methodologies to grow these tissues.[131] Trans epithelial electrical resistance (TEER) and mannitol permeability (typically 25 nm/s) are used as quality control parameters to validate the culturing procedure.[125,131]

Because Caco-2 cells are fully differentiated and functional, they are able to reproduce a variety of active and passive transport mechanisms. One of the features of Caco-2 monolayers is that they produce efflux proteins such as P-glycoprotein (P-gp), and multidrug resistant protein 2 (MDRP-2) that oppose the absorption of certain solutes. Caco-2 monolayers have been used to investigate the mechanism of transport of various nutrients and drugs. The active transport of calcium, iron, and other poorly absorbed metals has been investigated using these models.[133–135] However, in some cases the relative magnitude of drug transport through active and passive mechanisms, or the activity of efflux proteins in Caco-2 cells is different from the *in vivo* condition. In general Caco-2 monolayers tend to promote active transport mechanisms and hinder passive paracellular (between cell) transport.[131] Because of these limitations Caco-2 permeation studies are used as a secondary screening tool (after PAMPA screening) and not as an *in vitro* method for IVIVC. However, Caco-2 cell studies are recognized under the BCS classification method. A recent review suggests that values of permeability of 100 nm/s or more (measured using either PAMPA or Caco-2 cell models) can be considered highly permeable molecules that are fully absorbed by the body.[125]

There are other cell lines that have been used with similar purposes. Madin–Darby canine kidney cells (MDCK), despite being kidney and not epithelial intestinal cells, are suitable to evaluate selective active transport, and have the advantage of having a 3–7 day culturing time. The HT-29 is also a human colon carcinoma cell line which produces a variety of morphologies (depending on the culture media composition) that are useful to evaluate specific transport mechanisms. Some HT-29 sub-clones are capable of producing the mucus layer found in *in vivo* tissue, and that, as will be discussed later, is essential to evaluate the mucoadhesive properties of various controlled-release carriers. The cell line 2/4/A1 is derived from normal epithelial cells of fetal rat intestine. Monolayers of 2/4/A1 cells are more permeable than Caco-2 cells and are suitable to evaluate the paracellular and transcellular passive transport of active ingredients.[132]

Caco-2 cells have also been used to evaluate the toxicity of drug formulations. Some researchers use the change (increase) in permeability of certain markers such as rhodamine 123 or fluorescein as an indication of the toxicity of the formulation. However, the best approach is the use of biological markers such as MTT (3-(4,5-dimethylthiazol-2-yl)-2,5-diphenyltetrazolium bromide) which is transformed into an insoluble purple formazan by the enzymatic activity of living cells. The salt is later dissolved into suitable solvent (such as isopropyl alcohol) and the absorbance of this solution at 530 nm is used as a measure of the activity of living cells. By dividing this absorbance by the absorbance of a reference (untreated sample), the viability (fraction of cells that survived the treatment) of the formulation is calculated.[132,136–138]

Caco-2 cells and mucus producing HT-29 cell monolayers have been used as substrate to evaluate the adhesive properties of probiotics formulated in food products.[139] There are two versions of this technique, one that uses a radio-labeled bacterial suspension,[140] and another that uses Gram staining techniques.[141] In both techniques the bacterial suspension (in a simulated intestinal fluid) is incubated at 37 °C for 1.5 h (some researches use 0.5 h and 1 h) in well plates containing a monolayer of the cell line. After the incubation period is completed, the tissue is thoroughly rinsed. The number of remaining bacteria (after rinse) adhered to the monolayer is determined using scintillation counting or image analysis of micrographs of the stained tissue. The fractional adhesion of the bacterial strain is calculated as the ratio between the number of bacterial cells adhered and the initial (dosed) number of bacterial cells. It has been found that adhesion to HT-29 cells (with fully functional mucous layer) is typically two to three times greater than the adhesion to Caco-2 cells. However, both cell lines are suitable to identify differences in adhesion between different strains.[141] The composition of the simulated intestinal fluid has been found to affect the fractional adhesion of bacteria.[139] Based on the success of these studies, it has been proposed that cell culture models should be used to evaluate the effectiveness of controlled-release/encapsulation formulations for probiotics and as a quality assurance method.[142]

Caco-2 cell monolayers have also been used to evaluate the effectiveness of controlled-release formulations. Liu and Krishnan[143] evaluated an encapsulation vehicle formulated with alginate, polylysine and pectin, and mannitol as the active ingredient. These researchers found that the interaction between the encapsulation vehicle and the Caco-2 monolayer was important in determining the adhesion and release profile of this formulation. It has been proposed that Caco-2 permeability studies should be a part of the evaluation/screening methodologies for controlled-release formulas.[144] This view is certainly supported by the findings of Lameiro *et al.*[145] who evaluated the release profile of adenoviral vectors encapsulated using the complex coacervation between chitosan and bile salts. Lameiro *et al.* found that this coacervation method protected the viral vector but did not release the vector using conventional *in vitro* dissolution tests. When the suspension of the encapsulated vector was contacted with Caco-2 monolayers, this vector (in its bioactive form) was effectively released. The authors suggest that in this case the degradation of the shell might have been induced by biomolecules produced by the Caco-2 monolayers. Lameiro *et al.* referred to this delivery mechanism as a host-controlled release.[145]

Tao and Desai[146] reported the use of microelectromechanical systems (MEMS) with various permeation layers and coated with tomato lectin for increased adhesion to the epithelial cells. The authors found that the release profile of their device is influenced by the surface chemistry of the Caco-2 monolayers where they were deposited.

Ex vivo *intestinal models*

Everted intestinal sac models (typically obtained from rats) were first introduced in 1958 by Crane and Wilson[147] as a way to evaluate the permeation of sugars through the intestine. Before the introduction of the Caco-2 monolayers, everted intestinal (gut) sac was the method of choice to evaluate the permeability of active ingredients and the effect of excipients on permeability. The everted gut sac method is recognized by the FDA as a valid protocol to determine permeability.

In short, the everted gut sac method consists of excising the small intestine (ileum and jejunum) of a rat, and maintaining the tissue in culture media bubbled with 95% O_2–5% CO_2 mixture to keep the intestinal cells alive. A section of the gut (typically between 2.5 and 15 cm) is turned inside out and sealed at one of the ends. The setup is submerged into a simulated intestinal fluid containing the active ingredient. The open end of the gut is used to rinse the serosal side of the membrane with a buffer solution. The perfused buffer is collected and the concentration of the drug permeated from the mucosal side of the membrane is analyzed using liquid chromatography techniques. Details of construction can be found elsewhere.[148]

One of the advantages of using excised intestinal models is that their physiological and mechanical properties are similar to that of the *in vivo* tissue. However, the fact that the rat intestine is more permeable than the human intestine is one of the limitations of the methodology. Another limitation is that, as with the Caco-2 model, the complexity (various transport mechanisms acting at the same time) of the excised intestinal tissue represents a disadvantage in terms of investigating the mechanism of absorption of the active ingredient. In comparison with the Caco-2 model, the excised tissue does have the vili and mucus structure (which is an advantage in terms of simulating bioadhesion of controlled-release formulas), but it makes it extremely difficult (if not impossible) to obtain an accurate value of the surface area in order to calculate an absolute value of permeability. In most cases, these tests are carried out in a comparative form to evaluate the effect of surfactant and other formulation excipients on the rate of permeation.

El Gorab *et al.*[149] studied the effect of bile salts, nonionic surfactant Tween 20 and cationic surfactant hexadecyltrimethylammoniumbromide on the uptake of tritium-labeled β carotene and retinol. They found that the uptake of these antioxidants was greater when they were solubilized in micelles (compared to emulsions) and that in the case of β carotene the nonionic and cationic surfactants might have enhanced its uptake by modifying the interaction with the epithelial membrane and perhaps acting as a transport carrier. Lo[150] conducted a comparative study between Caco-2 monolayers and everted gut sac model and found that in both cases the use of nonionic surfactants (Span 80, Brij 30, Tween 20, Tween 80, Myrj 52) mitigated the efflux action of P-glycoprotein (P-gp) against epirubicin. Barthe *et al.*[151] introduced a modified (extended) incubation protocol for the excised tissue to enhance the action of P-gp activity in the excised tissues. Sharma *et al.*[152] introduced an ingenious optical microscopy method to evaluate the toxicity of permeation enhancers, finding that cyclodextrins (used in inclusion complexation formulas) are more toxic than bile salts and medium chain fatty acids. Naisbett and Woodley[153] used the everted gut sac model to evaluate the bioadhesive properties of tomato lectin, finding that this protein is absorbed (via transcytosis) by the enterocytes and that it tends to accumulate in these cells. The latter is an important mechanism of uptake for controlled-release formulas which will be further discussed in the next section.

It is interesting to note after this brief review that *in vitro* test methods are evolving towards more simplified models such as artificial membranes. More complex methodologies such as *in situ* studies in animal or humans are now less common. The topic of '*in situ*' studies was not covered in this review. However, information on these test methods can be found elsewhere.[154]

3.4.4 Adhesion/absorption tests

One of the issues with controlled (particularly extended)-release formulas is that the formulator has to ensure that the encapsulation vehicle remains in the intestinal track long enough to release its load before it leaves the small intestine (or the colon if the controlled release is designed for colonic delivery).

Smart *et al.*[155,156] introduced the use of tensile testing devices to measure the detachment force necessary to separate a disc of the polymeric coating of interest (e.g. carbopol, carboxymethyl cellulose, and others) from the mucosal surface of an excised rat intestine. Sam *et al.*[157] used the Wilhelmy plate method to measure the bioadhesive strength (detachment force) of polymer-coated glass slides submerged in mucus gel obtained from scraping freshly excised pig intestines. Mortazavi and Smart[158] used a modified rheometer to obtain the work of adhesion (the area under the force vs. displacement curve) of a pig mucus gel adhered to a disk of the assayed polymer, and found that the work of adhesion was directly proportional to the detachment force.

These and other tensile-test methods of bioadhesive properties are appropriate as a screening method to select a coating polymer for encapsulation formulas. However, the fact that the polymer has to be preformed into a disk (instead of the actual particle form in which the polymer is released into the intestine) represents a serious limitation of these methods. Hassan and Gallo introduced a simple rheological method where the viscosity of the mixture was equated to the sum of the contributions of the polymer assayed, mucin and the contribution due to the bioadhesive binding between the polymer and mucin.[159] This method was suitable to evaluate the adhesive force between a strong binding polymer such as chitosan and mucin. Tamburic and Craig[160] compared the storage modulus (obtained with oscillatory rheology) of polymer–mucin gels to the work of cohesion of these mixtures (obtained using penetrometry studies) and found that the measurements were consistent enough to rank the relative adhesive strength of the assayed polymers.

There are various lessons learned from these bioadhesion studies. First, that mucoadhesion is a phenomenon that involves electrostatic interactions, Van der Waals interactions, and hydrogen bonding. Second, that using cationic polymers such as chitosan, which binds to mucin via electrostatic interactions, it is possible to obtain high adhesive strength. Third, that the adhesive properties depend on the degree of hydration of mucin and the binding polymer, in which case more hydrophobic polymers such as hydroxypropyl cellulose are more resistant to these hydration effects and produce more stable bonds.

Despite the fact that these methodologies helped us understand the interactions between polymers and mucin, they are not suitable to describe the adhesion of micro- and nanoparticles (typically used in controlled-release formulas) to the surface of the intestine. To this end, particle counting methods are more appropriate for these applications. In 1998 Delie[161] reviewed the different methodologies available at that time to determine particle uptake by the gastrointestinal tract. Delie explained that, at the time, there were six different methodologies to determine the number of particles absorbed by intestinal tissue: optical microscopy, electron microscopy, fluorescence-based methods, gel permeation chromatography, radioiosotopes, and drug release methods. Out of these methods, optical (confocal) microscopy of fluorescence labeled particles was recommended as one of the best methods to determine the number of particles absorbed and the site of absorption. Delie highlighted that the inconsistency between different studies on micro- and nanoparticle uptake was a result of the different intestinal substrates (cultured cell tissue, excised tissue, *in situ* models, animal models), the different methods of tissue preparation, the different methods of exposure (ussing chambers, diffusion chambers, *in vivo* single/chronic uptake, *in vivo* closed intestinal loop), and the different concentrations of particles used by different researchers.

In a mini-review article Florence[162] also expressed concern about these inconsistencies in the literature. He explained that there is no dispute about the fact that particle uptake occurs, and that this uptake takes place through M cells, Peyer's patches and by transcytosis through enterocytes. He concluded that smaller particle size, positive surface charge of the particle, and the presence of ligands such as chitosan, lectin or invasin improved the adhesion properties of the nanoparticle system.

It is worth highlighting the work of Desai *et al.*[103] who studied the absorption of biodegradable polylactic-polyglycolic acid copolymer (PLAGA) particles of various sizes (ranging from 100 nm to 10 μm) into the small intestine of rats. These researchers used the *in situ* intestinal loop model. The particles were labeled with a fluorescence die (6-coumarin) which they use as a marker to determine particle uptake (using liquid chromatography) and to conduct histological studies using optical microscopy on selected patches in the tissue (to determine the particle uptake by epithelial and lymphatic tissue). These researchers observed that the uptake of 100 nm particles was 15–250 times higher than the uptake of particles with diameter 500 nm or larger, in both lymphatic tissue (Peyer's patch) and epithelial tissue. Furthermore, the histological studies revealed that 100 nm particles penetrated the submucosal layers of the intestine, while larger particles were trapped on the mucosal surface.

More recently, Takeuchi *et al.*[163] introduced two *in vitro* methods to evaluate the mucoadhesive properties of various polymers. The first method consists of evaluating the zeta potential of mucin particles in the presence of the assayed polymer. The change in this zeta potential (regardless of the 'sign' of the change) is considered a qualitative indication of the interaction between mucin and the polymer. In the case of chitosan, this change in zeta potential (from negative to near zero zeta potential as chitosan concentration increases) was accompanied by an increase in the size of the aggregate and an increase in the adhesiveness determined using particle counting methods. The second method consisted of the use of surface plasmon resonance to study the binding reaction between the polymer and a suspension of mucin particles. The polymer was bound to a Biacore® chip, and the kinetics of the polymer–mucin binding process was traced by the changes in the local refraction index on the surface of the chip. These techniques, similar to the ones based on tensile or rheological studies, serve as a pre-screening tool to select the polymer of interest, and as a way of exploring the molecular interactions involved in mucoadhesion.

An article by Kockisch *et al.*[164] presents evidence that *in vitro* mucoadhesion tests based on polymer–mucin or mucus interaction are sometimes misleading in the formulation of controlled-release formulations. In particular, these researchers found that carbopol outperformed the rest of the assayed polymer (including chitosan) in tensile adhesion tests. However, microparticles encapsulated with carbopol yield the lowest retention time in a flow-thru experiment that used esophageal mucosa as substrate. The flow-thru experiment revealed that the cationic polymer chitosan yielded the largest residence time for the encapsulated particles, even at high flow rate.

In our laboratory, we have used a double-chamber flow-thru diffusion cell (flow-thru DIALYZER™) to evaluate the adhesion/absorption of microemulsions on/into porcine small intestine mucosa in the presence of lectin and soluble mustard protein isolate (used as bioadhesive additives).[165] A schematic of this setup is shown in Fig. 3.10(a). In short, the microemulsion product is first diluted using an appropriate protocol (using a 1% solution of SDS to simulate the presence of bile salts). The microemulsion is then pumped through the retentate (mucosal) side of the pig intestine membrane, and the effluent is then passed through a quartz flow-thru cell where the red-dyed carrier oil is detected using visible spectrophotometry. The microemulsion system was circulated in a closed loop to evaluate the kinetic and equilibrium constants of the micelle-intestine binding for each formulation. The effluent of the serosal side of the membrane can be used to determine the fraction of the drug that has permeated the membrane, only that, in this case we were only interested in determining the effect of lectin and soluble mustard protein on the absorption of the microemulsified oil. Figure 3.10(b) presents the absorption of the microemulsified oil (isopropyl myristate) per unit area of intestine (assuming flat surface) as a function of time for three different formulations: lecithin microemulsion (LMe), LMe + 0.1 mg/ml soybean lectin, and LMe + 10 mg/ml mustard protein in the initial (concentrated) microemulsion.

The data in Fig. 3.10(b) reflects that LMe micelles are absorbed within 10 minutes of contact time at the conditions that the experiment was carried out (approximate surface velocity of 24 cm/min). This absorption data can be fitted

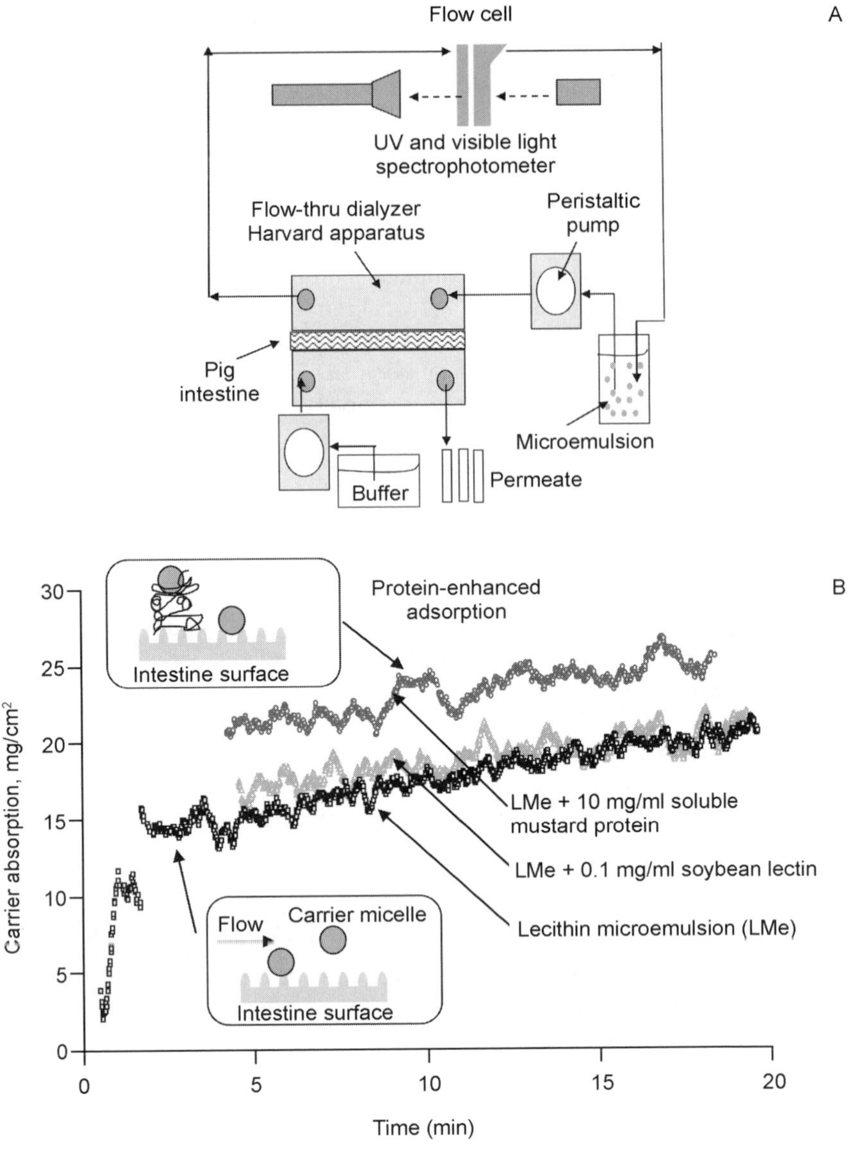

Fig. 3.10 Flow-thru double chamber diffusion cell studies. A: Experimental setup. B: Carrier absorption curve.

to a first-order reversible model that accounts for the initial rate of absorption, and the absorption/desorption equilibrium (binding constant) for this process. When the microemulsion was pre-incubated with a low dose of soybean lectin, no significant increase in absorption is obtained, suggesting that this lectin is not a suitable bioadhesive for microemulsions. This result may be explained by the fact that lectin is not conjugated to the microemulsion itself, and at this low dose tends to bind, preferentially, to the intestine. Additionally, the pig intestine tissue used in the study for Fig. 3.10 was not fully viable, thus some of the active mechanism of transport (especially lectin-mediated transcytosis) might not have been available in the model. Using the same setup we were able to determine that a relatively high concentration of soluble mustard protein does improve the absorption of the microemulsion. Other less soluble fraction of mustard and soybean isolates did not perform in the same way that the soluble mustard protein did.

Using simple diffusion/dissolution cells like the one presented in Fig. 3.10, it is possible to combine *in vitro* solubilization, permeation, and adhesion tests where the mass balances of Oh *et al.*[102] and Amidon *et al.*[118] (setup of Fig. 3.8) could be used to find the dissolution and absorption parameters necessary to evaluate the bioavailability even for complex controlled-release formulations.

3.5 Future trends

Before looking into future trends, it is important to recall the various constraints imposed on food and nutritional products: safety, low cost, appeal (taste, odor, color, and texture), stability, and nutritional value. In most cases, cost is the overriding factor that has hindered the introduction of sophisticated controlled-release technologies in food products. However, recent developments in microfluidics and self-assembled monolayer technologies are making micro- and nanoparticle-based controlled-release vehicles a cost-effective alternative to conventional spray drying and bed coating technologies. The confluence of these new technologies and market conditions are opening new opportunities for controlled-release formulations in food products.

The market for these formulations is fuelled by the increased awareness of the potential health benefits of nutraceuticals and probiotics. This situation has prompted various manufacturers to introduce these ingredients into their products, and as a part of their marketing strategy. The added value of nutraceutical ingredients justifies the additional cost of microencapsulation technology used to maintain the stability of these ingredients and mask their flavor (in the case of polyunsaturated fatty acids). However, food and nutritional supplement manufacturers should be aware that if these products do not produce real health benefits, consumer interest in these products may fade. Therefore it is important to ensure the bioavailability of the active ingredients, and perhaps use this fact as a marketing tool for the product. At the moment, nutritional supplements or food products containing nutraceutical additives are not held to the same standards as their pharmaceutical counterparts. However, the introduction of the 'USP-verified dietary supplement' seal of approval is a sign of the increasing pressure from regulatory agencies, nutritionists and consumer groups to discourage 'snake-oil' formulas from being marketed. Despite the comments offered above, the case of probiotic formulations deserves a special mention because these formulas are carefully designed and tested to make sure that the microorganism is protected from the action of gastric juices and that they are released at the site of action.

Another trend that is capturing the pharmaceutical industry and is now permeating the food industry is the use of an integrated formulation approach (also called retro-design, or formulation by design) whereby the selection of ingredient starts with the choice of release profile, as opposed to a linear formulation approach of trial and error. Certainly, integrated formulation is a challenging proposition because it requires a fundamental understanding of the mechanism of transport and release of the delivery vehicle, and hence the interest of numerous academic and industrial research groups on understanding the mechanism of mass transfer of the active ingredient at the cellular and molecular level.

Cell cultures (particularly Caco-2 and HT-29 cell lines) have shown tremendous potential as experimental tools to investigate the mechanisms of carrier transport and release, especially in the case of nanoparticle absorption, and host-controlled release of probiotics. The combination of cell cultures or *ex situ* tissue and flow-thru models (that simulate *in vivo* flow conditions) is gaining popularity as test protocols to evaluate the bioadhesive properties of extended-release formulations (especially those produced with chitosan or lectins).

3.6 Sources of further information and advice

To learn more about various encapsulation technologies and their relevance to the food industry, the 1999 review by Gibbs *et al.*[1] is highly recommended. A review article by Gouin on the industrial appraisal of microencapsulation technologies offers an updated view of the use of controlled-release technologies and future trends in the food industry.[20] The reader interested in learning more about 'retro-design' for the delivery of active ingredients (in particular probiotics) in food products should read the article by Ubbink and Krüger.[68]

The book on microencapsulation technologies edited by Simon Benita contains comprehensive reviews of the encapsulation technologies (specially those concerning nanoparticles and liposomes) introduced in the first part of this chapter.[5,27,28]

For additional information on mechanisms of controlled release and bioadhesion, Donald Wise's *Handbook of Pharmaceutical Controlled Release Technology* is a good source of information.[90,120]

For *in vivo* bioavailability studies, the review article by Fairweather-Tait and Teucher provides a good introduction to the topic of bioavailability of minerals.[85] However, the book by Abrams and Wong is strongly recommended for a more in-depth view of the growing field of stable isotopes applied to food and nutritional sciences.[82]

To learn more about *in vivo* and *in vitro* bioavailability studies the book on drug bioavailability edited by van de Waterbeemd, Lennernäs, and Artursson is one of the best sources, especially to understand the different transport pathways available for active ingredient transport.[121,131,132,154]

Finally, the review by Delie is recommended for those interested in nano- and microparticle uptake in delivery systems.[161]

3.7 References

1. GIBBS B.F., KERMASHA S., ALLI I., MULLIGAN C.N. (1999) Encapsulation in the food industry: a review. *Int. J. Food Sci. Nutr.*, 50 (3), 213–224.
2. DESAI K.G.H., PARK H.J. (2005) Recent developments in microencapsulation of food ingredients. *Drying Technol.*, 23 (7), 1361–1394.
3. AUGUSTIN M.A., SANGUANSRI L., BODE O. (2006) Maillard reaction products as encapsulants for fish oil powders. *J. Food Sci.*, 71 (2), E25–E32.
4. FORSSELL P., PARTANEN R., POUTANEN K. (2006) Microencapsulation – better performance of food ingredients. *Food Sci. Technol.*, 20 (3), 18–20.
5. GARTI N., ASERIN A. (1996) 'Pharmaceutical emulsions, double emulsions and microemulsions', in Benita, S. *Microencapsulation, Methods and Industrial Applications*, New York, Marcel Dekker, 411–534.
6. SALAGER J.L (2000) 'Emulsion properties and related know how to attain them', in Françoise N. and Gilberte M. *Pharmaceutical Emulsions and Suspensions*. New York, Marcel Dekker, 73–125.
7. MADENE A., JACQUOT M., SCHER J., DESOBRY S. (2006) Flavour encapsulation and controlled release – a review. *Int. J. Food Sci. Technol.*, 41 (1), 1–21.
8. REINECCIUS G.A. (2004) The spray drying of food flavors. *Drying Technol.*, 22 (6), 1289–1324.
9. KIM Y.D., MORR C.V. (1996) Microencapsulation properties of gum arabic and several food proteins: spray-dried orange oil emulsion particles. *J. Agric. Food Chem.*, 44 (5), 1314–1320.
10. McNAMEE B.F., O'RIORDAN E.D., O'SULLIVAN M. (1998) Emulsification and microencapsulation properties of gum arabic. *J. Agric. Food Chem.* 46 (11), 4551–4555.
11. LIU X.-D., ATARASHI T., FURUTA T., YOSHII H., AISHIMA S., OHKAWARA M., LINKO P. (2001) Microencapsulation of emulsified hydrophobic flavors by spray drying. *Drying Technol.*, 19 (7), 1361–1374.
12. KOSARAJU S.L., D'ATH L., LAWRENCE A. (2006) Preparation and characterisation of chitosan microspheres for antioxidant delivery. *Carbohydrate Polym.*, 64 (2), 163–167.
13. CHEN L., REMONDETTO G.E., SUBIRADE M. (2006) Food protein-based materials as nutraceutical delivery systems. *Trends in Food Sci. Technol.*, 17 (5), 272–283.
14. LEE S.J., ROSENBERG M. (2000) Whey protein-based microcapsules prepared by double emulsification and heat gelation. *Food Sci. Technol.*, 33 (2), 80–88.
15. ROSENBERG M., KOPELMAN I.J., TALMON Y. (1990) Factors affecting retention in spray-drying microencapsulation of volatile materials. *J. Agric. Food Chem.*, 38 (5), 1288–1294.
16. VAN DER GRAAF S., SCHROËN C.G.P.H., BOOM R.M. (2005) Preparation of double emulsions by membrane emulsification – a review. *J. Membrane Sci.*, 251 (1–2), 7–15.
17. CHO Y.-H., SHIM H.K., PARK J. (2003) Encapsulation of fish oil by an enzymatic gelation process using transglutaminase cross-linked proteins. *J. Food Sci.*, 68 (9), 2717–2723.

18. YANG Y.-Y., CHUNG T.-S., BAI X.-L., CHAN W.-K. (2000) Effect of preparation conditions on morphology and release profiles of biodegradable polymeric microspheres containing protein fabricated by double-emulsion method. *Chem. Eng. Sci.*, 55 (12), 2223–2236.
19. FORGIARINI A., ESQUENA J., GONZÁLEZ C., SOLANS C. (2001) Formation of nano-emulsions by low-energy emulsification methods at constant temperature. *Langmuir*, 17 (7), 2076–2083.
20. GOUIN S. (2004) Microencapsulation: industrial appraisal of existing technologies and trends. *Trends Food Sci. Technol.*, 15 (7–8), 330–347.
21. WEINBRECK F., MINOR M., DE KRUIF C.G. (2004) Microencapsulation of oils using whey protein/gum arabic coacervates. *J. Microencapsulation*, 21 (6), 667–679.
22. SCHROOYEN P.M.M., VAN DER MEER R., DE KRUIF, C.G. (2001) Microencapsulation: its application in nutrition. *Proc. Nutr. Soc.*, 60 (4), 475–479.
23. TAYLOR T.M., DAVIDSON P.M., BRUCE B.D., WEISS J. (2005) Liposomal nanocapsules in food science and agriculture. *Crit. Rev. Food Sci. Nutr.*, 45 (7–8), 587–605.
24. KIRBY C.J., GREGORIADIS G. (1984) Dehydration-rehydration vesicles – a simple method for high-yield drug entrapment in liposomes. *Biotechnology*, 2, 979–984.
25. MOZAFARI M.R., MORTAZAVI S.M. (2005) *Nanoliposomes: From Fundamentals to Recent Developments*. Oxford, Trafford.
26. LARIDI R., KHEADR E.E., BENECH R.-O., VUILLEMARD J.C., LACROIX C., FLISS I. (2003) Liposome encapsulated nisin Z: Optimization, stability and release during milk fermentation. *Int. Dairy J.*, 13 (4), 325–336.
27. MARGALIT R., YERUSHALMI N. (1996) 'Pharmaceutical aspects of liposomes: perspectives in, and integration of, academic and industrial research and development', in Benita S. (ed.), *Microencapsulation, Methods and Industrial Applications*. New York, Marcel Dekker, 259–296.
28. LASIC D.D. (1996) 'Stealth liposomes', in Benita, S. (ed.), *Microencapsulation, Methods and Industrial Applications*. New York, Marcel Dekker, 297–328.
29. LASIC D.D. (1990) On the thermodynamic stability of liposomes. *J. Colloid Interface Sci.*, 140, 302–304.
30. ÜNER M. (2006) Preparation, characterization and physico-chemical properties of solid lipid nanoparticles (SLN) and nanostructured lipid carriers (NLC): their benefits as colloidal drug carrier systems. *Pharmazie*, 61 (5), 375–386.
31. MÜLLER R.H., MÄDER K., GOHLA S. (2000) Solid lipid nanoparticles (SLN) for controlled drug delivery – a review of the state of the art. *Eur. J. Pharm. Biopharm.*, 50 (1), 161–177.
32. RUNGE S.A.F. (1998) *Lipid-Nanopartikel (SLN) als kolloidaler Arzneistoffträger für Cyclosporin A*, Ph.D. thesis, Free University of Berlin.
33. YANG S., ZHU J., LU Y., LIANG B., YANG C. (1999) Body distribution of camptothecin solid lipid nanoparticles after oral administration. *Pharm. Res.*, 16, 751–757.
34. MA Q.H., XIA Q., LU Y.Y., HAO X.Z., GU N., LIN X.F., LUO D. (2007) Preparation of tea polyphenols-loaded solid lipid nanoparticles based on the phase behaviors of hot microemulsions. *Solid State Phenom.*, 121–123, 705–708.
35. KRISTL J., VOLK B., GAŠPERLIN M., ŠENTJURC M., JURKOVIČ P. (2003) Effect of colloidal carriers on ascorbyl palmitate stability. *Eur. J. Pharm. Sci.*, 19 (4), 181–189.
36. POUTON C.W. (2000) Lipid formulations for oral administration of drugs: non-emulsifying, self-emulsifying, and self-microemulsifying drug delivery systems. *Eur. J. Pharm. Sci.*, 11 (2), S93–S98.
37. PORTER C.J.H., CHARMAN W.N. (2001) *In vitro* assessment of oral lipid based

formulations. *Adv. Drug Deliv. Rev.*, 50, S127–S147.
38. CUI S., ZHAO C., CHEN D., HE Z. (2005) Self-microemulsifying drug delivery systems (SMEDDS) for improving *in vitro* dissolution and oral absorption of Pueraria lobata isoflavone. *Drug Devel. Ind. Pharm.*, 31 (4–5), 349–356.
39. WU W., WANG Y, QUE L. (2006) Enhanced bioavailability of silymarin by self-microemulsifying drug delivery system. *Eur. J. Pharm. Biopharm.*, 63, 288–294.
40. BUSH A.J., BEYER R., TRAUTMAN R., BARBÉ C.J., BARTLETT J.R. (2004) Ceramic micro-particles synthesised using emulsion and sol-gel technology: an investigation into the controlled release of encapsulants and the tailoring of micro-particle size. *J. Sol-Gel Sci. Technol.*, 32 (1–3), 85–90.
41. CELLESI F., TIRELLI N. (2006) Sol-gel synthesis at neutral pH in W/O microemulsion: a method for enzyme nanoencapsulation in silica gel nanoparticles. *Colloids Surfaces A*, 288 (1–3), 52–61.
42. ZHU Y.-Y., ZHANG G.-Y., HONG X.-L., DONG J.-F., ZHANG X.-G., ZENG H. (2005) Synthesis of nanocapsules by complex coacervation in microemulsion. *Acta Chim. Sin.*, 63 (16), 1505–1509.
43. LESER M.E., SAGALOWICZ L., MICHEL M., WATZKE H.J. (2006) Self-assembly of polar food lipids. *Adv. Colloid Interface Sci.*, 123–126, 125–136.
44. FLANAGAN J., SINGH H. (2006) Microemulsions: a potential delivery system for bioactives in food. *Crit. Rev. Food Sci. Nutr.*, 46 (3), 221-237.
45. GARTI N., SPERNATH A., ASERIN A., LUTZ R. (2005) Nano-sized self-assemblies of nonionic surfactants as solubilization reservoirs and microreactors for food systems. *Soft Matter*, 1 (3), 206–218.
46. SABATINI D.A., ACOSTA E.J., HARWELL J.H. (2003) Linker molecules in surfactant mixtures. *Curr. Op. Coll. Interface Sci.*, 8 (4,5), 316–326.
47. ACOSTA E.J., NGUYEN T., WITTHAYAPANYANON A., HARWELL J.H., SABATINI D.A. (2005) Linker-based bio-compatible microemulsions. *Environ. Sci. Technol.*, 39(5), 1275–1282.
48. SALAGER J., ANTÓN R., SABATINI D., HARWELL J., ACOSTA E., TOLOSA L. (2005) Enhancing solubilization in microemulsions – state of the art and current trends. *J. Surfactant Detergents*, 8, 3–21.
49. EDWARDS M.F., INSTONE T. (2001) Particulate products – their manufacture and use. *Powder Technol.*, 119, 9–13.
50. ZHAO J., WHISTLER R.L. (1994) Spherical aggregates of starch granules as flavor carriers. *Food Technol.*, 48 (7), 104–105.
51. CONDE-PETIT B., ESCHER F., NUESSLI J. (2006) Structural features of starch-flavor complexation in food model systems. *Trends Food Sci. Technol.*, 17 (5), 227–235.
52. MASTERS K. (1991) *Spray Drying Handbook*, 5th edn, Harlow, Longman.
53. RÉ M.-I. (2006) Formulating drug delivery systems by spray drying. *Drying Technol.*, 24 (4), 433–446.
54. MOREAU D.L., ROSENBERG M. (1998) Porosity of whey protein-based microcapsules containing anhydrous milkfat measured by gas displacement pycnometry. *J. Food Sci.*, 63 (5), 819–823.
55. KIM E.H.-J., XIAO D.C., PEARCE D. (2005) Effect of surface composition on the flowability of industrial spray-dried dairy powders. *Colloid Surfaces B*, 46 (3), 182–187.
56. MUJUMDAR A.S. (1987) *Handbook of Industrial Drying*, 2nd edn, New York, Marcel Dekker.
57. FUCHS M., TURCHIULI C., BOHIN M., CUVELIER M.E., ORDONNAUD C., PEYRAT-MAILLARD M.N., DUMOULIN E. (2006) Encapsulation of oil in powder using spray drying and

fluidised bed agglomeration. *J. Food Eng.*, 75 (1), 27–35.

58. DIOSADY L.L., ALBERTI J.O., MANNAR M.G.V. (2002) Microencapsulation for iodine stability in salt fortified with ferrous fumarate and potassium iodide. *Food Res. Int.*, 35 (7), 635–642.
59. PELEG M. (1992) Disintegration and segregation kinetics of dry food particulates. *IFT Basic Symposium Series*, 7, 541–571.
60. MASSOL-CHAUDEUR S., BERTHIAUX H., DODDS J. (2003) The development and use of a static segregation test to evaluate the robustness of various types of powder mixtures. *Food and Bioproducts Processing*, 81, 106–118.
61. BLANCO M., GONZÁLEZ-BAÑO R., BERTRAN E. (2002) Monitoring powder blending in pharmaceutical processes by use of near infrared spectroscopy. *Talanta*, 56 (1), 203–212.
62. EL-HAGRASY A.S., MORRIS H.R., D'AMICO F., LODDER R.A., DRENNEN-III J.K. (2001) Near-infrared spectroscopy and imaging for the monitoring of powder blend homogeneity. *J. Pharm. Sci.*, 90 (9), 1298–1307.
63. LYON R.C., LESTER D.S., LEWIS E.N., LEE E., YU L.X., JEFFERSON E.H., HUSSAIN A.S. (2002) Near-infrared spectral imaging for quality assurance of pharmaceutical products: analysis of tablets to assess powder blend homogeneity. *AAPS Pharm. Sci. Tech.*, 3 (3): article 17.
64. EL-HAGRASY A.S., DRENNEN III J.K. (2006) Process analytical technology approach to near-infrared process control of pharmaceutical powder blending. Part III: Quantitative near-infrared calibration for prediction of blend homogeneity and characterization of powder mixing kinetics. *J. Pharm. Sci.*, 95 (2), 422–434.
65. BERMAN J., ELINSKI D.E., GONZALES C.R., HOFER J.D., JIMENEZ P.J., PLANCHARD J.A., TLACHAC R.J., VOGEL P.F. (1997) Blend uniformity analysis: validation and in process testing. *PDA Technical Rep.*, 25 (51), S1–S99.
66. CODE OF FEDERAL REGULATIONS [Title 21, Volume 4] [Revised as of April 1, 2006] [CITE: 21CFR211.110] Sec. 211.110 Sampling and testing of in-process materials and drug products.
67. INTERNATIONAL CONFERENCE ON HARMONIZATION WEBSITE (http://www.ich.org/cache/compo/276-254-1.html) accessed on January 2007.
68. UBBINK J., KRÜGER J. (2006) Physical approaches for the delivery of active ingredients in foods. *Trends in Food Sci. Technol.*, 17 (5), 244–254.
69. MATEEV Y.I., GRINBERG V.Y., TOLSTOGUZOV V.B. (2000) The plasticizing effect of water on proteins, polysaccharides and their mixtures. Glassy state of biopolymers, food and seeds. *Food Hydrocolloid*, 14, 425–437.
70. SOOTTITANTAWAT A., YOSHII H., FURUTA T., OHGAWARA M., FORSSELL P., PARTANEN R., POUTANEN K., LINKO P. (2004) Effect of water activity on the release characteristics and oxidative stability of D-limonene encapsulated by spray drying. *J. Agric. Food Chem.*, 52 (5), 1269–1276.
71. WEGMÜLLER R., ZIMMERMANN M.B., BÜHR V.G., WINDHAB E.J., HURRELL R.F. (2006) Development, stability, and sensory testing of microcapsules containing iron, iodine, and vitamin A for use in food fortification. *J. Food Sci.*, 71 (2), S181–S187.
72. PARTANEN R., HAKALA P., SJÖVALL O., KALLIO H., FORSSELL P. (2005) Effect of relative humidity on the oxidative stability of microencapsulated sea buckthorn seed oil. *J. Food Sci.*, 70 (1), E37–E43.
73. DRUSCH S., SERFERT Y., SCHWARZ K. (2006) Microencapsulation of fish oil with n-octenylsuccinate-derivatised starch: flow properties and oxidative stability. *Eur. J. Lipid Sci. Technol.*, 108 (6), 501–512.

74. RAPAKA R.S., COATES P.M. (2006) Dietary supplements and related products: a brief summary. *Life Sciences*, 78, 2026–2032.
75. FAIRWEATHER-TAIT S.J. (1992) Bioavailability of trace elements. *Food Chem.*, 43, 213–215.
76. SACKS P.V., HOUCHIN D.N. (1978) Comparative bioavailability of elemental iron powders for repair of iron deficiency anemia in rats. Studies of efficacy and toxicity of carbonyl iron. *Am. J. Clin. Nutr.*, 31 (4), 566–571.
77. NAVARRO M., WOOD R.J. (2003) Plasma changes in micronutrients following a multivitamin and mineral supplement in healthy adults. *J. Am. College Nutr.*, 22 (2), 124–132.
78. SIEGEL J.A., THOMAS S.R., STUBBS J.B., STABIN M.G., HAYS M.T., KORAL K.F., ROBERTSON J.S., HOWELL R.W., WESSELS B.W., FISHER D.R., WEBER D.A., BRILL A.B. (1999) MIRD pamphlet no. 16: Techniques for quantitative radiopharmaceutical biodistribution data acquisition and analysis for use in human radiation dose estimates. *J. Nuclear Med.*, 40 (2), 37S–61S.
79. HALLBERG L. (1981) Bioavailability of dietary iron in man. *Annual Rev. Nutr.*, 1, 123–147.
80. COOK J.D., LAYRISSE M., MARTINEZ-TORRES C., WALKER R., MONSEN E., FINCH C.A. (1972) Food iron absorption measured by an extrinsic tag. *J. Clin. Invest.*, 51 (4), 805–815.
81. CONSAUL J.R., LEE K. (1983) Extrinsic tagging in iron bioavailability research: a critical review. *J. Agric. Food Chem.*, 31 (4), 684–689.
82. ABRAMS S.A., WONG W.W. (2003) *Stable Isotopes in Human Nutrition.* Oxford, CABI International.
83. FAIRWEATHER-TAIT S.J., DAINTY J. (2002) Use of stable isotopes to assess the bioavailability of trace elements: a review. *Food Additives and Contaminants*, 19 (10), 939–947.
84. HALLBERG L., HULTHÉN L. (2000) Prediction of dietary iron absorption: an algorithm for calculating absorption and bioavailability of dietary iron. *Am. J. Clin. Nutr.*, 71 (5), 1147–1160.
85. FAIRWEATHER-TAIT S.J., TEUCHER B. (2002) Iron and calcium bioavailability of fortified foods and dietary supplements. *Nutr. Rev.*, 60 (11), 360–367.
86. RICKETTS C.D. (1993) Iron bioavailability from controlled-release and conventional iron supplements. *J. Appl. Nutr.*, 45 (1), 13–19.
87. BRONNER F., PANSU D. (1999) Nutritional aspects of calcium absorption. *J. Nutr.*, 129, 9–12.
88. ZHAO Y., MARTIN B.R., WEAVER C.M. (2005) Calcium bioavailability of calcium carbonate fortified soymilk is equivalent to cow's milk in young women. *J. Nutr.*, 135 (10), 2379–2382.
89. BACCIOTTINI L., TANINI A., FALCHETTI A., MASI L., FRANCESCHELLI F., PAMPALONI B., GIORGI G., BRANDI M.L. (2004) Calcium bioavailability from a calcium-rich mineral water, with some observations on method. *J. Clin. Gastroenterology*, 38 (9), 761–766.
90. PEPPAS N.A., LITTLE M. D., HUANG Y. (2000) 'Bioadhesive controlled release systems', in Wise D.L. (ed.), *Handbook of Pharmaceutical Controlled Release Technology*. New York, Marcel Dekker.
91. WILLIAMSON G., MANACH C. (2005) Bioavailability and bioefficacy of polyphenols in humans. II. Review of 93 intervention studies. *Am. J. Clin. Nutr.*, 81 (1), 243S–255S.
92. GÄRTNER C., STAHL W., SIES H. (1997) Lycopene is more bioavailable from tomato paste than from fresh tomatoes. *Am. J. Clin. Nutr.*, 66 (1), 116–122.

93. HOLLMAN P.C.H., VAN TRIJP J.M.P., BUYSMAN M.N.C.P., VAN DER GAAG M.S., MENGELERS M.J.B., DE VRIES J.H.M., KATAN M.B. (1997) Relative bioavailability of the antioxidant flavonoid quercetin from various foods in man. *FEBS Letters*, 418 (1–2), 152–156.
94. ZHANG K., ZUO Y. (2004) GC-MS determination of flavonoids and phenolic and benzoic acids in human plasma after consumption of cranberry juice. *J. Agric. Food Chem.*, 52 (2), 222–227.
95. TANG G., QIN J., DOLNIKOWSKI G.G., RUSSELL R.M., GRUSAK M.A. (2005) Spinach or carrots can supply significant amounts of vitamin A as assessed by feeding with intrinsically deuterated vegetables. *Am. J. Clin. Nutr.*, 82 (4), 821–828.
96. VAN LIESHOUT M., WEST C.E., VAN BREEMEN R.B. (2003) Isotopic tracer techniques for studying the bioavailability and bioefficacy of dietary carotenoids, particularly β-carotene, in humans: a review. *Am. J. Clin. Nutr.*, 77 (1), 12–28.
97. EMKEN E.A. (2001) Stable isotope approaches, applications, and issues related to polyunsaturated fatty acid metabolism studies. *Lipids*, 36 (9), 965–973.
98. MOZAFARI M.R., FLANAGAN J., MATIA-MERINO L., AWATI A., OMRI A., SUNTRES Z.E., SINGH H. (2006) Recent trends in the lipid-based nanoencapsulation of antioxidants and their role in foods. *J. Sci. Food Agric.*, 86 (13), 2038–2045.
99. GARTI N., SHEVACHMAN M., SHANI A. (2004) Solubilization of lycopene in jojoba oil microemulsion. *J. Am. Oil Chem. Soc.*, 81 (9), 873–877.
100. SHI J., LE MAGUER M. (2000) Lycopene in tomatoes: chemical and physical properties affected by food processing. *Crit. Rev. Food Sci. Nutr.*, 40 (1), 1–42.
101. JOHNSON, K.C., SWINDELL, A.C. (1996) Guidance in setting of drug particle size specifications to minimize variability in absorption. *Pharm. Res.*, 13 (12), 1795–1798.
102. OH D-M., CURL R.L., AMIDON G.L. (1993) Estimating the fraction dose absorbed from suspensions of poorly soluble compounds in humans: a mathematical model. *Pharm. Res.*, 10 (2), 264–270.
103. DESAI M.P., LABHASETWAR V., AMIDON G.L., LEVY, R.J. (1996) Gastrointestinal uptake of biodegradable microparticles: effect of particle size. *Pharm. Res.*, 13 (12), 1838–1845.
104. BIESA CH., LEHRA C-M, WOODLEY J.F. (2004) Lectin-mediated drug targeting: history and applications. *Adv. Drug Deliv.. Rev.*, 56, 425–435.
105. OHNO Y., NAGANUMA T., OGAWA T., MURAMOTO K. (2003) Effect of lectins on the transport of food factors in Caco-2 cell monolayers. *J. Agric. Food Chem.*, 51, 6056–6061.
106. PUSZTAI A. (1989) Transport of proteins through the membranes of the adult gastro-intestinal tract – a potential for drug delivery? *Adv. Drug Deliv. Rev.*, 3 (2), 215–228.
107. OHNO Y., NAGANUMA T., OGAWA T., MURAMOTO K. (2006) Effect of lectins on the transport of food factors in Caco-2 cell monolayers. *J. Agric. Food Chem.*, 54 (2), 548-553.
108. VIROONUDOMPHOL D., PONGPAEW P., TUNGTRONGCHITR R., CHANGBUMRUNG S., TUNGTRONGCHITR A., PHONRAT B., VUDHIVAI N., SCHELP F.P. (2003) The relationships between anthropometric measurements, serum vitamin A and E concentrations and lipid profiles in overweight and obese subjects. *Asia Pacific J. Clin. Nutr.*, 12(1), 73–79.
109. WORTSMAN J., MATSUOKA L.Y., CHEN T.C., LU Z., HOLICK M.F. (2000) Decreased bioavailability of vitamin D in obesity. *Am. J. Clin. Nutr.*, 72(3), 690–693.
110. BELL N.H., LIEL Y., HOLLIS B.W., EPSTEIN S. (1988) Evidence that alteration of the

vitamin D-endocrine system in obesity results from vitamin D deficiency. *Proc. 7th Workshop on Vitamin D*, 968–975.

111. ROIG M.J., ALEGRÍA A., BARBERÁ R., FARRÉ R., LAGARDA M.J. (1999) Calcium bioavailability in human milk, cow milk and infant formulas – comparison between dialysis and solubility methods. *Food Chem.*, 65, 353–357.
112. MILLER D.D., SCHRICKER B.R., RASMUSSEN R.R., VAN CAMPEN D. (1981) An *in vitro* method for estimation of iron availability from meals. *Am. J. Clin. Nutr.*, 34, 2248–2256.
113. LUTEN J., CREWS H., FLYNN A., VAN DAEL P., KASTENMAYER P., HURRELL R., DEELSTRA H., SHEN L.-H., FAIRWEATHER-TAIT S., HICKSON K., FARRÉ R., SCHLEMMER U., FRØHLICH W. (1996) Interlaboratory trial on the determination of the *in vitro* iron dialysability from food. *J. Sci. Food Agric.*, 72 (4), 415–424.
114. JUDPRASONG K., ORNTHAI M., SIRIPINYANOND A., SHIOWATANA J. (2005) A continuous-flow dialysis system with inductively coupled plasma optical emission spectrometry for *in vitro* estimation of bioavailability. *J. Anal. Atom. Spectrometry*, 20 (11), 1191–1196.
115. REBOUL E., RICHELLE M., PERROT E., DESMOULINS-MALEZET C., PIRISI V., BOREL P. (2006) Bioaccessibility of carotenoids and vitamin E from their main dietary sources. *J. Agric. Food Chem.*, 54 (23), 8749–8755.
116. TYSSANDIER V., REBOUL E., DUMAS J.F., BOUTELOUP-DEMANGE C., ARMAND M., MARCAND J., SALLAS M., BOREL P. (2003) Processing of vegetable-borne carotenoids in the human stomach and duodenum. *Am. J. Physiol. Gastrointest. Liver Physiol.*, 284 (6), G913–G923.
117. BAJPAI S.K., DUBEY S., SAXENA S. (2006) Flow through diffusion cell method: a better approach to study drug release behavior as compared to traditional dissolution test method. *J. Macromolecular Sci. – Pure Appl. Chem.*, 43 (4–5), 627–636.
118. AMIDON G.L., LENERNAS H., SHAH V.P., CRISON J.R. (1995) A theoretical basis for a biopharmaceutical drug classification: the correlation of *in vitro* drug product dissolution and *in vivo* bioavailability. *Pharm. Res.*, 12(3), 413–419.
119. ZAHIRUL M., KHAN I. (1996) Dissolution testing for sustained or controlled release oral dosage forms and correlation with *in vivo* data: challenges and opportunities. *Int. J. Pharm.*, 140 (2), 131–143.
120. QIU Y., SAMARA E.E., CAO G. (2000) '*In vitro–in vivo* correlations in the development of solid oral controlled release dosage forms', in Wise, D.L. (ed.), *Handbook of Pharmaceutical Controlled Release Technology*. New York, Marcel Dekker, 527–550.
121. AVDEEF A. (2003) 'High throughput measurement of permeability profiles', in van de Waterbeemd H., Lennernäs H., Artursson P. (eds), *Drug Bioavailability*. Weinheim (Germany), Wiley-VCH, 46–71.
122. KANSY M., SENNER F., GUBERNATOR K. (1998) Physicochemical high throughput screening: parallel artificial membrane permeation assay in the description of passive absorption processes. *J. Med. Chem.*, 41, 1007–1010.
123. KANSY M., AVDEEF A., FISCHER H. (2004) Advances in screening for membrane permeability: high-resolution PAMPA for medicinal chemists. *Drug Discovery Today*, 1 (4), 349–355.
124. KAZAN G. (2006) Improved capabilities of non-cell-based permeability assays. *Am. Lab.*, 38 (8), 26–29.
125. BALIMANE P.V., HAN Y.-H., CHONG S. (2006) Current industrial practices of assessing permeability and P-glycoprotein interaction. *AAPS J.*, 8 (1), E1–E13.

126. PINTO M., ROBINE-LEON S., APPAY M.D. (1983) Enterocyte-like differentiation and polarization of the human colon carcinoma cell line Caco-2 in culture. *Biology of the Cell*, 47 (3), 323–330.
127. VINCENT M.L., RUSSELL R.M., SASAK V. (1985) Folic acid uptake characteristics of a human colon carcinoma cell line, Caco-2. A newly-described cellular model for small intestinal epithelium. *Clin. Nutr.*, 39 (5), 355–360.
128. HIDALGO I.J., RAUB T.J., BORCHARDT R.T. (1989) Characterization of the human colon carcinoma cell line (Caco-2) as a model system for intestinal epithelial permeability. *Gastroenterology*, 96 (3), 736–749.
129. ARTURSSON P. (1990) Epithelial transport of drugs in cell culture. I: A model for studying the passive diffusion of drugs over intestinal absorptive (Caco-2) cells. *J. Pharm. Sci.*, 79 (6), 476–482.
130. ARTURSSON P., PALM K., LUTHMAN K. (2001) Caco-2 monolayers in experimental and theoretical predictions of drug transport. *Adv. Drug Deliv. Rev.*, 46 (1–3), 27–43.
131. ARTURSSON P., TAVELLIN S. (2003) 'Caco-2 and emerging alternatives for prediction of intestinal drug transport: a general overview', in van de Waterbeemd H., Lennernäs H., Artursson P. (eds), *Drug Bioavailability*. Weinheim (Germany), Wiley-VCH, 72–89.
132. UNGELL A.L., KARLSSON J. (2003) 'Cell cultures in drug discovery: an industrial perspective', in van de Waterbeemd H., Lennernäs H., Artursson P. (eds), *Drug Bioavailability*. Weinheim (Germany), Wiley-VCH, 90–131.
133. BOSSCHER D., VAN CAILLIE-BERTRAND M., ROBBERECHT H., VAN DYCK K., VAN CAUWENBERGH R., DEELSTRA H. (2001) *In vitro* availability of calcium, iron, and zinc from first-age infant formulae and human milk. *J. Ped. Gastroenterology Nutr.*, 32 (1), 54–58.
134. GARGARI B.P., RAZAVIEH S.V., MAHBOOB S., NIKNAFS B., KOOSHAVAR H. (2006) Effect of retinol on iron bioavailability from Iranian bread in a Caco-2 cell culture model. *Nutr.*, 22 (6), 638–644.
135. FAIRWEATHER-TAIT S., LYNCH S., HOTZ C., HURRELL R., ABRAHAMSE L., BEEBE S., BERING S., BUKHAVE K., GLAHN R., HAMBIDGE M., HUNT J., LONNERDAL B., MILLER, D., MOHKTAR N., NESTEL P., REDDY M., SANDBERG A.-S., SHARP P., TEUCHER B., TRINIDAD T.P. (2005) The usefulness of *in vitro* models to predict the bioavailability of iron and zinc: a consensus statement from the HarvestPlus expert consultation. *Int. J. Vit. Nutr. Res.*, 75 (6), 371–374.
136. LAHM H., AMSTAD P., WYNIGER J., YILMAZ A., FISCHER J.R., SCHREYER M., GIVEL J.-C. (1994) Blockade of the insulin-like growth-factor-I receptor inhibits growth of human colorectal cancer cells: evidence of a functional IGF-II-mediated autocrine loop. *Int. J. Cancer*, 58 (3), 452–459.
137. FRANCIS M.F., PIREDDA M., WINNIK F.M. (2003) Solubilization of poorly water soluble drugs in micelles of hydrophobically modified hydroxypropylcellulose copolymers. *J. Control. Rel.*, 93 (1), 59–68.
138. DIMITRIJEVIC D., SHAW A.J., FLORENCE A.T. (2000) Effects of some non-ionic surfactants on transepithelial permeability in Caco-2 cells. *J. Pharm. Pharmacol.*, 52 (2), 157–162.
139. BLUM S., RENIERO R., SCHIFFRIN E.J., CRITTENDEN R., MATTILA-SANDHOLM T., OUWEHAND A.C., SALMINEN S., VON WRIGHT A., SAARELA M., SAXELIN M., COLLINS K., MORELLI L. (1999) Adhesion studies for probiotics: need for validation and refinement. *Trends Food Sci. Technol.*, 10 (12), 405–410.
140. TUOMOLA E.M., SALMINEN S.J. (1998) Adhesion of some probiotic and dairy

Lactobacillus strains to Caco-2 cell cultures. *Int. J. Food Microbiol.*, 41 (1), 45–51.

141. GOPAL P.K., PRASAD J., SMART J., GILL H.S. (2001) *In vitro* adherence properties of *Lactobacillus* rhamnosus DR20 and *Bifidobacterium* lactis DR10 strains and their antagonistic activity against an enterotoxigenic *Escherichia coli*. *Int. J. Food Microbiol.*, 67 (3), 207–216.
142. TUOMOLA E., CRITTENDEN R., PLAYNE M., ISOLAURI E., SALMINEN, S. (2001) Quality assurance criteria for probiotic bacteria. *Am. J. Clin. Nutr.*, 73 (2), 393S–398S.
143. LIU P., KRISHNAN T.R. (1999) Alginate-pectin-poly-L-lysine particulate as a potential controlled release formulation. *J. Pharm. Pharmacol*, 51 (2), 141–149.
144. THOMBRE A.G. (2000) Feasibility assessment and rapid development of oral controlled release prototypes. *ACS Symposium Series*, 752, 69–77.
145. LAMEIRO M.H., MALPIQUE R., SILVA A.C., ALVES P.M., MELO E. (2006) Encapsulation of adenoviral vectors into chitosan-bile salt microparticles for mucosal vaccination. *J. Biotechnology,* 126 (2), 152–162.
146. TAO S.L., DESAI T.A. (2005) Micromachined devices: the impact of controlled geometry from cell-targeting to bioavailability. *J. Control. Rel.*, 109 (1–3), 127–138.
147. CRANE R.K., WILSON, T.H. (1958) *In vitro* method for the study of the rate of intestinal absorption of sugars. *J. Appl. Physiol.*, 12, 145–146.
148. KEARNEY P., MARRIOTT C. (1987) The effects of mucus glycoproteins on the bioavailability of tetracycline. III. Everted gut studies. *Int. J. Pharm.*, 38, 211–220.
149. EL GORAB M.I., UNDERWOOD B.A., LOERCH J.D. (1975) The roles of bile salts in the uptake of β carotene and retinol by rat everted gut sacs. *Biochim. Biophys. Acta*, 401 (2), 265–277.
150. LO Y.-L. (2003) Relationships between the hydrophilic-lipophilic balance values of pharmaceutical excipients and their multidrug resistance modulating effect in Caco-2 cells and rat intestines. *J. Control. Rel.*, 90 (1), 37–48.
151. BARTHE L., BESSOUET M., WOODLEY J.F., HOUIN, G. (1998) The improved everted gut sac: a simple method to study intestinal P-glycoprotein. *Int. J. Pharm.*, 173 (1–2), 255–258.
152. SHARMA P., VARMA M.V.S., CHAWLA H.P.S., PANCHAGNULA R. (2005) Absorption enhancement, mechanistic and toxicity studies of medium chain fatty acids, cyclodextrins and bile salts as peroral absorption enhancers. *Farmaco*, 60 (11–12), 884–893.
153. NAISBETT B., WOODLEY J. (1994) The potential use of tomato lectin for oral drug delivery: 2. Mechanism of uptake *in vitro*. *Int. J. Pharm.*, 110 (2), 127–136.
154. PETRI N., LENNERNÄS H. (2003) '*In vivo* permeability studies in the gastrointestinal tract of humans', in van de Waterbeemd H., Lennernäs H., Artursson P. (eds), *Drug Bioavailability*. Weinheim (Germany), Wiley-VCH, 155–188.
155. SMART J.D., KELLAWAY I.W., WORTHINGTON H.E.C. (1984) An *in vitro* investigation of mucosa-adhesive materials for use in controlled drug delivery. *J. Pharm. Pharmacol.*, 36, 295–299.
156. MORTAZAVI S.A., SMART J.D. (1994) An *in vitro* method for assessing the duration of micoadhesion. *J. Control. Rel.*, 31 (2), 207–212.
157. SAM A.P., VAN DEN HEUIJ J.T.M., TUKKER J.J. (1992) Mucoadhesion of both film-forming and non-film-forming polymeric materials as evaluated with the Wilhelmy plate method. *Int. J. Pharm.*, 79 (2–3), 97–105.
158. MORTAZAVI S.A., SMART J.D. (1995) An investigation of some factors influencing the *in vitro* assessment of mucoadhesion. *Int. J. Pharm.*, 116 (2), 223–230.

159. HASSAN E.E. GALLO J.M. (1990) A simple rheological method for the *in vitro* assessment of mucin-polymer bioadhesive bond strength. *Pharm. Res.*, 7, 491–495.
160. TAMBURIC S., CRAIG D.Q.M. (1997) A comparison of different *in vitro* methods for measuring mucoadhesive performance. *Eur. J. Pharm. Biopharm.*, 44 (2), 159–167.
161. DELIE F. (1998) Evaluation of nano- and microparticle uptake by the gastrointestinal tract. *Adv. Drug Deliv. Rev.*, 34 (2–3), 221–233.
162. FLORENCE A.T. (1997) The oral absorption of micro- and nanoparticulates: neither exceptional nor unusual. *Pharm. Res.*, 14 (3), 259–266.
163. TAKEUCHI H., THONGBORISUTE J., MATSUI Y., SUGIHARA H., YAMAMOTO H., KAWASHIMA Y. (2005) Novel mucoadhesion tests for polymers and polymer-coated particles to design optimal mucoadhesive drug delivery systems. *Adv. Drug Deliv. Rev.*, 57 (11), 1583–1594.
164. KOCKISCH S., REES G.D., YOUNG S.A., TSIBOUKLIS J., SMART J.D. (2003) Polymeric microspheres for drug delivery to the oral cavity: an *in vitro* evaluation of mucoadhesive potential. *J. Pharm. Sci.*, 92 (8), 1614–1623.
165. ACOSTA E., YUAN SH., SHARIFF A. (2006) Proteins as bioadhesives in microemulsions drug delivery vehicles. 97th AOCS Annual Meeting, 30 April–3 May.

4

Lyotropic liquid crystals as delivery vehicles for food ingredients

J. Barauskas, Institute of Biochemistry, Lithuania and T. Nylander, Lund University, Sweden

4.1 Introduction

The main challenges when designing delivery systems for sparingly water-soluble components in food and pharmaceuticals are to increase the solubility of the component and to prevent aggregation and crystallization. Other challenges are to protect the drug from degradation during storage until it reaches its targeted site in the body. Here different types of liquid crystalline phases are attractive as they can be prepared from food ingredients and they have large capacity to solubilize both hydrophilic and hydrophobic compounds. They can also be prepared to give intriguing nanostructures. An example of such structures, which will be highlighted throughout this review, are the bicontinuous cubic structures that also by virtue of their well defined porosity have a large potential use in drug delivery systems (Larsson *et al.*, 2006). The cubic liquid crystalline phases and their dispersions can also be used to achieve unique delivery functionalities in food systems, e.g. to solubilize nutrients and flavours as well as to control their release. Other applications in food systems can be to protect molecules from chemical degradation, or to increase the yield in Maillard reactions (Vauthey *et al.*, 2000; Sagalowicz *et al.*, 2006a,b). Also different types of liposomes can be used to achieve nanoencapsulation of antioxidants to increase their solubility, bioavailability and to prevent unwanted interactions with other food components (Mozafari *et al.*, 2006).

The monoolein-aqueous system, which is a thoroughly studied example of a nanostructured system, forms a cubic phase that exists in excess with water (Ericsson *et al.*, 1983; Hyde *et al.*, 1984; Landh, 1994; Briggs *et al.*, 1996; Qui and Caffrey, 2000). The bicontinuity of the cubic phase makes it possible to solubilize both hydrophobic and hydrophilic components. This means that lipophilic proteins like A-gliadin from wheat (Larsson and Lindblom, 1982) and bacteriorhodopsin (Landau and Rosenbusch, 1996) as well as relatively large amounts of membrane lipids (Nylander *et al.*, 1996; Razumas *et al.*, 1996b; Barauskas *et al.*, 1999; Engblom *et al.*, 2000) and other hydrophobic compounds of biological relevance (Caboi *et al.*, 1997, 2001; Barauskas *et al.*, 1999) can be included. These compounds are dispersed in the lipid bilayer region of the cubic phase, which can be verified by NMR self-diffusion measurements. For instance it was observed that the mobility vitamin K_1, dispersed in the lipid bilayer of monoolein-aqueous cubic phase follows that of monoolein, indicating complete dispersion of vitamin K_1 (Caboi *et al.*, 1997).

Hydrophilic compounds are included in the water channels in the bicontinuous cubic phases. The dimensions of these depend on the degree of swelling, the lipid composition and type of cubic phase, but are generally in the same range as the size of proteins (cf. Barauskas *et al.*, 2000; Clogston and Caffrey, 2005). Furthermore, being liquid crystalline phases they are quite flexible structures. These features have triggered a number of studies, which have shown that a large range of hydrophilic proteins with molecular weights up to 590 kD can be entrapped in the aqueous cavity of the monoolein-aqueous cubic phases (Razumas *et al.*, 1994, 1996a,b; Leslie *et al.*, 1996; Nylander *et al.*, 1996; Barauskas *et al.*, 2000). The entrapped proteins have been found to be protected in the cubic phase, with retained native confirmation (Ericsson *et al.*, 1983; Portmann *et al.*, 1991; Landau and Luisi, 1993; Leslie *et al.*, 1996; Razumas *et al.*, 1996b). Landau and Rosenbusch demonstrated that the bicontinuous phases based on monoolein and monopalmitolein could provide matrices for the crystallization of membrane proteins like bacteriorhodopsin (Landau and Rosenbusch, 1996). They pointed out that the use of these types of cubic phase is advantageous as they provide nucleation sites since the membrane proteins can be dissolved in the lipid bilayer. In addition, they support growth by allowing lateral diffusion of the protein molecules in the membrane.

Clogston and Caffrey (2005) investigated the transport properties of the cubic phase and were able to quantify the diffusion from the mesophase with respect to the size, shape and charge of the diffusing species. The size of the guest molecules ranged from a single amino acid to a multi-subunit protein consisting of 4176 residues and a 7 μm-long stretch of nuclear DNA. As expected, the size of the aqueous channel, as well as the interactions between the diffusing species and the lipid wall of the cubic phase, were found to affect diffusion rate. By controlling this interaction, they were able to obtain good control over the release.

Larsson and co-workers demonstrated nearly 20 years ago that it was possible to prepare stable particles of lipid-aqueous cubic phases, Cubosome® particles (Larsson, 1989, 2000; Landh, 1994; Gustafsson *et al.*, 1996, 1997; Larsson *et al.*, 2006). Several studies on different types of dispersed liquid-crystalline nanoparticles (LCNP) have pointed to the potential of using these systems for drug delivery as well as delivery of functionality to foods (Barauskas *et al.*, 2005a,b,

2006a,b; Esposito *et al.*, 2005; Spicer, 2005; Rangelov and Almgren, 2005; Angelova *et al.*, 2005; Boyd *et al.*, 2006; Johnsson *et al.*, 2006; Sagalowicz *et al.*, 2006a,b; Tamayo-Esquivel *et al.*, 2006; Vandoolaeghe *et al.*, 2006; Wörle *et al.*, 2006; Yaghmur *et al.*, 2006). This has been demonstrated with both model and *in vivo* studies for the drug substance propofol; a well-known anaesthetic agent currently used in clinical practice in the form of a stable emulsion (Johnsson *et al.*, 2006). The propofol-LCNP formulation shows several useful features including higher drug-loading capacity, lower fat-load, excellent stability, modified pharmacokinetics, and an indication of increased effect duration.

This review will focus on how the non-lamellar liquid crystalline phase can be turned into well-defined LCNP that can be used to entrap compounds with low aqueous solubility as well as hydrophilic compounds. We will also discuss the stability of the compounds in terms of hydrolysis as well as what happens when these particles interact with an interface.

4.2 Lyotropic lipid liquid crystals

From a physical point of view lipids can be naturally divided into two major groups: polar and nonpolar lipids. The nonpolar lipids, primarily triacylglycerols, diacylglycerols and sterols, have no, or small, polar groups, and hence show only limited interactions with aqueous systems. The polar lipids, with one or two hydrocarbon tails and large charged or uncharged polar head groups that control their amphiphilic nature, do interact with water and self-assemble into supramolecular liquid crystalline aggregates, also known as mesomorphic phases (Hyde, 2001; Larsson *et al.*, 2006). The common feature of lipid self-assembly in aqueous environment (lyotropic behaviour) is the formation of a polar interface, which separates layers of hydrocarbon and water. Consequently, the liquid crystalline phases formed by the addition of water are called lyotropic liquid crystals (lyotropic phases). The hydrocarbon chains in liquid crystal can exist either in a fluid state or a solid state. The latter is a characteristic property of lipid gel phases.

Polar lipids can exhibit a rich lyotropic phase behaviour, forming liquid crystalline structures with 1D, 2D or 3D periodicities. A milestone in the understanding of lyotropic liquid crystalline phases is the work of Luzzati and colleagues in France (Luzzati, 1968) and Fontell, Larsson and colleagues in Sweden (Fontell, 1974; Larsson, 1989) from the late 1960s, which clearly demonstrated that polar lipids can self-assemble into different phase structures, and established the canonical forms of lyotropic phase diagrams.

A summary of the most common lyotropic liquid crystalline phases that are formed by polar lipids and their mixtures is given in Fig. 4.1. The formation of a particular phase can in many cases be understood by looking at the geometric packing properties of the amphiphilic molecule in the particular environment (Israelachvili *et al.*, 1976; Mitchell and Ninham, 1981), that is the cross-section

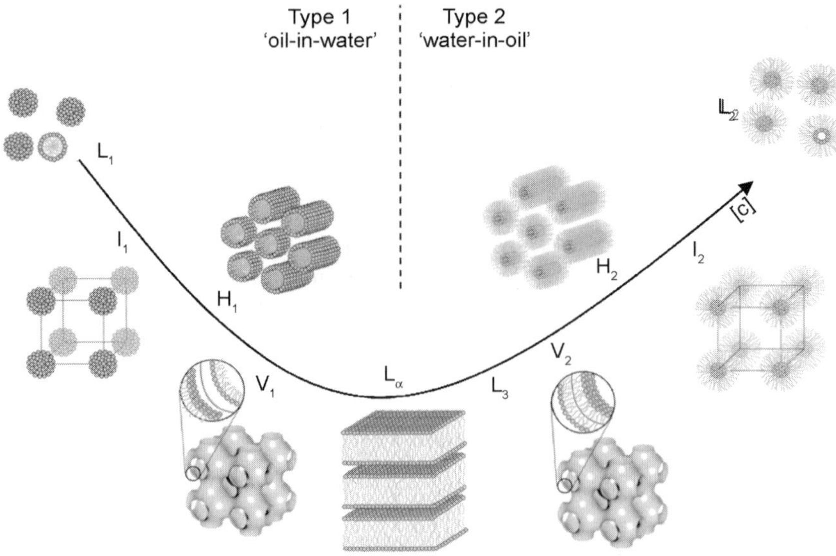

Fig. 4.1 Schematic illustrations of the structural units of different liquid crystalline phases including, from left to right, micellar cubic (I_1), hexagonal (H_1), bicontinuous cubic (V_1), lamellar (L_α), reversed bicontinuous cubic (V_2), reversed hexagonal (H_2), reversed micellar cubic (I_2). Also included are micellar (L_1), reversed micellar (L_2) as well as the typical location of the 'sponge' or L_3 phase.

area of the polar head group in relation to that of the acyl chain. This property can be expressed by the so-called packing parameter (*v/al*), which is defined as the ratio between the volume of the hydrophobic chain (*v*) and the product of the head group area (*a*) and the chain length (*l*). The packing parameter for a particular environment will determine the curvature of the interface and thus the particular phase. Generally speaking, a value of the packing parameter lower than unity (cone shaped lipids, high water solubility) facilitates the formation of structures where the polar interface is curved towards the hydrocarbon; i.e. 'oil-in-water' structures (type 1). On the other hand a value larger than one (reversed cone shaped lipids, low water solubility) will give the reverse curvature and favours 'water-in-oil' structures (type 2). With decreasing water content, the phase behaviour of polar lipids often follows the sequence shown in Fig. 4.1 (note that not all phases are found in particular lipid systems). Phase transitions can also occur with changes in temperature; with increasing temperature the sequence of thermal transitions is usually in the same direction as with decreased water content.

Lamellar phase (L_α) is the most commonly encountered mesophase in the majority of lipid/water systems. The mesostructure of L_α phase (α subscript refers to the molten lipid chains) consists of stacked infinite lipid bilayers separated by water layers, forming a 1D lattice. Another 1D mesophase, the gel phase (L_β), closely resembles L_α, but has a very much higher viscosity as the lipid bilayers in L_β phase have rigid, mostly all-*trans* alkyl chains. Another most common mesophase type is hexagonal phase. It consists of dense packing of cylindrical micelles, arranged on a 2D hexagonal lattice. The cylinders in this phase can either have a hydrocarbon core (H_1 phase) or a water core (H_2 phase). As shown in Fig. 4.1, the cubic phases (Q) can exist in several locations in the phase diagrams. Cubic phases exhibit the most complex spatial organization of all known lyotropic liquid crystals. As the name implies, the cubic phase structures are based on several possible 3D cubic lattices, namely primitive, body centered and face centered. Due to 3D periodicity they are optically isotropic and often highly viscoelastic. There are two distinct cubic phase aggregate structures. One is comprised of discrete globular micelles, normal (I_1) or reversed (I_2) (Seddon and Robins, 1999). So far, several types of space groups in the lipid micellar cubic phases are well established: $Fm\bar{3}m$ (Q^{225}), $Fd\bar{3}m$ (Q^{227}), $Im\bar{3}m$ (Q^{229}) and $Pm\bar{3}n$ (Q^{223}) (Seddon, 1996). Another cubic phase structure is based on 3D bicontinuous aggregates closely resembling triply periodic surfaces (minimal surfaces) with zero mean curvature (Hyde, 1996). To date, three bicontinuous phases are accepted, based on three surfaces: the P ($Im\bar{3}m$, Q^{229}), D ($Pn\bar{3}m$, Q^{224}), and G ($Ia\bar{3}d$, Q^{230}). Type 1 bicontinuous mesophases (V_1) consist of a water film folded on to the surface and separating two interwoven hydrocarbon continua, whereas type 2 (V_2) mesophases contain a lipid bilayer which follows minimal surface and separates two unconnected continuous systems of water channels (Hyde, 1996). In the phase diagrams micellar cubic phases are found between micellar solutions and hexagonal phases, while bicontinuous cubic phases usually occur between hexagonal and lamellar phases (Seddon, 1996).

Here it should be noted that many other more complex liquid crystalline mesophases may be formed by polar lipids, including non-cubic bicontinuous aggregates, branched and punctured polycontinuous structures, ribbon mesophases (Hyde and Schröder, 2003). In addition, there are a number of closely related disordered mesophases, i.e. 'sponge' (L_3) phase, which can be considered to be a melt of bicontinuous cubic phase (Anderson *et al.*, 1989).

4.3 Lamellar phase colloids

In many scientific and technological applications, such as delivery of pharmaceuticals or food ingredients, colloidal dispersions of lipid liquid crystals are more favourable than corresponding bulk mesophases. Realization that lipid phases can be dispersed into biomimicking particle structures was a breakthrough discovery. The simplest and most studied of such structures are the lamellar phase dispersions – vesicles and liposomes (in the literature, the term 'vesicle' most frequently refers to unilamellar aggregate, whereas the term 'liposome' refers to multilamellar structures) discovered by Bangham and colleagues in the 1960s (Bangham and Horne, 1964). The major requirement for liposome formation is the occurrence of a multiphase region that comprises the lamellar liquid crystal coexisting with an aqueous phase (Fig. 4.2(a)).

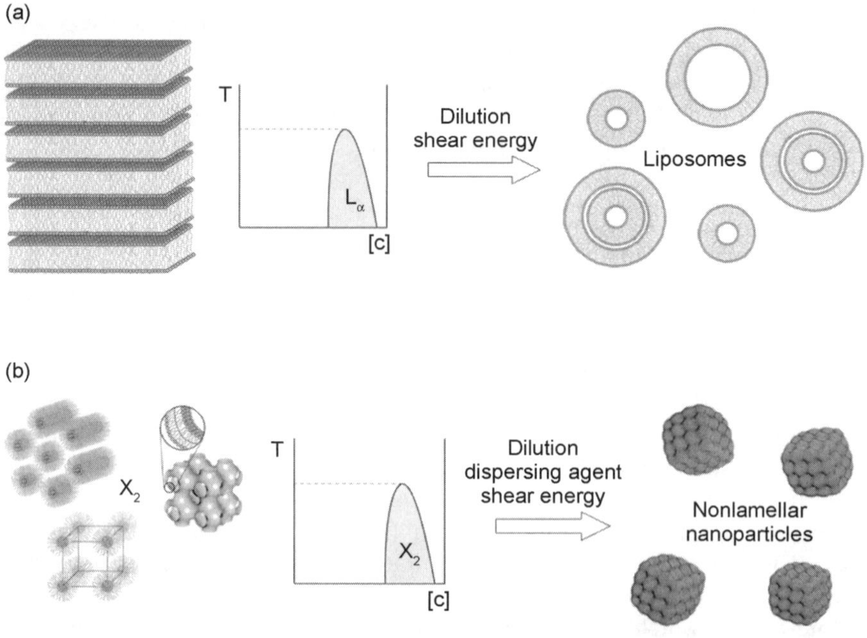

Fig. 4.2 (a) Schematic illustration of the L_α phase, phase diagram, and formation of liposomes in the diluted two-phase region. (b) Schematic illustration of the reversed phases, phase diagram, and formation of liquid crystalline nanoparticles in the diluted two-phase region.

Depending on lipid system there are several ways to produce vesicles and liposomes. The most frequently used is based on the formation of thin lipid films, hydration with excess water solution and subsequent dispersion by applying a shear force, i.e. vortexing. High-shear force methods, such as ultrasonication, homogenization or extrusion, can be further applied to reduce the size of the particles. As for all colloidal systems, stability issue of liposome dispersions is the major challenge. Since dispersions are typically not thermodynamically stable, kinetic stabilization by introducing electrostatic and steric repulsion is frequently used to prolong colloidal stability. More about vesicles and liposomes can be found in specialized literature (Lasic, 1993).

4.4 Nonlamellar liquid crystalline lipid-based nanoparticles

As discussed in the previous section, polar lipids can also self-assemble into various nonlamellar liquid crystals of different geometries. In analogy with liposomal dispersions, nonlamellar phase particles can be prepared by dispersion formation in a multiphase region comprising a type 2 ('water-in-oil') nonlamellar liquid crystal in coexistence with a dilute aqueous phase (Fig. 4.2(b)). *Nonlamellar phase coexistence with aqueous solution is an obligatory requirement*. Here it should be noted that type 1 liquid crystals ('oil-in-water') formed by water soluble polar lipids are not suitable since they lose their structure to micellar solution formation upon high dilution.

As noted in the previous section, with decreasing spontaneous curvature there is a variety of different type 2 liquid crystals: 'sponge' (L_3), reversed bicontinuous cubic (V_2), reversed hexagonal (H_2), and reversed micellar cubic (I_2). All these phases can be dispersed into nanoparticles with mean sizes in the range 50–500 nm featuring different characteristic attributes of internal nanostructure and functional properties. Virtually all possible nonlamellar liquid crystals can be dispersed into colloidal nanoparticles when using suitable lipid combinations and preparation techniques. Pending applications include utilization of lipid nonlamellar particle systems as drug and food ingredient delivery vehicles.

4.4.1 Formation mechanisms

Liquid crystalline nanoparticles (LCNPs) of nonlamellar geometry (only later was it realized that these aggregates had a bicontinuous cubic structure) were probably first observed in a light microscopy study of fat digestion in the late 1970s (Patton and Carey, 1979). When compared to liposome dispersions, nonlamellar liquid crystals can only be partially fractionated into particles that are small enough by using simple dilution of a given phase and subsequent dispersion. The cohesion forces within the aggregate are simple too strong to break up the structure into colloidal particles without dispersing agent.

The first versions of somewhat stabilized fragmented nonlamellar phase were recognized almost two decades ago when it was proposed that the cubic phase formed by unsaturated monoacylglycerols (uMGs) may be dispersed into micrometer-sized particles by mechanical breakup in the presence of micellar solutions of bile salts or caseins. The action of these agents was explained in terms of a formation of a lamellar envelope on the surface of the cubic phase particle (Bucheim and Larsson, 1987; Larsson, 1989). Later it was discovered that amphiphilic block copolymers can provide very powerful stabilization for the uMGs cubic phase dispersions (Landh, 1994). A number of further studies clearly demonstrated that poly(ethylene oxide)-poly(propylene oxide)-poly(ethylene oxide) triblock copolymers, such as Pluronic F127, are very efficient for the stabilization of uMG-based (mainly glycerol monooleate and glycerol monolinoleate) cubic LCNPs and simultaneous preservation of the inner cubic phase structure of the particles (Gustafsson *et al.*, 1996, 1997; Monduzzi *et al.*, 2000; Larsson, 2000; Nakano *et al.*, 2001; Yang *et al.*, 2004).

It has been shown that the choice of lipid-to-polymer ratio is crucial to produce colloidally stable LCNPs while simultaneously preserving inner liquid crystalline structure of the particles. As with many multicomponent lipid mixtures, ternary glycerol monooleate (GMO)/F127/water phase diagram is extremely complicated and shows the importance of the phase behaviour investigations in order to produce well-defined and stable nanoparticle dispersions (Landh, 1994). With increasing F127 concentration the cubic phase region

(Q^{224} and Q^{230} phases) originating from the binary GMO/water system is transformed into the cubic Q^{229} phase that exists over a large composition range with respect to both F127 (up to 20 wt%) and water (swelling limit about 65 wt%). At an approximate GMO/F127 ratio of 70/30 (w/w) and at a water content of 70 wt%, an L_α phase appears that transforms into a sponge phase (L_3) at even higher dilution. These data indicate that in order to produce colloidally stable cubic phase dispersions the GMO/F127 ratio should be kept in the range of about 94/6–80/20 (w/w) where the Q^{229} phase or phase mixture $Q^{229} + L_3$ (or L_α) phase is formed in equilibrium with excess solution at lower and higher polymer concentrations, respectively. Lower concentrations of polymer will simply lead to unstable dispersions due to non-sufficient density of sterically stabilizing polymer F127 at the particle surfaces. Higher concentrations will shift the system into a L_3 (or L_α) + solution phase region and induce the formation of vesicular structures only.

Pluronic F127 was also found to be an effective dispersing and stabilizing agent in some other acylglycerol systems, i.e. glyceryl monooleyl ether (Barauskas *et al.*, 2006a), mixtures of diglycerol monooleate and glycerol dioleate (Johnsson *et al.*, 2005a). However, for the majority of other lipid systems F127 does not provide either sufficient fragmentation ability or stabilization effect. It is therefore of importance to choose a suitable dispersing substance for each particular lipid or lipid mixture in order to facilitate fragmentation and provide steric stabilization of the particles. This is achieved by using dispersing agents which are amphiphilic polymeric substances with hydrophobic chain(s) for miscibility with particular lipid and long hydrophilic unit(s) for stabilizing barrier at the surface of the particles, i.e. PEGylated phospholipids, PEGylated poly(ethylene glycol) acylglycerol, polyoxyethylene 1,3-didodecyloxy-propane-2-ol, polyoxyethylene 1,3 didodecyloxy-2-glycidyl-glycerol, polyoxyethylene(20) sorbitan monooleate (Polysorbate 80), D-alpha-tocopheryl poly(ethylene glycol) 1000 succinate, hydrophobically modified cellulose (Johnsson and Edwards, 2001, Johnsson *et al.*, 2005b; Rangelov and Almgren, 2005; Barauskas *et al.*, 2005a, 2006b; Almgren *et al.*, 2007).

The studies of the aqueous phase behaviour have shown that all above-mentioned dispersing polymeric agents share similar features of the interaction with structure forming lipids. With increasing their concentration in respect to structure forming lipid the phase transition from reversed liquid crystal to lamellar phase is observed. In all successful cases the amount of polymer (lipid-to-polymer ratio) is balanced in a ‘window’ between sufficient stabilization of LCNPs and transformation into purely lamellar aggregates. However, fragmentation is facilitated by the coexistence of a lamellar or L_3 phase and some liposomal aggregates are usually present in the dispersion. For a complete characterization and control of the nonlamellar LCNPs the determination of phase diagrams including all components of the system is needed.

4.4.2 Preparation methods

The simplest way of making nonlamellar LCNPs is the mechanical agitation of all necessary components, i.e. structure forming lipid, polymeric stabilizer and excess aqueous solution (typically 90–99 wt% of water). Usually this results in a coarse dispersion with a particle size in the range of 1–100 μm (Ljusberg-Wahren *et al.*, 1997). As with liposomes, further reduction in size can be achieved by use of high-shear energy input techniques such as ultrasonication, homogenization and emulsification. However, it has also been shown that these high energy methods result not only in a desirable size reduction of lipid nanoparticles but also in the undesired formation of liposomes which in terms of number density typically dominate over the cubic phase particles (Nakano *et al.*, 2002; Siekman *et al.*, 2002; Esposito *et al.*, 2003). Since the stabilizing polymer slightly swells the nonlamellar liquid crystal and reduces cohesion forces, high-shear forces simply break down the equilibrium liquid crystalline structure (especially cubic and L_3 since their interfacial curvature is lowest) into nonequilibrium vesicular structures that later typically fuse into colloidally unstable large aggregates (Barauskas *et al.*, 2005b).

Some improvement of the above method in terms of particle yield can be achieved by executing homogenization procedure at elevated temperatures, i.e. 40–80 °C (Gustafsson *et al.*, 1997; Neto *et al.*, 1999; Nakano *et al.*, 2005). In this case higher temperature prevents the formation of nonequilibrium vesicular structures in large quantities.

Another method for producing GMO/F127 cubic phase dispersions was also introduced (Spicer *et al.*, 2001). It is based on mixing the structure forming lipid with ethanol in miscible proportions and dilution of the system with aqueous solution containing dispersing stabilizer, i.e. F127. If the dilution trajectory falls into a cubic phase region, the cubic phase particles are formed spontaneously due to molecular diffusion difference at the liquid/liquid interface. Later, minor modifications to the above preparation procedures have also been introduced (Chung *et al.*, 2002; Um *et al.*, 2003). Nevertheless, also with this preparation method, substantial amounts of vesicular material are typically obtained and the particle size distributions are rather broad.

Improvements in LCNP preparation schemes have been published recently (Barauskas *et al.*, 2005a,b; Wörle *et al.*, 2007). It was discovered that metastable liposome contaminants formed after particle size reduction with homogenization techniques can be converted back to equilibrium phase structures (back to non-lamellar aggregates) by a heat treatment step. In this approach the homogenized particle dispersion is heated to temperatures above the stabilizing polymer clouding point (i.e. F127 has a clouding point just above 100 °C) for a short period of time. Considering that the particles are sterically stabilized by adsorbed dispersing agent, very slow or even negligible fusion is expected below clouding point (Barauskas *et al.*, 2005b). Due to reduced solubility and stabilizing efficiency of dispersing agent a pronounced metastable liposome particle fusion occurs above the clouding point. When the dispersion is cooled back to room temperature the dispersing agent adsorbs back on to the nonlamellar particles.

Such a preparation technique facilitates the formation of well-defined, mostly monocrystalline particle dispersions with controlled and narrow size distribution (particle size in the range of 50–500 nm and polydispersity index of about 0.2) and high kinetic stability. Importantly this method is applicable to a large range of different nonlamellar particle compositions (Barauskas *et al.*, 2005a,b, 2006a; Johnsson *et al.*, 2005a, 2006; Popescu *et al.*, 2007). Furthermore, the size of the particles can be easily controlled just by changing the concentration of the dispersion. An example of such size distribution control is shown in Fig. 4.3. Note that after preparation the dispersion will retain the particular particle size with time or upon dilution.

Very recent findings have shown that the high-energy pre-treatment of nonlamellar dispersions is not always required to get well-defined nonlamellar LCNPs with narrow size distribution (Barauskas *et al.*, 2006b). As shown in this study, sponge phase nanoparticles can be easily prepared by self-dispersion (low energy mechanical agitation, i.e. on a mechanical mixing table) of aqueous diglycerol monooleate/glycerol dioleate/mixtures fortified with small amounts

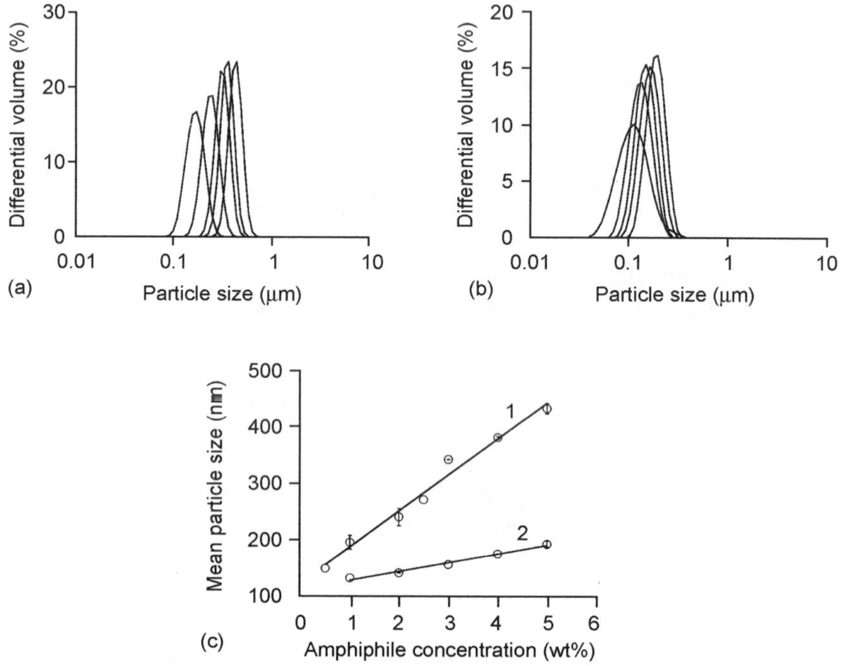

Fig. 4.3 Particle size distributions as a function of total amphiphile concentration (1, 2, 3, 4 and 5 wt%) of the homogenized and then heat-treated (125 °C) GMO/F127 (a) and GMO/oleic acid (OA)/F127 (b) dispersions. (c) Dependencies of the obtained mean particle size of the heat-treated GMO/F127 (1) and GMO/OA/F127 (2) dispersions on the total amphiphile concentration. Error bars represent the mean value ± StdErr of 1–5 separate experiments. Linear approximations are drawn to guide the eye. Reproduced with permission from *Langmuir* 2005, 21, 2569–2577. Copyright 1998 American Chemical Society.

of Polysorbate 80. The particles formed have a size of about 100–150 nm with intricate, disordered, and rather dense inner cores enclosed by a surface sponge-like region composed of intersecting lamellas. In contrast to Pluronic copolymers, Polysorbate 80 has a significantly lower molecular weight which makes it more miscible with structure forming lipids and more able to penetrate and swell the liquid crystal as well as promote dispersion formation. The surface coverage of the L_3 phase layer on L_2 droplets provides unique particle stabilization facilitating dispersion formation and long-term colloidal stability.

4.4.3 Nanostructure

The nonlamellar LCNPs are usually classified according to phase nanostructure as 'sponge' (L_3), reversed bicontinuous cubic (V_2), reversed hexagonal (H_2), or reversed micellar cubic (I_2) particles. Cubosome®, Hexosome® and Flexosome® are frequently used trade names of cubic, hexagonal and 'sponge' phase dispersions. The size of these particles can be controlled in the size range from about 50 to 500 nm, in some cases up to 1 μ or even 10 μ, without changes in the inner morphology. The main techniques to study lipid particle nanostructure are X-ray diffraction and electron microscopy. Other analytical methods such as NMR (Monduzzi *et al.*, 2000), atomic force microscopy (Neto *et al.*, 1999) and ellipsometry (Vandoolaeghe *et al.*, 2006) have also been applied.

Over the last ten years, the majority of structural and functional studies have been focused on bicontinuous cubic LCNPs formed in aqueous mixtures of uMGs and Pluronic triblock copolymers, namely GMO and F127 (Gustafsson *et al.*, 1997; Larsson, 2000; Monduzzi *et al.*, 2000; Nakano *et al.*, 2001; Siekman *et al.*, 2002; Barauskas *et al.*, 2005b). The Fourier transformations of cryo-TEM images and X-ray diffraction measurements reveal reflections characteristic of a body centered Q^{229} cubic phase structure of $Im\bar{3}m$ space group (or P-type in the minimal surface description) with a lattice parameter of about 130 Å (Barauskas *et al.*, 2005b). Due to internal cubic symmetry monocrystalline cubic LCNPs tend to maintain the shape of the cube.

Until recently only uMGs have been known to possess the aqueous phase behaviour suitable for the formation of bicontinuous LCNPs. Recent study has shown that 3,7,11,15-tetramethyl-1,2,3-hexadecane-triol which is widely used as an active ingredient in the cosmetics industry and commonly known as phytantriol (PtOH), exhibits a similar aqueous phase behaviour as uMGs (Barauskas and Landh, 2003). It forms V_2 phase in excess solution – the main criterion for preparing the corresponding nanoparticle structures. It was discovered that the PtOH-based cubic phase can be effectively dispersed into stable and reproducible Q^{229} structure-based nanoparticles by use of small amounts of D-alpha-tocopheryl polyethylene glycol 1000 succinate (Barauskas *et al.*, 2005a). In comparison to uMGs, PtOH-based nanoparticles have slightly smaller unit cell dimensions due to a shorter molecule and lower swelling limits in water.

Interesting examples of two coexisting different bicontinuous cubic LCNPs was recently published by Popescu *et al.* (2007). In this study the dispersions in

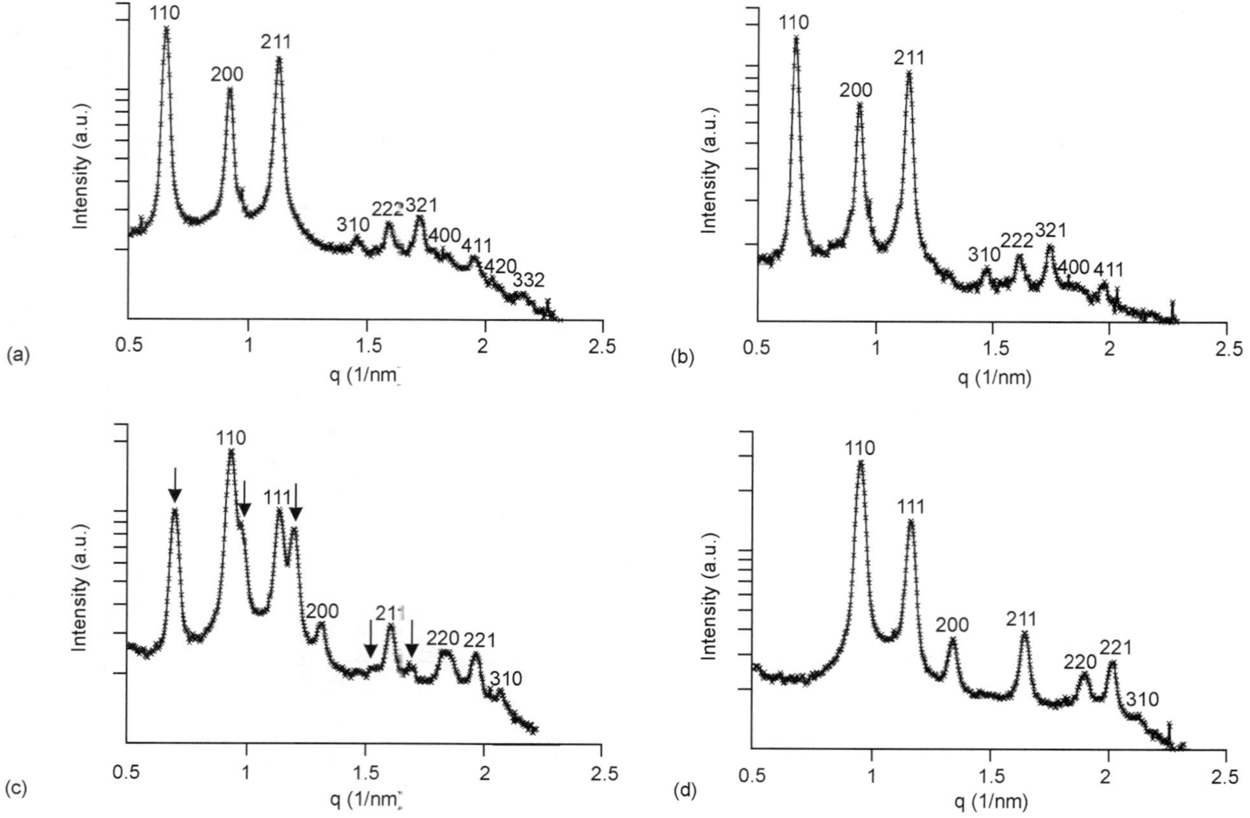

Fig. 4.4 Synchrotron X-ray diffractograms of prepared cubic phase nanoparticle dispersions in a quaternary GMO/GME/F127/water system. The GMO/GME (w/w) ratios are 95/5 (a), 90/10 (b), 85/15 (c), and 80/20 (d). All dispersions have lipid/polymer ratios of 9/1 and a water content of 95 wt%. The Miller indices in panel c stand for Q^{224} structure, whereas arrows denote the peak positions of the Q^{229} phase. Reproduced with permission from *Langmuir* 2005, 23, 496–503. Copyright 1998 American Chemical Society.

aqueous mixtures of GMO and glyceryl monooleyl ether (GME) are investigated. The results have shown that simply by changing the GMO/GME ratio it is possible to form different kinds of bicontinuous cubic nanoparticles, Q^{229} structure-based ($Im\bar{3}m$) at low GME concentration and Q^{224} structure-based ($Pn\bar{3}m$) at higher GME concentration. The example of this transformation is shown in Fig. 4.4.

Other lipids and mixtures for the preparation of V_2 structure based nanoparticles have been tried as well: glycerol monolinoleate (De Campo *et al.*, 2004; Yaghmur *et al.*, 2005), glycolopids (Abraham *et al.*, 2005), arginine- and glutamate-based surfactants (Rosa *et al.*, 2006). The formation of V_2 nanoparticles in dioleoylphosphatidylethanolamine and soy phosphatidylethanolamine-based systems was also shown (Johnsson *et al.*, 2005b; Barauskas *et al.*, 2005a). The last example is particularly interesting since natural phospholipid-based systems can be useful not only for the delivery of food ingredients but also for parenteral drug delivery.

In contrast to bicontinuous cubic particles the most common outer shape of the reversed hexagonal particles is the hexagonal prism. The fact that these prisms always appear projected along the [10] direction in cryo-TEM images suggested that they have a high aspect ratio and consequently are aligned in the relatively thin vitrified films of the experiment (Barauskas *et al.*, 2005a). To prove this hypothesis, angle-resolved cryo-TEM measurements were performed to visualize other facets of the particles and gain further insight into the 3D structure. The thickness of the reversed hexagonal particles was evaluated by measuring the projected width as a function of tilting angle. The results were then compared with the corresponding geometric model of a hexagonal prism. The fact that the unit cell dimensions of the inner particle structure does not change in the presence of polymeric stabilizer indicates its preferential location on the surface of LCNPs. Less ordered particles or particles prepared by high-shear force methods only usually adopt more spherical shapes (Gustafsson *et al.*, 1997; Sagalowicz *et al.*, 2005).

Several lipid and lipid mixture systems have been shown to be suitable to form reversed hexagonal LCNPs: mixtures of uMGs with oil (Gustafsson *et al.*, 1997; Yaghmur *et al.*, 2005), mixtures of diglycerol monooleate and glycerol dioleate (Johnsson *et al.*, 2005a), glyceryl monooleyl ether (Barauskas *et al.*, 2006a), and synthetic glycerate-based surfactants (Fong *et al.*, 2005; Boyd *et al.*, 2006).

Thermodynamically stable sponge phases can also be fragmented into LCNPs. Particles have been prepared in a three component lipid system composed of diglycerol monooleate, glycerol dioleate and Polysorbate 80 (Barauskas *et al.*, 2006b). This disordered phase dispersion forming system shows several advantageous features when compared to ordered liquid crystals: spontaneous dispersion formation and tunable particle phase structure. The relative thickness of the outer L3 phase layer can be tuned by only changing a fraction of Polysorbate concentration in the dispersion. Sponge phase layer consists of an inner network of interconnected lamellae, also referred to as molten bicontinuous cubic phase (Gustafsson *et al.*, 1999). Compared to other LCNPs, L_3 structure-based nanoparticles can incorporate much more water due to a high swelling and appearance in the water-rich region of the phase diagrams. The shape of the disordered L_3 phase particles is typically spherical.

Probably the least investigated LCNPs are dispersions of the reversed micellar cubic phase (I_2). This is due to the fact that only a limited number of lipid mixtures are shown to form I_2 liquid crystals in excess water solutions. I_2 structure-based ($Fd\bar{3}m$ space group) LCNPs have been prepared in mixtures of glycerol linoleate and tetradecane (Yaghmur *et al.*, 2006), and in mixtures of phosphatidylcholine and glycerol dioleate fortified with small amounts of Polysorbate 80 (Johnsson *et al.*, 2006).

4.4.4 Functionality – how to include the components, analyses of *in vitro* and *in vivo* release, and interaction with model membranes

Both liquid crystals and liquid crystalline nanoparticles formed through lipid self-assembly have a range of attractive properties as food delivery carriers. In particular they offer a wide solubilization and release properties range for different ingredients. Because of coexisting hydrophilic and hydrophobic domains and the very large interfacial area possessed by these structures (especially V_2 liquid crystals), they have a broad spectrum of applicability to different food substances. Both solubilization and release properties can be easily varied by changing the lipid composition and the nanostructure of liquid crystal. For example, the release properties of hydrophilic substance will be very different in a V_2 phase structure-based liquid crystal or nanoparticle, compared to an I_2 phase, which has discrete cage-like water domains. It is also obvious that if the hydrophobic species need to be incorporated the I_2 phase will be preferable to the others. In contrast, large water soluble molecules are more easily incorporated into sponge phase particles with a relatively high degree of swelling compared to, for example, reversed hexagonal liquid crystal with rather small water channels.

From a practical application point of view, an important feature of lipid-based liquid crystalline nanoparticles is the physical and chemical stability for extended periods of time. Using appropriate lipid compositions, the stability of these particles can be very good, from months to years at room temperature.

The primary issues addressed by nonlamellar particles are interaction with biological membranes. Very recent study has been addressed to an interfacial behaviour of GMO-based bicontinuous cubic nanoparticles at hydrophilic and hydrophobic surfaces studied by means of *in situ* null ellipsometry (Vandoolaeghe *et al.*, 2006). Depending on the surface properties and presence of electrolytes, different adsorption scenarios were discerned: at hydrophilic silica thick surface layers of lipid nanoparticles are generated by particle adsorption from dispersions containing added electrolyte. Insights into the structure formation and mechanisms of adsorption can be gained from analysis of the time evolution of the adsorbed layer properties. As illustrated in Fig. 4.5, the amount adsorbed at silica initially increases proportionally with time and

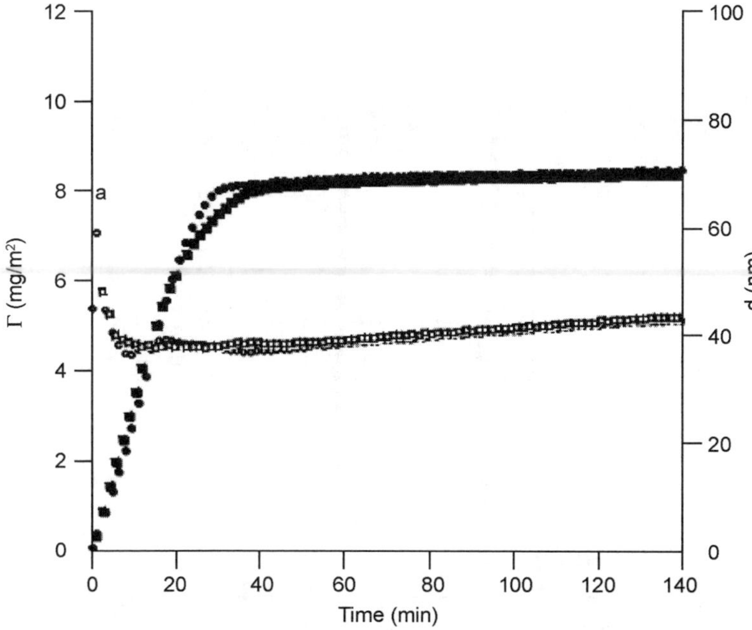

Fig. 4.5 Adsorbed amount (filled symbols) and layer thickness (open symbols) versus time after addition of bicontinuous cubic nanoparticle dispersion of GMO and F127 (9/1, w/w) to a final concentration of 0.05 mg/mL in the presence of 0.1 M NaCl and 0.017 M $CaCl_2$. The two curves (circles and squares) correspond to repeat experiments and show the good reproducibility of the experiment. Reproduced with permission from *Langmuir* 2005, 22, 9169–9174. Copyright 1998 American Chemical Society.

then levels off to reach a saturation value of around 8–9 mg/m^2. The layer thickness shows an initial decay but stabilizes rather quickly at a plateau value of about 40–50 nm, sometimes after a brief oscillation, which may reflect transient structuring at the interface. The analysis suggests an adsorption process during which the interfacial layer is successively being built up by attachment of lipid nanoparticles rather than their molecular components.

The adsorption at hydrophobic, silanized silica is dramatically different from that at bare silica. Adsorption is basically independent of the presence of electrolyte or pH and results in the formation of a thin and dense adsorbed layer, with a final thickness of 2–3 nm and an adsorbed amount of 1.7–2 mg/m^2. This shows a buildup of a monolayer and indicates that nanoparticles are dissociated at the interface.

By use of ellipsometry and quartz crystal microbalance with dissipation monitoring (QCM-D), the interaction of GMO-based bicontinuous cubic nanoparticles with supported lipid bilayers consisting of dioleoylphosphatidylcholine (DOPC) has been further investigated (Fig. 4.6) (Vandoolaeghe *et al.*, 2007). When the nanoparticles were injected, an initial fast adsorption at the bilayer covered surfaces is observed. The adsorbed amount reaches a maximum of 7.6 mg/m^2, but thereafter the adsorption decreases too and reaches a minimum of

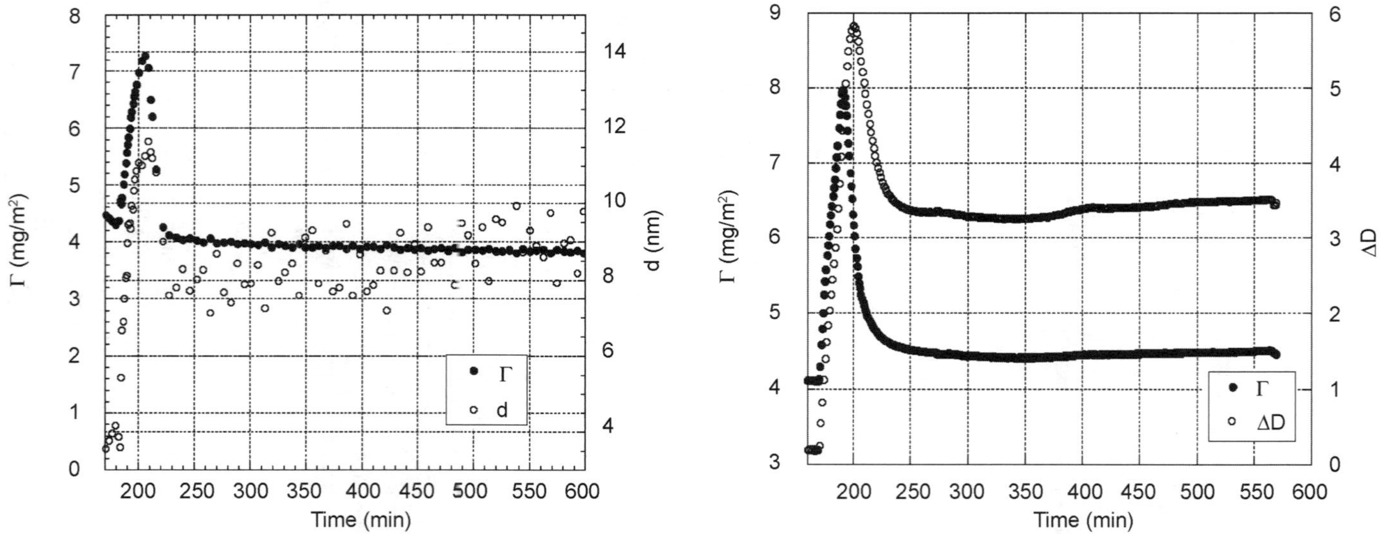

Fig. 4.6 (a) Adsorbed amount and layer thickness by ellipsometry calculated by using a four-layer model, and (b) adsorbed amount and dissipation difference by QCM-D as a function of time after the injection of GMO-based bicontinuous cubic nanoparticles at a concentration of 0.05 mg/mL on DOPC bilayer at pH 4. The dispersion has a GMO/F127 ratio of 9/1 and a water content of 95 wt%.

about 4 mg/m^2, i.e. slightly below or above the original DOPC bilayer adsorption value depending on the model used to evaluate the ellipsometry data. The increase in adsorbed amount just after the injection is indicative of an interaction between nanoparticles and the DOPC bilayer. The subsequent desorption evidenced by the decrease of the adsorbed amount is probably due to the mixing of nanoparticle components (GMO and F127) into the DOPC bilayer and a shift towards a new equilibrium surface structure. These findings are also supported by results from neutron reflectivity studies. The initial adsorption of nanoparticles and their molecular components subsequently results in partial desorption.

The QCM-D data are in agreement with the ellipsometry data, where the adsorbed amount increases drastically after the injection of the CPNP, to decrease but at a higher value than in the case of the ellipsometry, which could be explained by the presence of water in the formed layer (swelling of the layer) and then slightly increases. The same behaviour with the dissipation difference is observed. The layer becomes very viscoelastic just after the injection of CPNP to relax afterwards, but in a more viscous state than the initial behaviour. This is indicative of an initial adsorption of intact nanoparticles, or part of them, followed by a relaxation where an exchange between the lipid of the nanoparticles and the phospholipids of the bilayer could occur.

To summarize, it has been shown that there is an interaction between GMO-based LCNPs and its lipid constituents with model membranes (Vandoolaeghe *et al.*, 2007). This process involves initial adsorption of the particles, followed by a strong interaction between the GMO-based CPNP constituents and the phospholipid bilayer. This in turn leads to rearrangements within the surface layer and finally by the growth of a self-assembled structure on top of the bilayer.

Another important feature is the interaction of lipid nanoparticles with digestive enzymes, i.e. phospholipases and triacylglycerol hydrolases (lipases). It is obvious that the release of functional ingredients will also be affected by the rate of the lipid carrier hydrolysis. Most of the lipolysis takes place at interfaces and is extremely dependent on the structure and organization of the lipid/water boundary. During the reaction hydrolysis products interact with lipids, continuously change the interface characteristics and alter cleavage mechanisms from complex lipid aggregate.

GMO-based bicontinuous nanoparticles have been used as substrates for lipase activity investigations. The changes in lipid aggregate geometry as a function of a lipase action have been investigated by means of pH-stat titration, cryo-TEM and X-ray diffraction techniques (Barauskas *et al.*, 2007). From the pH-stat titration kinetic curves (Fig. 4.7), it is obvious that the reaction kinetics and end products (more than 50% of lipid hydrolyzed) of lipolytic process on GMO-based cubic lipid liquid crystalline nanoparticles is very pH dependent. The results presented in Fig. 4.7 clearly illustrate that kinetics of lipase induced hydrolysis at pH 8.4 is relatively simple. At pH 6.5 more complicated kinetics are observed and therefore were further analyzed in order to determine if intermediate aggregates are formed during the reaction. Time resolved

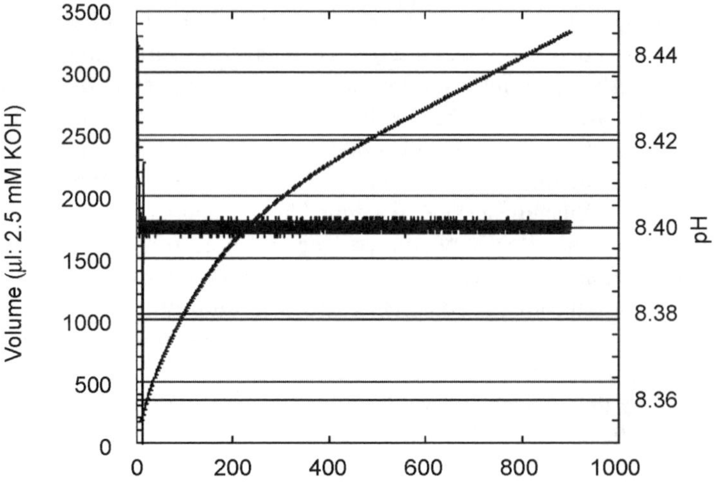

(a)

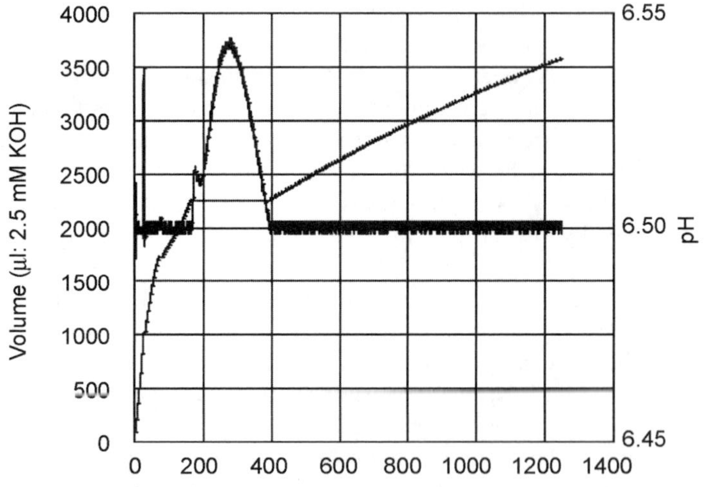

(b)

Fig. 4.7 pH-Stat titration curves of the GMO-based bicontinuous cubic nanoparticle hydrolysis by lipase at (a) pH 8.4 and (b) pH 6.5.

synchrotron X-ray diffraction and cryo-TEM have shown that during hydrolysis at pH 6.5 lipid nanoparticles undergo a phase transition sequence, bicontinuous cubic → reversed hexagonal → reversed micellar cubic, before final transformation into emulsion (reversed micellar solution). At the low pH the structural changes of the substrate were also found to affect the pKa of the generated acid, leading to an increase in pH.

4.5 Future trends

In order to prepare well-defined LCNP, one of the great challenges for the future is to obtain a fundamental understanding of the domain formation in liquid crystalline phases, that is regions with the same crystal lattice orientation. As discussed above, temperature cycling can be used to narrow the particle size distribution and therefore one can to some extent view the domain formation as a crystallization type (nucleation and growth) of process. Even in more simple systems such as mono- and bilayers the formation of different domains is quite complex (cf. the work of Mouritsen and coworkers: Mouritsen and Zuckermann, 1987; Mouritsen, 1991; Mouritsen *et al.*, 2006). For a 3D structure like the cubic liquid crystalline phases the challenges are even larger and we lack today good methods to study these liquid crystalline domains in 3D and the interface between them. Cryogenic transmission electron microscopy (cryo-TEM) images can be used to study morphological changes of liposomes induced by temperature as a function of lipid composition and one might analyze these in terms of domain formation. Edwards and co-workers report on the formation of bilayer discs, which can be related to the domain formation in the liposome bilayer (Ickenstein *et al.*, 2003, 2006; Johansson *et al.*, 2005). They observed that when lysolipid-containing thermosensitive liposomes (LTSL) were cycled through the gel ($L_{\beta'}$) to liquid crystalline (L_{α}) phase transition, open liposomes and bilayer disks appeared. The amount of bilayer disks was found to depend on the presence of both PEG-lipids and lysolipids in the membrane. Based on their results, they suggested that the micelle-forming membrane components stabilize the rim of bilayer openings as well as the membrane discs (Ickenstein *et al.*, 2003). Consequently they found that varying the content of the PEG-lipid, the average diameter of the disks can be changed in the interval from about 15 to 60 nm (Johansson *et al.*, 2005). Hydrolysis of DPPC in DPPC/DSPE-PEG$_{2000}$ liposomes, that is increasing the amount of palmitic acid, was also found to increase the disintegration of the liposomes into membrane disks (Ickenstein *et al.*, 2006). Again palmitic acid in its deprotonated form, the soap, is a micelle-forming lipid. These results indicate the importance of local effects for the domain formation. The development of single molecule techniques for studying these systems will in future deepen our understanding of lipid structure at the nanoscale.

In an *in vitro* study of lipolysis of triglycerides in a intestinal-like environment, Patton and Carey (1979) observed, apart from the initially occurring crystalline phase, a viscous isotropic phase composed of monoglycerides and fatty acids, which is identical to the one formed in monoglyceride systems. In excess of bile salts, the lipolysis products are rapidly solubilized in mixed micelles. However, the bile acid amounts *in vivo* are not sufficient to solubilize all lipids after a meal rich in fats, which implies that the liquid crystalline phases exist *in vivo* (Lindström *et al.*, 1981). Lipase and water must be free to diffuse through the phases formed by the lipolysis products, surrounding the diminishing fat droplet. Thus, the bicontinuity as well as the incorporation properties of the cubic monoglyceride phases are thought to be important features for the lipolysis process (Patton *et al.*, 1985). Various types of LCNP are therefore likely to be formed spontaneously in the GIT even if they are not administered as drug delivery vehicles or as food ingredients. This indicates a very important aspect of LCNP and liquid crystalline phases, namely their importance as carries of nutrients in food. This is important as many nutrients, such as vitamins, have a low aqueous solubility and their true nutritional value rests on their bioavailability. The analytically determined content of nutrients can therefore be irrelevant. In our opinion there is a lack of knowledge on how the structure and stability of the food that carries the nutrient affect the nutritional value. Here is a need for nutritionist, physical chemist and analytical chemists to work together. The food industry has just started to realize the potential of using LCNP as carriers of nutrients and antioxidants or to control release of flavours and to prevent unwanted interactions with other food components (Mozafari *et al.*, 2006). This type of added functionality also includes the protection of molecules from chemical degradation, or to increase the yield in Maillard reactions (Sagalowicz *et al.*, 2006a,b).

4.6 Acknowledgements

We are grateful to our colleagues Pauline Vandoolaeghe, Markus Johnsson, Fredrik Tiberg and Kåre Larsson for many stimulating discussions on lipid liquid crystalline nanoparticles and interfaces, as well as Allan Svendsen, Novozymes for discussions regarding lipases and lipase activity. I711 beamline staff at Max-Lab are acknowledged for beamtime and supporting discussions. Financial support was obtained from Camurus Lipid Research Foundation, Swedish Foundation for Strategic Research, Vinnova, Swedish Research Council, EU Marie Curie Research Training Network 'BIOCONTROL' (Contract no. MRTN-CT-2006-033439) and EU-STREP FP6 project BIOSCOPE (contract no. NMP4-CT-2003-505211).

4.7 References

ABRAHAM T, HATO M and HIRAI M (2005), 'Glycolipid based cubic nanoparticles: preparation and structural aspects', *Colloid Surface B*, 35, 107–118.

ALMGREN M, BORNÉ J, FEITOSA E, KHAN A and LINDMAN B (2007), 'Dispersed lipid liquid crystalline phases stabilized by a hydrophobically modified cellulose', *Langmuir*, 23, 2768–2777.

ANDERSON D, WENNERSTRÖM H and OLSSON U (1989), 'Isotropic bicontinuous solutions in surfactant-solvent systems: the L3 phase', *J Phys Chem*, 93, 4243–4253.

ANGELOVA A, ANGELOV B, PAPAHADJOPOULOS-STERNBERG B, BOURGAUX C and COUVREUR P (2005), 'Protein driven patterning of self-assembled cubosomic nanostructures: long oriented nanoridges', *J Phys Chem B*, 109, 3089–3093.

BANGHAM A D and HORNE R W (1964), 'Negative staining of phospholipids and their structural modification by surface-active agents as observed in the electron microscope', *J Mol Biol*, 8, 660–668.

BARAUSKAS J and LANDH T (2003), 'Phase behavior of the phytantriol/water system', *Langmuir*, 19, 9562–9565.

BARUSKAS J, RAZUMAS V and NYLANDER T (1999), 'Solubilization of ubiqinone-10 in the lamellar and bicontinous cubic phases of aqueous monoolein', *Chem Phys Lipids*, 97, 167–179.

BARAUSKAS J, RAZUMAS V and NYLANDER T (2000), 'Entrapment of glucose oxidase into the cubic Q^{230} and Q^{224} phases of aqueous monoolein', *Progr Colloid Polymer Sci*, 116, 16–20.

BARAUSKAS J, JOHNSSON M and TIBERG F (2005a), 'Self-assembled lipid superstructures: beyond vesicles and liposomes', *Nano Lett*, 5, 1615–1619.

BARAUSKAS J, JOHNSSON M, JOABSSON F and TIBERG F (2005b), 'Cubic phase nanoparticles (Cubosome): principles for controlling size, structure, and stability', *Langmuir*, 21, 2569–2577.

BARAUSKAS J, JOHNSSON M, NYLANDER T and TIBERG F (2006a), 'Hexagonal liquid-crystalline nanoparticles in aqueous mixtures of glyceryl monooleyl ether and Pluronic F127', *Chem Lett*, 35, 830–831.

BARAUSKAS J, MISIUNAS A, GUNNARSSON T, TIBERG F and JOHNSSON M (2006b), ' "Sponge" nanoparticle dispersions in aqueous mixtures of diglycerol monooleate, glycerol dioleate and Polysorbate 80', *Langmuir*, 22, 6328–6334.

BARAUSKAS J, ANDERBERG D, SVENDSEN A and NYLANDER T (2007), 'Lipase action on self-assembled lipid liquid crystalline nanoparticles'. Manuscript.

BOYD B J, WHITTAKER D V, KHOO S-M and DAVEY G (2006), 'Hexosomes formed from glycerate surfactants – formulation as a colloidal carrier for irinotecan', *Int J Pharm*, 318, 154–162.

BRIGGS J, CHUNG H and CAFFREY M (1996), 'The temperature-composition phase diagram and mesophase structure characterization of the monoolein/water system', *J Phys II France*, 6, 723–751.

BUCHEIM W and LARSSON K (1987), 'Cubic lipid-protein-water phases', *J Colloid Interface Sci*, 117, 582–583.

CABOI F, NYLANDER T, RAZUMAS V, TALAIKYTÉ Z, MONDUZZI M and LARSSON K (1997), 'Structural effects, mobility and redox behavior of vitamin K_1 hosted in the monoolein-water liquid crystalline phases', *Langmuir*, 13, 5476–5483.

CABOI F, AMICO G S, PITZALIS P, MONDUZZI M, NYLANDER T and LARSSON K (2001), 'Addition of hydrophilic and lipophilic compounds of biological relevance to the monoolein:water system. I. Phase behavior', *Chem Phys Lipids*, 109, 47–62.

CHUNG H, KIM J, UM J Y, KWON I C and JEONG S Y (2002), 'Self-assembled "nanocubicle" as a carrier for peroral insulin delivery', *Diabetologia*, 45, 448–451.

CLOGSTON J and CAFFREY M (2005), 'Controlling release from the lipidic cubic phase. Amino acids, peptides, proteins and nucleic acids', *J Contr Release*, 107, 97–111.

DE CAMPO L, YAGHMUR A, SAGALOWICZ L, LESER M E, WATZKE H and GLATTER O (2004), 'Reversible phase transitions in emulsified nanostructured lipid systems', *Langmuir*, 20, 5254–5261.

ENGBLOM J, MIEZIS Y, NYLANDER T, RAZUMAS V and LARSSON K (2000), 'On the swelling of monoolein liquid-crystalline aqueous phases in the presence of distearoyl-phosphatidylglycerol', *Progr Colloid Polym Sci*, 116, 9–15.

ERICSSON B, LARSSON K and FONTELL K (1983), 'A cubic protein-monoolein-water phase', *Biochim Biophys Acta*, 729, 23–27.

ESPOSITO E, EBLOVI N, RASI S, DRECHSLER M, DI GREGORIO G M, MENEGATTI E and CORTESI R (2003), 'Lipid-based supramolecular systems for topical application: a preformulatory study', *AAPS Pharm Sci*, 5, A30.

ESPOSITO E, CORTESI R, DRECHSLER M, PACCAMICCIO L, MARIANI P, CONTADO C, STELLIN E, MENEGATTI E, BONINA F and PUGLIA C (2005), 'Cubosome dispersions as delivery systems for percutaneous administration of indomethacin', *Pharm Res*, 22, 2163–2173.

FONG C, KRODKIEWSKA I, WELLS D, BOYD B J, BOOTH J, BHARGAVA S, McDOWALL A and HARTLEY P G (2005), 'Submicron dispersions of hexosomes based on novel glycerate surfactants', *Aust J Chem*, 58, 683–687.

FONTELL K (1974), 'X-ray diffraction by liquid crystals – amphiphilic systems', in Gray G W and Winsor P A, *Liquid Crystals and Plastic Crystals*, Chichester, Ellis Horwood, 80–109.

GUSTAFSSON J, LJUSBERG-WAHREN H, ALMGREN M and LARSSON K (1996), 'Cubic lipid-water phase dispersed into submicron particles', *Langmuir*, 12, 4611–4613.

GUSTAFSSON J, LJUSBERG-WAHREN H, ALMGREN M and LARSSON K (1997), 'Submicron particles of reversed lipid phases in water stabilized by a nonionic amphiphilic polymer', *Langmuir*, 13, 6964–6971.

GUSTAFSSON J, NYLANDER T, ALMGREN M and LJUSBERG-WAHREN H (1999), 'Phase behavior and aggregate structure in aqueous mixtures of sodium cholate and glycerol monooleate', *J Colloid Interface Sci*, 211, 326–335.

HYDE S T (1996), 'Bicontinuous structure in lyotropic liquid crystals and crystalline hyperbolic surfaces', *Curr Opin Solid State Mater Sci*, 1, 653–662.

HYDE S T (2001), 'Identification of lyotropic liquid crystalline mesophases', in Holmberg K, *Handbook of Applied Surface and Colloid Chemistry*, Chichester, John Wiley & Sons, 299–332.

HYDE S T and SCHRÖDER G E (2003), 'Novel surfactant mesostructural topologies: between lamellae and columnar (hexagonal) forms', *Curr Opin Colloid Interface Sci*, 8, 5–14.

HYDE S T, ANDERSSON S, ERICSSON B and LARSSON K (1984), 'A cubic structure consisting of a lipid bilayer forming an infinite periodic minimum surface of the gyroid type in the glycerolmonooleate-water system', *Z Crystallogr*, 168, 213–219.

ICKENSTEIN L M, ARFVIDSSON M C, NEEDHAM D, MAYER L D and EDWARDS K (2003), 'Disc formation in cholesterol-free liposomes during phase transition', *Biochim Biophys Acta*, 1614, 135–138.

ICKENSTEIN L M, SANDSTRÖM M C, MAYER L D and EDWARDS K (2006), 'Effects of phospholipid hydrolysis on the aggregate structure in DPPC/DSPE-PEG_{2000} liposome preparations after gel to liquid crystalline phase transition', *Biochim Biophys Acta*, 1758, 171–180.

ISRAELACHVILI J N, MITCHELL D J and NINHAM B W (1976), 'Theory of self-assembly of hydrocarbon amphiphiles into micelles and bilayers', *J Chem Soc Faraday Trans II*, 72, 1525–1568.

JOHANSSON E, ENGVALL C, ARFVIDSSON M, LUNDAHL P and EDWARDS K (2005), 'Development and initial evaluation of PEG-stabilized bilayer disks as novel model membranes', *Biophys Chem*, 113, 183–192.

JOHNSSON M and EDWARDS K (2001), 'Phase behavior and aggregate structure in mixtures of dioleoylphosphatidylethanolamine and poly(ethylene glycol)-lipids', *Biophys J*, 80, 313–323.

JOHNSSON M, LAM Y, BARAUSKAS J and TIBERG F (2005a), 'Aqueous phase behavior and dispersed nanoparticles of diglycerol monooleate/glycerol dioleate mixtures', *Langmuir*, 21, 5159–5165.

JOHNSSON M, BARAUSKAS J and TIBERG F (2005b), 'Cubic phases and cubic phase dispersions in a phospholipid-based system', *J Am Chem Soc*, 127, 1076–1077.

JOHNSSON M, BARAUSKAS J, NORLIN A and TIBERG F (2006), 'Physicochemical and drug delivery aspects of lipid-based liquid crystalline nanoparticles: a case study of intravenously administered propofol', *J Nanosci Nanotechnol*, 6, 3017–3024.

LANDAU E M and LUISI P L (1993), 'Lipidic cubic phases as transparent, rigid matrices for the direct spectroscopic study of immobilized membrane proteins', *J Am Chem Soc*. 115, 2102–2106.

LANDAU E M and ROSENBUSCH J P (1996), 'Lipidic cubic phases: a novel concept for the crystallization of membrane proteins', *Proc Natl Acad Sci*, 93, 14532–14535.

LANDH T (1994), 'Phase behavior in the system pine oil monoglycerides–Poloxamer 407-water at 20 °C', *J Phys Chem*, 98, 8453–8467.

LARSSON K (1989), 'Cubic lipid-water phases: structures and biomembrane aspects', *J Phys Chem*, 93, 7304–7314.

LARSSON K (2000), 'Aqueous dispersions of cubic lipid-water phases', *Curr Opin Colloid Interface Sci*, 5, 64–69.

LARSSON K and LINDBLOM G (1982), 'Molecular amphiphile bilayers forming a cubic phase in amphiphile-water systems', *J Disp Sci Technol*, 3, 61–66.

LARSSON K, QUINN P, SATO K and TIBERG F (2006), *Lipids: Structure, Physical Properties and Functionality*, Bridgewater, The Oily Press.

LASIC D D (1993), *Liposomes – from Physics to Applications*, New York, Elsevier.

LESLIE S B, PUVVADA S, RATNA B R and RUDOLPH A S (1996), 'Encapsulation of hemoglobin in a bicontinuous cubic phase lipid', *Biochim Biophys Acta*, 1285, 246–254.

LINDSTRÖM M, LJUSBERG-WAHREN H, LARSSON K and BORGSTRÖM B (1981), 'Aqueous lipid phases of relevance to intestinal fat digestion and absorption', *Lipids*, 16, 749–754.

LJUSBERG-WAHREN H, NYBERG L and LARSSON K (1997), 'Dispersion of the cubic liquid crystalline phase – structure, preparation and functionality aspects', *Chim Oggi*, 14, 40–43.

LUZZATI V (1968), 'X-ray diffraction studies of lipid-water systems', in Chapman D, *Biological Membranes*, New York, Academic Press, 71–123.

MITCHELL D J and NINHAM B W (1981), 'Micelles, vesicles and microemulsions', *J Chem Soc Faraday Trans II*, 77, 601–629.

MONDUZZI M, LJUSBERG-WAHREN H and LARSSON K (2000), 'A ^{13}C NMR study of aqueous dispersions of reversed lipid phases', *Langmuir*, 16, 7355–7358.

MOURITSEN O G (1991), 'Theoretical models of phospholipids phase transitions', *Chem. Phys. Lipids*, 57, 179–194.

MOURITSEN O G and ZUCKERMANN M J (1987), 'Model of interfacial melting', *Phys. Rev. Lett.* 58, 389–392.

MOURITSEN O G, ANDRESEN T L, HALPERIN A, HANSEN P L, JAKOBSEN A F, JENSEN U B, JENSEN M Ø, JØRGENSEN K, KAASGAARD T, LEIDY C, SIMONSEN A C, PETERS G H and WEISS M (2006), 'Activation of interfacial enzymes at membrane surfaces (Review)', *J Phys: Cond Matt*, 18, S1293–S1304.

MOZAFARI M R, FLANAGAN J, MATIA-MERINO L, AWATI A, OMRI A, SUNTRES Z E and SINGH H (2006), 'Recent trends in the lipid-based nanoencapsulation of antioxidants and their role in foods', *J Sci Food Agric*, 86, 2038–2045.

NAKANO M, SUGITA A, MATSUOKA H and HANDA T (2001), 'Small-angle X-ray scattering and ^{13}C NMR investigation on the internal structure of "Cubosomes" ', *Langmuir*,

17, 3917–3922.

NAKANO M, TESHIGAWARA T, SUGITA A, LEESAJAKUL W, TANIGUCHI A, KAMO T, MATSUOKA H and HANDA T (2002), 'Dispersions of liquid crystalline phases of the monoolein/oleic acid/Pluronic F127 system', *Langmuir*, 18, 9283–9288.

NAKANO M, KAMO T, SUGITA A and HANDA T (2005), 'Detection of bilayer packing stress and its release in lamellar-cubic phase transition by time-resolved fluorescence anisotropy', *J Phys Chem B*, 109, 4754–4760.

NETO C, ALOISI G, BAGLIONI P and LARSSON K (1999), 'Imaging soft matter with the atomic force microscope: Cubosomes and Hexosomes', *J Phys Chem B*, 103, 3896–3899.

NYLANDER T, MATTISSON C, RAZUMAS V, MIEZES Y and HÅKANSSON B (1996), 'A study of entrapped enzyme stability and substrate diffusion in a monoglyceride-based cubic liquid crystalline phase', *Colloids Surfaces A: Physicochem. Eng. Aspects*, 114, 311–320.

PATTON J S and CAREY M C (1979), 'Watching fat digestion', *Science*, 204, 145–148.

PATTON J S, VETTER R D, HAMOSH M, BORGSTRÖM B, LINDSTRÖM M and CAREY M C (1985), 'The light microscopy of fat digestion', *Food Microstructure*, 4, 29–41.

POPESCU G, BARAUSKAS J, NYLANDER T and TIBERG F (2007), 'Liquid crystalline phases and their dispersions in aqueous mixtures of glycerol monooleate and glyceryl monooleyl ether', *Langmuir*, 23, 496–503.

PORTMANN M, LANDAU E M and LUISI P L (1991), 'Spectroscopic and rheological studies of enzymes in rigid lipidic matrices: the case of a-chymotrypsin in a lysolecithin/water cubic phase', *J Phys Chem*, 95, 8437–8440.

QIU H and CAFFREY M (2000), 'The phase diagram of the monoolein/water system: metastability and equilibrium aspects', *Biomaterials*, 21, 223–234.

RANGELOV S and ALMGREN M (2005), 'Particulate and bulk bicontinuous cubic phases obtained from mixtures of glyceryl monooleate and copolymers bearing blocks of lipid-mimetic anchors in water', *J Phys Chem B*, 109, 3921–3929.

RAZUMAS V, KANAPIENIENÉ J, NYLANDER T, ENGSTRÖM S and LARSSON K (1994), 'Electrochemical biosensors for glucose, lactate, urea, and creatinine based on enzymes entrapped in a cubic liquid crystalline phase', *Anal Chim Acta*, 289, 155–162.

RAZUMAS V, LARSSON K, MIEZES Y and NYLANDER T (1996a), 'A cubic monoolein-cytochrome c-water phase: X-ray diffraction, FT-IR, differential scanning calorimetry and electrochemical studies', *J Phys Chem*, 100, 11766–11774.

RAZUMAS V, TALAIKYTÉ Z, BARAUSKAS J, LARSSON K, MIEZIS Y and NYLANDER T (1996b), 'Effects of distearoylphosphatidylglycerol and lysozyme on the structure of the monoolein-water cubic phase: X-ray diffraction and Raman scattering studies', *Chem Phys Lipids*, 84, 123–138.

ROSA M, INFANTE M R, DA MIGUEL M and LINDMAN B (2006), 'Spontaneous formation of vesicles and dispersed cubic and hexagonal particles in amino acid-based catanionic surfactant systems', *Langmuir*, 22, 5588–5596.

SAGALOWICZ L, MICHEL M, ADRIAN M, FROSSARD P, ROUVET M, WATZKE H J, YAGHMUR A, DE CAMPO L, GLATTER O and LESER M E (2005), 'Crystallography of dispersed liquid crystalline phases studied by cryo-transmission electron microscopy', *J Microsc*, 221, 110–121.

SAGALOWICZ L, LESER M E, WATZKE H J and MICHEL M (2006a), 'Monoglyceride self-assembly structures as delivery vehicles', *Trends Food Sci Techn*, 17, 204–214.

SAGALOWICZ L, MEZZENGA R and LESER M E (2006b), 'Investigating reversed liquid crystalline mesophases', *Current Opinion Colloid Interface Sci.*, 11, 224–229.

SEDDON J (1996), 'Lyotropic phase behaviour of biological amphiphiles', *Ber Bunsenges*

Phys Chem, 100, 380–393.

SEDDON J and ROBINS J (1999), 'Inverse micellar lyotropic cubic phases', in Sadoc J F and Rivier N, *Foams and Emulsions*, Dordrecht, Kluwer, 423–430.

SIEKMAN B, BUNJES H, KOCH M H J and WESTESEN K (2002), 'Preparation and structural investigations of colloidal dispersions prepared from cubic monoglyceride-water phases', *Int J Pharm*, 244, 33–43.

SPICER P T (2005), 'Progress in liquid crystalline dispersions: cubosomes', *Curr Opin Colloid Interface Sci*, 10, 274–279.

SPICER P T, HAYDEN K L, LYNCH M T, OFORI-BOATENG A and BURNS J L (2001), 'Novel process for producing cubic liquid crystalline nanoparticles (Cubosomes)', *Langmuir*, 17, 5748–5756.

TAMAYO-ESQUIVEL D, GANEM-QUINTANAR A, MARTINEZ A L, NAVARRETE-RODRIGUEZ M, RODRIGUEZ-ROMO S and QUINTANAR-GUERRERO D (2006), 'Evaluation of the enhanced oral effect of omapatrilat-monolein nanoparticles prepared by the emulsification-diffusion method', *J Nanosci Nanotechnol*, 6, 3134–3138.

UM J Y, CHUNG H, KIM K S, KWON I C and JEONG S Y (2003), '*In vitro* cellular interaction and adsorption of dispersed cubic particles', *In J Pharm*, 253, 71–80.

VANDOOLAEGHE P, TIBERG F and NYLANDER T (2006), 'Interfacial behavior of cubic liquid crystalline nanoparticles at hydrophilic and hydrophobic surfaces', *Langmuir*, 22, 9169–9174.

VANDOOLAEGHE P, TIBERG F, RENNIE A R, THOMAS R K, HÖÖK F, FRAGNETO G and NYLANDER T (2007), 'Adsorption of cubic liquid crystalline nanoparticles on model membranes'. Manuscript.

VAUTHEY S, MILO CH, FROSSARD PH, GARTI N, LESER M E and WATZKE H J (2000), 'Structures fluids as microreactors for flavour formation by the Maillard reactions', *J Agric Food Chem*, 48, 4808–4816.

WÖRLE G, SIEKMANN B, KOCH M H J and BUNJES H (2006), 'Transformation of vesicular into cubic nanoparticles by autoclaving of aqueous monoolein/poloxamer dispersions', *European J Pharm Sci*, 27, 44–53.

WÖRLE G, DRECHSLER M, KOCH M H J, SIEKMAN B, WESTESEN K and BUNJES H (2007), 'Influence of composition and preparation parameters on the properties of aqueous monoolein dispersions', *Int J Pharm*, 329, 150–157.

YAGHMUR A, DE CAMPO L, SAGALOWICZ L, LESER M E and GLATTER O (2005), 'Emulsified microemulsions and oil-containing liquid crystalline phases', *Langmuir*, 21, 569–577.

YAGHMUR A, DE CAMPO L, SALENTINIG S, SAGALOWICZ L, LESER M E and GLATTER O (2006), 'Oil-loaded monolinolein-based particles with confined inverse discontinuous cubic structure (*Fd3m*)', *Langmuir*, 22, 517–521.

YANG D, ARMITAGE B and MARDER S T (2004), 'Cubic liquid-crystalline nanoparticles', *Angew Chem Int Ed*, 43, 4402–4409.

Part II

Materials and techniques for controlled release and delivery of nutrients

5

Structured lipids as delivery systems

E. M. Hernandez, OmegaPure Technology and Innovation Center, USA

5.1 Introduction

Fat consumption in the Western diet has gone through a series of changes in recent years both in the amount consumed and the types of fats recommended. Dietary fat reductions were recommended in the 1980s and 1990s but in recent years wider fat dietary recommendations have been issued by official health institutions. Several new types of fats and oils have been introduced into foods as a response to the need by manufacturers to reformulate their products. These changes have been driven by new labeling laws and the demand by the consumer for healthier and more functional dietary fats.

It is well documented that lipids play an important role in the body's metabolic processes and are essential components of cell membranes and other biological structures. Figure 5.1 shows the major types of lipids with significant participation in metabolic processes. Major lipids are: triacylglycerols (TAG), free fatty acids, phospholipids, sphingolipoids, bile salts, steroids and sterols, cholesterol, eicosenoids and fat soluble vitamins. Lipids in general are an important source of essential fatty acids (omega-3 and omega-6), fat-soluble vitamins, cellular transport components and food/medical supplements (Dupont, 1996; Kinsella, 1988). More specifically, essential fatty acids are important components of cell structures and are the precursors of signaling molecules that control biological processes such as inflammation, immune response and also act as messengers in the central nervous system. The better understanding of the functionality of lipids is driving the development of a diverse number of applications in both the pharmaceutical and cosmetic fields (Hernandez, 2005).

Sales of lipid nutritional supplements have grown appreciably in the last ten years and their application has extended also to pharmaceutical and cosmetic

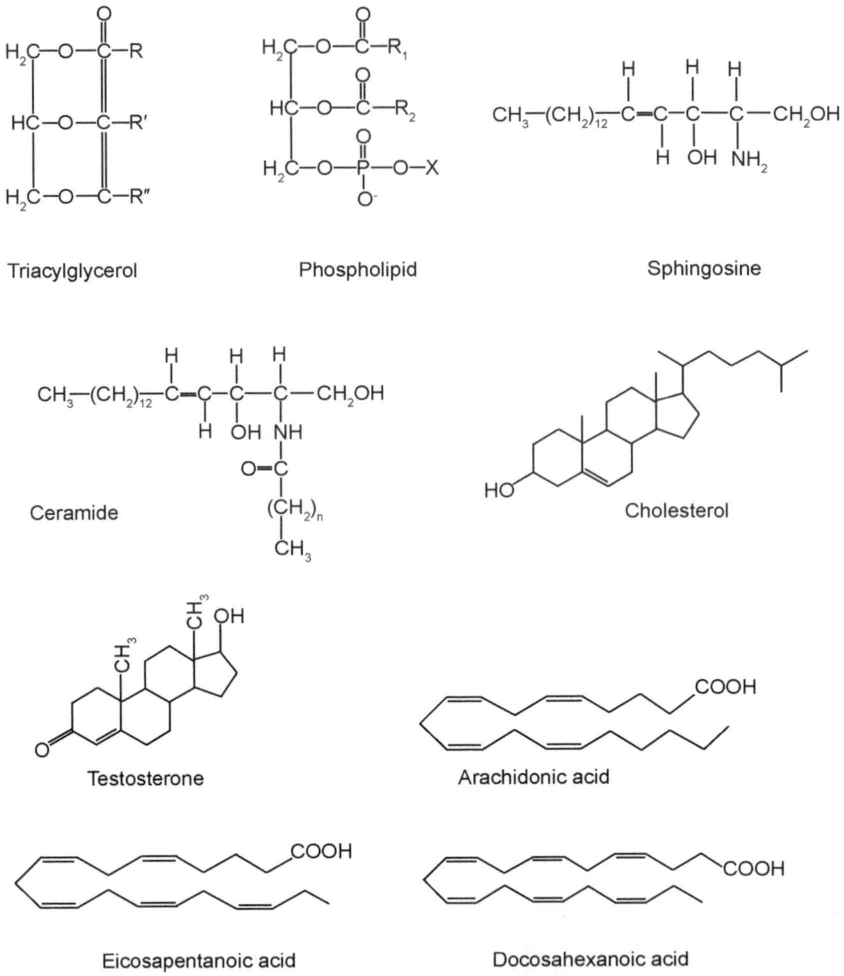

Fig. 5.1 Structures of major bioactive lipids.

fields such as disease prevention and treatment, excipients and coadjuvants, transdermal carriers and skin emolliency agents. These new functional lipids are being manufactured and supplied through several delivery media including the use of delivery systems in the form of structured lipids.

Currently structured lipids and phospholipids are used in food, cosmetic and pharmaceutical products. Lipids are being modified to enhance nutritional properties of foods, and improve the rheological and crystallization properties of some fats. They also play an important role in pharmaceutical applications like parenteral nutrition. Structured lipids are basically triglycerides whose fatty acid composition or distribution has been modified. In general modification of the triglyceride composition of vegetable oils can be done by hydrogenation, chemical and enzymatic interesterification.

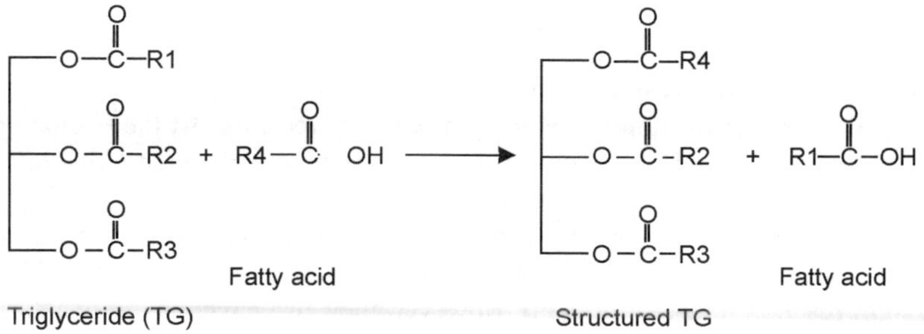

Fig. 5.2 Acidolysis reaction.

Interesterification of oils is generally known as the exchange of acyl group among triglycerides and is the most widely used method for the production of structured lipids. Acyl groups may exchange positions within a triglyceride or among triglyceride molecules. This process has found a wide variety of applications in the edible oils industry, for example to change the melting point profile of solid fats, to improve the compatibility of different tryglycerides, manipulate the plasticity of certain fats, to change the emulsifying properties of some oils and to change the nutritional properties of some edible oils (Willis *et al.*, 1998). When the interesterification is carried out to react the fat or oil with other fatty acids, it is called acidolysis, when the esterification is carried out with alcohols, it is called alcoholysis, and with other esters, transesterification. Figure 5.2 shows an example of an acidolysis reaction and Fig. 5.3 of an alcoholysis reaction.

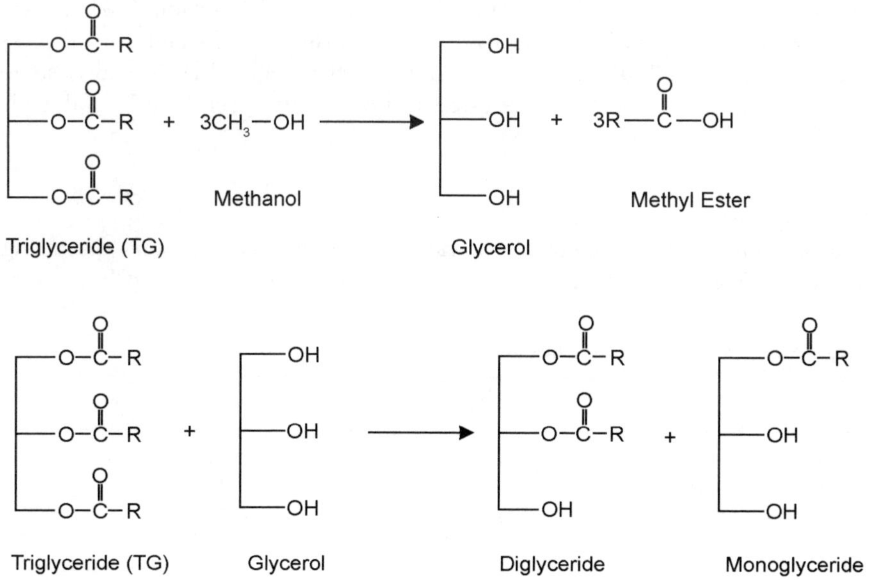

Fig. 5.3 Alcoholysis reaction.

5.2 Synthesis of structured lipids

5.2.1 Chemical interesterification

Chemical interesterification is generally a two-step reaction. At the beginning of the reaction the metal catalyst forms a transition complex with the original triacylglycerol that acts as the actual catalyst that forms the fatty acid esters. Sodium alkilates such as sodium ethoxide and sodium methoxide are the most common catalysts because of their relatively low reaction temperature (50–120 °C) and short reaction time (5–120 min). However, they also have the disadvantage of being inactivated easily in the oil by impurities such as water, free fatty acids, and peroxides. This means that extra expense has to be incurred to make up for decomposed catalyst during processing. Chemical interesterification is usually random.

5.2.2 Enzymatic interesterification

Under normal conditions lipases hydrolyze fatty acids from the triglyceride into free fatty acids. However, if the amount of water in the reaction is restricted, hydrolysis is minimized, the enzyme catalyzes the reaction of interesterification instead. A small amount of water needs to be present to maintain the integrity of the enzyme. Lipases can catalyze interesterification with triglycerides and also with free fatty acids. In order to optimize their use lipases have to be immobilized in microporous supports such as amberlite resins, celite, cellulose, silica gel, clay, carbon, or alumina.

Lipases are now commonly used for enzymatic interesterification of several types of trygliceride oils. The main source of lipases is from microorganisms and some are also obtained from animal pancreas and plant origin. Enzymes are usually position specific in the triglyceride (positions 1,3) and can also be specific to the length of the fatty acid in the triglyceride. Table 5.1 shows some types of enzymes used for interesterification of triglyceride oils and their specificity (Akoh and Min, 1998).

5.2.3 Structured phospholipids

Phospholipids are well-known natural emulsifiers, wetting and dispersing agents. They are used in many food and pharmaceutical applications because of

Table 5.1 Examples of enzymes and lipase activity

	Regiospecificity	Fatty acid specificity
Candida cylidracae	NPS	NFAS
Staphylococus aureus	NPS	NFAS
Aspergillus niger	1,3-	C:18
Candida rugosa	NPS	C:18
Mucor miehei	1,3-	—
Rhizopus arrhizus	1,3-	C:8, C:10

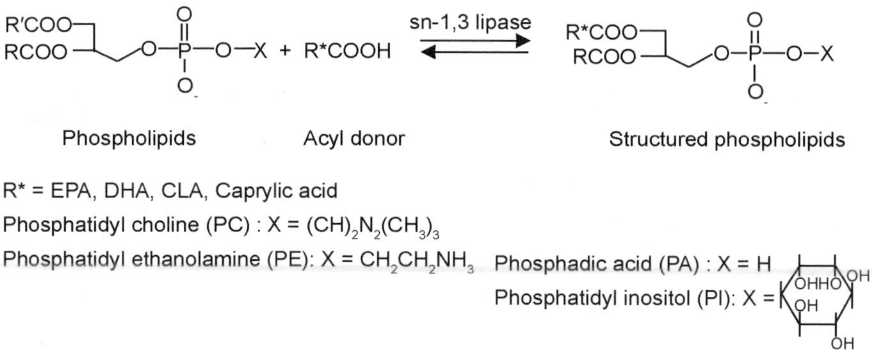

Fig. 5.4 Structured phospholipids.

their emulsifying and physiological properties. Phospholipids play an important role in the metabolism of lipids and other digestive fluids. Their emulsifying properties also make them part of metabolic systems at cellular level such as membrane structure and cell membrane transport systems (Totani and Hara, 1991; Haraldsson and Thorarenson, 1999). In fact, they ensure miscibility of hydrophilic and hydrophobic metabolites at liquid–liquid and liquid–solid interfaces in many biological pathways (Cevc and Paltauf, 2003). The most common types of phospholipids are phosphatidylcholine (PC), phosphatidylethanolamine (PE), phosphatidilinositol (PI), phosphatidic acid (PA), and phosphatidylserine (PS) (see Fig. 5.4). The major sources of phospholipids (PLs) in our diet are oilseeds and egg yolk.

Structured phospholipids can also be enzymatically manufactured to enhance the delivery of functional fats such as medium-chain fatty acids and omega-3 polyunsaturated fatty acids, such as eicosapentanoic acid (EPA) and DHA. By replacing the existing fatty acids asymmetrically in an original PL molecule with desired ones, new physical properties and special functions can also be achieved (Guo *et al.*, 2005; Clissold, 1999). Enzymatic production of both structured lipids and phospholipids are preferred to chemical methods, especially when the reaction is required to be position specific and a specific fatty acid is required to be incorporated.

The process of interesterification has been used to produce structured phospholipids enriched with phosphatidyl serine (PS) which has been shown to enhance neuronal membrane function and has been prescribed to the elderly to improve cognitive function (Crook *et al.*, 1991). PS is found at levels of up to 1% in commercial lecithin from vegetable oils and 3% in commercial lecithin from eggs. The Food and Drug Administration (FDA) authorized two health claims for PS: the first claim, 'Phosphatidylserine (PS) may reduce the risk of cognitive dysfunction in the elderly', and the second claim, 'Phosphatidylserine (PS) may reduce the risk of dementia in the elderly'. Phospholipids, as structured phospholipids, are also used to enhance the biological activity of other functional lipids such as conjugated linoleic acid (Hossen and Hernandez, 2005).

Other work on structuring phospholipids is to esterify nutraceutical components to hydrophilic part of the phospholipid molecule. Examples of these include the esterification of sterols to the phosphatidic head of the phospholipids. Sterols in general have been studied for a long time as cholesterol-reducing agents. Also several cholesterol absorption inhibitor drugs have been developed and cholesterol absorption by dietary plant sterols (PS) has received much research interest because of their claimed effectiveness in lowering plasma total and LDL cholesterol. Several studies have been reported to produce amphiphilic sterols by synthesizing phospholipid derivatives of PS using an enzymatic process catalyzed by phospholipase D (Hossen and Hernandez, 2004).

5.3 Nutrition, absorption, transport and metabolism of structured lipids

A classical example of *in vivo* structure lipid delivery takes place in the digestive tract. Ingested dietary lipids are absorbed in the intestine through a series of emulsification and chemical transformation reactions. Before enzymatic hydrolysis, lipids are emulsified with bile salts which are synthesized in the liver and secreted from the gall bladder following the ingestion of fat. The emulsified lipids are then hydrolyzed by pancreatic lipases and phospholipase (A_2). (Mu and Høy, 2001). The result of this enzymatic reaction are free fatty acids and a mixture of mono- and diacylglycerols. Since lipases are specific to the positions 1 and 3 of the triacylglycerides, the majority of the monoglycerides are position 2-acylglycerols. Phospholipids are hydrolyzed at the 2 position by pancreatic phospholipase A_2 releasing a free fatty acid and the lysophospholipids. The resulting fatty acids and 2-monoglycerides are then transported into the endoplasmic reticulum, where they are used to resynthesize triglycerols. These triacylglycerols then combine with lipoprotein and form chylomicrons. Chylomicrons basically consist of a core of hydrophobic lipids surrounded by a shell of proteins, phospholipids and cholesterol. Chylomicrons are first transported into the lymphatic vessel that penetrates into each villus. Chylomicron-rich lymph then flows into the blood vessels where they are disassembled and their constitutent lipids utilized by the various tissues for storage, production of energy through oxidation or participate in various metabolic pathways.

Short chain fatty acids (C:2 to C:12) fatty acids tend to be transported via the portal system and C12:0 to C24:0 via the lymphatic system. There is growing evidence that MCFA may indeed be absorbed as 2-monoacylglycerides, especially if they are esterified to the sn-2 position of the structured lipid due to the slow rate of hydrolysis at the sn-2 position of triacylglycerols, and as a result the fatty acid at this position remains intact as 2-monoacylglycerides during digestion and absorption. On the other hand, long chain triacylglycerides are partially hydrolyzed by pancreatic lipase and absorbed slowly as partial glycerides in mixed micelles. The resulting long chain fatty acids are re-esterified and incorporated into chylomicrons in the enterocyte, where they enter the lymphatics to reach the general circulation through the thoracic duct. In general, the metabolism of a structured lipid would be determined by the nature and position of the constituent fatty acids on the glycerol moiety which may account for the differences in the pathway of absorption: lymphatic vs portal. Free fatty acids, especially MCFA, tend to be absorbed directly from the intestine into the portal vein and sent straight to the liver where they are metabolized (Rubin *et al.*, 2000).

5.4 Foods applications of structured lipids

5.4.1 Specialty dietary fats

Due to the increasing awareness of the benefits of some bioactive lipids fats and the risks associated with excessive intake of some fats, there is a growing market for modified fats as vehicles of healthier and bioactive lipids in some foods. Some new classes of edible structured lipids include products such as lower calorie fats manufactured by interesterification of short and long chain fatty acids. Caprenin is a fat developed for use in a high melting temperature applications. It is made from the naturally occurring fatty acids capric (C6:0), caprylic (C8:0) and behenic (C22:0), and is metabolized like other triglycerides. Instead of 9 kcal/g, it yields only 5 kcal/g because the short chain fatty acids have lower energy values and behenic acid is not well absorbed. The structured lipid, SALATRIM, which is an acronym for short and long chain fatty acids consists of tributyrin, triacetin and tripropionin with hydrogenated canola and other fatty acids (mostly stearic with a small amount of palmitic on a glycerol backbone). This fat has fewer calories as a result of the oxidation of short chain fatty acids. Coupled with only 27–35% absorption of the stearate and other long chain fatty acids, the net energy contribution is 4.5–6 kcal/g. SALATRIM is used in filled dairy and reduced calorie bakery products (Akoh and Min, 1998).

The fat product Captex is produced via esterification to deliver C8 and C10 fatty acids. Captex products function as a readily available energy source for athletes and patients with special dietary needs. Applications for Captex products are snacks, main meals, frozen and refrigerated foods (Shahidi, 2006). Neobee is also a structured lipid that delivers short and medium chain fatty acids, i.e., capric and caprylic acids. This product may also serve as a flavor carrier and functions as a textural component for low-fat food products (Shahidi, 2005).

5.4.2 Infant formulas

The composition of human milk is now fairly well understood. Normalized data report some of the major fatty acids in human milk fat: lauric, C:12 (4.40%), myristic, C14:0 (6.27%), palmitic 16:0 (22.00%), stearic, 18:0 (8.06%), oleic acid, 18:1 (31.3%) and linolenic, C:2 (10.85%) plus small amounts of essential fatty acids such as arachidonic (0.46%), eicosapentanoic (EPA) (0.12%) and docohexanoic acids (DHA) (0.25%) (Jensen, 1996). Supplementation of infant formulas, and baby foods in general, with essential fatty acids has gained great interest. This is mostly due to the fact that term infants have limited capacity to desaturate short chain omega-3 fatty acids into long chain fatty acids such as EPA and DHA required by many metabolic functions. The two most abundant fatty acids in mothers' milk, palmitic and oleic acids and structured lipids, such as 1,3-dioleoyl-2-palmitoylglycerol (OPO), have been studied for their potential pharmaceutical and nutraceutical applications (Chen *et al.*, 2004).

Interesterification with lipases has been used to design human milk fat substitutes and deliver essential fatty acids. The use of lipases as biocatalysts allows for esterification of palmitic acid in the position 2 of triglyceride which, with the addition of essential fatty acids can mimic the composition of human milk. This has been used in the supplementation of infant formulas with essential fatty acids i.e, ARA, EPA and DHA (Shimada *et al.*, 2000; Sahin *et al.*, 2005, 2006).

5.4.3 Parenteral and enteral nutrition

Lipid malabsorption is a leading cause of essential fatty acid deficiencies, especially in recovering patients and injured people. Malabsorption can occur in any of the lipid absorption sequential steps, including lipolysis, solubilization by bile, uptake into the enterocyte, and chylomicron secretion into lymph. Treatments for essental fatty acid deficiencies include the use of structured lipids and phospholipids for the delivery, for example of medium chain fatty acids (MCFA) as an energy source and of essential fatty acids for patient's metabolic needs. Enzymatically modified lipids and physical mixtures of lipids have shown different metabolic pathways based on the structure and due to the difference in fatty acid composition at the sn-2 position of lipid molecules (Tso *et al.*, 2001).

Structured lipids composed of long chain fatty acids and medium chain fatty acids have been designed to provide more controlled release of the MCFA into the bloodstream (Bach and Babayan, 1982). One of the main advantages of enterally fed structured lipids is that MCFA provide a vehicle for rapid hydrolysis and absorption, mainly due to their smaller molecular size and greater water solubility in comparison to longer chain triacylglycerides. MCFA also play an important role in the metabolism of nitrogen. Besides faster oxidation rates, studies show that MCFAs improve the nitrogen balance. These properties make MCFA an even more important component of parenteral nutrition (Tao and Yoshimura, 1980). Studies also confirm that no alteration of liver function occurs in any of the patients treated with these structured lipids and thus they appear to be safe and may be associated with reduction in liver dysfunction (Rubin *et al.*, 2000).

5.5 Structured lipids in drug delivery

Vegetable oils in general are non-toxic which allows them to be used as reliable excipients or carriers in many pharmaceutical formulations. As a result many vegetable oils and fats have been approved as excipients to facilitate delivery of bioactive compounds and to act as fillers, binders, lubricants, solubilizers, emulsifiers and emollients in a variety of delivery forms including tablets, capsules, suppositories, emulsions (enteral/parenteral), ointments, creams and lotions. Other non-direct applications include artificial blood, gene delivery, diagnostic imaging, and medical devices (Charman, 1996).

Use of structured lipids in the development of controlled release of drugs and bioactive substances is also rapidly growing. Applications include delivery systems for cancer treatments, bacterial and fungal infections, and respiratory disease, as well as the delivery of drugs intravenously, orally and transdermally.

5.5.1 Lipids in transdermal applications

There are several advantages reported for transdermal drug delivery:

- bypassing of gastrointestinal and hepatic metabolism
- the potential for enhanced efficacy
- lower side effects through a more controlled delivery rate
- increased patient compliance through easier application and dosing frequency (Charman, 1996).

The skin can be used as a portal of entry for drugs, for both localized and systemic treatments and as a result transdermal delivery has become one of the fastest growing systems for some disease treatments. However, since the outer layer of the skin can be a difficult barrier to penetrate, permeation-enhancing agents are in some cases necessary to allow intake of some drugs.

The upper layer of the skin or stratum corneum has a very limited permeability and its basic function is to act as a barrier against outside elements. The human skin is composed of layers of differentiated strata (Stratum corneum, Stratum lucidum, Stratum lranolosum, Stratum spinosum, Stratum basale) that transdermal applications need to overcome in order to deliver bioactive elements. The lipid composition and distribution in stratum corneum has been used to elucidate and model transdermal transport phenomena of several substances through the skin (Wertz *et al.*, 1987). The sebaceous lipids in the skin consist mainly of 45–60% triacylglycerols, 25% wax and sterol esters, 12–15% squalene and 10% free fatty acids (Lampe *et al.*, 1983). This in turn has been used to design formulations with increased permeability and has been used in disease treatment through topical drug administration (Harding, 2004).

Some unsaturated fatty acids are well-known skin penetration enhancers. The addition of phosphatidylcholine (PC) to dermal dosage forms has been reported to increase percutaneous absorption. Lipid disperse systems (LDSs) containing polar lipids, such as PC and glycosylceramide, are also useful for increasing the percutaneous permeation of drug through rat abdominal skin in both *in vitro* and *in vivo* systems. Liposomes have been widely reported as an effective means of transdermal transport. The use of liposomes has been reported in applications for topical and intravenous uses (Egbaria and Weiner, 1990). Though the mechanism of this effect is not fully explained, it has been suggested that liposomes pass intact through the lipid rich outer layer of the skin to the dermis, where they become localized. There has also been the suggestion that the follicular pathway contributes to the liposomal delivery of drugs into the deeper skin strata (Ogiso *et al.*, 1996).

Liposomes have been extensively investigated in the last 40 years and extensively for their potential in the targeted delivery of chemotherapeutics applications. The unique structural properties of liposomes allow different methods to be used to entrap substances. Liposomal systems can also form an effective drug reservoir in the upper layers of the skin. This is particularly useful for local skin therapy. Ethosomal carriers composed of phospholipid vesicular systems with alcohols are also effective at enhancing transdermal delivery of both lipophilic and hydrophilic compounds. The use of these ethosomes has been used in the delivery of minoxidil to the pilosebaceous section of the skin with better results than conventional liposomes. Similar results are reported in clinical studies with acyclovir in a topical therapy treatment of recurrent herpes labialis. Other application reports with ethosomes are patches containing testosterone (Touitou *et al.*, 2000).

Topical applications in the form of spray have also been reported as vehicles for enhanced transdermal delivery of drugs such as testosterone, estradiol, progesterone, and norethindrone acetate. More effective drug penetration was reported with enhancers padimate and octyl salicylate and compared with laurocapram and oleic acid (Morgan *et al.*, 1998).

5.5.2 Lipids in oral drug delivery

In many cases drugs administered orally may degrade the digestive tract or have limited absorption. Lipid-based macromolecules are part of a growing group of new drugs with many clinical applications that can be used in oral delivery applications with enhanced absorption capabilities. In some cases some lipophilic drugs can be made more bioavailable by traveling though the processes of the digestive tract itself (Charman, 1996). It is widely documented that the addition of fatty acids can improve the bioavailabilty of drugs that are poorly absorbed by the gastrointestinal tract by increasing their uptake into the lymphatic system (Porter, 1997). For example the addition of fatty acyl groups also enhances permeability across intestinal mucosa as well as other epithelial barriers such as buccal, nasal, and pulmonary mucosae (Verhoef *et al.*, 1990). The use of triacylglycerol esters for delivery of peptides across the blood brain barrier was also reported.

The use of self-emulsifying drug delivery systems has been reported to produce stable microemulsions for the delivery of potent drugs. A successful example is the use of microemulsions for delivery of cyclosporine in transplant patients (Pouton, 2000). Self-emulsifying formulations, known as SEDDS A SEDDS formulation normally contains a lipid, surfactants, and co-solvents. The main objective is to facilitate the subsequent assimilation of a drug in the digestive tract without getting inactivated by the pH environment or in the bile salt micellar structures. Also it is desired that the emulsified drug be independent of the ingested food.

Besides emulsifying properties, another useful characteristic of solid fats are their malleable melting point profiles. Fat crystals have the structure-forming properties that allow the formation of a matrix for suspension of drug particles. When used simply as carriers some lipids can provide a matrix that can keep the bioactive components uniformly distributed and released in timely manner. The properties of the crystals such as melting time can be manipulated by structuring the lipids or phospholipid used in the delivery system (Porter *et al.*, 1997).

Biosomes made from phosphatidylcholine, and the structured lipids of medium-chain monoacylglycerol and low-molecular-weight heparin (LMWH), used in the treatment and prevention of thromboses was reported (Bohlinder *et al.*, 1994). Heparin is known as having poor oral absorption. The system is applicable for oral and parenteral administration as well as for enhancing dermal, rectal, and nasal absorption of other drugs.

The charge of the emulsion can also be manipulated and therefore enable the expansion of therapeutic applications to intravenous, ocular, and oral administration. It has been reported that positively charged submicron emulsions interact desirably with biological membranes that are negatively charged. The positive charge of submicron emulsions results from cationic excipients such as lipids, polymers, and surfactants in formulations. This difference in charge results in an increased drug uptake and can be designed for site-specific targeting. Use of this property has been shown in enhanced drug permeation through cornea, and drug targeting of the lung following IV injection (Yang and Benita, 2000). This also suggests a new therapeutic treatment for lung cancer chemotherapy.

The pharmaceutical industry is also using Tris, a compound regarded as relatively inexpensive, nontoxic, and readily available, because it has structural similarities to glycerol and has the advantage of being symmetrical and avoids structural isomerization. Tris can be used by readily attaching several types of drugs to the amine group. It also allows for the synthesis of symmetric tri-fatty acyl esters which is not possible with glycerol. Initially the technology was applied to the delivery of vaccines but it has since been used to link various drugs to fatty acids. It has also been used to generate a novel class of cationic lipid cytofectins. The synthesis of fatty acid conjugates of morphine, indomethacin (a nonsteroidal anti-inflammatory drug), AZT (an antiviral drug and methotrexate and chlorambucil (antineoplastics) was reported.

The formation of microemulsions of structured lipids and certain pharmaceuticals can also be applied intravenously, for example, a micromeulsion composed of medium-chain triacylglycerol (MCT), soybean phosphatidylcholine and poly(ethylene glycol)(660)-12-hydroxystearate (12-HSA- EO15) as amphiphiles, and poly(ethylene glycol) 400 (PEG 400) and ethanol as cosolvents (von Corswant *et al.*, 1998). The mean droplet diameter of the oil-in-water emulsion reported was between 60 and 200 nm. It was concluded that it is possible to administer it without producing any significant effect on acid–base balance, blood gases, plasma electrolytes, mean arterial blood pressure (MAP), heart rate (HR) (concentrations of up to 0.5 mL/kg of the microemulsion).

5.5.3 Other applications

New applications on the use of structured cationic lipids for delivery of plasmids DNA in genetic research have been reported in gene transfection (Tranchant *et al.*, 2004). Further work has also focused on enhancing and prolonging the stability of these cationic lipids in a physiological environment as well as increasing their site-specific and tissue-specific interactions. The use of synthetic vectors as compared with the use of viral vectors, and the recent progress realized in the conception of new vectors in terms of transfection efficiency, bioavailability, reduction of cytotoxicity and cellular targeting is very encouraging. For example, a liposomal preparation containing the cationic lipid N-[1-(2,3-dioleyloxy)propyl]-N,N,N-trimethylammonium chloride (DOTMA) was found to increase by at least 1000-fold the potency of an antisense oligonucleotide

The use of structuring triacylglycerols has been studied as a vehicle of other healthy fats. Figure 5.4 shows common bioactive lipids that can be used in structured lipids. Structured lipids of conjugated linoleic acid (CLA) and medium-chain triacylglycerols to produce new types of more healthful structured lipids (SLs) have been reported (Rocha and Hernandez, 2004). This work reported on the synthesis of structured lipids containing CLA, by the incorporation of different levels of CLA, as free fatty acid, into two triacylglycerols which are good sources of medium-chain fatty acids (i.e. coconut oil and tricaprylin) by enzymatic acidolysis in a solvent-free system, using Lipozyme IM (immobilized lipase from *Mucor miehei*) as biocatalyst.

Docosahexaenoic acid (DHA)-containing PLs has shown to have potential medical applications such as promoting cell differentiation in leukemia, enhancing survivals of tumor-bearing mice and preventing cerebral degeneration. Structured PLs, which contain medium-chain fatty acids also have similar positive effects on health (Guo *et al.*, 2005).

5.6 References

AKOH C and MIN D (1998). *Food Lipids: Chemistry, Nutrition and Biotechnology*. Marcel Dekker, New York.

BACH AC and BABAYAN VK (1982). *Am. J. Clinical Nutr.*, 36, 950–962.

BOHLINDER K, OLSSON B, RAGNARSSON G, SINGH SK, STJERNSTRÖM ÖHMAN G, WADSTEN T, CARLSSON A and HERSLÖF B (1994). *Eur. J. Pharm. Sci.*, 2 (4), 271–279.

CEVC G and PALTAUF F (EDS) (2003). *Phospholipids: Characterization, Metabolism, and Novel Biological Applications*. AOCS Press, Champaign IL.

CHARMAN WN (1996). *J. Pharm. Sci.*, 45 (3–4), 316–327.

CHEN ML, VALI SL, LIN J and JU YH (2004). *J. Am. Oil Chem. Soc.*, 81 (6), 525–532.

CLISSOLD D (1999). The potential for prostaglandin pharmaceuticals. In Tyman J.H.P (ed.), *Lipids in Health and Nutrition*, Royal Society of Chemistry, Cambridge, UK.

CROOK TH, TINKLENBERG J, YESAVAGE J, PETRIE W, NUNZI MG and MASSARI DC (1991). *Neurology* 41 (5), 644–649.

DUPONT J (1996). *Am. J. Clin Nutr.* 63 (6), 991S–993S.

EGBARIA K and WEINER N (1990). *Adv. Drug Del. Rev.*, 5, 287–300.

GUO Z, VIKBJERG A and XU X (2005). *Biotechnol. Adv.*, 23, 203–259.

HARDING CR (2004). *Dermatologic Therapy*, 17 (1), 6–15.

HARALDSSON GG and THORARENSON A (1999). *J. Am. Oil Chem. Soc.*, 76, 1143–1149.

HEIRD W, GRUNDY SM and VAN HUBBARD S (1986). *Am. J. Clin. Nutr.*, 43, 320–000.

HERNANDEZ E (2005). Pharmaceutical and cosmetic use of lipids. In Shahidi F (ed.), *Bailey's Industrial Oils and Fat Products*, John Wiley & Sons, Inc., New York.

HOSSEN M and HERNANDEZ E (2004). *Lipids,* 39 (8), 777–782.

HOSSEN M and HERNANDEZ E (2005). *Eur. J. Lipid Sci. Technol.*, 107, 730–736.

IP C, SCIMECA JA and THOMPSON HJ (1994). *Cancer*, 74, 1050–1054.

JENSEN RG (1996). *Prog. Lipid Research, Lipids in Human Milk.* 35 (1), 53–92.

KINSELLA JE (1998). *Food Technol.*, 42 (10), 124–142.

LAMPE MA, WILLIAMS ML and ELIAS PM (1983). *J. Lipid Research*, 24, 131–140.

MORGAN TM, REED BL and FINNIN BC (1998). *J. Pharm. Sci.*, 87 (10), 1213–1218.

MU H and HØY C (2001). *J. Lipid Research*, 42, 792–798.

OGISO T, NIINAKA N and IWAKI M (1996). *J. Pharm. Sci.*, 85 (1), 57–64.

PORTER ET AL. (1996) to be supplied.

PORTER CJ (1997). *Crit. Rev. Ther. Drug Carrier Syst.*, 14, 333–393.

POUTON CW (2000). *Eur. J. Pharm. Sci.*, 11, Suppl. 2, S93–S98.

ROCHA A and HERNANDEZ E (2004). *J. Am. Soc. Oil Chem.,* 81 (7).

RUBIN M, MOSER A, VASERBERG N, GREIG F, LEVY Y, SPIVAK H, ZIV Y and LELCUK S (2000). *Nutrition*, 16, 95–100.

SAHIN N, AKOH CR and KARAALI A (2005). *J. Am. Oil Chem. Soc.*, 82 (8), 549–557.

SAHIN N, AKOH CR and KARAALI A (2006). *J. Agr. Food Chem.*, 54, 3717–3722.

SHAHIDI F (2006). *Nutraceutical and Specialty Lipids and Their CO-Products*. CRC Press, Boca Raton.

SHIMADA Y, NAGAO T, HAMASAKI Y, AKIMOTO K, SUGIHARA A, FUJIKAWA S, KOMEMUSHI S and TOMINAGA Y (2000). *J. Am. Oil Chem. Soc.*, 77 (1), 89–93.

TAO RC and YOSHIMURA NN (1980). *J. Parenter Enteral Nutr.*, 4, 469.

TOTANI Y and HARA S (1991). *J. Am. Oil Chem. Soc.*, 68, 848–851.

TOUITOU E, GODIN B and WIESS C (2000). *Drug Development Research*, 50, 406–415.

TRANCHANT I, THOMPSON B, NICOLAZZI C, MIGNET N and SCHERMAN D (2004). *J. Gene Med.*, 6, S24–S35.

TSO P, LEE T and DEMICHELE SJ (2001). *Journal of Nutrition*, 131, 2157–2163.

VERHOEF JC, BODDE HE, DE BOER AG, BOUWSTRA JA, JUNGINGER HE, MERKUS FW and BREIMER

DD (1990). *Eur. J. Drug Metab. Pharmacokinet.*, 15, 83–93.

VON CORSWANT C, THORÉN P and ENGSTRÖM S (1998). *J. Pharm. Sci.*, 87 (2), 200–208.

WERTZ PW, SWARTZENDRUBER DC, KATHI CM and DOWNING DT (1987). *J. Invest. Dermatol.*, 89, 419–424.

WILLIS WM, LENCKI RW and MARANGONI AG (1998). *Crit. Rev. Food Sci. Nutr.*, 38 (8), 639–674.

YANG SC and BENITA S (2000). *Drug Development Research*, 50 (3–4), 476–486.

6

Micro- and nano-emulsions for delivery of functional food ingredients

N. Garti and I. Yuli-Amar, The Hebrew University of Jerusalem, Israel

6.1 Introduction

The term food colloid is used very often to describe the physical structure of common basic foods. The colloidal systems might be composed of a variety of complex dispersions, such as solid particles in solid matrix, liquid (or semi liquid) aqueous phases emulsified in an aqueous continuous phase, bubbles entrapped in complex solid or liquid matrices, etc. The colloidal dispersions are in most cases small particles embedded in a complex dispersion phase. Understanding how these systems work and learning how to control their properties is an enormous task. Even after years of very extensive studies we are far from truly understanding basic man-made food products such as baked goods (or just simple bread), dairy products, ice cream or meat products.

Some of the dispersed systems remain intact in the final food product due to the nature of the continuous phase (matrix). The matrix if solid or of high viscosity can immobilize the dispersion particles or droplets. Other systems (fluids, drinks) are stabilized by a set of interracially active compounds (surfactants) that adsorb onto the dispersed particles, droplets or bubbles, and prevent them from aggregating, clustering or coalescing. The stabilizers were given their names according to their obvious functionalities. These compounds were termed alternatively surfactants, wetting agents, stabilizers, spreading agents, foamers, emulsifiers, etc. The surfactants/emulsifiers can be amphiphilic monomeric or polymeric compounds and/or can be naturally occurring molecules.

The common goal for all those dealing with food colloids is to find the best performing compounds and to use them in minimum quantities in order to improve the cost/performance ratio. These additives have been in most cases synthetic compounds that revealed some hazardous health and environmental effects from minor byproducts that contaminated the active amphiphiles. Today we know that many of these compounds are considered to be unsafe and their use is restricted in most developed countries. The restrictions result in significant difficulties for food producers and are affecting the goodness of the products. The industry is searching for additives derived from natural sources, and/or extracted with permitted solvents. The major sources for such food additives are plants, fruits, vegetables, herbs, flowers, roots, tree barks, leaves, etc.

Utilization of a dispersed system is essential in many foods that contain multiple components with different characteristics. Adding lipophilic ingredients (fats, oils, oil-soluble vitamins, flavors, antioxidants, etc.) into an aqueous-based system is not an easy task and requires emulsification technology. Mixing lipophilic substances into fluid foods is an even more challenging task. The dispersed particles and droplets must remain stable for a long shelf-life and should not be too big to comply with taste and organoleptic requirements. It is therefore essential to learn how to properly disperse one phase into another. The dispersed systems once formed can also be utilized as carriers, vehicles, interfacial reactors, protective media, etc. for many other applications. For years, efforts have been made to design and stabilize food dispersions with specific properties and characteristics.

Consuming more healthy food has become a major trend in the last decade leading to another important trend of food additives with health benefits.[1–3] The old slogans of 'be healthier by consuming foods with less sugar, less fat and less salt' are now replaced by 'consume foods enriched with fibers, proteins, vitamins and other more sophisticated nutritional or healthy ingredients'. We are progressively introducing into our food lexicon the terms 'food supplements' or 'food nutraceuticals'. Food supplements such as minerals and vitamins (food additives) have been known for generations to provide some health 'maintenance' while nutraceuticals is a modern term to describe mainly plant, vegetable, fruit and root extracts that can provide some 'health benefits and even treat certain diseases'. Since no clear and official definition has been coined for this category of food additives, the term is widely used for almost any ingredient that someone insists can provide health benefits even without any scientific proof. It should also be stressed that in many cases there is insufficient data to support some of the health claims made by the producers of these extracts and no data to guide the consumer on daily dosage use, allergies, side effects, and other health hazards. The health authorities in many countries are helpless in attempts to control the use of these additives in our foods.

However, the extracted ingredients are very often complex mixtures of many organic compounds as well as biopolymers that are very difficult to characterize, separate or purify. In addition, most of these ingredients are water-insoluble and their incorporation into an aqueous food matrix might be a real challenge.

These ingredients are sometimes aqueous extracts and sometimes organic solvent extracts. In many cases the 'organic solvent extracts' are non-soluble in food-grade solvents or in simple triglyceride type oil phase (vegetable oils) and in some cases are insoluble in water and in oil. The poor solubility causes severe difficulties to the food product manufacturer. Moreover, the beneficial health or nutritional effects depend considerably on the bioavailability of the nutraceuticals since the transport of these ingredients across our gut membranes is practically negligible.

The food industry needs urgently carriers, vehicles and protective systems to improve nutraceutical performance. The search for such vehicles is the aim of many recent studies, a search that has become increasingly urgent.

One important trend to mention is the nanotechnology that can help food supplements or nutraceuticals that are added to our foods to become more efficient; i.e. more nutritional, healthier, tastier and safer. Nano-vehicles seem to be excellent candidates as solubilization carriers for insoluble active compounds.[4–5]

Recently, Weiss *et al.* published an excellent full review on 'Functional materials in food nanotechnology'.[6] The authors faced a dilemma in defining nanotechnology and nanostructures in foods. The authors defined nanotechnology as the 'science that focuses on characterization, fabrication and manipulation of biological and non-biological structures, smaller than 100 nanometers'. According to the National Nanotechnology Initiative (2006) 'nanotechnology is the understanding and control of matter at dimensions of roughly 1–100 nanometers, where unique phenomena enable application'. Weiss *et al.*[6] also stressed the fact that there is some criticism within the scientific community of the terminology and definitions of these systems. We, however, will use the term 'nano' only for structures in the length scale of 1–100 nm.

In this review we decided to concentrate on systems that are typically liquid dispersions in this length scale. We will discuss characteristics of liquid dispersions that provide the food industry unique advantages which cannot be obtained using other technologies.

Physical, chemical and biological properties of the nanoscale architectures that are substantially different from their macroscopic counterparts will be reviewed. The review will not discuss biopolymeric aggregated structures such as dispersed proteins (casein, whey proteins, soy proteins, egg albumins, etc.), polysaccharides, DNA, membranes and cells, etc., even if they are present in some foods. We will concentrate on the use of nano-sized dispersed liquid as vehicles for the solubilization of food additives, minor components such as flavors and antioxidants and mainly nutraceuticals.

Nanoscale structures can be solid particles, liquid droplets (microemulsions and nano-emulsions, or micelles) and gas bubbles (foams, aerosols), which dispersed in an aqueous continuous phase. This review is devoted only to the progress made in recent years in the preparation, understanding and manipulation of liquid-in-liquid dispersions (spontaneously formed, true microemulsions) to better serve food industry needs. Furthermore, we will scan the physical modifications that were made on lyotropic liquid crystals to make them more fluidic and to exhibit physical characteristics closer to liquids than to solids.

Nanoscale structures have the potential to impact many aspects of food systems such as food quality, texture, and flavor, food security (processing, shipping, and spoilage). We will give attention to the impact of the nanotechnology on food preservation (environmental and enzymatic protection), food quality (texture, flavor), and on nutrition and health aspects.

6.2 Microemulsions

6.2.1 Terminology

The term microemulsion is used for systems composed of oil phase (such as triglycerides or hydrocarbons) and an aqueous phase (water or water with some electrolytes or polyols) dispersed in each other to form systems with nanosized structures. However, Hoar and Schulman, being the first to describe microemulsions,[7] used the term 'microemulsions' for 'systems that are clear and solution-like in nature of aggregated droplets dispersed in each other'. A more recent and more generally accepted definition of microemulsion is: 'A spontaneous formed, clear, thermodynamically stable homogeneous dispersion of two immiscible liquids containing appropriate amounts of surfactants and co-surfactant'.[8–9] It should be noted that the interfacial tension of microemulsions is ultralow (*ca.* 10^{-2} mN/m) but not necessarily, or almost never zero.

Some researchers disagree with the term 'thermodynamically stable' and prefer to use only 'spontaneously formed'. However, we insist on defining microemulsions as systems that are entropy driven and thermodynamically stable. We can also use a very general definition of microemulsion for systems that are composed of oil and/or water nanodomains that coexist in thermodynamic equilibrium.[10,11]

In this broad definition we mean systems that are 'nano sized self-assembled liquid (NSSL)'. Any other attempt to term systems that are not self-assembled and not thermodynamically stable (but rather kinetically stable), as microemulsions is misleading. Submicronal droplets that are formed by high shear stresses are in most cases almost completely clear but not thermodynamically stable and therefore eventually they will separate. Real microemulsions are Newtonian and not shear-dependent.

6.2.2 Microemulsion preparation

The compositions that can produce isotropic regions with microemulsions structures were first described utilizing a phase diagram of three major components (emulsifier, oil and water) that were designed on the basis of titrating mixtures of surfactant and oil phase with water. This 'cook and look' approach led to the formation of isotropic regions formed in the presence of very large excess of the surfactant phase. Other preparations were obtained by titrating the two component system of oil and water with excess of surfactant phase until clear solution was obtained. This concept was based on the understanding that since microemulsions represent droplets of very small diameter, they form very high surfaces areas that will require huge amounts of surfactants. It is theoretically very simple to calculate, based on the tails and head groups' length and area, how much surfactant is needed for each droplet and for a given number of droplets. Such simple calculations led scientists to conclude that micro-emulsions are always rich in surfactant phase. Knowing that in food systems excess of surfactant (mainly hydrophilic in nature or ethoxylated) is not highly recommended led to the idea that microemulsions are not very suitable for food systems. Emulsions being poorer in the amounts of surfactants were always preferable in spite of their severe stability limitations. Yet, it is always important to bear in mind that the emulsions' smaller surface areas are also dictating lower solubilization capacities of bioactives at their interface.

General characterization

With years of research more sophisticated ternary or quaternary compositional phase diagrams of oil, surfactant, aqueous phase and cosolvent or cosurfactant were constructed and the identified clear isotropic phases were modified micro-emulsions both from a structural and composition viewpoint. Microemulsion of oil phase entrapped into an aqueous phase was termed L_1 isotropic phase or oil-in-water (o/w) microemulsion. Such systems are basically micelles swollen with oil. Similarly if the water was entrapped in the core of the micelles, L_2 isotropic region was obtained (termed water-in-oil, or w/o, microemulsion). A typical phase diagram of the three components is shown in Fig. 6.1. The L_2 micro-emulsion boundaries are termed as the A_T isotropic region and they indicate the compositions of the three components that will form microemulsion spontaneously, remain clear, and be thermodynamically stable. A type I (Winsor I) microemulsion is swollen micelles (o/w) surrounded by water surfactant micelles coexisting with excess oil. A type II (Winsor II) microemulsion is swollen reverse micelles (w/o) surrounded by oil where the reverse micelles coexist with excess of water. A type III microemulsion consists of oil, water, and a middle bicontinuous microemulsion phase coexisting in three-phase equilibrium. A type IV microemulsion is defined as a single-phase microemulsion system where both oil and water are completely solubilized in the surfactant microemulsion phase. Microemulsion transitions can be achieved in several ways, depending on the type of surfactants. For example, for ionic surfactants, a type I-II-III transition can be obtained by increasing the electrolyte concentration, whereas for nonionic surfactants temperatures can change the hydrophilicity of the surfactant and can allow similar transitions. It should be noted that some consider as optimum the type III microemulsion into which the water and the oil are added in equal amounts.

The aim of every formulator is to enlarge the one phase region and to expand it so that minimal levels of surfactant will be required. If w/o microemulsions are required for some spreads or margarine-like products it is important to maximize the amount of water that is solubilized within the oil and surfactant mixtures while if the application is of aqueous continuous phase such as beverages it is

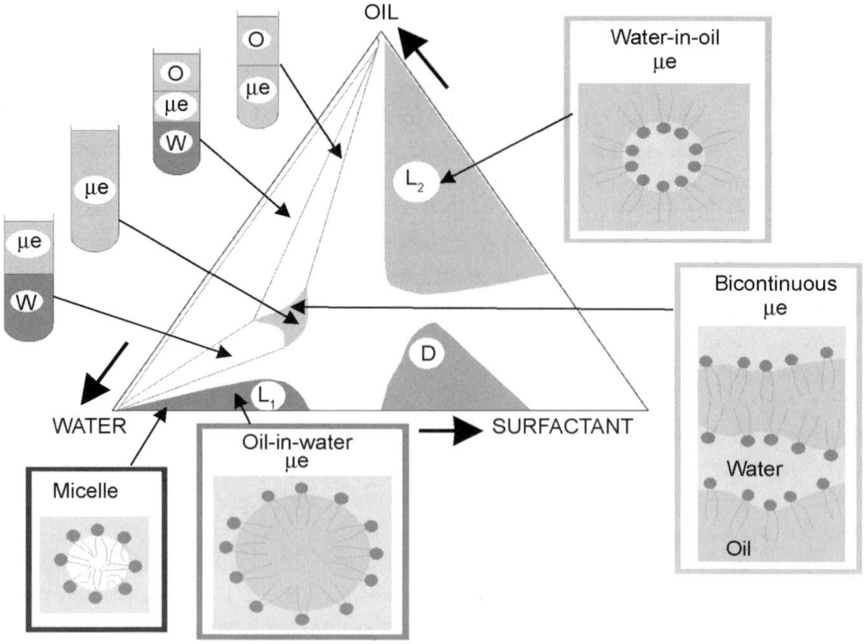

Fig. 6.1 Ternary phase diagram.

important to solubilize maximum amounts of an oil phase within water and surfactant mixtures and maximal levels of oil phase will be solubilized. In both cases it should be achieved with minimum amounts of surfactants. However, in some of the new applications it is essential to prepare an oil concentrate (oil and surfactants) that can be diluted with an aqueous phase so that the Winsor I will be transformed into Winsor IV type structure progressively.[12]

In order to control the behavior of the surfactant in entrapping oil or water, it is essential to control two important parameters: the curvature of the microemulsion, and the flexibility of the surfactant film. An elastic and flexible film is a prerequisite for the formation of microemulsion. In order to improve the flexibility some additional components such as alcohols can be added to the three component system. Similarly one can manipulate the curvature of the oil and/or the water droplets by various additives such as cosurfactants, electrolytes, etc.

The composition that will solubilize the maximum water will be termed the maximum solubilization dilution line and will be termed W_T. The dilution line represents the oil/surfactant weight ratio. Since the main application of such microemulsion is not just to solubilize water but to dissolve, into the water, hydrophilic active matter, it is essential to obtain maximum A_T and dilution lines with maximum amount of water. If, for example, maximum 40 wt% of water can be solubilized along dilution line 73, we will indicate it as W_T (73) = 40 wt%.

Similarly, if one is interested in solubilizing oil into an aqueous phase, an o/w NSSL has to be formed with minimum amounts of surfactant and maximum amounts of solubilized water. As one can see from Fig. 6.1 the two regions o/w and w/o are not connected and in between the isotropic region a thermodynamically non-stable region of emulsion is formed. In some cases, an additional isotropic region can be formed, termed by most scientists 'the bicontinuous mesophase' which is a transitional phase between the two types of droplets and is actually composed of elongated nanodomains of interwoven water and the oil phase.

Microemulsions have numerous potential applications in detergency,[13,14] pharmaceuticals,[15,16] cosmetics[17,18] and agrochemical.[19,20] Depending on the specific applications, attempts were made to increase the amount of the entrapped inner phase. The most common variations were to utilize (mostly by cook-and-look) various types of lipophilic compounds to serve as the oil phases and various types and combinations of surfactants. Numerous combinations of surfactants and oils were used.[13–20]

Thermodynamic properties of microemulsions

Some of the advantages of microemulsions are derived from the intrinsic thermodynamic properties. The free energy of microemulsion formation is considered to depend on the extent to which surfactant lowers the interfacial tension of the oil–water interface and the gain in entropy of the system

$$\Delta G_f = \gamma \Delta A - T\Delta S,$$

where ΔG_f is the free energy of formation, γ is the interfacial tension of the oil–water interface, ΔA is the change in interfacial area upon microemulsification, ΔS is the change in entropy of the system, and T is the temperature. It should be noted that when a microemulsion is formed the change in ΔA is very large due to the large number of small droplets formed which dictates large amounts of surfactant that will be sufficient to cover all the interface, γ is very small, (of the order of fractions of mN/m but not zero) and the entropic component is the major factor dictating spontaneous formation and thermodynamic stability. The dominant favorable entropic contribution is the very large dispersion entropy arising from the mixing of one phase in the other in the form of large numbers of small droplets. However, there are also expected to be entropic contributions arising from other dynamic processes such as surfactant diffusion in the interfacial layer. Thus a negative free energy of formation is achieved, and the microemulsification is spontaneous.[21]

6.2.3 Microemulsion composition-structure dependence

Microemulsions containing aliphatic or aromatic hydrocarbons have been extensively studied. Data on composition, properties, and structure are widely available.[22–24] Most of the surfactants and the oils are synthetic and aimed at applications like ternary oil recovery, agricultural pesticide emulsion, etc.

Food formulations containing lipids and oils with limited solubility were in demand for the preparation of low-fat products such as sauces, salad dressings, and drinks. Attempts to solubilize long-chain triacylglycerols (soybean, peanut oil, etc.) into microemulsions was a difficult task and in many cases only liquid crystalline (gel-like) structures were formed instead of the Newtonian microemulsions.

Few attempts at solubilizing water into food-grade microemulsions were made in early studies but only few of these studies dealt with the problems related to the formation of such microemulsions.[25–30] Only relatively small and limited isotropic regions of w/o microemulsions within the phase diagrams could be obtained.[31] In most cases authors were not able to solubilize more than 10–20% water within the vegetable oil with any mixture of surfactants, and the best results were obtained with polyglycerolesters where the surfactant to water ratio was 10 (Table 6.1). As a result authors replaced the triglycerides with flavor or essential oils as the oil phase.[32,33] No significant attempts were made in those days to explain why essential oils or long-chain hydrocarbons of alkyl esters are so different in their behavior from the triacylglycerols.

Table 6.1 Surfactants forming w/o microemulsion in Crisco oil at s/w = 12 or less

Surfactant trade name	Composition	S/W wt/wt
A. Glycerides		
Dimodan LS (G)	Sunflower (80% C18:2; 20% C18:1)	10:1*
AM #505 (G)	No saturated FA	10:1*
Dimodan O (G)	Soybean oil (51% C18:2; 23% C18:1)	None
AM #489 (C)	Rapeseed oil (40% di-) (21% C18:2; 62% C18:1)	None
AM #490 (G)	Sunflower oil (39% di-)	None
Monooleate (P)	Monooleate	None
Aldo MD (L)	Mono- and dioleate	None
Aldo MLD (L)	Monolaurate	None
B. Polyglycerol esters of fatty acids		
Caprol 3GO (C)	Triglycerol monoleate	None
Polyaldo 4-2-0 (L)	Tetraglycerol dioleate	10:0.71
Caprol 6G20 (C)	Hexaglycerol dioleate	None
Polyaldo 2010 (L)	Decaglycerol dioleate	None
Am #506 (G)	Polyglycerol oleate**	None
Experimental (L)	Tetraglycerol laurate	None
Polyaldo 4-2-L (L)	Tetraglycerol dilaurate	None
93-919 (L)	Polyglycerol monolaurate	None
AM #507 (G)	Polyglycerol mono-dilinoleate**	9:1
Triodan 20 (G)	Polyglycerol esters	14:1
Homodan PT (G)	Polyglycerol ester of dimerized soybean oil	None

* Microemulsion formed
** Polyglycerol: mainly tetra
G = Grindsted
L = Lonza/Glyco
C = Capitol City
P = Pflatz and Baur

Several different approaches have been tested with the idea of preventing triglycerides from forming liquid crystalline structures (derived from the fact that the surfactant lipophilic tail cannot easily penetrate within the large and rigid triglyceride molecules at the water interface). The idea was to disorder the triglyceride (triolein) film at the interface. Such disorder could be obtained by various methods.

1. Adding alcohol as cosolvent and cosurfactant with a strong tendency to disorder the triglycerides film.
2. Changing the temperature. It was also shown that the extension of the w/o microemulsion regions were strongly dependent on temperature. This is an expected behavior since ethyleneoxide-based surfactants are known to become more hydrophobic with temperature and it imparts strong film disorder and flexibility.
3. Increasing the surfactant levels.
4. Adding cosurfactant with different chain length or adding branched surfactants.
5. Decreasing or manipulating salinity.
6. Use of co-oils of different structures.

Some of the early attempts to form triglyceride-based food microemulsions were in fact experimental efforts (mostly based on experimental attempts) to enlarge the isotropic region (A_T) of w/o microemulsions by adding additional components to the formulations such as sucrose, short-chain alcohols, and other polyols.[29,30] The ethanol and the polyols served as hydrotropes as well. It was shown that small amounts of added sucrose (*ca* 10 wt% of the aqueous phase) enlarged the L_2 and decreased the L_1 isotropic regions (Fig. 6.2). On the other hand, higher levels (20–40 wt%) of sucrose reduced the two isotropic regions. It was also concluded that triglycerides with short chain length can solubilize more water (L_2) than those with long-chain fatty acids, and vice versa when oil was solubilized in an aqueous phase (L_1).[29] Close examination of the scanned attempts reveal that, in spite of all the efforts, the isotropic regions in the phase diagram were very small (Fig. 6.2).

The effect of alcohol was studied by Leser *et al.*[34] and Van Corswant *et al.*[35] Leser *et al.*[34] studied the phase diagrams of phosphatidylcholine, MCT and alcohols with different chain-length. However, none of these systems could be made with food-grade emulsifiers or food-grade alcohols. Van Corswant *et al.*[35] studied the microstructure of the PC-based microemulsions in the presence of 1-propanol as well, and learned that extended isotropic regions can be obtained with the mixture of 1-propanol/PC/MCT/water. Leser *et al.* concluded that propanol disorders the liquid crystalline structure, enhances the flexibility of the surfactant film and enables the MCT to penetrate the surfactant 'brush'. It seems that ethanol that is food-permitted cannot fulfill all of these requirements.

Aboofazeli *et al.* examined the oil effect on the isotropic L_2 region.[36] A partial phase diagram composed of water, egg lecithin, propanol and different polar oils such as MCT and soybean oil were reported. It was showed that long-

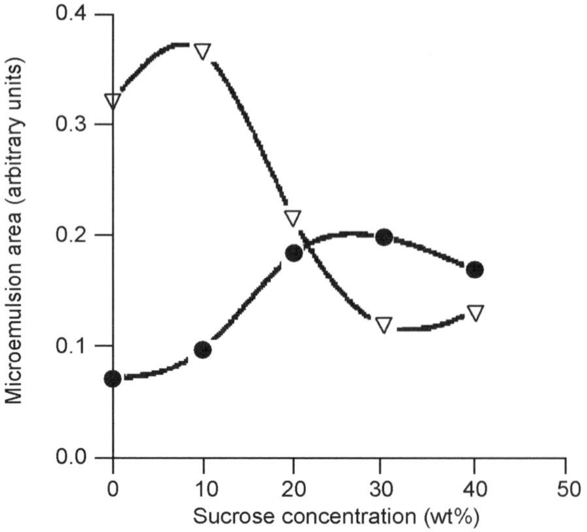

Fig. 6.2 Effect of sucrose concentration on the area of the microemulsion regions for a system containing soybean oil, ethoxylated monodiglyceride/monoglyceride (75/25 wt%), and aqueous phase containing 20 wt% 96 ethanol solution and various amounts of sucrose: (●) L1; (∇) L2.

chain triglycerides were capable of solubilizing less water than any other lipophilic molecules.

The only study that managed to solubilize up to 40 wt% water in food-grade ingredients was that of Costantinides and Scalart[37] that was composed of a complex mixture of medium-chain triglycerides (capric and caprilic) with mono- and diglycerides of the same fatty acids as the oil phase, and Tween 60 as the surfactant phase, while the aqueous phase was a saline mixture (ratio of 65/22/10/3 wt/wt). The blend seems to be obtained mostly by 'trial and error' ('cook-and-look') since no structural data were provided. The isotropic region that solubilized 40 wt% water is very rich in surfactant (*ca.* 90 wt% Tween from the oil phase). The authors have shown that at 40 wt% water the microemulsion already inverted into oil-in-water droplets but the system remains clear. However, once soybean oil was used as the oil phase the total solubilization capacity dropped to *ca.* 10 wt%.

All the previously prepared formulations suffered from a few major drawbacks:

- Oil-based concentrates could not be diluted with water above 40–50 wt%. Any addition of water caused discontinuity in the phase diagram with regions of two phases.
- There were significant difficulties in solubilizing large amounts of water or oil if the oil phase was based on triacylglycerols (triglycerides).
- Almost no formulations based only on food-grade components could be prepared.

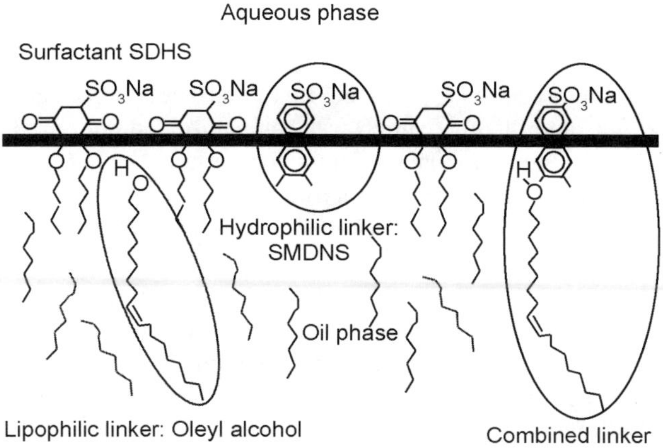

Fig. 6.3 Schematic of the linker effect, showing the surfactant, lipophilic, and hydrophilic linker at the oil–water interface. SMDNS, sodium mono- and dimethylnaphthalene sulfonate.

Graciaa and coworkers[38,39] were the first to introduce the concept of lipophilic linker to help enhance surfactant–oil interaction and thus improve the solubilization capacity of the hydrocarbon and polar oils. The lipophilic linkers were long-chain alcohols that have the tendency to segregate near the oil side of the water–oil interface, close to the tails of the surfactants[40] as depicted in Fig. 6.3. The linker, dodecanol, promotes the local order and increases the interaction between the surfactant tail and the oil phase. Similarly a hydrophilic linker such as sodium mono and dimethylnaphthalene sulfonate (SMDNS) increased the total interfacial area and the overall interaction between the surfactant layer and the aqueous phase.[41,42]

Komesvarakul *et al.*[43] recently published a modified approach to form microemulsions based on triglycerides. They incorporated a combined linker of the hydrophilic heads and the lipophilic tails. They managed to tune the curvature of the surfactant film by adding co-oil to the triglycerides. These formulations have some advantages since they can reduce the amount of surfactant and completely remove the alcohol from the formulation. However, in spite of the fact that the concept was proved to work, the components that were used as hydrophilic and lipophilic linkers as well as the co-oil are cosmetic grade but not permitted for any food application. The concept should be studied further and the search for food-grade components should continue.

New emulsifier mixtures to form food-grade microemulsions

As one can see from the number of publications, most efforts in the last three decades have been to increase the amounts of solubilized water in the oil phase. These efforts were mainly with the idea of making dietetic liquid foods (such as margarines or dressings) where the oil phase will be replaced by a mixture of water and oil without affecting the major physical and organoleptic properties of the final product. However, it seems that this goal was only partially achieved and only very few products that are based on w/o microemulsions exist in the marketplace.[44–46]

In recent years, researchers realized that microemulsions can be used in water-based drinks and that their interfaces can be utilized not only to solubilize water or oil, but to serve as nanoreactors for certain organic and enzymatic reactions. Furthermore, the very large surface areas of the microemulsions can be utilized to control kinetics of chemical reactions as well as to dictate regioselectivity of some processes.[47–49]

Therefore, the importance of constructing phase diagrams with large o/w isotropic regions gained significant interest by technologists as well as by scientists. The producer of the final product would like to have the *liberty of diluting the microemulsion with water to infinity*. Therefore, the idea of making 'concentrates' of reverse micelles that can be progressively diluted with aqueous phases caught the imagination of many researchers. The term 'U-type' or Winsor IV type microemulsion gained special attention. U-type microemulsion is a system that can be progressively diluted with an aqueous phase without phase separation. Such microemulsions were made with non-food ingredients but were not available with food-grade ingredients.

Microemulsion based on five-component mixtures of R(+)-limonene, ethanol, ethoxylated sorbitan ester, polyol and water for food applications have been studied in the authors' laboratory. The five-component systems first explored by Garti and Yaghmur[50] and Yaghmur *et al.*[51] are described on phase diagrams in Figs 6.4 and 6.5. In microemulsion containing R(+)-limonene/ethanol (1/1), Tween 60, and water/propylene glycol (PG) (1/1) (Fig. 6.4), the total monophasic area is 64% (relatively high water solubilization capacity). Modification of the aqueous phase by replacing some of the water by glycerol to

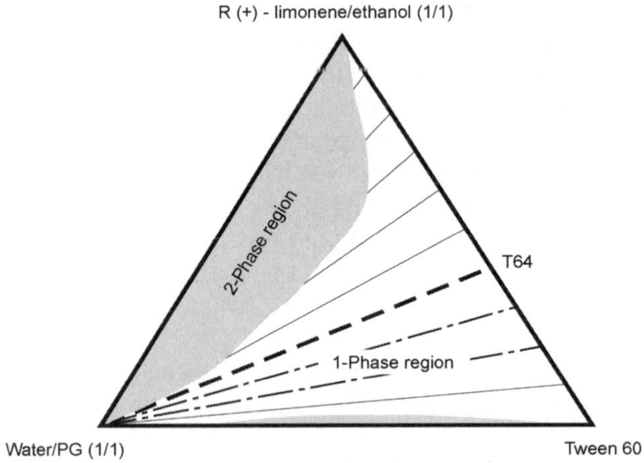

Fig. 6.4 Phase diagram (25 °C) and dilution lines (T64 means 60 wt% surfactant and 40 wt% oil phase in the micellar solution, so-called concentrate), composed of: R(+)-limonene/EtOH (1:1 wt/wt) as the oil phase; Tween 60 as the emulsifier phase; water/propylene glycol (1:1 wt/wt) as the aqueous phase.

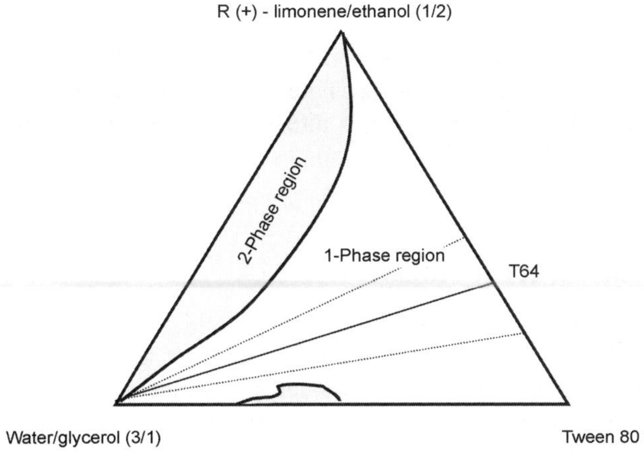

Fig. 6.5 Phase diagram (25 °C) and dilution lines, consists of: R(+)-limonene/EtOH (1:2 wt/wt) as the oil phase, Tween 80 as the emulsifier and water/glycerol (3:1 wt/wt) as the aqueous phase. The dilution line 64 is of 60 wt% surfactant and 40 wt% oil phase.

a ratio of water/glycerol (3/1) and substituting the surfactant by Tween 80 reduces the total monophasic area to 52% (as can be seen in Fig. 6.5). According to Shiao *et al.*[52] the mutual miscibility between the hydrophobic part of the surfactant and the oil affects the degree of the oil penetration into the amphiphilic film. The solubilization mechanism of perfume oils such as R(+)-limonene by nonionic surfactants has been studied by Tokuoka *et al.*[53] and Kanei *et al.*[54] They found that perfume molecules tend to be solubilized in the vicinity of the interface of the water–hydrocarbon moieties of the surfactant or the surfactant palisade layer. These results are in agreement with the results of Garti and Yaghmur and Yaghmur *et al.*[51] and show that the improved water solubilization in R(+)-limonene is related to the oil compatibility with the surfactant and hence its partial penetrates into the surfactant film at the interface.

The effects of polyols (propylene glycol and glycerol) and a short-chain alcohol (ethanol) on the phase behavior of nonionic surfactants and food-grade oils have been studied by Garti and Yaghmur.[50] The aim was to find mixtures of ingredients that will impart a positive interfacial curvature of the surfactant film in order to form o/w microemulsion, inhibit liquid crystal formation, and maximize mutual dissolution of the ingredients. They found that in order to formulate o/w microemulsion the addition of polyols and short-chain alcohols is needed. The phase behavior of the system of R(+)-limonene, ethanol, water/PG (1/1) and Tween 60 containing a 1:1:3 R(+)-limonene/ethanol/Tween 60 weight ratio is characterized by a single continuous microemulsion region starting from a pseudo binary solution (surfactant/oil phase) to the microemulsion water PG (1:1) corner. Due to the addition of the short-chain alcohol and polyol, this dilution line inverts from w/o into o/w via a bicontinuous region continuously, without any phase separation. Microemulsions based on glycerol and polyoxyethylene sorbitan monooleate (Tween 80) show similar behavior to that of those made of PG and Tween 60. However, one can note (Fig. 6.4) that microemulsion based on Tween 60 and PG has larger isotropic area, formed with broader dilution channels. It is possible that the PG (also considered to be cosolvent) is practically propagating into the interface while glycerol remains mostly in the continuous phase.

6.3 Loaded microemulsions

Several researchers discovered that the microemulsions can be good reservoirs for entrapment of bioactive molecules in the microemulsion's core or at its interface. Microemulsions were mentioned as vehicles for drug delivery, since they have a higher drug loading capacity than other similar micellar solutions.

Two types of microemulsions can be suitable for food applications: those based on solubilization of water and water-soluble bioactive molecules in an oil continuous phase and vice versa. The w/o microemulsion can be used for the preparation of dietetic products where part of the oil phase, rich in calories, will be replaced by water. Yet, most of the potential applications are for foods where the continuous phase is water, such as beverages. In such applications the idea is to solubilize oil-soluble bioactives or bioactive molecules that are insoluble in water and/or in oil.[22–24]

Along with the new diet-health pattern motivated by 'prevention rather than disease', functional foods have shown promise at a time when consumer interest in diet and health is at an all-time high.[1,2] The new health pattern emphasizes the role of food components as essential nutrients for preventing or delaying premature onset of chronic diseases later in life, beyond its simple nutritional value for sustaining life and growth. Functional foods are typically rich in phytochemicals, which are derived from natural plant products. Bioactive phytochemicals, known as nutraceuticals,[3,4] providing medical or health benefits including the prevention and treatment of disease, have attracted much interest in the last decade, as constituents of functional foods. These include the major ten categories of strategic nutraceuticals such as: vitamins (vitamin E, D, A, K), phytosterols, arytenoids (β-carotene, lutein, lycopene), fibers (inulin, fenugreek, guar gum) and flaxseed lignans, isoflavones (Soya) citrus limonoides, omega fatty acids, chemoprevention compounds (detoxification enzymes, inhibiting nitrosamine formation, antineoplastic agents, diluting carcinogens in digestive tract, anti-altering hormone metabolism, antioxidants), organosulfur compounds (garlic), polyphenols (tea catechins), phospolipids (PS), terpenoids (vitamin E, carotenoids, phytosterols, etc.) and other phenolic constituents (vitamin C, flavonoids), etc.[5]

Since some of the nutraceuticals are not soluble in water their transport via the digestive tract membrane is very limited hence their bioavailability is restricted. Therefore, attempts were made to search for new vehicles that will enhance solubility, transport the nutraceuticals, and improve bioavailability. There is a growing interest in using microemulsions[8,9,55] as vehicles for nutraceuticals and pharmaceutical formulations due to their physicochemical properties, such as spontaneous formation, clear appearance, low viscosity, thermodynamic stability, and high solubilization capacity.[8,9,55,56] Moreover, microemulsions have the potential to enhance nutraceutical absorption in the digestive system, as promoted by surfactant-induced permeability changes.[57,58] The solubilized drug can be widely distributed in the gastrointestinal tract, by the small droplets of microemulsion ($<5\,\mu m$), following oral administration.[59]

6.3.1 Solubilization of flavors, aromas, and dyes

The lipophilic nature of many flavors and aromas makes them excellent candidates for incorporation within o/w microemulsions. Most of the information was patented but close examination reveals that the direct application of these systems seems to be very limited.

Rosano issued a patent as early as 1979[60] disclosing a method for the preparation of oil-in-water microemulsions via a four-step process:

1. a surfactant is selected which is just barely soluble in the oil phase;
2. the surfactant thus selected is dissolved in the oil to be emulsified in an amount effective to yield a fine emulsion of the emulsified oil in an aqueous phase;
3. the oil, together with its dissolved surfactant is added to the water phase and shaken or stirred; and
4. finally there is provided a second surfactant in the water phase which is somewhat more soluble in water than the first surfactant to produce a substantially clear microemulsion of oil in water.

Wolf and Havekotte issued a different patent[32] in 1989 which discloses microemulsions of edible oils in a matrix of water and certain alcohols which are prepared using certain edible surfactants for use in various products such as beverages. El-Nokaly *et al.* were granted a different patent,[61] which discloses microemulsions which are thermodynamically stable, clear, and homogeneous. These microemulsions were made from a polar solvent, a specific polyglycerol mono- and diester (of branched or unsaturated fatty acids having from 12 to 24 carbon atoms) and a lipid.

El-Nokaly *et al.*[61] discloses that these microemulsions are edible, have good flavor and can be used to disperse water soluble nutrients, vitamins, flavors and flavor precursors in oils. Tabibi and Siciliano[62] issued a patent in 1992 for oral cavity and dental products prepared by microemulsifying an adsorptive oil in an aqueous medium to produce uniform submicron sized droplets. Nothing in the prior art discloses or implies a high flavor or fragrance loading, e.g. 25% flavor or fragrance oil in a microemulsion which has unexpectedly and advantageously high stability yet is free of lower alkanols, e.g. ethyl alcohol or contains less than about 1% of lower alkanols, e.g, ethyl alcohol.

A few years later, an o/w microemulsion was prepared from a flavor or fragrance oil, water and a surfactant with or without a co-surfactant in order to stabilize the microemulsion.[63] The objective of this research was to prepare a mouthwash or fragrance composition containing little or no alkanol (such as ethyl alcohol) but that would still maintain its stability and yield a transparent product. The flavor or fragrance oil which was used may be one or a mixture of oils. For example, the fragrance oil may be a mixture of geranium oil and lavender oil and the flavor oil may be a mixture of peppermint and spearmint. Both the fragrance oil and the flavor oil are insoluble in water and hence they are considered to be the oil phase where water is the continuous phase in the microemulsion. The microemulsion contains at least 30 wt% of flavor oil and at least two nonionic food grade surfactants with a defined hydrophilic lipophilic balance, and an optimized ratio of surfactant versus a co-solvent.

Slocum *et al.*[64] found that greatly enhanced solubility of a range of flavor elements, including alcohols, 2-ketones, 1-alcohols, and ethyl esters, could be achieved through solubilization in a variety of food-grade surfactants. Tween 20 (food-grade surfactant) was able to solubilize relatively large quantities of flavors, up to 3 moles of flavor solubilized per mole of surfactant – a quite attractive result considering the very small quantities of flavor component that are generally needed in foods. Several of the microemulsions discussed by Wolf and Havekotte[32] contained essential oils as their organic phase. In some of these studies the flavor is incorporated directly in the micelle (at the interface or at the micelle's core).[65] Note also that flavor incorporation is not limited to oil-soluble flavors; El-Nokaly *et al.*[33] reported the incorporation of water-soluble flavors within food-grade w/o microemulsions. It should be noted that in none of the early reports was the solubilization achieved by incorporating the nutraceuticals into the oil concentrate and in none of the reports could the systems be diluted to infinity with aqueous phase.

6.3.2 Solubilization of nutraceuticals

Nutraceuticals are a new class of food supplement with biological added values. These molecules are usually not soluble in water or aqueous body fluids, and usually are not soluble in oils as well. Microemulsions can be very useful vehicles for the solubilization and enhance the bioavailability of nutraceuticals.

One of the most obvious applications for microemulsions in foods is to facilitate their incorporation of food ingredients into final food products.

Microemulsions can provide many advantages for the incorporation of food ingredients. The transparent quality of microemulsions can provide a host of new options for altering the appearance of food, as well as being a useful quality when employing these phases in food analysis. The small droplet sizes in microemulsions provide excellent contact between the lipid and aqueous phases – particularly advantageous when solubilizing preservatives within the droplets. There is also some indication that microemulsions or micelles may protect solubilized components from unwanted degradative reactions.[66,67] Finally, the stability characteristics of microemulsions could be a real benefit during food processing and storage.

The pharmaceutical and medical literature is replete with studies of enhanced micellar delivery of vitamins, in particular vitamin E, vitamin K_1 and β-carotene.[68–70] Chiu and Yang[68] claim that solubilization within nonionic surfactants protects vitamin E from oxidative degradation. The *in vivo* effectiveness of the antioxidants vitamins E and C has been shown to be enhanced in micellar solutions, with important applications for cancer prevention.[68]

Fat-soluble vitamins, including vitamin E, are known to be better absorbed in the presence of surfactants from emulsified vehicles than from oily preparations.[71] Emulsification of vitamin E has been reported[72] but emulsions do not possess thermodynamic stability and have to be shaken before use; hence, a thermodynamic stable microemulsion is an appropriate delivery system.

Studies in the authors' laboratory[70] have shown that the oil-soluble vitamin E (tocopherol and tocopheryl acetate) can be solubilized in continuously dilutable w/o and o/w microemulsion, and can serve as part or all of the oil phase (R(+)-limonene). The basic system in which the solubilization was explored consists of: R(+)-limonene and ethanol as the oil phase, water/PG (1/1) as the aqueous phase and Tween 60 as the surfactant (Fig. 6.4). The authors showed that when d-α-tocopheryl acetate (TocAc) was incorporated into the oil phase (Fig. 6.6(a)), the two-phase region of the blank microemulsion, made of (R)-(+)-limonene alone as the oil at the phase diagram, was enhanced. They detected that the TocAc incorporation decreased the solubilization capacity of water and oil, by modifying the interactions with the surfactant and thus increasing the interfacial tension between water and oil. On the other hand, when free tocopherol (TocOH) was introduced into the oil phase (Fig. 6.6(b)), it enhanced the solubilization capacity of water in the oil-rich region; however, it reduced the solubilization of oil in water closer to the aqueous corner, even to a greater extent than tocopheryl acetate.

One should note that upon increasing the concentration of the surfactant, with TocOH being a part of the oil phase, the dilution capacity of the surfactant-oil solution decreased. This behavior may be attributed to the polar nature of the tocopherol which accounts for its penetration ability into the interfacial palisade layer closer to the hydrophilic heads of the surfactant. In the w/o droplet's structure, there was enough space for the TocOH to penetrate into the reversed micelles. The hydrophobic tails of the surfactant moved freely, hence the TocOH could be distributed at the interface and provided some flexibility to the surfactant film. As a result, a larger amount of the aqueous phase was solubilized.

In the aqueous-rich region, the hydrophobic tails entangled inside the micelles and hence the TocOH penetration into the interface may result in decreased flexibility of the surfactant film. For surfactant-concentrated compositions near the aqueous corner (Fig. 6.6(b)), the formulations tend to be more translucent, indicating larger structures. This may be ascribed to the localization of TocOH within the interface and the possible hydration of the polar hydroxyl group on the chroman ring giving rise to more swollen micelles.

In o/w microemulsions, R(+)-limonene appears to be a better oil than both tocopherols, due to its smaller molecular volume and the more favorable

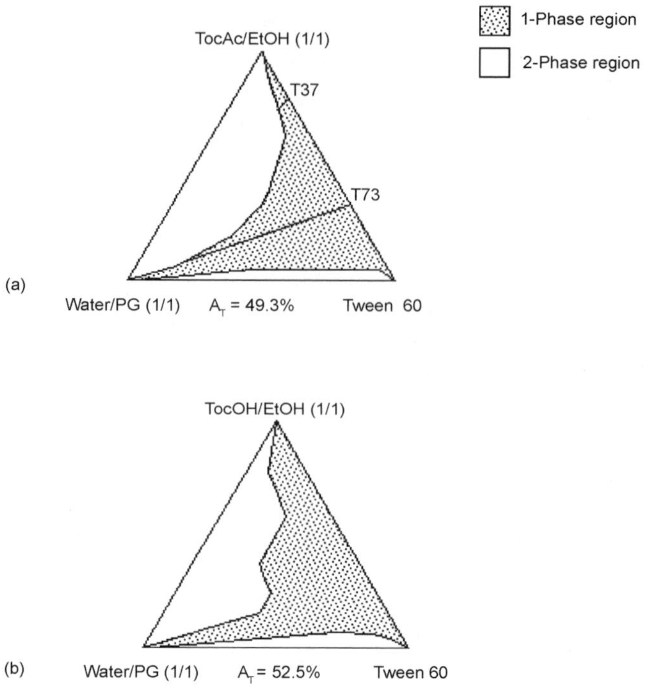

Fig. 6.6 Phase diagrams of the system water/PG/oil phase/Tween 60 at 25 °C with constant weight ratio of water/PG (1:1). The oil phase contain: (a) TocAC/ethanol (1:1) (b) TocOH/ethanol (1:1).

interactions it undergoes with the hydrophobic portions of the surfactant, leading to a dispersed oil phase with a higher thermodynamic stability.

Spernath *et al.* used the same food-grade microemulsion system for the solubilization of another very important nutraceuticals, lycopene.[73] Lycopene is an essential carotenoid that provides the characteristic red color of tomatoes. It is a lipophilic compound (Fig. 6.7) insoluble in water and in most food-grade oils. Several recent studies have indicated the important role of lycopene in reducing risk factors of chronic diseases such as cancer, coronary heart disease and aging.[74–76] In turn, this has led to the idea of studying the effect of lycopene uptake on human health.

Lycopene, a non-soluble powdered nutraceutical, with solubility of 700 ppm in R(+)-limonene, was solubilized in the core, and at the interface, of w/o,

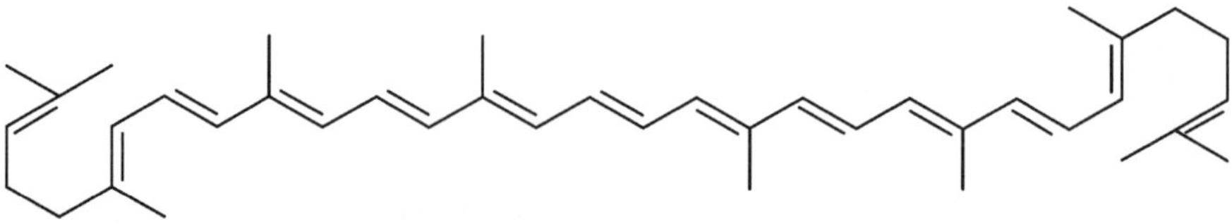

Fig. 6.7 Molecular structure of lycopene.

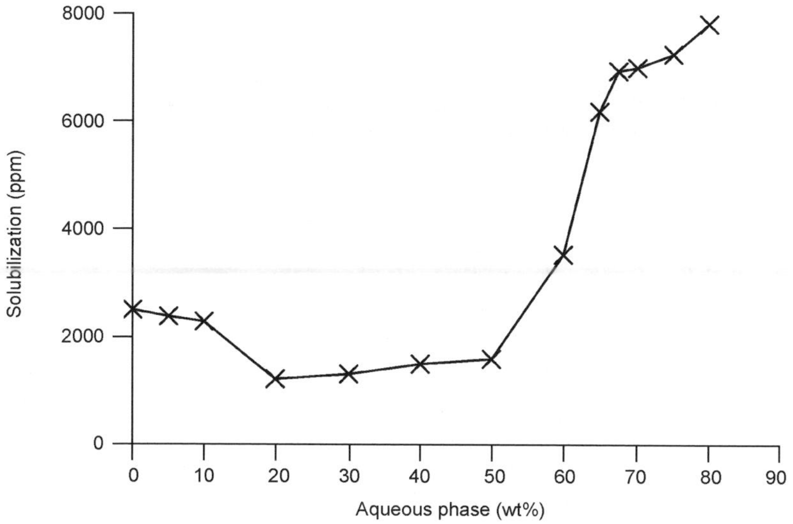

Fig. 6.8 The solubilization capacity of lycopene, mg lycopene normalized to the amount of oil phase at each dilution point tested. The solubilization was plotted against the aqueous phase content along the dilution line 64, at 25 °C.

bicontinuous and o/w microemulsions.[73] The solubilization capacity of lycopene along dilution line T64 (60 wt% surfactant and 40 wt% of oil phase in the concentrate) that was measured by the authors is shown in Fig. 6.8. Four different regions can be identified along this dilution line. At 0–20 wt% of aqueous phase (region I), the solubilization capacity of lycopene decreases dramatically from 500 to 190 ppm (reduction of 62%). In this region, most of the water is bound to the polyol, alcohol and to the hydroxyl groups of the surfactant at the interface. When water is introduced to the core, the micelle swells, and more surfactant and co-surfactant are participating at the interface, replacing the lycopene and therefore decreasing its solubilization.

At 20–50 wt% aqueous phase (region II) the solubilization capacity remains almost unchanged (decreased only by an additional 7%). The system transforms gradually into a bicontinuous phase, and the interfacial area remains almost unchanged when the aqueous phase concentration increases.

In region III (50–67 wt% aqueous phase), the solubilization capacity increases from 160 to 450 ppm and in region IV decreases to 312 ppm.

Spernath and Garti[73] found that lycopene cannot serve as an oil phase, yet it can be solubilized up to 10 times better than its dissolution capacity in R(+)-limonene or any other edible oil (Fig. 6.8).

The authors also studied the effect of the HLB of the nonionic surfactants, ethoxylated sorbitan esters homologies (Tween 20, 40, 60, 80) on the solubilization capacity of lycopene. The authors showed that Tween 60, being the most hydrophobic surfactant solubilized the maximum amount of lycopene. The authors postulate that lycopene is sensitive to the hydrocarbon chain length, and thus favors surfactants with long hydrocarbon chain length. Since an optimal

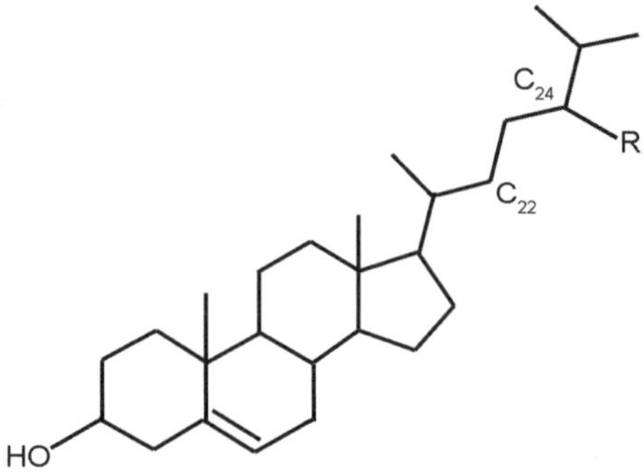

Fig. 6.9 Structures of cholesterol and some abundant phytosterols molecular structure ($R = H$ – cholesterol; $R = CH_2CH_3$ – β-sitosterol; $R = CH_2CH_3$ and additional double bond at C_{22} – stigmasterol; $R = CH_3$ – campasterol; $R = CH_3$ and additional double bond at C_{22} – brassicasterol).

interaction of the solubilized oil with the hydrophobic part of the surfactant is very important in the formation of microemulsions,[51,77,78] it seems that an optimal interaction of lycopene with the hydrophobic part of the surfactant would be just as important.

Spernath *et al.* also investigated the solubilization capacity of phytosterols and cholesterol, in the U-type food-grade microemulsion.[79] Phytosterols (plant sterols) are steroid alcohols. Their chemical structure resembles human cholesterol, as can be seen in Fig. 6.9. Both sterols are made up of a tetracyclic cyclopenta[α]phenanthrene ring system and a long flexible side chain at the C-17 carbon atom. The four rings have *trans* configurations, forming a flat α system.[80,81] Moreover, the sterols create planar surfaces, at both the top and the bottom of the molecules, since the 20R conformation is preferred in the side chain. This allows for multiple hydrophobic interactions between the rigid sterol nucleus (the polycyclic component) and the membrane matrix.[80,82,83] Numerous clinical investigations have indicated that administration of phytosterols to human subjects reduces the total plasma cholesterol and LDL cholesterol levels.[84,85]

The authors investigated the solubilization capacity (SC) of phytosterols and cholesterol along dilution line 64 (Fig. 6.10). When comparing the solubilization capacity of phytosterols and cholesterol, one can easily see the difference, especially in the w/o and bicontinuous microemulsions. The SC of cholesterol is 25–55% higher than phytosterol's SC in systems containing up to 40 wt% of aqueous phase (Fig. 6.10). Increasing the aqueous phase concentration over 50 wt% causes an inversion in the SC behavior of cholesterol and phytosterols. The phytosterol's SC is higher by 10–35% than that of cholesterol in micro-emulsions containing more than 50 wt% of aqueous phase. Thus, the authors postulate that, in order for phytosterols to block cholesterol incorporation (by competition mechanism) into bile-salt micelles effectively, one should use o/w microemulsions. The solubilization capacity of phytosterols was increased by 10-fold in comparison to its solubility in R(+)-limonene.

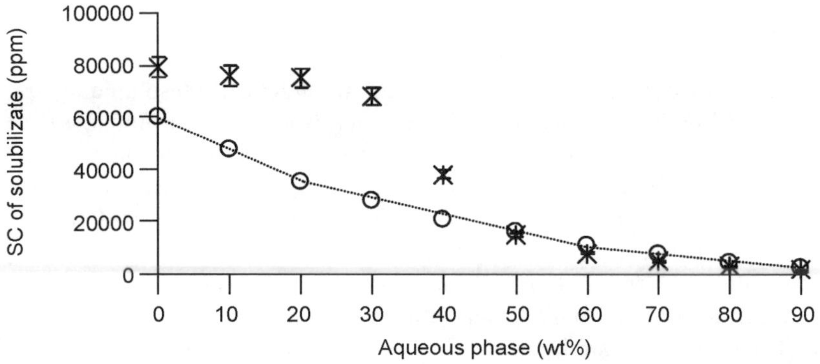

Fig. 6.10 Solubilization capacity (SC) of cholesterol (×) and phytosterols (◯) along the dilution line T64 at 25 °C.

Amar *et al.* used some different U-type microemulsion system for the solubilization of lutein (Fig. 6.5).[65] The lutein (Fig. 6.11) play an important role in the prevention of age-related macular degeneration, cataracts, and other blinding disorders. Lutein, a naturally-occurring carotenoid, is widely distributed in fruits and vegetables and is particularly concentrated in the *Tagetes erecta* flower. Epidemiological studies suggest that high lutein intake (6 mg/day), increases serum levels that are associated with a lower risk of cataract and age-related macular degeneration.[86,87] Lutein can either be free or esterified (miristate, palmitate or stearate). The carotenoids are situated in the macula (*macula lutea*, yellow spot) between the incoming photons and the

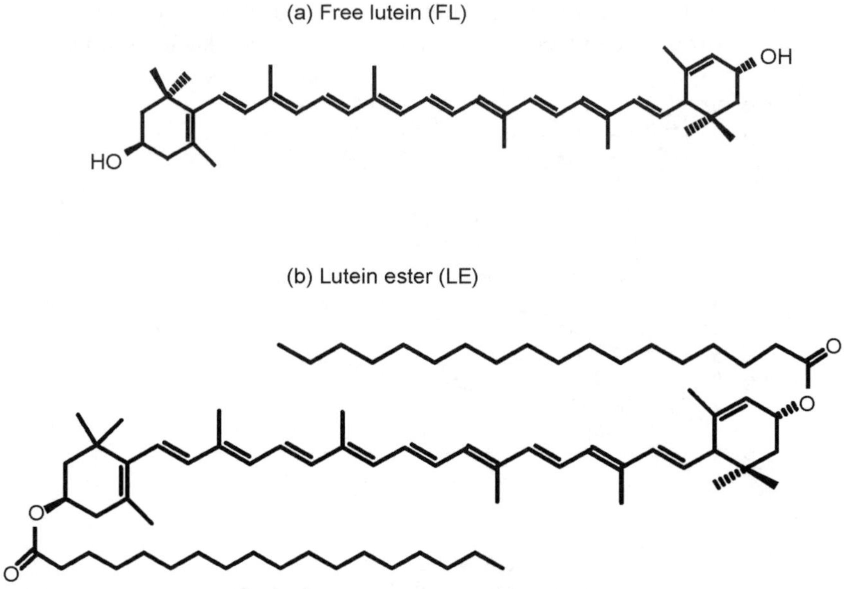

Fig. 6.11 Molecular structures of (a) free lutein, (b) lutein ester.

photoreceptors. As a result, lutein can function as a blue light filter (400–460 nm).[86,87]

Both forms of lutein (free and esterified) are practically insoluble in aqueous systems and their solubility in food-grade oils (such as R(+)-limonene) is very limited, 1200 and 7000 ppm, respectively.[65] Free lutein and lutein ester behave somewhat like phytosterol and lycopene, respectively. They cannot serve as an oil phase as the vitamin E did, and their solubilization was found to be surfactant-dependent in all mesophase structures.

The authors used microemulsion composed of water, Tween 80, R(+)-limonene, ethanol and polyol, but with somewhat different compositions (Fig. 6.5). The differences are in the oil/alcohol and in the water/polyol weight ratios. The solubilization along the T64 dilution line of both lutein and lutein mono- and diesters was explored and the differences between the two have been stressed.[65,88]

From Fig. 6.12 one can note that at L_2, which consists of oil/alcohol/surfactant at a 1:2:4.5 weight ratio and 0% aqueous phase, 5600 ppm of lutein can be solubilized. This is higher than its solubility in each of the components. An improved solubilization is observed once water was added (up to 40 wt% of aqueous phase) despite the dilution effect. This suggests that the interface, which becomes richer in alcohol/surfactant, can solubilize some lutein that under these conditions will be less soluble in the continuous phase (less alcohol and less surfactant). Once the w/o droplets start to invert into a bicontinuous micro-structure (40 wt% aqueous phase), the free lutein is better accommodated at the interface and its solubilization capacity increases significantly (up to 10 000 ppm). Once the droplets invert into o/w, a strong decrease occurs. It seems that the *R*-(+)-limonene participates, in part, at the o/w interface, which leads to a lower free oil content in the o/w microemulsion. Similarly, the

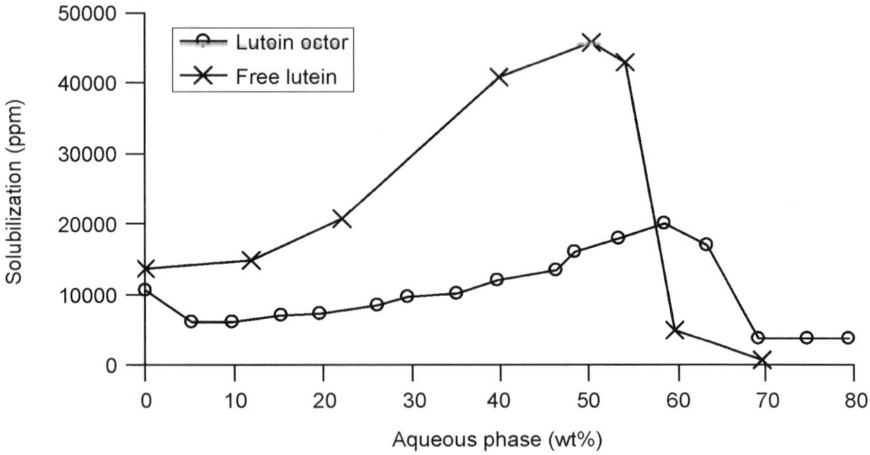

Fig. 6.12 The solubilization capacity of lutein ester (◯) and free lutein (×), mg lutein normalized to the amount of oil phase at each dilution point tested. The parameters were plotted against the aqueous phase content along dilution line 64 at 25 °C.

surfactant (hydrophilic nature) is efficiently consumed by the interface and is no longer present in the core of the microemulsion.[65,88]

The strongest solubilization effect was in the bicontinuous phase, where free and esterified lutein were solubilized up to ten and eight times higher than their dissolution capacity in the oil phase (R(+)-limonene/ethanol 1/2).

Comparison between lutein and lutein esters revealed that:

- Reverse micellar and w/o compositions solubilized both luteins better than o/w microemulsion, and maximum solubilization was obtained within the bicontinuous phase.
- Free lutein was solubilized better than the esterified one, in the w/o microemulsions, whereas the esterified lutein was better accommodated within the o/w microemulsion.
- Vegetable oils decrease the solubilization of free lutein.
- Glycerol and alcohol enhance the solubilization of both luteins.

Recently, the solubilization of CoQ10 was tested using a unique and new U-type food-grade microemulsion.[89] Coenzyme Q (CoQ), also known as ubiquinone (Fig. 6.13), is a naturally occurring substance classified as a fat-soluble quinone with characteristics that are vitamin-like. Its chemical structure is similar to that of vitamin K, and it is found naturally in the tissues of animals and plants. Coenzyme Q is one of the substances in the chain of reactions which produces energy in the metabolism of food. Due to its necessity for energy production, almost every cell of a living organism contains it. CoQ10 assists in driving the mitochondrial energy production vital for all body functions. The functioning of all organs depends on each cell having adequate levels of CoQ10 to provide life-sustaining energy.

CoQ10 is also a powerful antioxidant both on its own and in combination with vitamin E and is vital in the body's energy production (ATP) cycle. CoQ10 is found throughout the body in cell membranes, especially in the mitochondrial membranes and is particularly abundant in the heart, lungs, liver, kidneys, spleen, pancreas, and adrenal glands. The total body content of CoQ10 is only about 500–1500 mg and decreases with age.[90]

Like other vitamins discovered in the early part of the twentieth century, CoQ10 is an essential element of food that can now be used medicinally to support the sick host in conditions where nutritional depletion and cellular dysfunction occur. It would seem that the combination of disease attacking

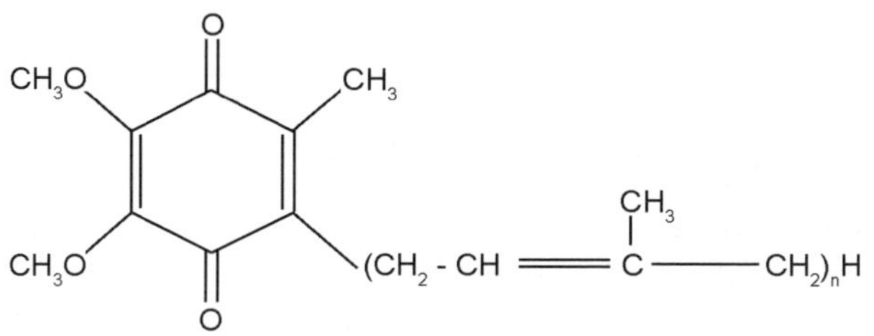

Fig. 6.13 Molecular structures of CoQ_{10}.

strategy and host supportive treatments would yield much better results in clinical medicine.

Since CoQ10 is essential to the optimal function of all cell types, it is not surprising to find a seemingly diverse number of disease states which respond favorably to CoQ10 supplementation. One of the disease states which has received attention is cancer. Low levels of CoQ10 in the blood of some cancer patients have been noted,[59] but overall, there is little data regarding cancer. All metabolically active tissues are highly sensitive to a deficiency of CoQ10. CoQ10's function as a free radical scavenger only adds to the protean manifestations of CoQ10 deficiency. Preliminary observations in a wide variety of disease states have already been published.[91–95]

The antioxidant or free radical quenching properties of CoQ10 serve to greatly reduce oxidative damage to tissue as well as significantly inhibit the oxidation of LDL cholesterol (much more efficiently than vitamin E).[96,97]

Being a lipophilic compound, CoQ10 is not soluble in water and therefore cannot be used in soft clear drinks. In addition, it was found that its lipophilicity slows down its ability to be absorbed from the guts into the blood stream. It was also claimed and demonstrated in previous studies that once nutraceuticals are added to our diet, the bioavailability of the lipophilic vitamins was damaged since the nutraceuticals compete with the vitamins for the solubilization locus within the giant micelles in the digestive tracts and thus less vitamin will be absorbed.

Attempts were made by several investigators to formulate the CoQ10 in a powdered form do be dispersed in water as cold water dispersible material and to incorporate it into emulsions. Yet, the stability and bioavailability was very limited.

In order to form microemulsions with large isotropic regions where we can optimally solubilize CoQ10, we have built phase diagrams with several nonionic surfactants and cosolvents.[98] After many attempts using microemulsions with a large isotropic region, a large solubilization capacity was obtained. The stability of the concentrate in different temperatures and the system's aqueous progressive dilution ability were tested. In the concentrate up to 8 wt% of CoQ10 was solubilized and it can be further diluted to 95 wt% water without any phase separation.[89] Physical properties including viscosity and thermal stability were tested. The microstructure of the system was tested with SAXS and SD-NMR and the boundaries at which transition from bicontinuous to o/w droplets occurs was determined.

6.4 Bioavailability of nutraceuticals

When individuals are given the same amount of drug or nutraceutical, the degree of physiological/biochemical response may not be the same. Degree of duration may vary greatly. The physical form of the drug/food may lead to partial absorption of the nutrients depending on its physical formulation.

Some scientists will define bioavailability as the proportion of a drug or nutrient that is digested, adsorbed, and utilized in normal metabolism. However, a more practical definition is usually related to estimates of amounts of nutrient absorbed into the serum. Some scientists have also coined 'bioactivity' as a term of significance which includes the journey of a compound from food to its functional site. The pharma definition of bioavailability is a working concept, meaning the rate and extent to which a drug reacts in its site of action. This parameter is difficult to follow because of ethnic differences between individuals. A pharmacological definition calls for the fraction of an oral dose from a particular preparation that trenches the systemic circulation.[99] The most accurate definition calls for bioavailability as a sum of events that includes absorption and bioactivity or tissue distribution and the functional consequences of absorption.[98]

Bioavailability will be defined in this review as 'the fraction of an ingested nutrient that is available for utilization in normal physiological functions or for storage', or 'the amount of drug transported through the digestive membranes to the serum'. Published information on nutraceutical (i.e. carotenoids, vitamins) bioavailability is based mainly on measurement of the nutrient in serum or plasma after investigation.[100,101] In the small intestine (Fig. 6.14), the nutraceutical is associated with bile salts where it is solubilized and absorbed at the brush border membrane of the mucosa by passive diffusion, along with other lipids in the micelles formed there. It is then incorporated into lipoprotein known as chylomicron and secreted into the intestinal lymph on its way to the different organs and tissues via the bloodstream.[102]

Factors that may interfere with the rate of each of the absorption steps will affect the overall bioavailability of the nutrient.

- Molecular configuration (*cis*/*trans*) of the nutraceutical molecules: the bioavailability of *cis*-isomer is higher than the bioavailability of the *trans*-isomer. This may result from the greater solubility of *cis*-isomers in mixed micelles and lower tendency of *cis*-isomers to aggregate.[100,101]
- Particle size,[103] e.g. tomatoes converted into tomato paste enhance the bioavailability of lycopene, as the processing includes particle size reduction and heat treatment.
- The matrix in which the nutraceutical is incorporated strongly affects its bioavailability. For example, the relative bioavailability of lutein appears to be lower from green leafy vegetables (45% from spinach) than from other vegetables (67% from a variety of vegetables).[102] It may be attributed to its entrapment and complexing to proteins in chloroplasts and within cell structures.
- Studies have shown that intake together with vegetable oils improved bioavailability.[102,104] As the bile salts are required in the nutraceuticals' absorption, the nutraceutical (such as lycopene, lutein and vitamin E) that are administered as a dietary supplement should preferably be taken with food to induce bile secretion,[105] and not on an empty stomach.

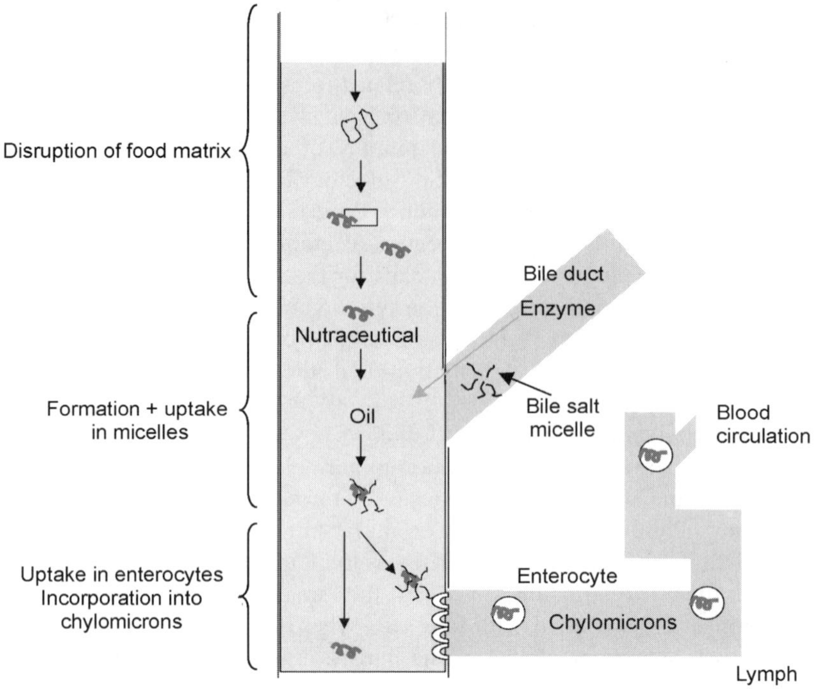

Fig. 6.14 Steps of oil-soluble nutraceutical absorption.

Figure 6.15 shows the bioavailability results that were obtained from our preliminary study carried out in Japan on seven volunteers. We found that the uptake of vitamin E was not damaged as a result of introducing the CoQ10 in the nanovehicles. As can be seen, the CoQ10 that was taken in our formulations was always more bioavailable than the commercial CoQ10 in a liquid form that is claimed by the producers to be 275% more bioavailable. It indicates that the fact that CoQ10 is solubilized at the o/w nano droplets interface allows obtaining a very stable and clear water-dilutable formulation with very high solubilization capacity and very significant bioavailability advantages.

It is therefore, important to provide new possible liquid vehicles for improved bioavailability and for possible incorporation of the nutraceutical into aqueous-based drinks or foods.

6.5 Microemulsions as microreactors

There has been a constant interest in exploiting self-assembled systems such as micellar solutions, microemulsions, lipid-bilayers, etc., as reaction media.[106–108] Among these systems, microemulsions have emerged as a popular choice for a number of diverse applications.[109] The microemulsion droplets can be considered as swollen nanoreactors that provide very small domains within which

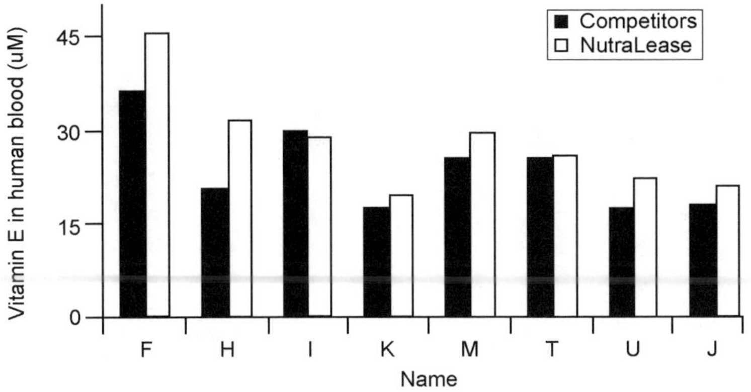

Fig. 6.15 Bioavailability test of uptake of vitamin E in nanovehicles (NutraLease) in comparison to others.

certain chemical reactions, including organic synthesis, formation of inorganic particles, polymerization, bioorganic synthesis, and electrochemical catalysis, take place.[109,110] In preparative organic chemistry, the constrained environments, the reactant compartmentalization, and the large interfacial area per unit volume of these 'organized structured' fluids allow the possibility of controlling the reaction, pathways, rates, yields, and regioselectivity.[109–113] This effect is called 'microemulsion catalysis'.[106,107]

The Maillard model system involving ribose (or furfural) and cysteine has been used widely in the generation of sulfur-containing flavors.[114–117] These flavors play an important role in contributing meaty, savory, roast flavors to cooked and roasted foods, due to their low odor thresholds.[114–117] The Maillard reaction-based flavors have been detected in many different thermally processed foods as well as in model Maillard reaction systems.[114–117] The best-known examples of these model compounds produced by furfural (or ribose)-cysteine interactions are 2-furfurylthiol (FFT) and 2-methyl-3-furanthiol (MFT).

Vauthey *et al.*[118] described the application of two-component structured fluids (cubic phases) as microreactors for the thermal generation of flavors from cysteine/furfural and cysteine/ribose model systems. The cubic phases led to the formation of two new sulfur-containing flavors. Similarly, generation of 2-furfurylthiol (FFT) and 2-methyl-3-furanthiol (MFT) was strongly enhanced in cubic phases (Fig. 6.16).

In our previous reports, we have examined the potential of two complex food-grade w/o and o/w microemulsion systems in the Maillard process of furfural-cysteine and have studied the activity of the two different curvatures on the regioselectivity and kinetics of the reactions.[119,120] These systems included oil and other ingredients that can participate and partition at the interface (short- and medium-chain alcohols) along with cosolvents such as polyols (water-soluble polyol for modifying the nature of the water phase). We found that the Maillard reactions occur at lower temperatures than in water-based systems and advance much faster. The interfaces of both w/o and o/w microemulsions are

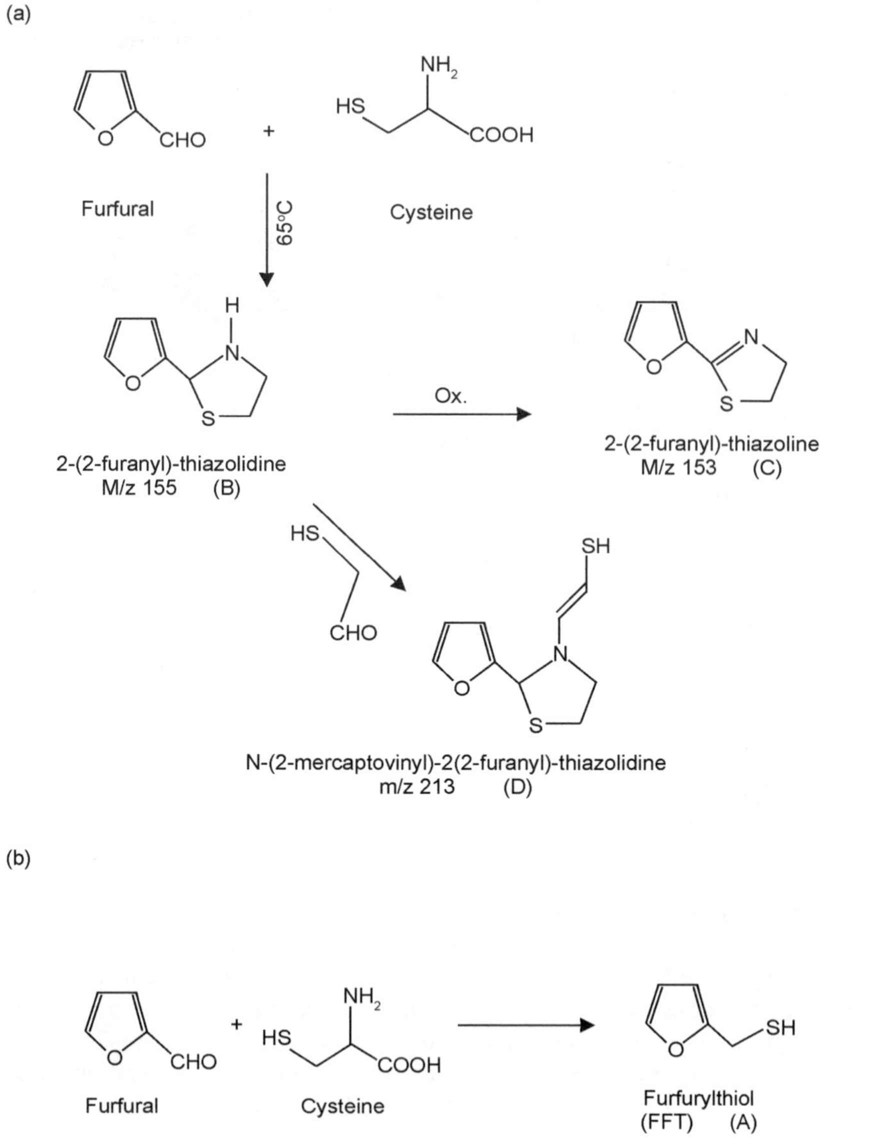

Fig. 6.16 Formation of sulfur-containing flavors in microemulsion reaction system (a) and in water (b).

capable of enhancing the Maillard reactions in which the selectivity and the reactivity are controlled by the composition of the interface and its curvature.[119,120] In the w/o microemulsions, the Maillard reaction was controlled and enhanced by the interfacial concentration of the coemulsifier such as butanol and are restricted by the concentration of the core water reservoir.[120] In contrast, in the o/w microemulsions, where the water is the continuous phase, the reaction

Table 6.2 Pseudo-first-order rate constants, k_{obs}, and linear regression (r^2) for water and o/w microemulsions

Temperature	Microemulsion (o/w) k_{obs} $(hr^{-1}) \times 10^2$	r^2	Water k_{obs} $(hr^{-1}) \times 10^2$	r^2	Aqueous solution k_{obs} $(hr^{-1}) \times 10^2$
40	20.1	0.99	3.2	0.97	1.7
50	29.1	0.97	6.7	0.96	4.7
65	48.6	0.98	12.6	0.94	9.3
70	59.8	0.97	16.0	0.95	12.2

rates will be enhanced by the increase in the water content and the microemulsion curvature.[119]

In another study we extended our knowledge on the reactivity of the five-component microemulsion systems and have investigated the influence of the following effects: temperature, pH, alcohol chain length, and the surfactant/oil weight ratio on the generation of sulfur-containing flavors from the model furfural-cysteine reaction.[121] In addition, a SD-NMR study on the dynamics of the reactants in the microemulsion systems was studied. Maillard furfural/cysteine reaction was chosen for exploring and understanding the different aspects that control the reactivity in five-component fully dilutable U-type food-grade oil/water (o/w) microemulsions of R(+)-limonene/ethanol/water/propylene glycol/Tween 60. It was found that these self-assembled nanosized structured systems are capable of enhancing the rate of Maillard reactions in which the selectivity and the reactivity are controlled by the composition of the interface and its curvature (Table 6.2). The sulfur-containing flavor formation is very efficient if carried out in those o/w microemulsions (modified oil-swollen nanoreactors) and the reaction yields high amounts of key aroma compounds. The proposed Tween 60-based o/w systems shows strong preference for the formation of 2-(2-furanyl)-thiazolidine (the main product) (Fig. 6.17). The o/w reaction medium enhances furfural-cysteine reaction rate (microemulsion catalysis) in comparison with the water-based reaction.[121]

It was found that the initial reaction rate (V_0) is affected by the solubilized aqueous phase content. At the w/o region (region I) where the two reagents are in the inner phase and remain totally immobilized (water activity is zero) the initial reaction rate is small and constant. In the bicontinuous region (region II), where water and oil form a bicontinuous water and oil domain, the reaction rate increases gradually since the mobility of the reactants increases, and once the reagents are located at the external interface in the o/w interface (region III) the reaction rate increases significantly (Fig. 6.18). The furfural-cysteine reaction is also affected by the process conditions (temperature, pH, time of reaction).

Similar reactions of glucose and cysteine in o/w microemulsions were also studied and the regioselectivity and the reaction rate were found similarly to be strongly affected by the nature of the interface and the mobility of the reactants at the interface.[121,122]

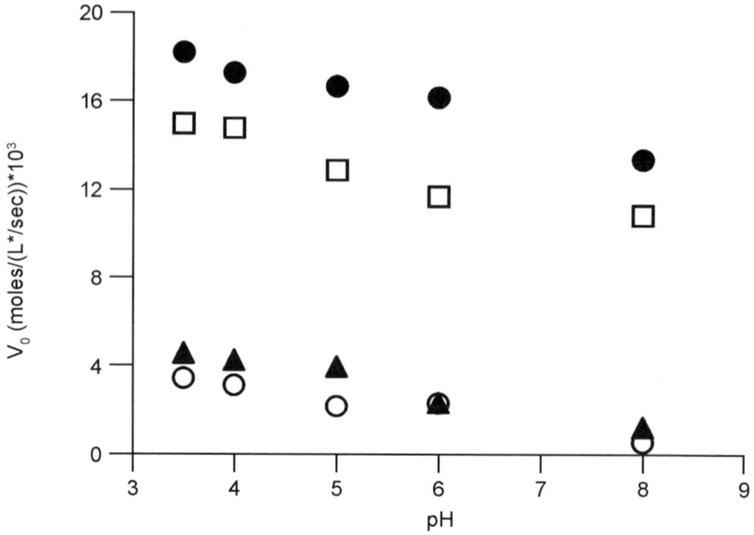

Fig. 6.17 Initial velocity (rate) of the disappearance of furfural (V_0) as function of the value of pH in two Tween 60-based O/W microemulsions, water (▲), and aqueous binary system (○, water/PG at a constant weight ratio of 1/1). The microemulsion systems composed of buffered water/PG/R(+)-limonene/EtOH/Tween 60 and contained 75 (□), and 90 wt% (●) aqueous phase (buffered water/PG with a weight ratio of 2/1). The reaction was carried out at 65 °C.

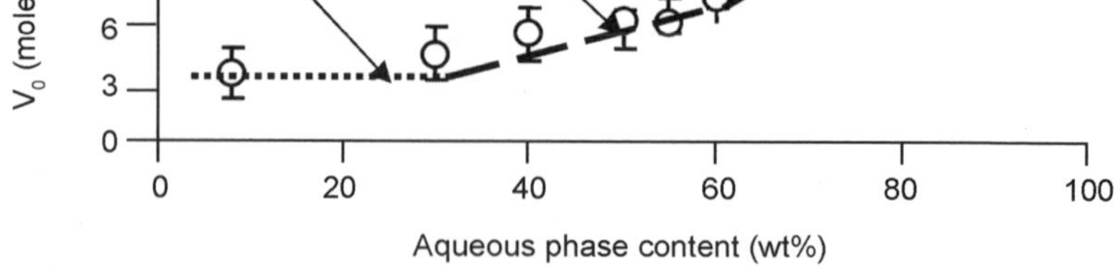

Fig. 6.18 Initial velocity (rate) of the disappearance of furfural (V_0) as function of aqueous phase content in microemulsion system containing buffered water (pH 5.0)/PG/R(+)-limonene/EtOH/Tween 60 along a dilution line with the same composition of the dilution line T64 (the Tween 60/R(+)-limonene/EtOH at a constant weight ratio of (3/1/1)). In these systems, the aqueous phase is water/PG with a weight ratio of 3/1 and its contents varying from 10 to 90 wt%). The reaction was carried out at 65 °C.

These results suggest that o/w microemulsions are a promising choice for generation of flavors and food products.

6.6 Conclusions

The aim of this review was to demonstrate that modified microemulsions can serve as new vehicles for the solubilization and transport of various bioactive molecules into food systems as well as regioselective microreactors, and as vehicles to deliver bioactives across the digestive membrane to the blood stream. Such microemulsions can help to improve the activity and biological efficacy of nutraceuticals.

In order to be able to utilize these systems in food systems we have to learn how to prepare them both from essential oils and vegetable oils and to be able to dilute them progressively with aqueous phase. The solubilization capacity must be very high and therefore the nutraceuticals must be solubilized at the oil phase and at the interface without precipitating out.

Advanced new techniques must be applied in order to study the nature of the microstructures, phase transitions upon dilution and the locus of the solubilized guest molecule at the interface or within the inner core of the microstructure. Many parameters play a significant role in stabilizing the loaded microemulsions and quantitative considerations related to nucleation growth and interpenetration of the emulsions must be quantified in order to be able to correlate the nature of the ingredients that are composing the microemulsion and its solubilization capacity at the concentrate and along the dilution line.

It seems that we still have long way to go in order to better understand how to control the structures of microemulsions and how to enhance their solubilization capacities for the inner phase as well as for any guest molecule. Although the microemulsions seem to be relatively easy to make, nanostructures require deeper understanding of the nature of the film, its curvature, and elasticity along with geometrical and packing parameters and entropy considerations. The complexity of microemulsions has yet to be disentangled.

6.7 References

1. BIDLACK WR, WANG W. In: Bidlack WR, Omaye ST, Meskin MS, Topham DKM, Eds, *Phytochemicals as Bioactive Agents*. Lancaster: Technomic Publishing Co, 2000.
2. BIDLACK WR, OMAYE ST, MESKIN MS, JAHNER D, *Phytochemicals: A New Paradigm*. Lancaster: Technomic Publishing Co, 1998.
3. CHERUVANKY R. In: Bidlack WR, Omaye ST, Meskin MS, Topham DKM, Eds. *Phytochemicals as Bioactive Agents*. Lancaster: Technomic Publishing Co, 2000.
4. DEFELICE SL, *Trends in Food Science and Technol.*, 1995, 6, 59–61.
5. DILLARD CJ, GERMAN JB, *J. Sci. Food Agric.*, 2000, 80, 1744–1756.
6. WEISS J, TAKHISTOV P, MCCLEMENTS J, *J. Food Sci.*, 2006, 71, R107–R116.

7. HOAR TP, SCHULMAN JH, *Nature*, 1943, 152, 102–103.
8. SOLANS C, PON R, KUNIEDA H, In: Solans C, Kunieda H, Eds, *Industrial Applications of Microemulsions*. New York: Marcel Dekker, 1997.
9. ZANA R, *Heterogeneous Chem. Rev.*, 1994, 1, 145–157.
10. GARTI N., ASERIN A, In Benita, S. *Microencapsulation, Methods and Industrial Applications*, New York: Marcel Dekker, 1996, 411–534.
11. SALAGER JL, In Françoise N, Gilberte M, *Pharmaceutical Emulsions and Suspensions*, New York, Marcel Dekker, 2000, 73–125.
12. DAICIC J, OLSSON U, WENNERSTROM H, *Langmuir*, 1995, 11, 2451–2459.
13. AZEMAR N, In: Solans C, Kunieda H, Eds, *Industrial Applications of Microemulsions*. New York: Marcel Dekker, 1997, 375–388.
14. SALAGER JL, In: Broze G, Ed., *Handbook of Detergents. Part A: Properties*, New York, Marcel Dekker, 1999, 253–302.
15. GASCO MR, In: Solans C, Kunieda H, Eds, *Industrial Applications of Microemulsions*. New York: Marcel Dekker, 1997, 97–121.
16. VAN CORSWANT C, SODERMAN O, *Langmuir*, 1998, 14, 3506–3511.
17. NAKAJIMA H, In: Solans C, Kunieda H, Eds, *Industrial Applications of Microemulsions*. New York: Marcel Dekker, 1997, 175–198.
18. FRIBERG SE, YANG J, Surfactants in cosmetics, in: Rieger MM, Rhein LD, (Eds.). *Surfactant Science Series*, 2nd edition 1997, 225–236
19. TADROS TF, In: Tadros TF, Ed., *Surfactants in Agrochemicals*. New York: Marcel Dekker, 1995, 183–205.
20. TADROS TF, In: Solans C, Kunieda H, Eds, *Industrial Applications of Microemulsions*. New York: Marcel Dekker, 1997, 199–208
21. LAWRENCE MJ, REES GD, *Adv. Drug Delivery Rev.*, 2000, 45, 89–121.
22. SHINODA K, FRIBERG S, *Emulsions and Solubilisation*, New York: John Wiley & Sons, 1986.
23. AVEYARD R, BINKS BP, MEAD J, *J. Chem. Soc. Faraday Trans.*, 1987, 83, 2347–2357.
24. AVEYARD R, BINKS BP, CLARK S, FLETCHER PDI, *J. Chem. Tech. Biotechnol.*, 1992, 54, 231–236.
25. DUNN RO, SCHWAB AW, BAGBY MO, *J. Dispersion Sci. Technol.*, 1993, 14, 1–16.
26. FRIBERG S, *J. Am. Oil Chem. Soc.*, 1971, 48, 578.
27. ALANDER J, WARNHEIN W, *J. Am. Oil Chem. Soc.*, 1989, 66, 1656–1660.
28. SCHWAB AW, NIELSEN HC, BROOKS DD, PRYDE EH, *J. Dispersion Sci. Technol.*, 1983, 4, 1–17.
29. PARRIS N, JOUBRAN RF, LU DP, *J. Agric. Food Chem.*, 1994, 42, 1295–1299.
30. JOUBRAN RF, CORNELL DG, PARRIS N, *Colloid Surf. A: Physicochem. Eng. Aspects*, 1993, 80, 153–160.
31. LINDSROM M, LJUSBERG-WAHREN H, LARSSON K, *Lipids*, 1981, 16, 749–754.
32. WOLF PA, HAVEKOTTE MJ, US Patent 4835002, 1989.
33. EL-NOKALY M, HILER G, MCGRADY J, In: El-Nokaly M, Cornell D, Eds, *Microemulsions and Emulsions in Foods*, American Chemical Society, 1991.
34. LESER ME, VANEVERT WC, AGTEROF WGM, *Colloids Surf. A*, 1996, 116, 293–308.
35. VAN CORSWANT C, ENGSTRöM S, SöDERMAN O, *Langmuir*, 1997, 13, 5061–5070.
36. ABOOFAZELI R, PATEL N, THOMAS M, LAWRENCE MJ, *Int. J. Pharm.*, 1995, 125, 107–116.
37. CONSTANTINIDES PP, SCALART JP, *Int. J. Pharm.*, 1997, 158, 57–68.
38. GRACIAA A, LACHAISE J, CUCUPHAT C, BOURREL M, SALAGER JL, *Langmuir*, 1993, 9, 669–672.

39. GRACIAA A, LACHAISE J, CUCUPHAT C, BOURREL M, SALAGER JL, *Langmuir*, 1993, 9, 3371–3374.
40. ACOSTA E, MAI PD, HARWELL JH, SABATINI DA, *J. Surfact. Deterg.*, 2003, 6, 1–12.
41. ACOSTA E, UCHIYAMA DA, SABATINI DA, HARWELL JH, *J. Surfact. Deterg.*, 2002, 5, 151–157.
42. ACOSTA E, TRAN S, UCHIYAMA DA, SABATINI DA, HARWELL JH, *Environ. Sci. Technol.*, 2002, 36, 4618–4624.
43. KOMESVARAKUL N, SANDRES MD, SZEKERES E, ACOSTA EJ, FALLER JF, MENTLIK T, FISHER LB, NICOLL G, SABATINI DA, SCAMEHORN JF, *J. Cosmet. Sci.*, 2006, 55, 309–325.
44. GARTI N, ASERIN A, SPERNATH A, AMAR I, US Patent 7182950, 2007.
45. BORDEN RC, MICHAEL DL, US Patent 6398960, 2002.
46. VAUTHEY S, LESER M, CHRISTIAN M, US Patent 6562391, 2003.
47. LUTZ R, ASERIN A, GARTI N, *J. Dispersion Sci. Technol.*, 2005, 26, 535–547.
48. YAGHMUR A, ASERIN A, ABBAS A, GARTI N, *Colloids and Surf. A*, 2005, 253, 223–234.
49. GARTI N, *Curr. Opin. Colloid & Interface Science*, 8, 2003, 197–211.
50. GARTI N, YAGHMUR A, *Colloids and Surfaces A*, 2002, 209, 71–81.
51. GARTI N,. YAGHMUR A, LESER ME, CLEMENT V, WATZKE HJ, *J. Agric. Food Chem.*, 2001, 49, 2552–2562.
52. SHIAO SY, CHHABRA V, PATIST A, FREE ML, HUIBERS PDT, GREGORY A, PATEL S, SHAH DO, *Colloid. Interface Sci.*, 1998, 74, 1–29.
53. TOKUOKA Y, UCHIYAMA H, ABE, M CHRISTIAN SD, *Langmuir*, 1995, 11, 725–729.
54. KANEI N, TAMURA Y, KUNIEDA H, *J. Colloid Interface Sci.*, 1999, 218, 13–22.
55. DUNGAN SR, In: Solans C, Kunieda H, Eds, *Industrial Applications of Microemulsions*, New York: Marcel Dekker, 1997.
56. KAWAKAMI K, YOSHIKAWA T, MOROTO Y, KANAOKA E, TAKAHASHI K, NISHIHARA Y, MASUDA K, *J. Controlled Rel.*, 2002, 81, 65–74.
57. CONSTANTINIDES PP, *Pharm. Res.*, 1995, 12, 1561–1572.
58. SHAH NH, CARVAJAL MT, PATEL CI, INFELD MH, MALICK AW, *Int. J. Pharm.*, 1994, 106, 15–23.
59. KIM CK, CHO YJ, GAO ZG, *J. Controlled Rel.*, 2001, 70, 149–155.
60. ROSANO HL, US Patent 4146499, 1979.
61. EL-NOKALY M, HILER G, MCGRADY J, US Patent 5045337, 1991.
62. TABIBI E, SICILIANO AA, US Patent 5130122, 1992.
63. CHUNG SL, TAN CT, TUHILL, IM, SCHARPF LG, US Patent 5283056, 1994.
64. SLOCUM SA, KILARA A, NAGARAJAN R, In: Charalambous G, Ed., *Flavors and Off-flavors*, Amsterdam: Elsevier Science Publishers, 1989.
65. AMAR I, AERIN A, GARTI N, *J. Agric. Food Sci.*, 2003, 51, 4775–4781.
66. KRASOWSKA H, *Int. J. Pharm.*, 1979, 4, 89–97.
67. MORO ME, NOVILLOFERTRELL J, VELAZQUEZ MM, RODRIGUEZ LJ, *J. Pharm. Sci.*, 1991, 80, 459–468.
68. CHIU YC, YANG WL, *Colloids Surfaces*, 1992, 63, 311–322.
69. LIU ZL, HAN ZX, CHEN P, LIU YC, *Chem. Phys. Lipids*, 1990, 56, 73–80.
70. GARTI N, ZAKHARIA I, SPERNATH A, YAGHMUR A, ASERIN A, HOFFMAN RE, *Prog. Colloid Poly. Sci.*, Received.
71. BATEMAN NE, UCCELLINI DA, *J. Pharm. Pharmacol.*, 1984, 36, 461–464.
72. JULIANTO T, YUEN KH, NOOR AM, *Int. J. Pharm.*, 2000, 200, 53–57.
73. SPERNATH A, YAGHMUR A, ASERIN A, HOFFMAN RE, GARTI N, *J. Agr. Food Chem.*, 2003, 50, 6917–6922.
74. RAMLEY PM, *Phytochemistry*, 2000, 54, 233–236.

75. RAO AV, AGARWAL S, *Nutr. Res.*, 1999, 19, 305–323.
76. COOKE JP, *Am. J. Cardiol.*, 1998, 82, 43S–46S.
77. CHEN SJ, EVANS DF, NINHAM BW, MITCHELL DJ, BLUM FD, PICKUP S, *J. Phys. Chem.*, 1986, 90, 842–847.
78. HOU MJ, SHAH DO, *Langmuir*, 1987, 3, 1086–1096.
79. SPERNATH A, YAGHMUR A, ASERIN A, HOFFMAN RE, GARTI N, *J. Agr. Food Chem.*, 2003, 51, 2359–2364.
80. PIIRONEN V, LINDSAY DG, MIETTINEN TA, TOIVO J, LAMPI AM, *J. Sci. Food Agric.*, 2000, 80, 939–966.
81. IUPAC, International Union of Pure and Applied Chemistry and International Union of Biochemistry and Molecular Biology. Available at: http://www.chem.qmw.ac.uk/iupac/steroid/ (1989).
82. NES WR, In: Stumpf PK, Mudd BJ, Nes WR, Eds, *The Metabolism, Structure and Function of Plant Lipids*. New York: Plenum Press, 1987, 3–9.
83. BLOCH K, *J. Am. Oil Chem. Soc.*, 1988, 65, 1763–1766.
84. PELLETIER X, BELBRAOUET S, MIRABEL D, MORDRET F, PERRIN JL, PAGES X, DEBRY G, *Ann. Nutr. Metab.*, 1995, 39, 291–295.
85. JONES PJH, MacDOUGALL DE, NTANIOS F, VANSTONE CA, *Can. J. Physiol. Pharmac.*, 1997, 75, 217–227.
86. LANDRUM JT, BONE RA, *Arch. Biochem. Biophys.*, 2001, 385, 28–40.
87. JUNGHANS A, SIES H, STAHL W, *Arch. Biochem. Biophys.*, 2001, 392, 160–164.
88. AMAR I, ASERIN A, GARTI N, *Colloids and Surfaces B: Biointerfaces*, 2004, 33, 143–150.
89. CHAGI SHARABI, MSc Thesis, private communication.
90. MELLORS A, TAPPEL AL, *J. Biol. Chem.*, 1966, 241, 4353–4356.
91. HANSEN IL, *Res Comm Chem Pathol Pharmacol*, 1976, 14, 729–738.
92. IWAMOTO Y, WATANABE T, OKAMOTO H, OHATA N, FOLKERS K, In: Folkers K, Yamamura Y, Eds, *Biomedical and Clinical Aspects of Coenzyme Q10*, Amsterdam: Elsevier, 1981, 109–119.
93. FOLKERS K, LANGSJOEN PH, *Biochem. Biophys. Res. Com.*, 1988, 153, 888–896.
94. LANGSJOEN PH, FOLKERS K, RICHARDSON P, In: Folkers K, Littarru GP, Yamagami T, Eds., *Biomedical and Clinical Aspects of Coenzyme Q*, Amsterdam: Elsevier, 1991, 409–415.
95. CORTES EP, MOHINDER G, PATEL M, MUNDIA A, FOLKERS K, In: Folkers K, Yamamura Y Eds., *Biomedical and Clinical Aspects of Coenzyme Q*, Amsterdam: Elsevier, 1977, 267–273.
96. BOWRY VW, MOHR D, CLEARY J, STOCKER R, *J. Biol. Chem.*, 1995, 17, 5756–5763.
97. INGOLD KU, BOWRY VW, STOCKER R, WALLING C, *Proc. Natl. Acad. Sci.*, 1993, 90, 45–49.
98. Patent Application, 2007.
99. SCHUMANN K, CLASSEN HG, HAGES M, PRINZLANGENOHL R, PIETRZIK K, BIESALSKI HK, *Arzneimittel-Forschung/Drug Research*, 1997, 47, 369–380.
100. CASTENMILLE JJM, WEST CE, *Annu. Rev. Nutr.*, 1998, 118, 19–38.
101. BRAMLEY PM, ELMADFA I, KAFATOS A, KELLY FJ, MANIOS Y, ROXBOROUGH HE, SCHUCH W, SHEEHY PJA, WAGNER KH, *J. Sci. Food Agric.*, 2000, 80, 913–938.
102. VAN HET HOF KH, GÄRTNER C, WEST CE, TIJBURG LBM, *Int. J. Vitam. Nuter. Res.*, 1998, 68, 366–370.
103. PAIVA SAR, RUSSELL RM, *J. Am. Coll. Nutr.*, 1999, 18, 426–433.
104. VAN HET HOF KH, WEST CE, WESTSTRATE JA, HAUTVAST GAJ, *J. Nutrition*, 2000, 130, 503–506.
105. GALLO TORRES HE, *Lipids*, 1970, 5, 379–384.

106. BUNTON CA, In: Grätzel M, Kalyanasundaram K Eds., *Kinetics and Catalysis in Microheterogeneous Systems*, New York: Marcel Dekker, 1991, 13.
107. BUNTON CA, ROMSTED LS, In: Kumar P, Mittal KL Eds., *Handbook of Microemulsion Science and Technology*, New York: Marcel Dekker, 1999, 457.
108. GARTI N, HOLMBERG K, *Curr. Opin. Colloid Interface Sci.*, 2003, 8, 135–136.
109. SJÖBLOM J, LINDBERG R, FRIBERG SE, *Adv. Colloid Interface Sci.*, 1996, 65, 125–287.
110. HOLMBERG K, *Curr. Opin. Colloid Interface Sci.*, 2003, 8, 187–196.
111. CURRIE F, HOLMBERG K, WESTMAN G, *Colloids Surfaces A: Physicochem. Eng. Aspects*, 2001, 182, 321–327.
112. GARCÍA-RÍO L,, LEIS JR, REIGOSA C, *J. Phys. Chem. B*, 1997, 101, 5514–5520.
113. GUTFELT S, KIZLING J, HOLMBERG K, *Colloids Surfaces A: Physicochem. Eng. Aspects*, 1997, 128, 265–271.
114. MOTTRAM DS, *Food Chem.*, 1998, 62, 415–424.
115. HOFMANN T, SCHIEBERLE P, *J. Agric. Food Chem.*, 1995, 43, 2187–2194.
116. HOFMANN T, SCHIEBERLE P, *J. Agric. Food Chem.*, 1998, 46, 235–241.
117. HOFMANN T, SCHIEBERLE P, *Lebensm Z, Unters. Forsch. A*, 1998, 207, 229–236.
118. VAUTHEY S, MILO CH, FROSSARD PH, GARTI N, LESER ME, WATZKE HJ, *J. Agric. Food Chem.*, 2000, 48, 4808–4816.
119. YAGHMUR A, ASERIN A, GARTI N, *J. Agric. Food Chem.*, 2002, 50, 2878–2883.
120. FANUN M, LESER M, ASERIN A, GARTI N, *Colloids Surfaces A: Physicochem. Eng. Aspects*, 2001, 194, 175–187.
121. LUTZ R, ASERIN A, GARTI N, *J. Dispersion Sci. Technol.*, 2005, 26, 535–547.
122. GARTI N, SPERNATH A, ASERIN A, LUTZ R, *Soft Matter*, 2005, 1, 206–218.

7

Emulsion droplet interfacial engineering to deliver bioactive lipids into functional foods

L. A. Shaw, H. Faraji, T. Aoki, D. Djordjevic, D. J. McClements and E. A. Decker, University of Massachusetts, USA

7.1 Introduction

ω-3 Polyunsaturated fatty acids are an essential part of the diet since humans cannot synthesize these fatty acids. α-Linolenic acid (ALA, 18:3 n-3) is recognized as the essential omega-3 fatty acid because it is a potential substrate for the synthesis of all the other ω-3 fatty acids. However, humans are very inefficient at converting ALA to long chain ω-3 fatty acids (Francois *et al.*, 2003; Ismail, 2005), eicosapentaenoic acid (EPA, 20:5 n-3) and docosahexaenoic acid (DHA, C2:6 n-3) indicating that the diet is the most important source of these long chain ω-3 fatty acids. Over the past decade, the health benefits of dietary ω-3 fatty acids and in particular EPA and DHA have been widely published for diseases such as coronary heart disease, stroke, mental illness and immune response disorders (Simopoulos, 1991; Harris, 1997; Kris-Etherton *et al.*, 2003; Siddiqui *et al.*, 2004; Ismail, 2005; Macchia *et al.*, 2005). Despite the evidence supporting the health benefits of ω-3 fatty acids, the consumption of these important fatty acids is still very low. Therefore, new strategies are needed to increase the amounts of ω-3 fatty acids in the diet.

The major sources of dietary EPA and DHA are seafood and oils from marine or algal sources. Unfortunately for food processors, marine and algal oils are very unstable to oxidative rancidity, making it difficult to incorporate them into food products. Recently, studies have shown that it is possible to stabilize fish and marine oils by their incorporation into stable emulsion systems. This strategy utilizes emulsion droplet interfacial engineering to minimize interactions between the ω-3 oils and water-soluble prooxidants (Djordjevic *et al.*, 2004a,b). An example of this technique is to coat emulsion droplets with multiple layers of emulsifiers resulting in an emulsion with improved physical and chemical stability. These multiple layered interfacial membrane emulsions systems can be spray dried to further microencapsulate the oxidatively sensitive oil. The microencapsulated fish and marine oil powders are oxidatively more stable than their bulk oil counterparts and are suitable for food application.

Owing to the potential health benefits of ω-3 fatty acids, there are an ever-increasing number of fortified food products being introduced to the worldwide market. These products typically have the fortified ω-3 fatty acids dispersed into the product in some form of an emulsion. Examples include milk and dairy products, spreadable fats, salad dressings and meat products. However, challenges remain to produce a broader array of ω-3 fatty acid fortified foods with high consumer acceptance. In order to overcome these challenges, new strategies are needed to inhibit omega-3 fatty acid oxidation.

7.2 Dietary ω-3 fatty acids

Current diets of Western culture contain high amounts of saturated fatty acids and ω-6 fatty acids and a very low amount of ω-3 fatty acids. Our ancestors had diets containing high amounts of ω-3 fatty acids due to high intake of fruits, vegetables, whole grains and wild and free-range animals (Horrocks and Yeo, 1999; Shahidi and Finley, 2001; Siddiqui *et al.*, 2004). Wild and free-range animals contain more ω-3 fatty acids in comparison to farmed animals today that are fed diets high in protein and ω-6 fatty acids (e.g. linoleic acid). ω-6 Fatty acids in our current diets are commonly consumed in the form of vegetable oils, namely soybean, sunflower and safflower, while saturated fatty acids originate primary from animal products. As more ω-6 oils are being used in food products to replace saturated and *trans* fatty acids the dietary ω-6 to ω-3 fatty acids ratio has increased. Leaf and Weber (1987) estimated that humans living in the year 1800 BC had relatively the same amounts of ω-6 to ω-3 fatty acids in their diet, and had a lower incidence of coronary heart disease (CHD). By the 1980s, the ω-6: ω-3 ratio was approximately 11:1 during which time the rate of CHD had risen approximately 40-fold (Leaf and Weber, 1987). The dietary ratio of ω-6 to ω-3 is still on the rise and is now estimated to be approximately 20-30:1 (Siddiqui *et al.*, 2004). As more ω-6 fatty acids are being consumed, the ratio of ω-6: ω-3 seems to be too high to maintain optimal health benefits. This off balance can lead to alterations of several metabolic functions within the body in addition to coronary heart disease.

Humans and most mammals do not produce α-linolenic acid (ALA, 18:3 n-3) and linoleic acid (18:2 n-6) and therefore, dietary sources of these fatty acids are essential for health. These fatty acids play an important role in skin, brain and retina development and are important precursors for numerous enzymes (Innis, 1991; Horrocks and Yeo, 1999; Francois *et al.*, 2003; Siddiqui *et al.*, 2004). ω-6 and ω-3 Fatty acids function differently in the body due to the positioning of the first double bond from the methyl end of the carbon chain. α-Linolenic acid and linoleic acid can be manipulated within the body through elongation and desaturation enzymes from the liver to produce longer chain fatty acids including docosahexaenoic acid (DHA, 22:6 n-3), eicosapentaenoic acid (EPA, 20:5 n-3) and arachidonic acid (AA, 20:4 n-6) (Innis, 1991; Siddiqui *et al.*, 2004).

Even though the desaturation and elongation enzymes for conversion of ALA to EPA and DHA are found in humans, the conversion rate is limited. In a study completed by Pawlosky and coworkers (2001), the conversion rate of α-linolenic acid to EPA was only 0.2%. From the 0.2% of EPA produced, only 63% of that is converted into DPA (docosapentaenoic acid, 22:5, n-6), and only 37% of that is desaturated to DHA (Pawlosky *et al.*, 2001). This data strongly suggest that dietary α-linolenic acid is not converted into significant amounts of EPA and DHA. Since dietary EPA and DHA are thought to have greater bioactivity than ALA (Horrocks and Yeo, 1999; Ismail, 2005), it is necessary to include EPA and DHA in the diet to obtain maximal health benefits.

7.3 Dietary ω-3 fatty acids and health

There are several important molecular mechanisms by which ω-3 fatty acids can impact health. Dietary ω-3 fatty acids can affect the fatty acid composition of cellular membranes by displacing other less unsaturated fatty acids. Increasing membrane polyunsaturation can aid in increasing metabolic rates, therefore increasing the efficiency of energy consumption (Hulbert *et al.*, 2005). EPA and DHA incorporation into cell membranes has also been linked to improved membrane fluidity, ion channel function, enzyme regulations and gene expression, all qualities that are associated with normal cell function (Young and Conquor, 2005). Increasing EPA and DHA consumption will also decrease AA concentrations in the cell membranes of erythrocytes, neutrophils, monocytes, platelets and liver cells (Simopoulos, 1999). These increased concentrations of EPA and DHA into cellular membranes decreases platelet aggregation and vasoconstriction as well as inflammatory responses (Simopoulos, 1999). By altering the chemical and physical properties of cells, ω-3 fatty acids are able to positively impact disease. The following subsections will review the relationship between dietary omega-3 fatty acids and coronary heart disease, immune response disorders and infant development.

7.3.1 Coronary heart disease

According to the Centers for Disease Control (CDC), coronary heart disease is the leading cause of death in the United States, accounting for about 40% of all deaths. More than 70 million Americans live with some type of coronary heart disease. Research has provided strong evidence that consumption of DHA and

EPA at levels of greater than 1 g per day will provide cardio-protective benefits such as reduced plasma triacylglycerols and reduced risk of sudden cardiac death. Benefits of DHA and EPA consumption include: reduction of serum triglyceride (TG) levels (Harris, 2004), anti-inflammatory/anti-arrhythmic properties (Kris-Etherton *et al.*, 2003; Leaf *et al.*, 2003, 2005), reduction of smooth muscle proliferation leading to less chance of atherosclerosis (Horrocks and Yeo, 1999), and control of platelet aggregation (Horrocks and Yeo, 1999; Kris-Etherton *et al.*, 2003; Vanschoonbeek *et al.*, 2003).

ω-3 Fatty acids are known to be hypotriglyceridemic when consumed at approximately 4 g EPA and/or DHA per day according to human studies surveyed by Harris (1997). Studies have shown that consuming high amounts of EPA and DHA will decrease serum triacylglycerol concentrations by 25 to 30% (Harris, 1997). The primary mechanism of serum triacylglycerol reduction is through a decrease in triacylglycerol production within the body (Nestel, 2000). It has also been shown that while EPA and DHA decrease serum triacylglycerol levels, they increase low density lipoprotein (LDL) levels in hypotriglyceridemic patients by 5 to 10% (Harris, 1997; Nestel, 2000). The mechanisms for an increase in LDL are not completely understood. It has been suggested that the increased intake of ω-3 fatty acids results in the production of VLDL that is susceptible to lipoprotein and/or hepatic lipase type, leading to increases in conversion of VLDL and LDL (Roche, 2000).

Disorders of the rhythmic beating of the heart are called arrhythmias. If arrhythmias are not corrected they can lead to heart disease, stroke or sudden cardiac death. Sudden cardiac death occurs when patients die soon after the onset of the symptoms of a heart attack. ω-3 Fatty acids have anti-arrhythmic properties, meaning that they stabilize myocardium tissues and help the heart return to a rhythmic beating pattern. DHA and EPA therefore have been found to decrease mortality rates in post myocardial infarction patients due to arrhythmic problems (Kris-Etherton *et al.*, 2003; Leaf *et al.*, 2003). An Italian study with over 11 000 subjects showed that 1 g ω-3 fatty acids from fish oil per day decreased cardiovascular and sudden cardiac death risk by 30 and 45%, respectively (Macchia *et al.*, 2005).

7.3.2 Inflammation

Inflammatory disorders are connected with heart disease, rheumatoid arthritis, bowel diseases and asthma. EPA and DHA have been linked to decreased inflammatory problems in numerous diseases (Horrocks and Yeo, 1999; Siddiqui *et al.*, 2004). Anti-inflammatory mechanisms involve the ability of EPA and DHA to alter the ω-6 to ω-3 ratio, thereby decreasing the formation of eicosanoid metabolic products that cause high levels of inflammatory response. High amounts of AA in cell membranes will result in increased production of leukotrienes, thrombonane A2 and prostoglandins by lipoxygenase (LOX) and cyclooxygenase (COX) resulting in an elevated inflammatory response. When DHA and EPA are consumed, they displace AA in cellular membranes

(Simopoulos, 1991). Prostoglandins and leukotrienes formed from EPA and DHA by lipoxygenase and cyclooxygenase are structurally different and less bioactive therefore leading to less inflammatory response (Simopoulos, 2002; Ismail, 2005).

7.3.3 Infant health and lactating women

DHA is one of the major fatty acids in nervous tissues and the retina. Rapid developmental changes in the brain and eyes of infants therefore require large amounts of DHA. However, as with adults, infants have limited pathways to convert ALA to DHA. Therefore, it is important for infants to receive DHA from maternal or supplementary means. Human breast milk contains DHA. Breast-fed infants tend to have higher cognitive development and visual acuity than infants fed formula not supplemented with DHA (Fleith and Clandinin, 2005). A study completed by Francois and coworkers (2003) showed that supplementing lactating women with high ALA flaxseed oil did not increase DHA concentration in breast milk, presumably due to the low rate of ALA to DHA conversion. In addition, during breastfeeding women undergo a decrease in plasma DHA levels unless they are consuming dietary DHA (Innis, 2004; Lauritzen *et al.*, 2005). Therefore, it is important for lactating women to consume adequate amounts of DHA through food and/or supplemental means to provide optimum nutrition for the infant.

7.4 Potential for ω-3 fatty acids in functional foods

ω-3 Fatty acid consumption in an average Western diet is around 0.1 g/day (Siddiqui *et al.*, 2004). This is a very small amount considering that many nutritionists are recommending 0.5–1.0 g/day for maximal health benefits (Djordjevic *et al.*, 2004a,b). The American Heart Association suggests different ω-3 fatty acid intake levels for different populations. The AHA suggests that normal people without CHD should consume at least two servings of fatty fish a week (0.3–0.5 g ω-3 fatty acid per day). For people with CHD, 1 g per day of EPA and DHA is recommended, with sources mainly coming from fatty fish. For people trying to lower serum triglyceride levels, 2–4 g per day of EPA and DHA is recommended, mainly from supplemental sources.

ω-3 Fatty acids are found mainly in nuts, seeds and marine sources. The nut and seed sources of ω-3 fatty acids are exclusively found as ALA. Flaxseeds contain about 18.1% ALA, whereas flaxseed oil contains around 53% ALA. Walnuts and canola oils contain around 9% ALA (DeFilippis and Sperling, 2006). EPA and DHA are generally limited to seafood sources. ω-3 Fatty acids are not produced by fish, but originate instead from marine microorganisms that produce EPA and DHA (Harris, 2004). The ω-3 fatty acids from these microorganisms then move through the food chain and accumulate in fish. Total ω-3 levels in fish are very dependent on fat content with high fat fish containing higher EPA and DHA concentrations compared to fish with low fat contents. Some examples of fatty fish that typically contain higher amounts of EPA and DHA are salmon, herring, fresh tuna, swordfish and mackerel.

A problem with consuming large amounts of high fat marine fish is that these fish are often contaminated with methyl mercury. Methyl mercury is a problem if consumed in high amounts, especially for pregnant women and children, because it is neurotoxic. Methyl mercury, originating primarily from pollution, is initially incorporated into microorganisms. Small marine animals then consume the microorganisms and the methyl mercury moves up the food chain and accumulates in large predatory fish. Therefore, fish that contain the higher amounts of EPA and DHA often also contain high amounts of methyl mercury. The exception to this is aquaculture fish (e.g. salmon) because their diet does not contain methyl mercury. Due to high amounts of methyl mercury in fish, consumers are looking for alternative sources of EPA and DHA to obtain the health benefits without the health risk. EPA and DHA, fish and algae oil supplements are increasing in popularity. These supplements come from refined oils where methyl mercury is removed during the processing of the oil. Although supplements in correct dosage can provide a safe way of consuming EPA and DHA, many consumers would rather consume EPA and DHA directly from food rather than taking supplements.

Decreasing consumption of ω-3 fatty acids and the increasing evidence of the health benefits suggest that these bioactive lipids are ideal candidates for functional foods. Functional foods containing EPA and DHA would have many advantages because oil processes can provide ω-3 oils free of methyl mercury, unpleasant taste and odors. In addition, functional foods containing EPA and DHA could provide an ω-3 fatty acid source for people who dislike seafood or have difficulty obtaining high quality, low cost fish. Unfortunately, ω-3 oils are oxidatively unstable and their decomposition will lead to the development of fishy off-flavors that result in consumer rejection of the food product. The oxidation of ω-3 fatty acids is much quicker than that of mono or saturated fatty acid. For example, DHA (22:6) is about five times more oxidizable than linoleic acid (18:2) (Frankel, 1998). An additional problem with oxidation of omega-3 fatty acids is that the oxidation products produced can contain multiple double bonds to produce compounds that interact strongly with olfactory tissues resulting in very low sensory thresholds. This means that off-aromas and flavors can be detected very early in the oxidation pathway. Another contributing factor to rapid oxidation of fish and marine oils is that they are naturally low in antioxidants such as tocopherols. Because of the high susceptibility to oxidative deterioration of ω-3 fatty acids, not all foods are good candidates for fortification. Parameters for ideal conditions in foods suitable for ω-3 fortification include short shelf life, low storage temperature, limited oxygen and light exposure, low levels of endogenous prooxidants and high levels of endogenous antioxidants.

ω-3 Oils are commonly provided to food processors as bulk oil. Bulk oils can be a problem because they can be difficult to incorporate into foods. An additional problem is that the antioxidants used to stabilize bulk oils are not effective when the oils are dispersed into foods (see discussion below on polar antioxidant paradox). An alternative to the addition of bulk oils to foods is to add bioactive lipids in the form or an oil-in-water emulsion. An oil-in-water emulsion contains small spherical oil droplets dispersed into a continuous aqueous phase. In food emulsion systems, oil droplets typically range from 0.1–100 μm (McClements, 2005). Owing to the large free energy at the droplet surfaces, oil-in-water emulsions are very thermodynamically unstable, and thus cannot be produced without emulsifiers or surfactants that lower the interfacial tension between oil and water. An oil-in-water emulsion consists of three specific regions within emulsions: the lipid core of the emulsion droplet, the interfacial material that is the barrier between the aqueous phase and the oil, and the aqueous phase (McClements, 2005; McClements and Decker, 2000; Friberg *et al.*, 2004). The interfacial region of emulsion droplet is very important to the physical and chemical stability of the lipid as outlined below.

7.5 Mechanisms of lipid oxidation

Oxidation of unsaturated lipids occurs through free radical reactions. Lipid oxidation proceeds through three steps: initiation, propagation and termination.

7.5.1 Initiation

The initiation step of lipid oxidation begins with an abstraction of hydrogen (H^+) from the methylene interrupted carbon of a polyunsaturated fatty acid (LH) (Fig. 7.1). This hydrogen abstraction forms an alkyl free radical (L•) that is stabilized via dislocation of its energy through double bonds within the fatty acids resulting in rearrangement of the double bonds to form a conjugation system. Lipid free radicals are not formed through hydrogen abstraction by oxygen bi-radicals because it is thermodynamically difficult (Nawar, 1996). Therefore, free radical formation must occur through presence of an initiator, such as UV light, irradiation and metal ions.

As the bond dissociation energies of hydrogens adjacent to the double bonds of unsaturated fatty acids decrease, hydrogen abstraction becomes easier, resulting in faster initiation. Polyunsaturated fatty acids will oxidize much quicker than monounsaturated fatty acids. Polyunsaturated fatty acids have a methylene interrupted carbon that has lower hydrogen–carbon bond dissociation energy than monounsaturated fatty acids making hydrogen abstraction to form a free radical much easier. This difference in hydrogen–carbon bond dissociation energy makes polyunsaturated fatty acids over 10-fold more susceptible to oxidation than monounsaturated fatty acids. The relative oxidizability of polyunsaturated fatty acids will also depend upon the number of methylene interrupted carbons within the fatty acid chain. Oxidation rates will double with each additional methylene interrupted carbons (Frankel, 1998).

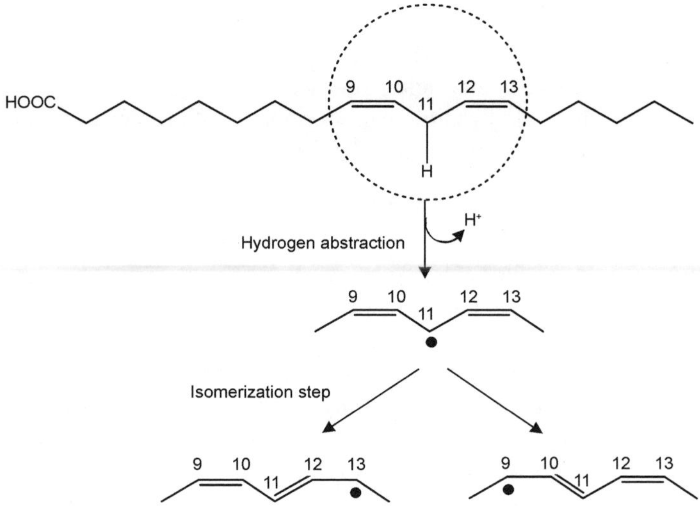

Fig. 7.1 The initiation step of lipid oxidation.

7.5.2 Propagation

The propagation step of lipid oxidation occurs upon the addition of oxygen (O_2) to an alkyl radical (L•) (Fig. 7.2). Since atmospheric oxygen is a bi-radical, this is a very rapid radical–radical reaction, which is limited only by diffusion rates. The result of this reaction is a highly reactive peroxyl radical (LOO•). Peroxyl

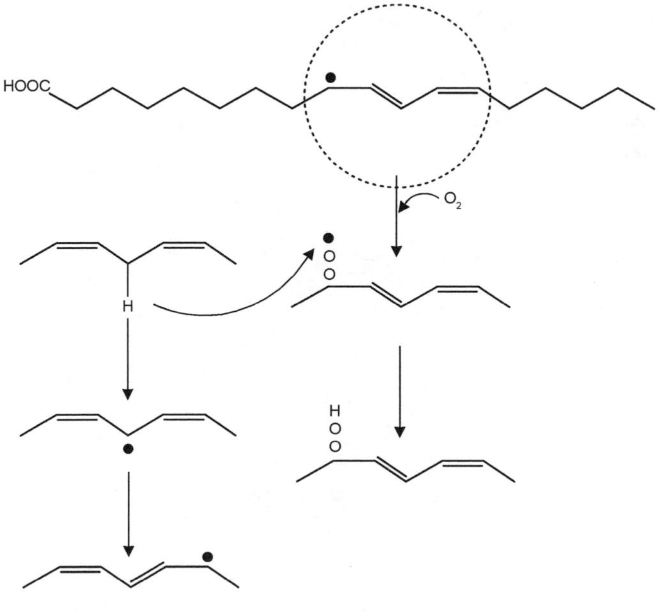

Fig. 7.2 The propagation step of lipid oxidation.

radicals can easily abstract hydrogen from methylene interrupted groups upon other fatty acid molecules (LH). This reaction results in hydrogen addition to the peroxyl radical, forming a fatty acid hydroperoxide (LOOH) and new alkyl free radical (L•). The new alkyl radical can then react with oxygen and continue a cyclic chain of self-propagating reactions (Nawar, 1996).

7.5.3 Termination

Free radical reactions can be halted through termination reactions (Fig. 7.3). These reactions occur when two radicals (e.g. LOO•, L•) collide with each other to form a non-radical product (Erickson, 2002).

Lipid hydroperoxides are one of the most important intermediate products of the lipid oxidation reaction. Lipid hydroperoxides are formed via the propagation step described above or via direct addition of oxygen to the fatty acids by singlet oxygen or lipoxygenases. Hydroperoxides are unstable and can decompose into free radicals especially in the presence of transition metals that are found in essentially every food source. Transition metals break down lipid hydroperoxides into alkoxy radicals through scission reactions at the oxygen–oxygen bond. Formation of alkoxyl radicals represents the formation of a second radical species that can attack unsaturated fatty acids (the first being the peroxyl), formation of the second radical resulting in an exponential increase in oxidation rates (Nawar, 1996; McClements and Decker, 2000; Reische *et al.*, 2002; Genot *et al.*, 2003). The alkoxyl radicals can also undergo β-scission reactions that break down the fatty acid chain through a complex set of reactions that yield volatile aldehydes, ketones, hydrocarbons and alcohols that change the sensory properties of oil.

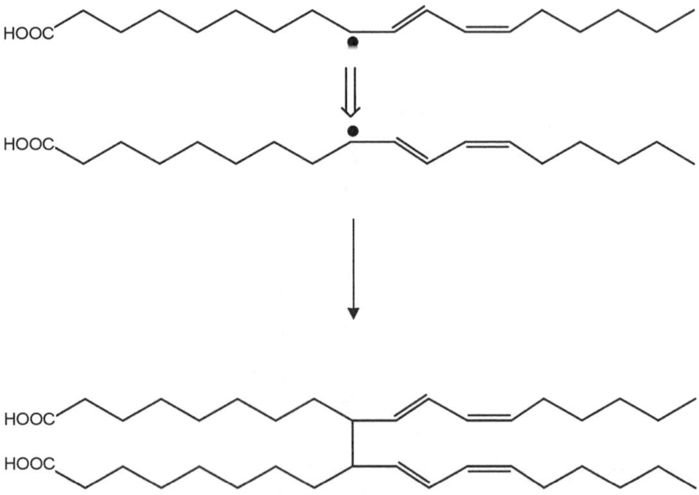

Fig. 7.3 The termination step of lipid oxidation.

7.6 Lipid oxidation in emulsions

Lipid hydroperoxides are surface active molecules and thus accumulate at the interface of an emulsion droplet (McClements and Decker, 2000). Lipid hydroperoxides at the emulsion droplet surface can interact with iron to produce free radicals, which in turn can interact with the oil within the droplet core. In emulsions, the rate of oxidation is therefore dependent upon the surface properties of the emulsion droplet which could alter the ability of iron to decompose hydroperoxides and the resulting free radicals to reach the lipid core (McClements and Decker, 2000). Understanding how the properties of the emulsion droplet surface impacts the interactions between iron and hydroperoxide as well as free radicals and lipids is important for manipulating emulsion systems to inhibit the oxidative deterioration of the lipids.

The ability of transition metals to break down lipid hydroperoxides is one of the main pathways promoting lipid oxidation in food emulsions. Iron is found in food emulsions at concentrations sufficient to promote lipid oxidation (Mancuso *et al.*, 1999a). Both ferric and ferrous states of iron can accelerate the breakdown of lipid hydroperoxides, leading to formation of radicals (Dunford, 1987; Mei *et al.*, 1998b). When transition metals are in close proximity to the unsaturated oil–water interface of the emulsion droplet, their ability to promote and accelerate oxidation increases (McClements and Decker, 2000). Therefore, one of the mechanisms by which chelators inhibit lipid oxidation is through their ability to physically remove metals from the emulsion droplet surface. In addition, chelators can bind metals and prevent their redox cycling thus decreasing their reactivity (Reische *et al.*, 2002). Ethylenediaminetetraacetic acid (EDTA) is a commonly used chelator in the food industry. This chelator will bind metals and form a thermodynamically stable complex. The addition of EDTA in oil-in-water emulsion systems will provide increased oxidative stability (Mei *et al.*, 1998b; Mancuso *et al.*, 1999b; Hu *et al.*, 2004a). Citric acid and polyphosphates also bind metals; however, they are not effective antioxidants in oil-in-water emulsions (Hu *et al.*, 2004a) most likely owing to their lower binding constants in comparison to EDTA. Proteins and peptides can also be effective metal chelators (Diaz *et al.*, 2003; Faraji *et al.*, 2004).

Lipid oxidation can also be inhibited by free radical scavengers (FRS). FRS inhibits lipid oxidation by having a liable hydrogen that can donate an electron to a free radical before the free radical can oxidize unsaturated fatty acids. Effective FRS can then decrease the energy of the radical by resonance delocalization so that the resulting free radical scavenger radical is unable to oxidize unsaturated fatty acids. Antioxidant effectiveness is not only dependent on the chemical properties of the free radical scavenger but also on the physical nature of the lipid (Porter, 1993; Frankel, 1998). Polar antioxidant paradox is a term that describes the observation that hydrophilic antioxidants are often less effective in oil-in-water emulsions than lipophilic antioxidants, whereas lipophilic antioxidants are less effective in bulk oils than hydrophilic antioxidants (Porter, 1993; Frankel, 1998). Hydrophilic antioxidants are good free radical scavengers in bulk oils because they accumulate at the air–oil interface while lipophilic antioxidants are more effective in emulsions because they are retained in the oil droplets. The tocopherol homologs are one of the most effective lipophilic antioxidants in oil-in-water emulsions. Many hydrophilic antioxidants can increase lipid oxidation rates in oil-in-water emulsions presumably through their ability to reduce iron to its more prooxidative ferrous state. Proanthocyanidins in grape seed extract are one of the few groups of water-soluble antioxidants found to be effective in oil-in-water emulsions (Hu *et al.*, 2004b). This could be because of their ability to both scavenge free radicals and chelate prooxidant metals.

7.7 Interfacial engineering

As discussed above, lipid oxidation in oil-in-water emulsions is influenced by properties of the interfacial membrane region of emulsion droplets, since several important oxidative interactions occur at the surface of the emulsion droplet. Certain oils, such as fish and marine oils are very difficult to stabilize through traditional antioxidant technologies. Manipulation of the emulsion droplet interfacial region could provide an additional technique to increase protection for these highly sensitive oils. Engineering of the interfacial region upon oil droplets may not only provide an effective way to protect oils from oxidation but could also allow for the product of a food ingredient (oil-in-water emulsion) that could be easily incorporated into foods.

Research has shown that a thick interfacial membrane on emulsion droplets may act as a physical barrier to decrease the interaction between prooxidative species (e.g. aqueous phase metals) with lipids in the emulsion droplet core. Silvestre and coworkers (2000) completed studies on oxidative stability of salmon oil-in-water emulsions stabilized by polyoxyethylene (10) stearyl ether (Brij 76) and polyoxyethylene (100) stearyl ether (Brij 700), both non-ionic surfactants with the same nonpolar tail group. These surfactants were tested because of the size of their polar head groups, with Brij 700 containing ten times more polyoxyethylene groups than Brij 76, thus producing large differences in the size of the hydrophilic head groups at the interfacial membrane. Emulsions were prepared with similar droplet size distributions to minimize the effect of surface area on oxidation kinetics. Lipid hydroperoxides decomposed faster and salmon oil oxidized sooner in oil-in-water emulsions stabilized with Brij 76 compared to the Brij 700 stabilized emulsions. Chaiyasit *et al.* (2000) found that increasing the length of the surfactant tail's group from 12 to 16 carbons decreased lipid oxidation in oil-in-water emulsions slightly. The thickness of the interfacial membrane may also explain differences in the oxidative stability of oil-in-water emulsions stabilized by proteins where casein, which forms an interfacial membrane thickness of approximately 10 nm, decreases lipid oxidation more effectively than whey protein isolate which produces an interfacial thickness of 1–2 nm (Mackie *et al.*, 1991; Dalgleish, 1993; Fang and

Dalgleish, 1993; Hu *et al.*, 2003). This research suggests that the thickness of the interfacial layer of oil-in-water emulsion droplets could inhibit lipid oxidation by sterically hindering interactions between the lipids in the emulsion droplet with prooxidants in the continuous phase.

The electrical charge of emulsion droplets will depend upon the type and characteristics of the emulsifier or surfactant that is adsorbed to the interface of the oil droplet and environmental factors such as pH and ionic strength. Emulsifiers and surfactants that can produce charged emulsion droplets in foods include proteins, polysaccharides and certain small molecule surfactants such as lecithin (McClements, 2005). Changing the pH of oil-in-water emulsions containing proteins adsorbed to the interfacial region of the droplets will change the overall charge of the emulsion droplets producing a cationic charge at pH values below the pI and an anionic charge at pH values above the pI of the protein (Donnelly *et al.*, 1998; Mancuso *et al.*, 1999b).

Physical stability can be dependent on the charge density of an emulsion droplet since droplets with like charges will repel each other and decrease collision, which cause flocculation or coalescence (McClements, 2005). Electrically charged droplet surfaces will also impact the ability of the droplet to interact with ions in the aqueous phase of the emulsion. Interactions between emulsion droplets with aqueous phase ions can impact both the physical and chemical stability of emulsions. Counter ions can decrease physical stability by shielding the surface charge or by causing bridging flocculation. Electrostatic interactions between prooxidative ions (iron, copper) and the emulsion droplets can also impact the chemical stability of an emulsion (McClements and Decker, 2000).

Mei *et al.* (1998a) found that the oxidative stability of sodium dodecyl sulfate (SDS, anionic)-, polyoxyethylene(23) lauryl ether (Brij 35, nonionic)- or dodecyltrimethylammonium bromide (DTAB, cationic)-stabilized corn oil-in-water emulsions in the presence of iron was in the order DTAB > Brij 35 > SDS. Donnelly *et al.* (1998) investigated the oxidative stability of menhaden (fish oil) oil-in-water emulsion stabilized with whey protein isolate as a function of pH in the presence of added iron. When whey protein isolate (WPI)-stabilized emulsion droplets were at pH values below the pI of WPI ($\approx$ 5.1), the emulsion droplets were cationic compared to pH values above 5.1 where the droplets were anionic. Under these conditions, the oxidative stability of the cationic droplet (pH 3 and 4) was much greater than the anionic droplets (pH 6 and 7). These observations suggest that cationic emulsion droplets are able to electrically repel iron and decrease lipid–iron interactions thus inhibiting lipid oxidation since iron is such an important prooxidant in oil-in-water emulsion. The fact that lipid oxidation can be inhibited by both thick and cationic emulsion droplet interfacial membranes suggests that engineering the physical properties of emulsion droplet could be used to inhibit oxidative reactions.

7.8 Electrostatic layering in oil-in-water emulsions

A novel interfacial engineering technology has been developed recently, which can be used to improve the stability of food emulsions to environmental stresses, such as pH extremes, high mineral contents, thermal processing, freeze-thaw cycling, and drying. The same technology can also be used to encapsulate and protect various lipophilic substances from chemical degradation. This technology utilizes food-grade ingredients (e.g., phospholipids, proteins, polysaccharides) and simple unit operations (e.g., homogenization, stirring) that are commonly used in the food industry. Consequently, the commercial implementation of this technology is practical and economically feasible. The underlying principle of interfacial engineering technology is electrostatic deposition of charged biopolymers onto oppositely charged emulsion droplets to build up multilayered interfacial membranes around the droplets (Fig. 7.4). This relatively simple technology enables one to systematically control interfacial characteristics such as composition, electrical charge and layer thickness, which in turn allows one to create emulsions with improved stability or novel properties, e.g., better stability to environmental stresses or the ability to protect and release functional components.

The physicochemical principle of the interfacial engineering technology is illustrated schematically in Fig. 7.4, which shows how oil-in-water emulsions containing droplets stabilized by interfacial membranes consisting of two layers (i.e., emulsifier-biopolymer) can be formed using a two-step process. First, a primary emulsion containing electrically charged droplets surrounded by a layer of emulsifier is prepared by homogenizing oil, aqueous phase and a charged emulsifier together. Second, incorporating a biopolymer into the primary emulsion forms a secondary emulsion containing droplets stabilized by emulsifier-biopolymer membranes. The biopolymer should have an opposite electrical

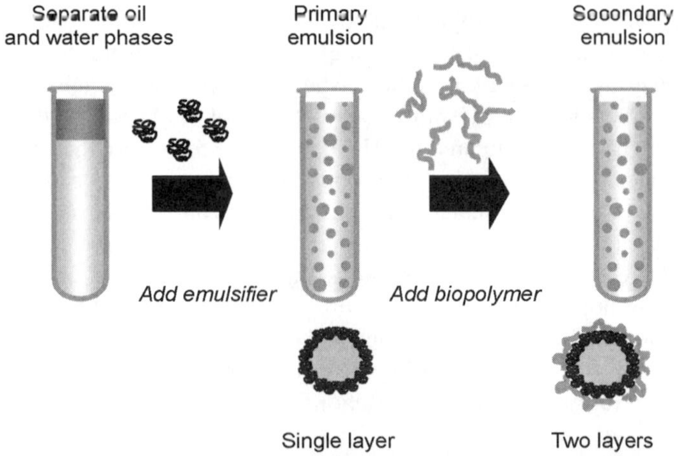

Fig. 7.4 Schematic representation of production of oil-in-water emulsions containing droplets stabilized by two layers (emulsifier-biopolymer).

charge to the droplets in the primary emulsion. If necessary mechanical agitation is applied to the secondary emulsion to disrupt any flocs formed. This procedure can be continued to add more layers to the interfacial membrane, but from a practical standpoint it would probably be economically unfeasible to exceed three layers. The emulsifier used to form the primary emulsion could be any food-grade surface-active ingredient that produces electrically charged droplets, e.g., phospholipids, small molecule surfactants, fatty acids, proteins, polysaccharides, and biopolymer complexes. The biopolymer used to form the secondary emulsion could be any food-grade biopolymer that has an electrical charge, such as proteins and polysaccharides.

The major advantage of this technology is that it can be used to systematically control the characteristics of the interfacial layers surrounding emulsified materials, e.g., the electrical charge, thickness, permeability, and responsiveness of the layer to the environment. Consequently, it is possible to design emulsions that have improved or novel functional characteristics by carefully manipulating the properties of the interfacial layers. Nevertheless, the multilayer emulsions are often difficult to form because of the tendency for bridging and depletion flocculation to occur (McClements, 2005). It is therefore important to carefully control the preparation conditions, such as composition, pH, ionic strength, mixing conditions, and order of addition.

In previous studies, we have utilized the interfacial electrostatic deposition technique to create multilayer emulsions using a variety of different emulsifiers and biopolymers, *i.e.* lecithin/chitosan (Ogawa *et al.*, 2003a,b, 2004), SDS/chitosan (Mun *et al.*, 2005), β-Lg/pectin (Moreau *et al.*, 2003), lecithin/chitosan/pectin (Ogawa *et al.*, 2004), β-Lg/carrageenan (Gu *et al.*, 2004a,b), and β-Lg/carrageenan/gelatin (Gu *et al.*, 2005). Under certain conditions, these multilayer emulsions have been shown to have better stability to environmental stresses than conventional single-layer emulsions, such as pH extremes, high ionic strengths, elevated temperatures or freeze-thaw cycling. It has also been proposed that these multilayer emulsions could be used to create emulsion-based delivery systems that could encapsulate and release active agents (such as flavors, antimicrobials, or enzymes) in response to environmental triggers (such as pH, salt concentration, or temperature).

Emulsions stabilized by multiple layers of emulsifiers have the potential to increase the oxidative stability of lipids. As mentioned previously, emulsion droplets with thick interfacial membranes exhibit increased oxidative stability (Silvestre *et al.*, 2000), as do emulsion droplets with a cationic surface (Donnelly *et al.*, 1998; McClements and Decker, 2000). Stabilizing emulsion droplets with multiple layers of emulsifiers has the potential to inhibit lipid oxidation by forming both thick and cationic emulsion droplet interfacial membranes by using two or more emulsifier layers to increase thickness and by using a cationic biopolymer for the outermost layer to produce a positive charge.

In studies on the oxidative stability of tuna oil-in-water emulsions stabilized by a multilayer system consisting of lecithin and chitosan, it was found that emulsion droplets coated with only lecithin oxidized quicker than the

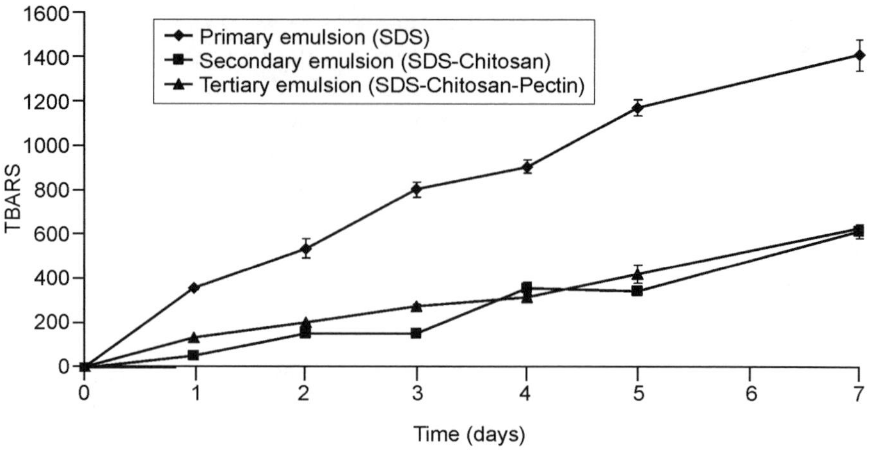

Fig. 7.5 Formation of thiobarbituric acid reactive substance (TBARS) in multilayered emulsion systems stabilized by sodium dodecyl sulfate (SDS; primary emulsion), SDS-chitosan (secondary emulsion) or SDS-chitosan-pectin (tertiary emulsions).

combination of lecithin-chitosan as determined by measuring both lipid hydroperoxide and thiobarbituric acid reactive substances (Klinkesorn *et al.*, 2005a). The improved stability of the lecithin-chitosan emulsion compared with the lecithin emulsion is possibly due to cationic repulsion of iron and other prooxidative metals by the positively charged lecithin-chitosan emulsion droplet interfacial membrane. This trend was also seen in an emulsion stabilized by SDS and chitosan where the formation of thiobarbituric acid reactive substance (TBARS) was faster in the anionic SDS-stabilized emulsion compared to a cationic SDS-chitosan multilayer emulsion. Addition of pectin to create a third layer on to the emulsion droplets resulted in the formation of an anionic interfacial membrane. The oxidative stability of this tertiary emulsion was similar to the secondary SDS-chitosan emulsion (Fig. 7.5). These data suggest that by increasing the thickness of the interfacial membrane of the emulsion droplet, it can be possible to overcome the prooxidative effects of an anionic emulsion droplet interface.

7.9 Microencapsulation of oil-in-water emulsion systems

Microencapsulation is a technique by which solids, liquids, or gases are sealed within a solid material that can be triggered to release their contents under specific environmental conditions (Re, 1998; Augustin *et al.*, 2001; Hogan *et al.*, 2001; Desai and Park, 2005; Kolanowski *et al.*, 2006). Through microencapsulation, sensitive materials known as core materials can be protected within an encapsulant, known as the wall matrix. The core material is typically a sensitive material, needing further protection from the surrounding environment. Typically, low molecular weight carbohydrates, such as maltodextrins and corn syrup solids, are used as the wall matrix, and have been shown to be effective for providing microencapsulation (Klinkesorn *et al.*, 2005b). In order to micro-encapsulate, water must be removed. Typically within the food industry, spray and freeze-drying are common means of removing water. When an oil-in-water emulsion containing maltodextrins or corn syrup solids is dried, the emulsifier-stabilized oil droplets are entrapped within a carbohydrate matrix. This technology may make it easier to store sensitive ingredients for long periods of time and can provide a simple means for the addition of the sensitive material into food products.

ω-3 Fatty acid containing oils are very oxidatively unstable making them good candidates for microencapsulation. However, when dried microencapsulated oils are added to a liquid-based food, the solid wall material dissolves with the encapsulated emulsion droplets being dispersed into the aqueous phase. The loss of wall material means that the emulsion droplets are now exposed to the environment where they can interact with prooxidant factors such as iron. Therefore, if the physical characteristics of the emulsion droplet favor iron–lipid interactions (e.g. the emulsion droplets are anionic), the oxidative stability of the oil can be very low upon addition of the encapsulated oil into the food product since the wall material will dissolve and expose the emulsion droplets. In order to produce microencapsulated ω-3 fatty acids with maximal oxidative stability in the dried form as well as when they are dispersed into a water-based food, oxidative protection must be provided by both the wall material and the emulsifiers coating the oil. This could be accomplished by microencapsulation of emulsions coated with multiple layers of emulsifiers.

Tuna oil-in-water emulsions stabilized with a lecithin and chitosan multilayer system have good oxidative stability (Klinkesorn *et al.*, 2005b). Addition of corn syrup solids to these emulsions had no impact on the oxidative stability of the emulsions during storage. When the lecithin-chitosan-stabilized tuna oil oil-in-water emulsion containing corn syrup solids was then spray dried, the oxidative stability of the resulting powder was greater than the bulk oil during storage at 37C and 33% R.H. Even when mixed tocopherols were added to the bulk oil, the powders without mixed tocopherols still had greater oxidative stability (Klinkesorn *et al.*, 2005c). Upon dispersion of the spray dried emulsion into water, the droplets retained their positive charge indicating that the chitosan remained absorbed onto the emulsion droplets. This spray dried lecithin-chitosan-stabilized oil-in-water emulsion had greater oxidative stability than a spray dried lecithin-stabilized oil-in-water emulsion after dispersion of the powder into an aqueous phase at pH 3.0 (Fig. 7.6).

This research suggests that the oxidative stability of oils containing ω-3 fatty acids can be improved by coating with multiple layers of emulsifiers and then microencapsulating the emulsion in solid matrix of corn syrup solids by spray drying. This system would not only protect the ω-3 fatty acids oils in the microencapsulated powder but would also provide protection against oxidation upon dispersion into a water-based food.

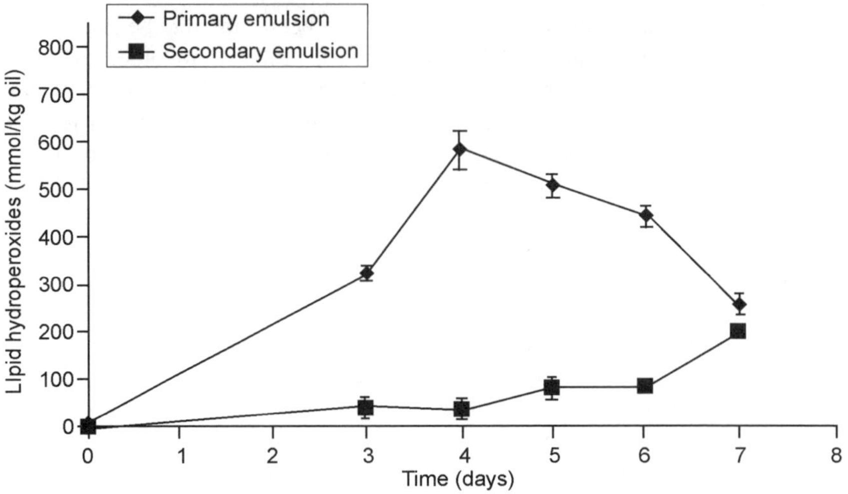

Fig. 7.6 Formation of lipid hydroperoxides in spray dried emulsions menhaden oil-in-water emulsion stabilized by lecithin (primary emulsion) or lecithin and chitosan (secondary emulsion) after dispersion into an aqueous phase at pH 3.0.

7.10 Emulsion droplet engineering to stabilize ω-3 fatty acids

When bulk ω-3 oils are added to water-based foods, the oil will be dispersed into small droplets. To stabilize bulk oil during storage, most ω-3 oil processors add polar antioxidants. Unfortunately, these polar antioxidants are not effective when the oil is dispersed into foods since the polar antioxidants will partition into the aqueous phase where they are ineffective and can be potentially prooxidative because of their ability to reduce transition metals into a more prooxidative state. An oil-in-water emulsion is a potential vehicle for the delivery of bioactive lipids into water-based food products such as dairy products, salad dressings and meat products. An advantage of using an oil-in-water emulsion delivery system is that the emulsion droplet interface could be engineered to inhibit oxidation of the bioactive lipids, which would protect the lipids both in the delivery systems as well as in the food in which they are incorporated. In addition, nonpolar antioxidants that are used to stabilize the oil in the emulsion would also be effective upon addition of the emulsion to the foods since the antioxidants would be retained in the lipid.

To evaluate the potential of an oil-in-water system to deliver bioactive lipids into functional foods, ω-3 fatty acids from algae oil were emulsified with WPI at pH 3.0 to produce a cationic emulsion droplet interfacial layer (Djordjevic *et al.*, 2004a,b). Low viscosity emulsions that are easy to handle (e.g. pumping) could be prepared at lipid concentrations $\leq$35 wt%. The emulsion's mean particle size, apparent viscosity, and creaming were low and did not change during storage in the presence of $\leq$150 mM NaCl (Djordjevic *et al.*, 2004a). This emulsion could be thermally processed up to 70 °C without altering

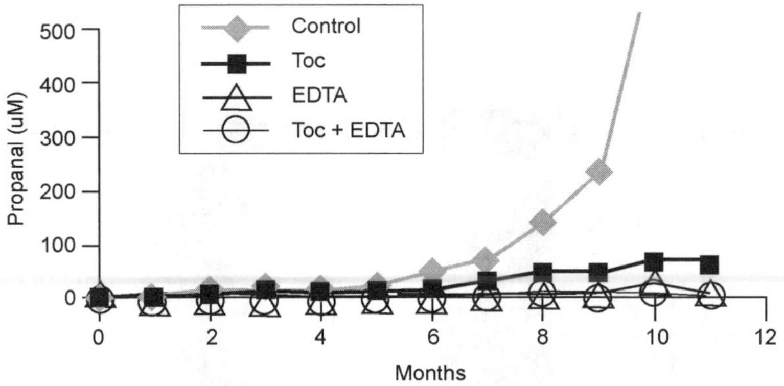

Fig. 7.7 Formation of propanal in whey protein isolated stabilized oil-in-water emulsion containing 25% algal oil at pH 3.0. Treatments include control (no added antioxidants), tocopherols (500 ppm), ethylenediamine tetraacetic acid (EDTA, 100 μM) and tocopherols + EDTA. Emulsion also contained potassium sorbate (0.2%) and were pasteurized at 75 °C for 30 min.

physical stability. The oxidative stability of the emulsion could be increased by addition of mixed tocopherols to the lipid phase and EDTA to aqueous phase (Djordjevic *et al.*, 2004b). An ω-3 oil-in-water emulsion delivery system for fortification of ω-3 fatty acids into functional foods was prepared with 25% algal oil stabilized by whey protein isolate at pH 3.0. Mixed tocopherols (500 ppm), EDTA (50 μM) and potassium sorbate (0.2%) were added to increase oxidative stability and inhibit microbial growth. The final emulsions were pasteurized at 75 °C for 30 min (Djordjevic *et al.*, 2004b). Emulsions containing this combination of ingredients had no detectable oxidation after 11 months of storage (Fig. 7.7).

The algal oil-in-water emulsion delivery system was then added to salad dressings, meats and dairy products to determine its potential to fortify functional foods. Strawberry yogurt was produced with 500 mg of ω-3 fatty acid per serving, 180 mg from DHA and 120 mg from EPA (Chee *et al.*, 2005). The ω-3 fatty acids within the yogurt oxidized slowly within the samples as determined by lipid hydroperoxides, but were not correlated to any off-flavors when evaluated by a consumer sensory panel with over 230 participants (Chee *et al.*, 2005). A consumer panel (85 participants) was also not able to distinguish the difference between control (added corn oil emulsion) or fortified (350 mg ω-3 fatty acid per serving) ranch or Italian salad dressing (Fig. 7.8). When the algae oil-in-water emulsion delivery system was added to fresh ground turkey patties, restructured ham or fresh ground pork sausages at 500 mg of ω-3 fatty acid per serving, additional antioxidants (rosemary extract, sodium citrate and sodium erythorbate) had to be added to inhibit oxidation since the pH of the meat products was above the pI of the WPI (5.1) and therefore the emulsion droplets would be anionic and susceptible to oxidation. Consumer sensory panels (>90 participants) could not determine a difference between ω-3 fatty acid fortified ground turkey and pork sausage compared to the control. However,

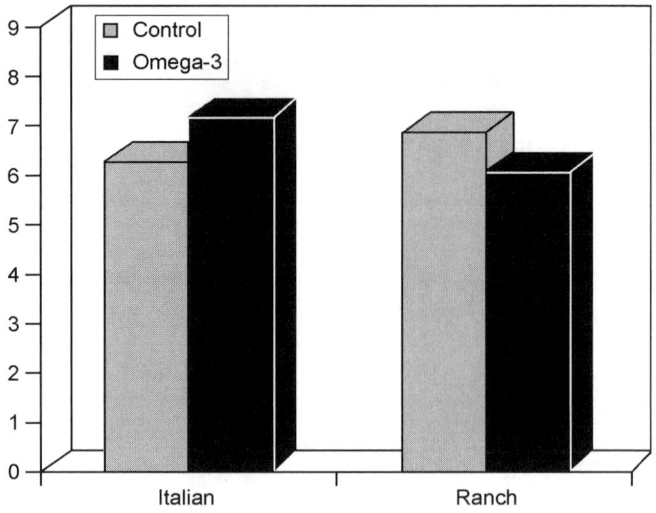

Fig. 7.8 Consumer sensory evaluation of salad dressings fortified with emulsion containing either corn oil or algal oil. Salad dressings were fortified at 350 mg ω-3 fatty acid per serving.

the presence of the ω-3 fatty acid delivery system could be detected in the restructured ham (Lee *et al.*, 2006).

7.11 Conclusions

When bulk ω-3 oils are added to foods, the oil will no longer be in a bulk form but will instead exist as an oil-in-water or water-in-oil dispersion. Most omega-3 oil processors add polar antioxidants to protect their bulk oil against oxidation during storage. However, when the oil is then dispersed into foods, these antioxidants can partition into the aqueous phase where they are ineffective and can be potentially prooxidative due to their ability to make transition metals more prooxidative. Use of oil-in-water emulsions to deliver ω-3 fatty acids into functional foods may provide an alternative to bulk oils. An oil-in-water emulsion ω-3 fatty acid delivery system would have several advantages over bulk oils because the non-polar antioxidants effective in the emulsion would also be effective once the emulsion is added to the food, and because the interface of the emulsion droplet can be engineered to inhibit lipid oxidation by decreasing interactions between lipids and aqueous phase prooxidants. In addition, these emulsions can be microencapsulated in solid wall materials to further enhance their oxidative stability. Use of emulsion in combination with microencapsulation technologies could be extremely effective in the fortification of functional foods if the emulsion droplet retains its antioxidant properties after the wall material is lost upon dispersion into foods. Additional research is needed to understand the physical and chemical properties of emulsion droplets upon dispersion into foods in order to maintain an emulsion droplet interface that inhibits lipid oxidation during processing and storage.

7.12 References

AUGUSTIN, M. A., SANGUANSRI, L., MARGETTS, C., YOUNG, B. Microencapsulation of food ingredients. *Food Australia*. **2001**. 53(6): 220–223.

CHAIYASIT, W., SILVESTRE, M. P. C., McCLEMENTS, D. J., DECKER, E. A. Ability of surfactant tail group size to alter lipid oxidation in oil-in-water emulsions. *Journal of Agricultural and Food Chemistry*. **2000**. 48: 3077–3080.

CHEE, C. P., GALLAHER, J. J., DJORDJEVIC, D., FARAJI, H., McCLEMENTS, D. J., DECKER, E. A., HOLLENDER, R., PETERSON, D. G., ROBERTS, R. F., COUPLAND, J. N. Chemical and sensory analysis of strawberry flavoured yogurt supplemented with an algae oil emulsion. *Journal of Dairy Research*. **2005**. 72(3): 311–316.

DALGLEISH, D. G. The sizes and conformations of the proteins in adsorbed layers of individual caseins on lattices and oil-in-water emulsions. *Colloids and Surfaces B – Biointerfaces*. **1993**. 1(1): 1–8.

DEFILIPPIS, A. P., SPERLING, L. S. Understanding omega-3's. *American Heart Journal*. **2006**. 151(3): 564–570.

DESAI, K. G. H., PARK, H. J. Recent developments in microencapsulation of food ingredients. *Drying Technology*. **2005**. 23(7): 1361–1394.

DIAZ, M., DUNN, C. M., McCLEMENTS, D. J., DECKER, E. A. Use of caseinophosphopeptides as natural antioxidants in oil-in-water emulsions. *Journal of Agricultural and Food Chemistry*. **2003**. 51(8): 2365–2370.

DJORDJEVIC, D., KIM, H. J., McCLEMENTS, D. J., DECKER, E. A. Physical stability of whey protein-stabilized oil-in-water emulsions at pH 3: potential omega-3 fatty acid delivery systems (Part A). *Journal of Food Science*. **2004a**. 69(5): C351–C355.

DJORDJEVIC, D., MCCLEMENTS, D. J., DECKER, E. A. Oxidative stability of whey protein-stabilized oil-in-water emulsions at pH 3: potential omega-3 fatty acid delivery systems (Part B). *Journal of Food Science*. **2004b**. 69(5): C356–C362.

DONNELLY, J. L., DECKER, E. A., McCLEMENTS, D. J. Iron-catalyzed oxidation of menhaden oil as affected by emulsifiers. *Journal of Food Science*. **1998.** 63(6): 997–1000.

DUNFORD, H. B. Free-radicals in iron-containing systems. *Free Radical Biology and Medicine*. **1987**. 3(6): 405–421.

ERICKSON, M. *Food Lipids*. New York: Marcel Dekker, **2002**.

FANG, Y., DALGLEISH, D. G. Dimensions of the adsorbed layers in oil-in-water emulsions stabilized by caseins. *Journal of Colloid and Interface Science*. **1993**. 156(2): 329–334.

FARAJI, H., McCLEMENTS, D. J., DECKER, E. A. Role of continuous phase protein on the oxidative stability of fish oil-in-water emulsions. *Journal of Agricultural and Food Chemistry*. **2004**. 52(14): 4558–4564.

FLEITH, M., CLANDININ, M. T. Dietary PUFA for preterm and term infants: review of clinical studies. *Critical Review in Food Science and Nutrition*. **2005**. 45(3): 205–229.

FRANCOIS, C. A., CONNOR, S. L., BOLEWICZ, L. C., CONNOR, W. E. Supplementing lactating women with flaxseed oil does not increase docosahexaenoic acid in their milk. *American Journal of Clinical Nutrition*. **2003**. 77(1): 226–233.

FRANKEL, E. N. *Lipid Oxidation*. Dundee, Scotland: The Oily Press, **1998**.

FRIBERG, S., LARSSON, K., SJOBLOM, J. *Food Emulsions*. New York: Marcel Dekker; **2004**.

GENOT, C., MEYNEIER, A., RIAUBLANC, A. Lipid oxidation in emulsions. In *Lipid Oxidation Pathways*; A. Kamal-Eldin (ed.); Champaign, IL: AOCS Press, 190–244. **2003**.

GU, Y. S., DECKER, E. A., McCLEMENTS, D. J. Influence of pH and iota-carrageenan concentration on physicochemical properties and stability of beta-lactoglobulin-stabilized oil-in-water emulsions. *Journal of Agricultural and Food Chemistry*. **2004a**. 52(11): 3626–3632.

GU, Y. S., DECKER, E. A., McCLEMENTS, D. J. Influence of iota-carrageenan on droplet flocculation of beta-lactoglobulin-stabilized oil-in-water emulsions during thermal processing. *Langmuir*. **2004b**. 20(22): 9565–9570.

GU, Y. S., DECKER, E. A., McCLEMENTS, D. J. Production and characterization of oil-in-water emulsions containing droplets stabilized by multilayer membranes consisting of beta-lactoglobulin, iota-carrageenan and gelatin. *Langmuir*. **2005**. 21(13): 5752–5760.

HARRIS, W. S. n-3 Fatty acids and serum lipoproteins: human studies. *American Journal of Clinical Nutrition*. **1997**. 65: 1645S–1654S.

HARRIS, W. S. Fish oil supplementation: evidence for health benefits. *Cleveland Clinic Journal of Medicine*. **2004**. 71(3): 208–221.

HOGAN, S. A., MCNAMEE, B. F., O'RIORDAN, E. D., O'SULLIVAN, M. Emulsification and microencapsulation properties of sodium caseinate/carbohydrate blends. *International Dairy Journal*. **2001**. 11(3): 137–144.

HORROCKS, L. A., YEO, Y. K. Health benefits of docosahexaenoic acid (DHA). *Pharmacological Research*. **1999**. 40(3): 211–225.

HU, M., McCLEMENTS, D. J., DECKER, E. A. Lipid oxidation in corn oil-in-water emulsions stabilized by casein, whey protein isolate and soy protein isolate. *Journal of Agricultural and Food Chemistry*. **2003**. 51: 1696–1700.

HU, M., McCLEMENTS, D. J., DECKER, E. A. Impact of chelators on the oxidative stability of whey protein isolate-stabilized oil-in-water emulsions containing omega-3 fatty acids. *Food Chemistry*. **2004a**. 88(1): 57–62.

HU, M., McCLEMENTS, D. J., DECKER, E. A. Antioxidant activity of a proanthocyanidin-rich extract from grape seed in whey protein isolate stabilized algae oil-in-water emulsions. *Journal of Agricultural and Food Chemistry*. **2004b**. 52(16): 5272–5276.

HULBERT, A. J., TURNER, N., STORLIEN, L. H., ELSE, P. L. Dietary fats and membrane function: implications for metabolism and disease. *Biological Reviews*. **2005**. 80(1): 155–169.

INNIS, S. M. Essential fatty-acids in growth and development. *Progress in Lipid Research*. **1991**. 30(1): 39–103.

INNIS, S. M. Polyunsaturated fatty acids in human milk – an essential role in infant development. *Protecting Infants Through Human Milk*. **2004**. 554**:** 27–43.

ISMAIL, H. M. The role of omega-3 fatty acids in cardiac protection: an overview. *Frontiers in Bioscience*. **2005**. 10: 1079–1088.

KLINKESORN, U., SOPHANODORA, P., CHINACHOTI, P., McCLEMENTS, D. J., DECKER, E. A. Increasing the oxidative stability of liquid and dried tuna oil-in-water emulsions with electrostatic layer-by-layer deposition technology. *Journal of Agricultural and Food Chemistry*. **2005a**. 53(11): 4561–4566.

KLINKESORN, U., SOPHANODORA, P., CHINACHOTI, P., McCLEMENTS, D. J., DECKER, E. A. Encapsulation of emulsified tuna oil in two-layered interfacial membranes prepared using electrostatic layer-by-layer deposition. *Food Hydrocolloids*. **2005b**. 19(6): 1044–1053.

KLINKESORN, U., SOPHANODORA, P., CHINACHOTI, P., MCCLEMENTS, D. J., DECKER, E. A.

Stability of spray-dried tuna oil emulsions encapsulated with two-layered interfacial membranes. *Journal of Agricultural and Food Chemistry*. **2005c**. 53(21): 8365–8371.

KOLANOWSKI, W., ZIOLKOWSKI, M., WEISSBRODT, J., KUNZ, B., LAUFENBERG, G. Microencapsulation of fish oil by spray drying-impact on oxidative stability. Part 1. *European Food Research and Technology*. **2006**. 222(3–4): 336–342.

KRIS-ETHERTON, P. M., HARRIS, W. S., APPEL, L. J. Fish consumption, fish oil, omega-3 fatty acids, and cardiovascular disease. *Arteriosclerosis Thrombosis and Vascular Biology*. **2003**. 23(2): E20–E31.

LAURITZEN, L., JORGENSEN, M. H., OLSEN, S. F., STRAARUP, E. M., MICHAELSEN, K. F. Maternal fish oil supplementation in lactation: effect on developmental outcome in breast-fed infants. *Reproduction Nutrition Development*. **2005**. 45(5): 535–547.

LEAF, A., WEBER, P. C. A new era for science in nutrition. *American Journal of Clinical Nutrition*. **1987**. 45(5): 1048–1053.

LEAF, A., KANG, J. X., XIAO, Y. F., BILLMAN, G. E. Clinical prevention of sudden cardiac death by n-3 polyunsaturated fatty acids and mechanism of prevention of arrhythmias by n-3 fish oils. *Circulation*. **2003**. 107(21): 2646–2652.

LEAF, A., XIAO, Y. F., KANG, J. X., BILLMAN, G. E. Membrane effects of the n-3 fish oil fatty acids, which prevent fatal ventricular arrhythmias. *Journal of Membrane Biology* **2005**. 206(2): 129–139.

LEE, S., FAUSTMAN, C., DJORDJEVIC, D., FARAJI, H., DECKER, E. A. Effect of antioxidants on stabilization of meat products fortified with n-3 fatty acids. *Meat Science*. **2006**. (72): 18–24.

MACCHIA, A., LEVANTESI G., FRANZOSI, M. G., GERACI, E., MAGGIONI, A. P., MARFISI, R., NICOLOSI, G. L., SCHWEIGER, C., TAVAZZI, L., TONGNONI, G., VALAGUSSA, F., MARCHIOLI, R. Left ventricular systolic dysfunction, total mortality, and sudden death in patients with myocardial infarction treated with n-3 polyunsaturated fatty acids. *European Journal of Heart Failure*. **2005**. 7(5): 904–909.

MACKIE, A. R., MINGINS, J., NORTH, A. N. Characterization of adsorbed layers of a disordered coil protein on polystyrene latex. *Journal of the Chemical Society*. **1991**. 87(18): 3043–3049.

MANCUSO, J. R., McCLEMENTS, D. J., DECKER, E. A. Ability of iron to promote surfactant peroxide decomposition and oxidize α-tocopherol. *Journal of Agricultural and Food Chemistry*. **1999a**. 47: 4146–4149.

MANCUSO, J. R., McCLEMENTS, D. J., DECKER, E. A. The effects of surfactant type, pH, and chelators on the oxidation of salmon oil-in-water emulsions. *Journal of Agricultural and Food Chemistry*. **1999b**. 47(10): 4112–4116.

McCLEMENTS, D. J. *Food Emulsions: Principles, Practice, and Techniques*. Boca Raton, FL: CRC Press. **2005**.

McCLEMENTS, D. J., DECKER, E. A. Lipid oxidation in oil-in-water emulsions: impact of molecular environment on chemical reactions in heterogeneous food systems. *Journal of Food Science*. **2000**. 65(8): 1270–1282.

MEI, L. Y., McCLEMENTS, D. J., WU, J. N., DECKER, E. A. Iron-catalyzed lipid oxidation in emulsion as affected by surfactant, pH and NaCl. *Food Chemistry*. **1998a**. 61(3): 307–312.

MEI, L. Y., DECKER, E. A., McCLEMENTS, D. J. Evidence of iron association with emulsion droplets and its impact on lipid oxidation. *Journal of Agricultural and Food Chemistry*. **1998b**. 46(12): 5072–5077.

MOREAU, L., KIM H. J., DECKER, E. A., McCLEMENTS, D. J. Production and characterization of oil-in-water emulsions containing droplets stabilized by beta-lactoglobulin-pectin membranes. *Journal of Agricultural and Food Chemistry*. **2003**. 51(22): 6612–6617.

MUN, S., DECKER, E. A., McCLEMENTS, D. J. Influence of droplet characteristics on the formation of oil-in-water emulsions stabilized by surfactant-chitosan layers. *Langmuir*. **2005**. 21(14): 6228–6234.

NAWAR, W. W. Lipids. In *Food Chemistry*, Fennema, O. R. (ed.). New York: Marcel Dekker, **1996**.

NESTEL, P. Fish oil and cardiovascular disease. *American Journal of Clinical Nutrition*. **2000**. 71: 288S–331S.

OGAWA, S., DECKER, E. A., McCLEMENTS, D. J. Production and characterization of O/W emulsions containing cationic droplets stabilized by lecithin-chitosan membranes. *Journal of Agricultural and Food Chemistry*. **2003a**. 51(9): 2806–2812.

OGAWA, S., DECKER, E. A., McCLEMENTS, D. J. Influence of environmental conditions on the stability of oil in water emulsions containing droplets stabilized by lecithin-chitosan membranes. *Journal of Agricultural and Food Chemistry*. **2003b**. 51(18): 5522–5527.

OGAWA, S., DECKER, E. A., McCLEMENTS, D. J. Production and characterization of O/W emulsions containing droplets stabilized by lecithin-chitosan-pectin mutilayered membranes. *Journal of Agricultural and Food Chemistry*. **2004**. 52(11): 3595–3600.

PAWLOSKY, R. J., HIBBELN, J. R., NOVOTNY, J.A., SALEM, N. Physiological compartmental analysis of alpha-linolenic acid metabolism in adult humans. *Journal of Lipid Research*. **2001**. 42(8): 1257–1265.

PORTER, W. L. Paradoxical behavior of antioxidants in food and biological-systems. *Toxicology and Industrial Health*. **1993**. 9: 93–122.

RE, M. I. Microencapsulation by spray drying. *Drying Technology*. **1998**. 16(6): 1195–1236.

REISCHE, D., LILLARD, D. A., EITENMILLER, R. R. Antioxidant. In *Food Lipids: Chemistry, Nutrition, and Biotechnology;* Akoh C. C. and Min D. B. (eds). New York: CRC Press, **2002:** 489–516.

ROCHE, H., GIBNEY, M. Effect of long-chain n-3 polyunsaturated fatty acids on fasting and postprandial triacylglycerol metabolism. *American Journal of Clinical Nutrition* **2000**. 71: 232S–237S.

SHAHIDI, F., FINLEY J. W. *Omega-3 Fatty Acids: Chemistry, Nutrition, and Health Effects.* Oxford: Oxford University Press, **2001**.

SIDDIQUI, R. A., SHAIKH, S. R., SECH, L. A., YOUNT, H. R., STILLWELL, W., ZALOGA, G. P. Omega 3-fatty acids: health benefits and cellular mechanisms of action. *Mini-Reviews in Medicinal Chemistry*. **2004**. 4(8): 859–871.

SILVESTRE, M. P. C., CHAIYASIT, W., BRANNAN, R. G., McCLEMENTS, D. J., DECKER, E. A. Ability of surfactant headgroup size to alter lipid and antioxidant oxidation in oil-in-water emulsions. *Journal of Agricultural and Food Chemistry*. **2000**. 48(6): 2057–2061.

SIMOPOULOS, A. P. Omega-3 fatty acids in health and disease and in growth and development. *American Journal of Clinical Nutrition*. **1991**. 54(3): 438–463.

SIMOPOULOS, A. P. Essential fatty acids in health and chronic disease. *American Journal of Clinical Nutrition* **1999**. 20: 560S–569S.

SIMOPOULOS, A. P. The importance of the ratio of omega-6/omega-3 essential fatty acids. *Biomedicine & Pharmacotherapy*. **2002**. 56(8): 365–379.

VANSCHOONBEEK, K., DE MAAT, M. P. M., HEEMSKERK, J. W. M. Fish oil consumption and reduction of arterial disease. *Journal of Nutrition*. **2003**. 133(3): 657–660.

YOUNG, G., CONQUER J. Omega-3 fatty acids and neuropsychiatric disorders. *Reproduction Nutrition Development*. **2005**. 45(1): 1–28.

8

Lipid self-assembled particles for the delivery of nutraceuticals

O. Ramon and D. Danino, Technion – Israel Institute of Technology, Israel

8.1 Introduction

Supplementing staple foods with physiologically active components such as vitamins, minerals, phyto-chemicals, probiotics, and unsaturated fatty acids raises considerable technological challenges concerning the application of the active molecules to the food formula in uniform distribution, assuring their stability during product processing, distribution and storage, and, most importantly, retaining the additive's bioactivity.

Lipid-based structures at the micro- or nano-scale are most promising systems for encapsulation and delivery of such sensitive nutrients, and as bio-reactors for the release of aroma compounds and flavors. Their prominent advantages are related to their being mostly natural food ingredients, and to their ability to incorporate and target high loads of guest functional molecules with different solubility, and to release them at specific sites.[1–3] Lipidic carriers can be targeted to the specific site via active (e.g. antibodies in drug delivery applications) or passive (particle size) mechanisms.[4] Release at a controlled rate and time considerably increases their effectiveness by providing an optimal dosage. Lipid vehicles can accommodate several nutrition agents simultaneously, providing a synergistic effect. They protect the sensitive nutrients from temperature and oxidative stresses, free radicals, pH, and metal ions, as well as from degradation by enzymes and loss of bioactivity.[1,2]

In this chapter we review the principles of lipid self-assembling systems, their integration in the mainframe of the packing concept of Israelachvili and coworkers,[5,6] thermodynamic constraints,[7] and physico-chemical properties.

Common industrial techniques of liposome formation are described as well as characterization methods of the vehicles' properties and functionality. We then discuss the effects of introducing guest molecules on the self-assembly behavior and structure, the release of entrapped molecules from the lipid vehicles, and the application of liposomes at micro- and nano-scales for the delivery of nutrients, nutraceuticals, and food additives. The last part of the chapter focuses on non-parallel liquid-crystalline phases, cubic and hexagonal, and their dispersed nanoparticles. We review basic properties, methods for preparation, and characterization of the bulk phases and particles, and parameters controlling the structure. Prospective future applications in food science and technology are also delineated. We conclude with a concise exposition of future trends that will enable larger scale applications of these lipid vehicles in the area of food nutrition.

8.2 Structure and properties of lipids

Lipids are essential components of biological membranes and living cells. Polar lipids (synthetic and natural) are amphiphilic molecules composed of both hydrophilic (water-soluble) and hydrophobic (lipophilic, oil-soluble) moieties, called 'head' and 'tail', respectively. The polar head-group can be charged, zwitterionic, or uncharged. The hydrophobic lipid part consists of one or two hydrocarbon chains, which can be saturated or unsaturated to different degrees. Natural phospholipids typically have asymmetric hydrocarbon chains, containing an even number of carbon atoms. The chain length and degree of saturation are key parameters, determining many physical parameters of lipids and self-assembled lipid structures, such as their phase behavior, stability, permeability, and order of the chains.[8,9] Sterols (e.g. cholesterol, ergosterol, sitosterol) constitute another family of lipids.[8] Sterols have a rigid, planar hydrophobic portion which can be linked to other groups. They normally cannot form bilayers, but they are widely present in biological membranes formed by other lipids, and affect their properties. Cholesterol, for example, decreases the mobility of the membrane molecules, increases membrane stability, and reduces membrane permeability.

The general chemical structure of polar lipids is based on either a glycerol or a spingosine backbone, most frequently connected to a phospho or glyco polar group.[8] One or two fatty acids may be attached to the backbone, forming lysolipids and lipids, respectively. Thus, four groups of double-tail polar lipids are defined. An example is given to clarify the general terminology. In glycerolipids (i.e. lipids with a glycerol backbone), two of the hydroxyl groups (normally at the *sn1* and *sn2* positions) are connected with long chain fatty acids by an ester link. In glycerophospholipids, also called phospholipids, the hydroxyl group at the *sn3* position (the other end of the glycerol backbone) is esterified to phosphoric acid. One of the phosphoric acid oxygen groups is further covalently bound to an organic molecule such as ethanolamine, choline, glycerol, serine, or inositol, and other oligosaccharides. The two hydrophobic chains make up the molecule hydrophobic tail, while the phosphate with the attached group makes up the phospholipid hydrophilic head-group.[8,9]

Double-chain lipids normally self-organize into bilayers. The bilayer is composed of two opposing lipid monolayers arranged so that their hydrophobic tails face one another to form an oily core, while their charged head-groups face the aqueous phase on either side of the bilayer.

A key parameter of lipids is the gel to liquid-crystalline transition temperature, T_m, which represents the melting temperature of the hydrophobic chains. Above this temperature the bilayer loses its solid order packing, and its fluidity and permeability are highly increased. The gel to liquid-crystalline transition is a first-order transition, hence it involves a sharp change in membrane properties.[1,3] T_m increases with the chain length,[1,10,11] branching, degree of chain saturation, and head-group interactions. In natural phospholipids, however, the transition normally occurs over a range of temperatures due to chain asymmetry.

8.3 Self-assembly and microstructure of lipid systems

Lipids, like surfactants, display complex phase behavior that is strongly dependent on composition and temperature. Their dual hydrophilic-hydrophobic nature drives the spontaneous self-assembly into a variety of micellar structures and ordered liquid-crystalline (LC) phases when dispersed in a polar solvent. The driving force for the assembly is the hydrophobic effect,[7] which involves attraction between hydrophobic moieties (hydrophobic interaction) and the structure of water around nonpolar molecules (hydrophobic hydration). The size and shape of the spontaneously formed aggregates is determined by a delicate balance between inter- and intramolecular interactions that include the tendency to minimize the unfavorable contact of the hydrophobic tails with the polar solvent, repulsive forces between neighboring head-groups, and steric interactions.[7,12] Two types of structures are formed – normal and inverted. In the normal assemblies the hydrophobic moieties point inwards and form a hydrophobic core, while the polar head-groups are in contact with the polar environment, shielding the tails from the surrounding water. Additionally, one dimension of the assemblies must be smaller than the tail length. In the inverted structures the polar heads point inwards towards water tubes, while the tails face the surrounding nonpolar environment. A general representation of the common structures and phases that form in aqueous solutions of amphiphiles is presented in Fig. 8.1.

The size and shape of the assemblies can be predicted using the packing parameter, a dimensionless criterion developed by Israelachvili.[5,6] This concept relates molecular parameters that dictate the geometry of the molecule (chain length, chain volume, and head-group area) with intensive parameters (such as temperature and ionic strength) and the equilibrium aggregates nanostructure.[5–7,12–14] The packing parameter is defined by the ratio of the

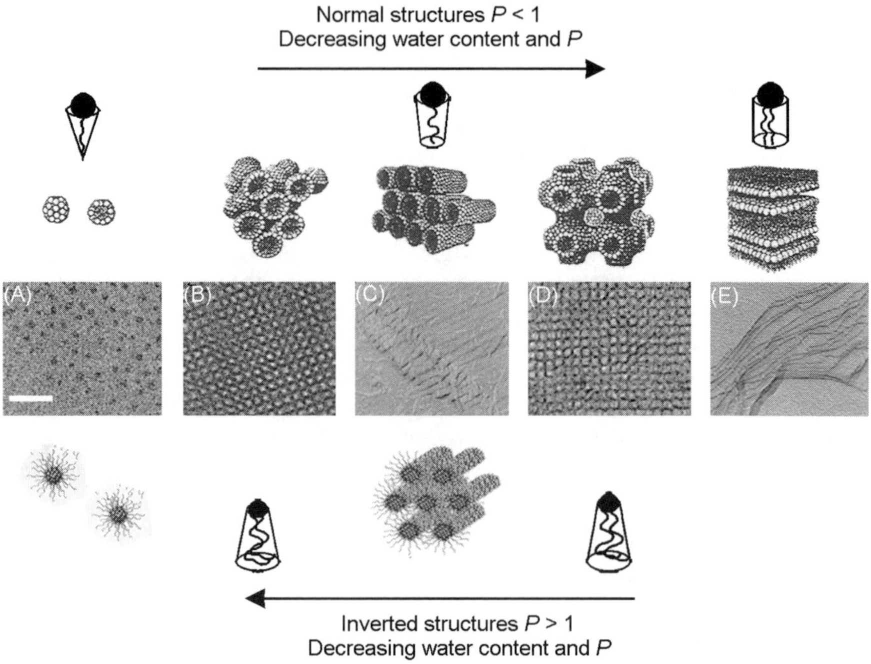

Fig. 8.1 Schematic representation of lipid self-assembled structures and mesophases, and their respective cryo-TEM images. Upper drawings show the normal structures and the representative geometrical packing of the molecules; the lower drawings correspond to the inverted structures: (A) spherical micelles, (B) discrete cubic mesophase, (C) hexagonal mesophase, (D) bicontinuous cubic mesophase and (E) lamellar mesophase. A, B and D are direct-imaging cryo-TEM images, C and E are freeze-fracture cryo-TEM images.

hydrophobic amphiphile chain volume (V_t) to the product of the equilibrium head-group area of the molecule (a_h) and its effective tail length in solution (l_t):

$$P = \frac{V_t}{a_h \times l_t} \tag{8.1}$$

Small, highly-curved assemblies like spherical micelles form when $P < 1/3$, typically by single-tailed surfactants; elongated micelles are created when $P \sim 1/2$. Double-chain amphiphiles such as lipids are characterized with P close to or greater than one. These molecules form planar bilayered structures and various types of lyotropic liquid-crystalline phases (mesophases). When $P < 1$, the mean curvature of the aggregate surface is positive and the structures are curved towards the hydrophobic parts (the normal structures), while $P > 1$ relates to a negative curvature, in which the molecules create structures curved towards the hydrophilic parts (the inverted structures).

Similarly, the ratio of the cross-sectional area of the lipophilic moiety (A_l) to that of the polar part (A_p) is defined. Highly curved structures form for $A_p > A_l$, and bilayer structures for molecules with comparable hydrophilic and hydrophobic cross-sectional areas ($A_p \sim A_h$). For multicomponent systems, average values are defined, $\langle A_l \rangle$ and $\langle A_p \rangle$, which are the linear combination of the individual A_i multiplied by the mole fraction of each molecule.[1] Minute changes in molecular or environmental properties may affect the morphology and phase transition temperatures, and induce structural and phase transformations.[15,16]

The most common liquid-crystalline phases are lamellar, and to a lesser extent hexagonal and cubic mesophases form, possessing orientation in one, two, and three dimensions, respectively. These mesophases can be dispersed in aqueous solutions into colloidal particles that retain the short-range symmetry of their bulk mesophases.[8,17] The dispersed particles are called liposomes, cubosomes, and hexosomes. These types of assemblies will be discussed individually in the following sections.

Microemulsions constitute another family of lipid assemblies which are formed in ternary lipid-water-oil solutions. Chapter 6 in this book is dedicated to micro- and nanoemulsions; hence, these assemblies will not be discussed further here.

8.4 Liposomes

8.4.1 Liposome formation

Liposomes are lipid vesicles, supramolecular assemblies made of closed bilayers surrounding an inner volume of solvent (Fig. 8.2). The polar head-groups in the inner and outer surfaces (leaflets) of the membrane point towards the inner and outer solvent phase, and the hydrophobic tails point towards the center of the membrane, forming a hydrophobic core. The solvent in the interior volume is polar, but may differ from the continuous phase in which the liposomes are suspended.[18]

The packing parameter concept predicts that vesicles will form when $1/2 \leq P \leq 1$. For common amphiphiles the ratio V_t/l_t is a constant, ~21 Å^2 for a single tail and ~42 Å^2 for double tails.[12] Thus, the specificity of an amphiphile and its packing parameter are independent of the tail length and are essentially determined by the surface area, i.e. the area per molecule in the aggregate a_n (a_n is a thermodynamic quantity obtained by equilibrium considerations of minimum free energy for the lipid in an aggregated state, not the geometrical head-group area of the individual molecule). However, Nagarajan showed that the packing parameter criterion is insufficient and spherical liposomes will not form unless the inner volume per molecule is larger than the head-group volume.[19,20] This criterion is satisfied only when the hydrophobic tail exceeds a critical length, defined by linear relations between the aqueous volume per molecule and the tail length ($V_{\text{cavity}} = 46.2\ l_t$).

8.4.2 Liposome classification

Liposomes can be classified according to their preparation method, by their number of bilayers, or by their size. The most common classification is by

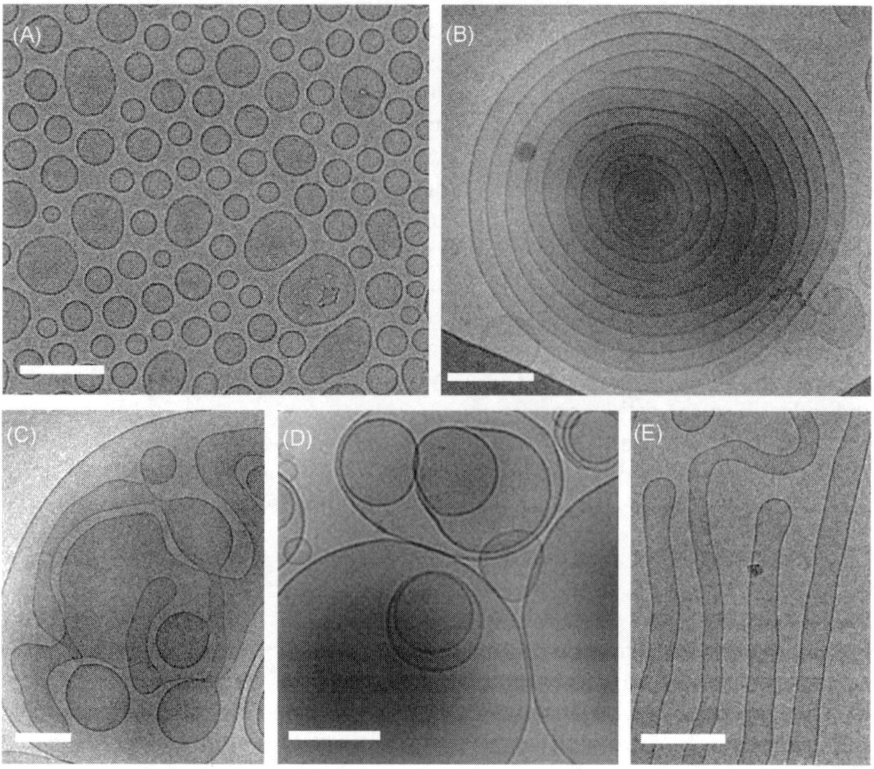

Fig. 8.2 Direct-imaging cryo-TEM images of liposomes: (A) small unilamellar vesicles, (B) multilamellar vesicles composed of concentric bilayers, (C, D) multivesicular vesicles and (E) tubular vesicles. The vesicles shown in (A) form in the MO/ethanol/water system. All the structures shown in panels B to E form spontaneously in binary phosphatidylserine/water solutions. Bars equal 100 nm in all panels.

structure (lamellarity) and size, which in turn depend on the preparation method.[21,22] Vesicles can be unilamellar (ULVs) – consisting of a single bilayer (Figs 8.2(a) and 8.3(a)), multilamellar (MLVs) – consisting of concentric bilayers (Figs 8.2(b) and 8.3(a)), or multivesicular (MVVs) – composed of many small non-concentric vesicles encapsulated within a single lipid bilayer (Figs 8.2(c), 8.2(d) and 8.3(b)).[11] Vesicles can be further classified by their size, as small unilamellar vesicles (SUVs) characterized by diameters ranging between 20 and 100 nm and small size distribution (Fig. 8.2(a)), and large unilamellar vesicles (LUVs) of sizes reaching up to tens of microns (Fig. 8.3(a)). SUVs have a low aqueous core volume to lipid ratio (0.2–1.5 l/mole lipid); thus they are not efficient encapsulants of functional foods and nutraceuticals.[23] On the other hand, LUVs have a large aqueous volume to lipid ratio (7 l/mole lipid),[24] and they can carry large loads of encapsulated hydrophilic molecules in their internal volume. Multilamellar vesicles are large aggregates, with diameters reaching up to 500 μm. They may contain hundreds of concentric bilayers, hence they are

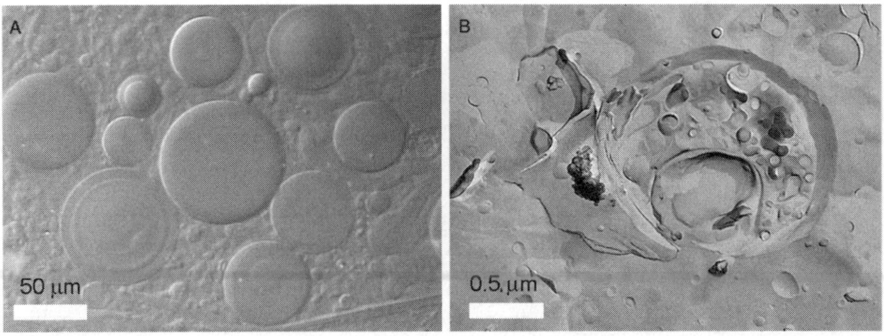

Fig. 8.3 (A) Light microscopy image at Nomarski optics showing large unilamellar, multilamellar and tubular liposomes of phosphatidylserine in water. (B) Freeze-fracture image of the same sample disclosing unilamellar and multivesicular liposomes. Note the different scales in A and B.

characterized by moderate aqueous volume to lipid ratio (1–4 l/mole lipid), but high entrapment capacity for lipophilic and/or hydrophobic nutraceuticals and functional foods.[23,24]

8.4.3 Methods for liposome preparation

Technological developments during the last ten years make liposome microencapsulation and nanoencapsulation more routine processes in the food industry.[25] Because liposomes are thermodynamically unstable assemblies,[1,8,26] their structure and characteristics, such as size, polydispersity, surface charge, degree of ionization, thermal phase behavior, permeability, and physical stability, depend on their preparation methods.[27–30] The first step in liposome preparation involves mixing in an organic solvent, followed by evaporation, lyophilization or spray drying. MLVs readily form upon resuspension of the dry bilayer-forming lipid in an aqueous phase during agitation.[1,31] Highly charged lipids, especially with low ionic strength, swell and disperse better.[32] However, MLVs are typically too heterogeneous for industrial applications. In order to form unilamellar vesicles and reduce their size, methods requiring substantial energy input such as sonication or extrusion (on a small scale) or homogenization techniques (on a large scale) are used.[32,33] The amount of energy required may vary significantly with the type of liposome produced. Non-mechanical methods include reverse phase evaporation, removal of detergents from mixed detergent/lipid micelles, and freeze-drying followed by rehydration and freeze-thawing. Additional information on these manufacturing techniques can be found in several reviews.[3,25,32]

Some methods enable continuous production of liposomes on a large scale and at acceptable costs, without the use of sonication[22,34–37] or organic solvents.[25,38–40] Microfluidization technology, for example, is a continuous, cost-effective, solvent-free method that can process high lipid concentrations of uni- or multilamellar vesicles with a good encapsulation efficiency (>20%).[39–44] Liposome formulations are generally kept in dilute aqueous suspensions. Under these conditions liposomes can undergo fusion, thus stability (physical, chemical, and biological) and shelf-life become important and must be considered.[1] Stabilization with antimicrobials can affect liposome structure, storage, and delivery and adds extra cost; hence, other techniques such as freeze-drying/rehydration and freeze-thawing are employed to improve liposome stability.[45] Sugars such as trehalose, sucrose, or lactose are added in these methods to protect the lipid membrane from damage during the freezing processes, and to increase the retention of the encapsulated compounds.[46] Rehydration of the lipid above the gel to liquid-crystalline phase transition initiates fusion of SUVs to MLVs, and the formation of large (>1000 nm) liposomes displaying a high entrapment efficiency (up to 50%).[35,45,46]

8.4.4 Controlling liposome self-assembly and bilayer properties for improved entrapping and delivery of functional foods and nutraceuticals

The self-assembly into liposomes and the bilayer properties are dictated by the chemical nature of the lipid (type, head-group charge, tail length, and asymmetry) and the solution properties (solvent type and quality, ionic strength, temperature, composition, and presence of additives). The application and functionality of liposomes as delivery vehicles are in turn dependent on their characteristics (liposome size, surface properties, lipid layer consistency, fluidity) and their interaction with other materials.[3] Vesicles can be programmed to respond to specific stimuli that will induce changes in the bilayer properties and facilitate the release of encapsulated active compounds. The gel-to-liquid-crystalline temperature, T_m, can be tuned by controlling the lipid composition to facilitate release from the liposomes at a desired temperature.[3,47] Changing the average area of the polar head-group ($\langle A_p \rangle$) will lead to transitions into other phases and mesophases, easing release as well.[1] Increasing $\langle A_p \rangle$ of the outer layer due to hydrolysis or protonation of lipids decreases the effective packing parameter, thus leading to transitions from bilayers into micelles (dissolution of the membrane) and as a result to leakage of the encapsulated guest molecules. Decreasing $\langle A_p \rangle$ triggers a lamellar-to-inverted hexagonal phase transition and liposome fusion, thereby releasing the encapsulated components.[1] In pH-sensitive liposomes, lipid bilayers with protonable weak groups can induce a phase transition due to the decrease of $\langle A_p \rangle$ upon lowering of the pH.[48]

Colloidal instability and aggregation of liposomes can be overcome by increasing the charge on their surface, or by decreasing the ionic strength of the medium. Alternative approaches include sterically increasing the repulsion forces between liposomes (which is salt independent and effective at high liposome concentrations) or coating of the liposomes with inert hydrophilic polymers.[1,8,26,49–56]

8.4.5 Liposome characterization and evaluation of performance

Adequate quality control of the liposomes during and immediately after their preparation, and during their storage, requires reproducible, accurate, and fast characterization methods. The liposome size and size distribution can be characterized by numerous techniques. Transmission electron microscopy (TEM) methods, such as freeze-fracture and cryogenic-TEM are especially advantageous as investigative tools of liposome morphology and membrane properties. TEM methods disclose details regarding the size and degree of lamellarity (see Figs 8.2 and 8.3), as well as the shape (spherical, tubular, or oblate), fluidity (below T_m, for example, vesicles will appear faceted), co-existence of various types, and transitions to other phases.[57] Photon correlation spectroscopy is an accurate method for measuring the size of the vesicles.[24] Stability of liposomes can be determined by dynamic light scattering[58–61] and the amount of sensitive ingredients entrapped in the liposomes can be quantified by column chromatography.[46] The latter technique enables discrimination between the unencapsulated nutraceuticals (measured after separation from the liposome fraction) and the amount delivered in the lipid vesicles (determined after solubilization of the membrane and release of the entrapped molecules).[32,33,62]

8.4.6 Applications of liposomes in food systems

Implementation of liposomes in medicine (such as in cancer treatments and therapy) and in many pharmaceutical, drug delivery, and cosmetic applications, combined with considerable development in formulation technologies at an industrial scale, enhanced the use of micro- and nanoliposome encapsulation in the food industry.[41,63] The lipophilic/hydrophilic properties, colloidal size, stability, encapsulation potential at reasonable efficiencies (>20%), and protection of the additive bioactivity make liposomes efficient delivery vehicles of sensitive nutraceuticals such as proteins, enzymes, probiotics, and vitamins in various applications of the food industry. One of their main advantages is the ability to entrap hydrophilic molecules in their interior volume, and hydrophobic compounds in their lipophilic bilayer core. Liposomes can also be designed to deliver, uniformly disperse, and release their load in a controlled manner and at specific areas, whether within a food matrix or in a biological system.[2] Additionally, liposomes provide higher protection and stability than other encapsulation technologies to sensitive water-soluble materials such as ascorbic acid (vitamin C) at high water-activity conditions.[64]

Applications in dairy products

Liposomes have been used in dairy processes for many years.[65–67] Early studies involved entrapment of enzymes such as proteases in MLVs and their use in cheese production to accelerate the ripening process and protect essential substrates in the milk until the cheese is formed.[65–69] The entrapment of proteases also prevents bitterness from developing during the ripening process, increases cheese firmness, and improves the texture, homogeneity and yield of the end products, at 100-fold lower enzyme requirement.[64,70–72] Kheadr *et al.* reported that liposome-encapsulated lipases increased cheddar cheese cohesiveness and elasticity, but reduced the cheese firmness.[73] Appropriate control of the lipase concentration also enabled the development of a desirable flavor profile.[73] For a detailed summary of the use of enzyme-loaded liposomes in cheese ripening and production the readers are referred to the review by Fresta and Puglisi.[74]

The successful use of liposome-entrapped enzymes in milk and cheese production is associated with relatively high encapsulation efficiency and stability of the formulations. In milk, higher encapsulation efficiencies were found at milk pH conditions with positively charged liposomes encapsulating negatively charged enzymes. The highest encapsulation efficiency was found by freeze-thaw preparation of MLVs and LUVs.[75] The stability of liposome systems can be enhanced by incorporation of cholesterol.[70] Cholesterol imparts rigidity to the bilayer membrane by changing interactions between the polar head-groups and the hydrophobic tail, and decreasing the phase transition temperature, T_m, and the transition enthalpy.[76] For example, β-galactosidase was encapsulated in soy-PC-cholesterol liposomes for control of lactose digestion.[77–80] Liposome systems were also employed to fortify dairy products with vitamins and thereby increase their nutritional potential. Vitamin D entrapment in liposomes is a good example of such an application.[81]

Liposome-encapsulation of vitamins, minerals, and antioxidants in food

Vitamins are important nutraceuticals with established therapeutic properties. Many of the vitamins' bioactivity is related to their antioxidative properties, i.e. their ability to delay or inhibit oxidation of substrates at low concentrations.[2] Food antioxidants are divided into two main groups: (1) acids and their salts and esters (e.g. ascorbic acid (vitamin C), citric acid) and (2) synthetic and natural phenol compounds (e.g. α-tocopherol (Vitamin E)). Both types are non-enzymatic antioxidants. Vitamins are encapsulated in liposomes to enhance their retention. Liposome-entrapped vitamin systems include vitamins A and E[2,82–85] and β-carotenes.[86] The attractiveness of these systems is related to the protection provided by the liposome bilayer, which extends the half-life of the entrapped antioxidants and facilitates their intracellular uptake.[87–89] For example, better bioavailability of the hydrophobic Coenzyme Q_{10} (CoQ_{10}, ubiquinone) was found upon encapsulation in a liposomal spray, compared with entrapment in a gelatin capsule. Entrapment of the enzyme bromelain (used as a meat tenderizer) in liposomes was also found to significantly increase enzyme stability and bioavailability.[90]

Bifunctional liposome carriers can simultaneously incorporate and release materials of various solubilities.[91] Good examples are entrapment of two antioxidant agents – α-tocopherol, a lipid-soluble molecule, with the water-soluble molecules ascorbic acid or glutathione.[64,91,92] In such formulations, a synergistic increase in the antioxidative effect of vitamin E was reported.

A new type of lipid-based nanoliposome prepared from polar ether lipids and named archaeosomes was recently introduced by Mozafari.[2] Compared with liposomes (that are prepared from ester phospholipids), archaeosomes display higher resistance to a low pH environment and bile salts, and better thermo-stability against chemical and enzymatic oxidation and hydrolysis. These properties make archaeosomes promising vehicles for antioxidants and temperature-sensitive nutraceuticals such as unsaturated long chain PUFA (polyunsaturated fatty acids) during food processing.[2,4,93,94]

Liposomes are also used to deliver minerals such as Ca^{+2}, Mg^{+2}, and Fe^{+3} into foods.[1,2,95–97] An interesting application involves the use of liposome-entrapped calcium in soy-milk products to prevent the gelling of soy protein that occurs when the free mineral (Ca^{+2}) is added and is in direct contact with the soy milk products.[95,96] The key factor for successful use of liposome-based vehicles is the ability to maintain the bioactivity of entrapped nutraceuticals and functional foods in the liposomes, by preventing degradation processes, and the potential of providing superior targeted delivery at a specific site.

Several liposome-based nutritional products are available in the market, including formulations with vitamins A, B_2, B_{12}, and E, Coenzyme Q_{10}, melatonin, and zinc compounds. A detailed list is provided in the review by Keller.[98]

8.5 Cubic and hexagonal mesophases and their dispersed nanoparticles

8.5.1 Structure and properties of non-parallel (cubic and hexagonal) mesophases

In addition to lamellar phases, cubic and hexagonal mesophases form in lipid systems. Cubic phases consist of either discrete micellar aggregates (I phases) or bicontinuous bilayer elements (V phases), arranged in a three-dimensional long-range order. Normal (Ii and Vi) and inverted (Iii and Vii) cubic mesophases form, located in the phase diagram at $P < 1$ and $P > 1$, respectively (see Fig. 8.1). Bicontinuous cubic mesophases are made of two intercrossing lipid and solvent networks; bicontinuous means that the hydrophilic and hydrophobic bilayer domains are extended continuously, and do not intersect throughout the phase (see Fig. 8.1). The lipid bilayers form surfaces called infinite periodic minimal surfaces (IPMS), characterized by periodicity in three dimensions, and zero mean curvature everywhere.[99] Three bicontinuous inverted cubic phases with minimal surfaces were identified in lipid systems – C_G, C_D, and C_P – corresponding to the IPMS gyroid (G surface), diamond (D surface), and primitive (P surface), and associated with the space groups *Ia3d* (Q_{230}), *Pn3m* (Q_{224}), and *Im3m* (Q_{229}), respectively,[100,101] Micellar cubic mesophases also form in lipid systems,[102] but less frequently. They may exist as simple cubic, body centered cubic (bcc), or face centered cubic (fcc). The most common form is the bcc, which consists of close-packed identical quasi-spherical micelles obeying the *Im3m* (Q_{229}) symmetry.[103] The fcc and simple cubic mesophases appear to contain two types of micelles, and are typically associated with the *Fd3m* (Q_{227}) and *Pm3n* (Q_{223}) space groups,[104–106] though additional symmetries were proposed by Hyde *et al.*[99]

Hexagonal mesophases are made of infinitely long micellar aggregates packed in a hexagonal array. The normal hexagonal mesophase (Hi) exists between the Ii and Qi liquid-crystalline phases and has a positive curvature, while the inverted hexagonal mesophase (Hii) forms between the Qii and Iii mesophases, and is made of negatively curved cylinders, in which the hydrophobic parts point towards the solvent (Fig. 8.1).

Monoglycerides are important food emulsification compounds characterized by poor water solubility, which drives their self-assembly into various inverted liquid-crystalline phases in aqueous solutions. The cubic and hexagonal mesophases of unsaturated monoglycerides are the most studied[101,107–119] (and additional references mentioned later in the text). The binary phase diagram of glycerol monooleate (GMO) in water shows one lamellar and two inverted bicontinuous cubic mesophases (diamond Q_{230} and gyroid Q_{224}) at room temperature,[107] and an inverted hexagonal phase at elevated temperatures. In recent years, several tertiary mixtures containing GMO, water, and various polyols and hydrotropes have been studied.[119–121] The addition of the third component was shown to shift the phase boundaries, promote the formation of additional phases that are not present in the binary lipid-water mixtures, and affect the phase properties.[117,118,122,123] The hydrotropes (cosurfactants) can be localized both in the continuous phase where they will have no effect, and at the interface where they will affect the structure and the characteristic d-spacing of the ordered mesophases. The GMO-water-ethanol phase diagram constructed by Spicer and colleagues shows the presence of the same liquid-crystalline phases as those found in the binary system.[119] Our recent studies with the GMO-water-ethanol system revealed the formation of a new cubic mesophase at room temperature, in addition to the *Pn3m* and *Ia3d* cubic mesophases reported previously.[117,118] The new mesophase, which we termed Q_L, exists in a small isotropic island within a two-phase region connecting the cubic, lamellar, and micellar phases. Like other cubic mesophases, Q_L displays three-dimensional long-range order – it is transparent, non-birefringent, and stable. However, it is also non-viscous, while other cubic mesophases are highly viscous. Our analysis showed that Q_L is made of discrete micellar aggregates arranged in a primitive lattice. Determination of the space group of this mesophase is not definite; we suggest it could belong to the *Pm3n* (Q_{223}), *P4₃32* (Q_{212}), or *P4₂32* (Q_{208}) space groups.[117,118]

Formation of lyotrophic phases similar to those created by GMO was reported for a few other lipid systems. Examples include phytantriol (PtOH or PHYT), a lipid comprised of a highly branched phytanyl tail with a tri-hydroxy head-group,[124] dioleoyl phosphatidylethanolamine (DOPE) containing small amounts of pegilated GMO,[125] mixtures of anionic and cationic lipids,[126] and unsaturated phosphatidylcholines with cholesterol.[127] The phase behavior and properties of these systems are much less characterized at present.

The cubic and hexagonal mesophases are thermodynamically stable systems that often exist in equilibrium with excess water.[128,129] The great interest in these phases, especially in the bicontinuous cubic mesophases for encapsulation and delivery applications, is linked to the extremely high surface area of the bilayer, and its ability to solubilize guest molecules of hydrophilic, lipophilic, and amphiphilic natures. However, except for Q_L, these phases are characterized by a gel-like appearance and high viscosity,[129] which complicate their use in commercial applications. To overcome this, cubic and hexagonal mesophases are dispersed into low-viscosity nanostructured aggregates that remain ordered and preserve many of the valuable lipid characteristics.

8.5.2 Dispersion of the non-parallel mesophases into cubosomes and hexosomes

Following the pioneering work of Luzzati on LC phases of amphiphiles,[15] cubosomes were first presented in 1989 by Larsson.[130] Due to their low solubility in water, cubosomes retain the high degree of internal order and structural regularity of the bulk mesophases from which they were formed (Fig. 8.4). Hence, many of the attractive physico-chemical characteristics such as high surface-area-to-volume ratio and the bilayer ability to solubilize molecules of various types are preserved. The viscosity of the dispersed phases is considerably lower, making them more usable in technological applications and attractive vehicles for the transport of bioactive compounds.

The dispersion can be done at high temperatures in an excess of water, or at low temperatures using high energy devices (mechanical or ultrasonic) or high pressure homogenizers.[111,112,114,131,132] Simple dispersion yields unstable particles that aggregate over time. To prevent aggregation, the dispersed particles are stabilized by surfactants, proteins, lipids, or polymers.[130,133] It was found that amphiphilic block copolymers such as the tri-block copolymer Pluronic F127 can preserve the inner cubic structure and stabilize cubosomes created from GMO.[101,109,110,112,114,115,131,132] F127 adsorbs onto the particle surface and reduces the interfacial energy. The adsorption provides a coating that stabilizes the particles and prevents their aggregation. Stabilization with F127 was found to be concentration dependent: a low quantity of stabilizer may lead to swelling of inverted cubic mesophases, followed by transitions into other symmetries[101,134] or into hexagonal mesophases. Addition of high F127 concentrations results in the formation of numerous vesicles that coexist with the cubosomes.[112] Small and stable cubosomes coexisting with many vesicles also formed upon dispersion of the cubic mesophase in the GMO-alcohol-water system.[119] Ethanol was originally added to GMO to allow spontaneous formation of cubosomes, since hydrotropes are generally considered as compounds that destroy LC phases.[119] Though cubosomes did not form spontaneously, the presence of ethanol was found to considerably ease the dispersion of the cubic phase[119] and cubosome formation. The mechanism of the alcohol intervention is not fully clear,[38,119,129] but was explained by the chaotropic effect of the alcohol, which increases lipid solubility. We recently found that the inverted micellar Q_L mesophase can also be easily dispersed into stable, nanosized cubosomes (unpublished data). A new protocol for producing GMO-based and PtOH/vitamin E-based cubosomes and other lipid nanoparticles with good colloidal stability was developed by the group of Barauskas and

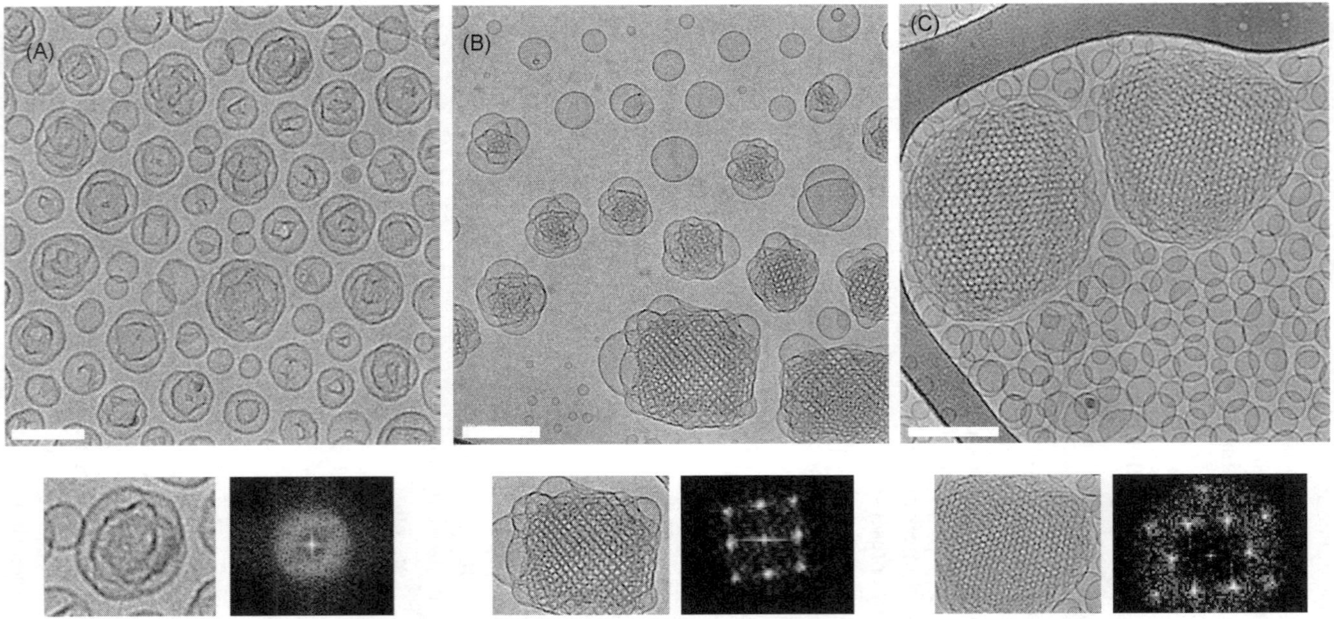

Fig. 8.4 Direct-imaging cryo-TEM pictures revealing the distribution of sizes, shapes and types of nanoparticles that form upon dispersion of a bicontinuous cubic monoolein/ethanol/water phase. The symmetry of the nanoparticles can be determined by applying the fast Fourier transformation on the nanoparticles. Bars equal 100 nm.

coworkers.[128] Using high energy microfluidization and heat treatment they were able to minimize the quantity of residual vesicles and to obtain good control of the particle size distribution (100–500 nm).

Hexosomes can be created by heating GMO cubosomes[135] or by dispersion of a hexagonal phase. In both routes the hexosomes are stabilized with F127. As for cubosomes, the competition between the rate of hydration of the lipid head-groups due to the dispersion (which decreases the particle size), and the polymer adsorption to the surface (surface coating) required the addition of a critical, relatively high concentration of stabilizer in order to form stable and ordered nano-sized hexosomes.[116] Only large hexosomes, of lower order and lower stability, were created at low F127 concentrations.

8.5.3 Characterizing the non-parallel bulk phases and their dispersed nanoparticles

Increasing our understanding of the bulk two-dimensional and three-dimensional mesophases requires extensive basic research. Insight into these systems can be achieved using numerous indirect (scattering, spectroscopy) and direct (microscopy) methods that can reveal structural details on the formation, morphology, shape, size, internal structure, and stability of the bulk mesophases and their dispersed nanoparticles.

Polarized light microscopy is used to construct the phase diagrams.[136] The cubic phases are isotropic and thus show no birefringence under cross-polarized light. In contrast, the lamellar and hexagonal LC phases are identified by their characteristic textures. Normal vs inverted mesophases are identified by their location in the phase diagram, following the packing parameter concept introduced in equation (8.1), with inverted structures forming at higher lipid content and higher temperatures. Small angle X-ray scattering (SAXS) is essential for identifying the type, space-group, and dimensions of the bulk liquid crystalline phases. SAXS analysis is performed by calculating the relative position of the sharp successive peaks in the scattering diffraction curves.[106,118,137] NMR[109,138] and differential scanning calorimetry (DSC)[131,139,140] are good techniques to follow phase transitions, and are used to determine the phase boundaries. Discriminating the mesophases, even the bicontinuous cubic phases from each other, is also possible using rheological measurements.[141]

Analyzing the structure and internal order of the dispersed particles is more complex. Because the particles are small they do not show birefringence; thus polarized light microscopy is not very useful. Scattering from the dispersed particles is rather weak due to their low concentration, small size, and heterogeneous shape. Moreover, the number of diffraction peaks from the dispersed nanoparticles is small, and they become wider and not well resolved. These parameters, combined with coexistence of various types of assemblies that often occurs following dispersion, make symmetry and space-group determination from SAXS data complex. Scattering methods, however, remain very useful for size and particle polydispersity determination. SANS (small-angle neutron scattering), which is sensitive to the isotopic composition of the scattering material, and SAXS can additionally provide information on the localization of solubilized molecules within the ordered lipid structures.

Transmission electron microscopy techniques (freeze-fracture and direct-imaging cryo-TEM) play a central role in the study of the bulk mesophases and predominantly their dispersions. Direct imaging cryo-TEM is especially advantageous since it provides direct structural details on shape, size, and internal ordering at a high resolution of ~1–2 nm that is comparable with the assembly's dimensions (a bilayer thickness is ~4 nm, thus the bilayer leaflets can be readily resolved). Moreover, the information obtained by cryo-TEM is detailed and aggregate specific, and is not averaged over the whole volume as with the indirect methods; thus coexistence of various structures is readily recognized. In addition to disclosing micro- and nanostructural characteristics such as the shape and size of the particles, direct-imaging cryo-TEM can probe the order, periodicity, and symmetry, and it often allows space-group determination by applying the fast Fourier transform algorithm (FFT) on the ordered regions. This approach has been demonstrated in numerous studies of cubic and hexagonal mesophases and their dispersions.[109,118,119,128,135,137,142,143] (and additional references mentioned elsewhere in the text). Direct-imaging cryo-TEM is also very powerful in revealing new morphologies and short-lived intermediates that form during phase transitions, and in detecting how incorporation of guest molecules affects the structure. Digital-imaging using CCD cameras makes direct-imaging cryo-TEM even more practical, efficient, and reliable for such investigations.[144,145]

Freeze-fracture cryo-TEM (FF-TEM) has also been frequently applied to study cubic and hexagonal mesophases of lipids and their dispersed particles. This technique directly discloses the shape, size, morphology, and heterogeneity of the particles, their surface topology, and lateral periodicity. Fractures often form along different crystallographic directions, thereby exposing different internal planes that allow conclusive determination of the periodicity, symmetry, and space-group.[104,146,147] FF is also very useful in exploring the effect of additives on the structure (e.g. changes in the lattice parameter, defects in the structure, and transitions into other symmetries)[148,149] as well as in studying temperature-induced phase transitions and formation of intermediate structures. Complementary FF and SAXS experiments,[104] or FF with modeling and computational analysis[148,149] have been shown to be effective approaches for quantitative analysis.

8.5.4 Introducing guest molecules in non-parallel mesophases and their nanoparticles

Cubic and hexagonal phases and their dispersed nanoparticles are attractive vehicles for controlled release of bioactive molecules because of their small pore size (<10 nm), their ability to solubilize hydrophobic, hydrophilic, and amphiphilic molecules, and their easy biodegradability.[108] They can potentially provide extremely large surfaces for encapsulation of active molecules such as nutraceuticals. For example, a surface of 400 m^2 g^{-1} of cubic phase was reported for the inverted bicontinuous gyroid cubic mesophase.[150] Additionally, the dispersed particles offer high bilayer area-to-particle ratio (volume ratios) compared to liposomes, as well as higher resistance to rupture due to their high internal viscosity.[151]

However, introducing guest molecules often modifies the mesophase and nanoparticle internal order and stability due to changes introduced in the packing, thickness, curvature, and rigidity of the bilayer interface. The additive character determines its loading capacity and location in the aggregate and the resulting changes in the structure. Lipophilic compounds are expected to penetrate to the hydrophobic core of the inverted mesophases, thereby increasing the volume and promoting transitions towards structures of higher negative curvature. Indeed, addition of oils,[152] diglycerides,[108] some sugars,[153] and hydrophobic compounds[136] enhanced the formation of Hii at the expense of Qii. It was also shown that hydrophilic molecules like diglycerol monooleate[154] and sodium oleate,[109,138] which likely reside in the polar domains, induced transitions in the other direction, namely, from inverted bicontinuous cubic to lamellar mesophases. Analogous behavior to that reported for the bulk mesophases was also found for monoglyceride cubosomes: transitions from inverted bicontinuous cubic to inverted hexagonal symmetry were initiated upon the addition of hydrophobic compounds such as triglycerides,[112] and further transitions into an inverted cubic phase of discrete micelles were generated when oleic acid[114] or tetradecane[115] were added.

Only a few studies explored the introduction of food additives into non-lamellar mesophases. Despite the enormous surface of cubic structures, investigations show that the entrapment capacity of nutraceuticals is relatively low. In recent investigations, Efrat and coworkers solubilized different hydrophobic nutraceuticals in the bulk Q_L mesophase.[155] The solubilization capacities were found to be highly dependent on the specific nature of the active molecules. For example, relatively high levels of phytosterols could be incorporated, though the order of the mesophase decreased continuously with the increase in phytosterol concentration. On the other hand, only traces of the unsaturated antioxidant lycopene could be incorporated before precipitation occurred. However, the authors reported on synergistic effects when the two active molecules were solubilized together, reflected in an increase in lycopene-loading as well as an increase in the order and stability of the nutraceutical-loaded mesophase.

The tortoise-like structure of the cubic phase allows a slow diffusive release of solubilized nutraceuticals and functional ingredients from the bulk phases.[156] The rate of release and delivery of nutraceuticals can potentially be tuned via appropriate formulation of the lipids, stabilizers, and external parameters (temperature, pH, lipids modifications, additives) to match specific applications in food and nutrition. Recently, cubic phases were used to enhance flavor formation in a model Maillard reaction. High release of aroma from cubic phases was found, likely due to the large internal interface of the cubic phase and controlled-release associated with the low diffusivity of the compounds inside the cubic phase.[157] The production was found to be more effective than in microemulsions and several-fold higher than in water. However, despite their large bilayer area, cubosomes display a burst release of both hydrophilic and hydrophobic compounds,[158] which may be associated with their nanometric size.

8.6 Future trends

While liposomes are employed commercially in many food applications, no commercial product exists that utilizes cubosomes or hexosomes. Developing advanced technologies enabling the use of these nanometric lipid carriers to deliver, target, and release entrapped nutraceuticals in a controlled manner requires a thorough understanding of the fundamental characteristics of the mesophases and their dispersions. To gain better insight into these systems, comprehensive investigation of the complex links between the molecular characteristics, self-assembly, phase behavior, and the properties of the formed structures and their interaction with the guest molecules is required. Future research should continue to focus on producing lipid-based vehicles from non-purified polar lipid sources to reduce the cost of raw materials.[3] Establishment of new dispersion protocols that avoid the use of non-food-grade solvents and, in addition, will better preserve the internal order of the assemblies, minimize formation of residual vesicles, and improve the stability of the lipid supra-molecular structures in their dispersed phase are needed. This line of research should eventually yield reproducible, large-scale preparations at low costs.

Lipid carriers are mainly utilized in drug delivery applications, where small quantities of a bioactive drug are sufficient to resolve various health-related problems. Successful application of lipid self-assembled superstructures in food nutrition on a large scale will often require higher loads and possibly other delivery approaches. The interactions of the lipid assemblies with the entrapped guest molecules and with the environment as well as the stability of the loaded vehicles need to be addressed in more detail. Obtaining controlled release profiles of encapsulated nutraceuticals at appropriate rates, at specific loci, will be major issues toward implementation of vehicles such as cubosomes and hexosomes in food systems.[106,151] Pioneer reports on the use of nonlamellar lipid mesophases to release flavor and aroma compounds in foods in a controlled manner have shown great potential. More work is required to explore the capabilities of lipid mesophases and their dispersed phases as reactors in Maillard reactions.[135]

Dietary supplements are now considered a primary source for preventing or reducing the risk factors of several diseases; however, data regarding their bioactivity and bioavailability are scarce. Proper evaluation of the bioactivity and bioavailability of sensitive food additives when they are released is required, as well as investigation of the clinical efficiency and mechanism of action of the released components.

New concepts based on nanotechnology are being explored to improve the functionality and delivery efficiency of the lipid vehicles. One such approach involves encapsulation of wet or dry forms of lipid complexes inside other nanoparticles. Such multi-structures can be used to develop high performance delivery vehicles to target the lipid carriers to specific sites. The challenges of this form of delivery are to achieve better protection of the sensitive ingredients, maintain their concentration at desired levels and in a stable form, prevent their degradation, and increase their bioactivity.[159]

8.6 References

1. LASIC D D, 'Novel applications of liposomes', *Trends in Biotechnology*, 1998 **16**(7) 307–21.
2. MOZAFARI M R, FLANAGAN J, MATIA-MERINO L, AWATI A, OMRI A, SUNTRES Z E and SINGH H, 'Recent trends in the lipid-based nanoencapsulation of antioxidants and their role in foods', *Journal of the Science of Food and Agriculture*, 2006 **86**(13) 2038–45.
3. TAYLOR T M, DAVIDSON P M, BRUCE B D and WEISS J, 'Liposomal nanocapsules in food science and agriculture', *Critical Reviews in Food Science and Nutrition*, 2005 **45**(7–8) 587–605.
4. MOZAFARI M R and MORTAZAVI S M, *Nanoliposomes: From Fundamentals to Recent Developments*, Oxford, Trafford, 2005.
5. ISRAELACHVILI J N, MITCHELL D J and NINHAM B W, 'Theory of self-assembly of hydrocarbon amphiphiles into micelles and bilayers', *Journal of the Chemical Society – Faraday Transactions II*, 1976 **72** 1525–68.
6. ISRAELACHVILI J N, MITCHELL D J and NINHAM B W, 'Theory of self-assembly of lipid bilayers and vesicles', *Biochimica et Biophysica Acta*, 1977 **470**(2) 185–201.
7. TANFORD C, *The Hydrophobic Effect: Formation of Micelles and Biological Membranes*, New York, Wiley Intersience Publications, 1980.
8. LASIC D D, *Liposomes: From Physics to Applications*, Amsterdam, Elsevier, 1993.
9. VOET D and VOET G J, *Biochemistry*, New York, John Wiley & Sons, 1995.
10. MARCELJA S and WOLFE J, 'Properties of bilayer membranes in the phase-transition or phase-separation region', *Biochimica et Biophysica Acta*, 1979 **557**(1) 24–31.
11. NEW R R C, 'Preparation of liposomes', in New R R C, *Liposomes: A Practical Approach*. Oxford, IRL Press, 1990, 33–104.
12. TANFORD C, 'Monolayers, micelles, lipid vesicles and biomembranes', in *Physics of Amphiphiles: Micelles, Vesicles and Microemulsions,* Degiorgio V, Ed. Amsterdam, Elsevier Science, 1995, 547–54.
13. ISRAELACHVILI J N, 'Measurements of hydration forces between macroscopic surfaces', *Chemica Scripta*, 1985 **25**(1) 7–14.
14. EVANS D F and WENNERSTROM H, *The Colloidal Domain: Where Physics, Chemistry, Biology, and Technology Meet*, New York, Wiley-VCH, 1999.
15. LUZZATI V, GULIK-KRZYWICKI T and TARDIEU A, 'Polymorphism of lecithins', *Nature*, 1968 **218**(5146) 1031–4.
16. SEDDON J M, 'Structure of the inverted hexagonal (Hii) phase, and non-lamellar phase-transitions of lipids', *Biochimica et Biophysica Acta*, 1990 **1031**(1) 1–69.

17. LARSSON K, 'Colloidal dispersions of ordered lipid-water phases', *Journal of Dispersion Science and Technology*, 1999 **20**(1–2) 27–34.
18. CHEN H D, WEISS J C and SHAHIDI F, 'Nanotechnology in nutraceuticals and functional foods', *Food Technology*, 2006 **60**(3) 30-6.
19. NAGARAJAN R, 'Molecular packing parameter and surfactant self-assembly: the neglected role of the surfactant tail', *Langmuir*, 2002 **18**(1) 31–8.
20. NAGARAJAN R, 'Theory of micelle formation', in *Structure-performance Relationships in Surfactants*, Seumi K and Ueno M, Eds. New York, Marcel Dekker, 2003, 1–82.
21. DEAMER D W and NICHOLS J W, 'Proton hydroxide permeability of liposomes', *Proceedings of the National Academy of Sciences of the United States of America – Biological Sciences*, 1983 **80**(1) 165–8.
22. GREGORIADIS G, 'Old drugs in new clothing', *Nature*, 1984 **310**(5974) 186–7.
23. SHARMA A and SHARMA U S, 'Liposomes in drug delivery: progress and limitations', *International Journal of Pharmaceutics*, 1997 **154**(2) 123–40.
24. AUGUSTIN M A, SANGUANSRI L, MARGETTS C and YOUNG B, 'Microencapsulation of food ingredients', *Food Australia*, 2001 **53**(6) 220–3.
25. GOUIN S, 'Microencapsulation: industrial appraisal of existing technologies and trends', *Trends in Food Science & Technology*, 2004 **15**(7–8) 330–47.
26. LASIC D D and PAPAHADJOPOULOS D, 'Liposomes revisited', *Science*, 1995 **267**(5202) 1275–6.
27. ANGELOVA M I, SOLEAU S, MELERD PH and FAUCON J I, 'Preparation of giant vesicles by external AC electric fields. Kinetics and applications', *Prog. Colloid Polym. Sci*, 1992 **89** 127–31.
28. FEITOSA E, BARRELEIRO P C A and OLOFSSON G, 'Phase transition in dioctadecyldimethylammonium bromide and chloride vesicles prepared by different methods', *Chemistry and Physics of Lipids*, 2000 **105**(2) 201–13.
29. JOANNIC R, AUVRAY L and LASIC D D, 'Monodisperse vesicles stabilized by grafted polymers', *Physical Review Letters*, 1997 **78**(17) 3402–5.
30. LASIC D D, VALLNER J J and WORKING P K, 'Sterically stabilized liposomes in cancer therapy and gene delivery', *Current Opinion in Molecular Therapeutics*, 1999 **1**(2) 177–85.
31. BANGHAM A D, HILL M W and HILLER N G A, 'Preparation and use of liposomes as models of biological membranes', in *Methods in Membrane Biology*, Korn E D, Ed. New York, Plenum Press, 1974, 1–68.
32. VEMURI S and RHODES C T, 'Preparation and characterization of liposomes as therapeutic delivery systems: a review', *Pharm Acta Helv*, 1995 **70**(2) 95–111.
33. VEMURI S and RHODES C T, 'Encapsulation of a water-soluble drug in a liposome preparation – removal of free drug by washing', *Drug Development and Industrial Pharmacy*, 1995 **21**(11) 1329–38.
34. BATZRI S and KORN E D, 'Single bilayer liposomes prepared without sonication', *Biochimica et Biophysica Acta*, 1973 **298**(4) 1015–19.
35. KIRBY C and GREGORIADIS G, 'Dehydration-rehydration vesicles – a simple method for high-yield drug entrapment in liposomes', *Bio-Technology*, 1984 **2**(11) 979–84.
36. PEEL M, 'Liposomes produced by combined homogenization/extrusion', *GIT Laboratory Journal*, 1999 **3** 37–8.
37. ZHANG L, HU J and LU Z, 'Preparation of liposomes with a controlled assembly procedure', *Journal of Colloid and Interface Science*, 1997 **190**(1) 76–80.
38. FREDERIKSEN L, ANTON K, VAN HOOGEVEST P, KELLER H R and LEUENBERGER H,

'Preparation of liposomes encapsulating water-soluble compounds using supercritical carbon dioxide', *Journal of Pharmaceutical Sciences*, 1997 **86**(8) 921–8.

39. ZHENG L, HUANG X L, FAN Z, BOROWSKI L, WILSON C C and RINALDO C R, 'Delivery of liposome-encapsulated HIV type 1 proteins to human dendritic cells for stimulation of HIV type 1-specific memory cytotoxic T lymphocyte responses', *Aids Research and Human Retroviruses*, 1999 **15**(11) 1011–20.
40. ZHENG S, ALKAN-ONYUKSEL H, BEISSINGER R L and WASAN D T, 'Liposome microencapsulations without using any organic solvent', *Journal of Dispersion Science and Technology*, 1999 **20**(4) 1189–203.
41. KIM H H Y and BAIANU I C, 'Novel liposome microencapsulation techniques for food applications', *Trends in Food Science & Technology*, 1991 **2** 55–61.
42. KOIDE K and KAREL M, 'Encapsulation and stimulated release of enzymes using lecithin vesicles', *International Journal of Food Science and Technology*, 1987 **22** 707–23.
43. MAA Y F and HSU C C, 'Effect of primary emulsions on microsphere size and protein-loading in the double emulsion process', *Journal of Microencapsulation*, 1997 **14**(2) 225–41.
44. VUILLEMARD J C, 'Recent advances in the large-scale production of lipid vesicles for use in food-products – microfluidization', *Journal of Microencapsulation*, 1991 **8**(4) 547–62.
45. GREGORIADIS G, BACON A, CAPARROS-WANDERLEY W and McCORMACK B, 'Plasmid DNA vaccines: entrapment into liposomes by dehydration-rehydration', *Liposomes, Pt A*, 2003 **367** 70–80.
46. MADDEN T D, HARRIGAN P R, TAI L C L, BALLY M B, MAYER L D, REDELMEIER T E, LOUGHREY H C, TILCOCK C P S, REINISH L W and CULLIS P R, 'The accumulation of drugs within large unilamellar vesicles exhibiting a proton gradient – a survey', *Chemistry and Physics of Lipids*, 1990 **53**(1) 37–46.
47. PAPAHADJ D, 'Studies on mechanism of action of local anesthetics with phospholipid model membranes', *Biochimica et Biophysica Acta*, 1972 **265**(2) 169–86.
48. YATVIN M B, KREUTZ W, HORWITZ B A and SHINITZKY M, 'pH-sensitive liposomes – possible clinical implications', *Science*, 1980 **210**(4475) 1253–4.
49. EMANUEL N, KEDAR E, BOLOTIN E M, SMORODINSKY N I and BARENHOLZ Y, 'Preparation and characterization of doxorubicin-loaded sterically stabilized immunoliposomes', *Pharm Res*, 1996 **13**(3) 352–9.
50. CEVE G, BLUMEB G and SCHTITZLEIN A, 'The skin: pathway for systemic treatment with patches and lipid based agent carries', *Advanced Durg Delivery Reviews*, 1996 **18** 349–78.
51. KLIBANOV A L, BOGDANOV A A, LUKYANOV A N, TONEVITSKY A G, TRAKHT I N, TIKHONENKO T I, ZAVIZION B A and TORCHILIN V P, 'Biotinylated pH-sensitive liposomes, a container for targeting of biologically-active substances', *Vestnik Akademii Meditsinskikh Nauk Sssr*, 1990(8) 50–3.
52. KLIBANOV A L, MARUYAMA K, TORCHILIN V P and HUANG L, 'Amphipathic polyethylene-glycols effectively prolong the circulation time of liposomes', *FEBS Letters*, 1990 **268**(1) 235–7.
53. LASIC D D, 'Sterically stabilized vesicles', *Angewandte Chemie – International Edition in English*, 1994 **33**(17) 1685–98.
54. SENIOR J H, TRIMBLE K R and MASKIEWICZ R, 'Interaction of positively-charged liposomes with blood – implications for their application *in vivo*', *Biochimica et Biophysica Acta*, 1991 **1070**(1) 173–9.

55. WOODLE M C, NEWMAN M S, COLLINS L R and MARTIN F J, 'Efficient evaluation of long circulating or stealth liposomes by studies of *in vivo* blood-circulation kinetics and final organ distribution in rats', *Biophysical Journal*, 1990 **57**(2) A261.
56. PERRY D G and MARTIN W J, 'Fluorescent liposomes as quantitative markers of phagocytosis by alveolar macrophages', *Journal of Immunological Methods*, 1995 **181**(2) 269–85.
57. SIEGEL D P, GREEN W J and TALMON Y, 'The mechanism of lamellar-to-inverted hexagonal phase transitions: a study using temperature-jump cryo-electron microscopy', *Biophys J*, 1994 **66**(2 Pt 1) 402–14.
58. BEZOT P, OSTROWSKY N and HESSEBEZOT C, 'Light-scattering data-analysis for samples with large polydispersities', *Optics Communications*, 1978 **25**(1) 14–18.
59. BROWN J C, PUSEY P N, GOODWIN J W and OTTEWILL R H, 'Light-scattering study of dynamic and time-averaged correlations in dispersions of charged-particles', *Journal of Physics A – Mathematical and General*, 1975 **8**(5) 664–82.
60. KOPPEL D E, 'Analysis of macromolecular polydispersity in intensity correlation spectroscopy – method of cumulants', *Journal of Chemical Physics*, 1972 **57**(11) 4814–20.
61. SELSER J C, YEH Y and BASKIN R J, 'Light-scattering measurement of membrane vesicle permeability', *Biophysical Journal*, 1976 **16**(12) 1357–71.
62. VEMURI S and RHODES C T, 'Development and validation of a drug-release rate method for a water-soluble drug in a liposome preparation', *Drug Development and Industrial Pharmacy*, 1995 **21**(11) 1353–64.
63. GREGORIADIS G, DAVIS D and DAVIES A, 'Liposomes as immunological adjuvants – antigen incorporation studies', *Vaccine*, 1987 **5**(2) 145–51.
64. KIRBY C J, 'Controlled delivery of functional food ingredients: opportunities for liposomes in the food industry', in *Liposome Technology*, Gregoriadis G, Ed., London, CRC Press, 1993, 215–32.
65. GREGORIADIS G and RYMAN B E, 'Liposomes as carriers of enzymes or drugs – new approach to treatment of storage diseases', *Biochemical Journal*, 1971 **124**(5) P58.
66. LAW B A and WIGMORE A, 'Accelerated cheese ripening with food grade proteinases', *Journal of Dairy Research*, 1982 **49**(1) 137–46.
67. LAW B A and WIGMORE A S, 'Accelerated ripening of cheddar cheese with a commercial proteinase and intracellular enzymes from starter streptococci', *Journal of Dairy Research*, 1983 **50**(4) 519–25.
68. DUFOUR P, LALOY E, VUILLEMARD J C and SIMARD R, 'Liposomes in cheesemaking', in *Handbook of Nonmedical Applications of Liposomes,* Lasic D and Barenholz Y, Eds, Boca Raton, FL, CRC Press, 1996, 158–64.
69. ALKHALAF W, PIARD J C, ELSODA M, GRIPON J C, DESMAZEAUD M and VASSAL L, 'Liposomes as proteinase carriers for the accelerated ripening of Saint-Paulin type cheese', *Journal of Food Science*, 1988 **53**(6) 1674–9.
70. KIRBY C J, BROOKER B E and LAW B A, 'Accelerated ripening of cheese using liposome-encapsulated enzyme', *International Journal of Food Science and Technology*, 1987 **22**(4) 355–75.
71. BERGSTRAND N, ARFVIDSSON M C, KIM J M, THOMPSON D H and EDWARDS K, 'Interactions between pH-sensitive liposomes and model membranes', *Biophysical Chemistry*, 2003 **104**(1) 361–79.
72. BOOMER J A, INEROWICZ H D, ZHANG Z Y, BERGSTRAND N, EDWARDS K, KIM J M and THOMPSON D H, 'Acid-triggered release from sterically stabilized fusogenic liposomes via a hydrolytic DePEGylation strategy', *Langmuir*, 2003 **19**(16) 6408–15.

73. KHEADR E E, VUILLEMARD L C and EL-DEEB S A, 'Acceleration of cheddar cheese lipolysis by using liposome-entrapped lipases', *Journal of Food Science*, 2002 **67**(2) 485–92.
74. FRESTA M and PUGLISI G, 'Enzyme loaded liposomes for cheese ripening', *Microsperes, Microcapsule and Liposomes,* 1999 **2** 639–70.
75. FRESTA M, WEHRLI E and PUGLISI G, 'Neutrase entrapment in stable multilamellar and large unilamellar vesicles for the acceleration of cheese ripening', *Journal of Microencapsulation*, 1995 **12**(3) 307–25.
76. KIRBY C, CLARKE J and GREGORIADIS G, 'Effect of the cholesterol content of small unilamellar liposomes on their stability *in vivo* and *in vitro*', *Biochemical Journal*, 1980 **186**(2) 591–8.
77. LALOY E, VUILLEMARD J C, DUFOUR P and SIMARD R, 'Release of enzymes from liposomes during cheese ripening', *Journal of Controlled Release*, 1998 **54**(2) 213–22.
78. MATSUZAKI M, McCAFFERTY F and KAREL M, 'The effect of cholesterol content of phospholipid-vesicles on the encapsulation and acid resistance of beta-galactosidase from *Escherichia Coli*', *International Journal of Food Science and Technology*, 1989 **24**(4) 451–60.
79. RAO D R, CHAWAN C B and VEERAMACHANENI R, 'Liposomal encapsulation of beta-galactosidase – comparison of two methods of encapsulation and *in vitro* lactose digestibility', *Journal of Food Biochemistry*, 1995 **18**(4) 239–51.
80. ANNESINI M C, BRAGUGLIA C M, MEMOLI A, PALERMITI L G and DISARIO S, 'Surfactant as modulating agent of enzyme-loaded liposome activity', *Biotechnology and Bioengineering*, 1997 **55**(2) 261–6.
81. BANVILLE C, VUILLEMARD J C and LACROIX C, 'Comparison of different methods for fortifying Cheddar cheese with vitamin D', *International Dairy Journal*, 2000 **10**(5–6) 375–82.
82. TESORIERE L, BONGIORNO A, PINTAUDI A M, D'ANNA R, D'ARPA D and LIVREA M A, 'Synergistic interactions between vitamin A and vitamin E against lipid peroxidation in phosphatidylcholine liposomes', *Archives of Biochemistry and Biophysics*, 1996 **326**(1) 57–63.
83. WATERS R E, WHITE L L and MAY J M, 'Liposomes containing alpha-tocopherol and ascorbate are protected from an external oxidant stress', *Free Radical Research*, 1997 **26**(4) 373–9.
84. YIN M C and CHENG W S, 'Oxymyoglobin and lipid oxidation in phosphatidylcholine liposomes retarded by alpha-tocopherol and beta-carotene', *Journal of Food Science*, 1997 **62**(6) 1095–7.
85. EASTMAN S J, SIEGEL C, TOUSIGNANT J, SMITH A E, CHENG S H and SCHEULE R K, 'Biophysical characterization of cationic lipid:DNA complexes', *Biochimica et Biophysica Acta – Biomembranes*, 1997 **1325**(1) 41–62.
86. STAHL W, JUNGHANS A, DE BOER B, DRIOMINA E S, BRIVIBA K and SIES H, 'Carotenoid mixtures protect multilamellar liposomes against oxidative damage: synergistic effects of lycopene and lutein', *FEBS Letters*, 1998 **427**(2) 305–8.
87. MUZYKANTOV V R, 'Targeting of superoxide dismutase and catalase to vascular endothelium', *Journal of Controlled Release*, 2001 **71**(1) 1–21.
88. SMITH T, 'Liposomes in cell culture', in *Physical Chemistry of Food Processes*, Baianu I C, Pesse H and Kumosinski T F Eds. New York, Van Nostrand Reinhold, 1996, 554–87.
89. STONE W L and SMITH M, 'Therapeutic uses of antioxidant liposomes', *Molecular Biotechnology*, 2004 **27**(3) 217–30.

90. LEE D H, JIN B H, HWANG Y I and LEE S C, 'Encapsulation of bromelain in liposome', *Journal of Food Science and Nutrition*, 2000 **5** 81–5.
91. SUNTRES Z E and SHEK P N, 'Alleviation of paraquat-induced lung injury by pre-treatment with bifunctional liposomes containing alpha-tocopherol and glutathione', *Biochem Pharmacol*, 1996 **52**(10) 1515–20.
92. MOZAFARI M R, 'Micro and nano carrier technologies: high quality production within pharmaceutical standards', *Cellular & Molecular Biology Letters*, 2004 **9**(Suppl. 2) 44–5.
93. KIKUCHI H, CARLSSON A, YACHI K and HIROTA S, 'Possibility of heat sterilization of liposomes', *Chemical & Pharmaceutical Bulletin*, 1991 **39**(4) 1018–22.
94. PATEL G B, AGNEW B J, DESCHATELETS L, FLEMING L P and SPROTT G D, '*In vitro* assessment of archaeosome stability for developing oral delivery systems', *International Journal of Pharmaceutics*, 2000 **194**(1) 39–49.
95. ARNAUD J P, 'Liposomes in the agro/food industry', *Hi-Tech*, 1995 **6**(5) 30–6.
96. ARNAUD J P, 'Pro-liposomes for the food industry', *Food Technol Eur*, 1995 **2** 30–4.
97. ARNAUD J P, 'Liposome-based functional drinks', in *Food Ingredients Europe Conference Proceedings 1996*, Maarssen Miller Freeman plc, 1997.
98. KELLER B C, 'Liposomes in nutrition', *Trends in Food Science & Technology*, 2001 **12**(1) 25–31.
99. HYDE S T, ANDERSSON S, LARSSON K, BLUM Z, LANDH T, LIDIN S and NINHAM B W, *The Language of Shape*, New York, Elsevier, 1997.
100. SEDDON J M, TEMPLER R H, LIPOWSKY A R and SACKMANN R, 'Polymorphism of lipid-water systems', in *Handbook of Biological Physics*, Lipowsky A R and Sackmann R, Eds, New York, Elsevier, 1995, 97–153.
101. LANDH T, 'Phase-behavior in the system pine oil monoglycerides-Poloxamer-407-water at 20ºC', *Journal of Physical Chemistry*, 1994 **98**(34) 8453–67.
102. HYDE S T, 'Identification of lyotropic liquid crystalline mesophases', in *Handbook of Applied Surface and Colloid Chemistry*, Holmberg K, Ed., New York, John Wiley & Sons Ltd., 2001, 299–329.
103. LUZZATI V, VARGAS R, MARIANI P, GULIK A and DELACROIX H, 'Cubic phases of lipid-containing systems – elements of a theory and biological connotations', *Journal of Molecular Biology*, 1993 **229**(2) 540–51.
104. DELACROIX H, GULIK-KRZYWICKI T and SEDDON J M, 'Freeze fracture electron microscopy of lyotropic lipid systems: quantitative analysis of the inverse micellar cubic phase of space group Fd3m (Q(227))', *Journal of Molecular Biology*, 1996 **258**(1) 88–103.
105. LUZZATI V, DELACROIX H and GULIK A, 'The micellar cubic phases of lipid-containing systems: analogies with foams, relations with the infinite periodic minimal surfaces, sharpness of the polar apolar partition', *Journal De Physique Ii*, 1996 **6**(3) 405–18.
106. YAGHMUR A, DE CAMPO L, SALENTINIG S, SAGALOWICZ L, LESER M E and GLATTER O, 'Oil-loaded monolinolein-based particles with confined inverse discontinuous cubic structure (*Fd3m*)', *Langmuir*, 2006 **22**(2) 517–21.
107. LARSSON K, 'Two cubic phases in monoolein water-system', *Nature*, 1983 **304**(5927) 664.
108. BORNE J, NYLANDER T and KHAN A, 'Microscopy, SAXD, and NMR studies of phase behavior of the monoolein-diolein-water system ', *Langmuir* 2000 **16**(26) 10044–54.
109. BORNE J, NYLANDER T and KHAN A, 'Phase behavior and aggregate formation for the

aqueous monoolein system mixed with sodium oleate and oleic acid', *Langmuir* 2001 **17**(25) 7742–51.

110. DE CAMPO L, YAGHMUR A, SAGALOWICZ L, LESER M E, WATZKE H and GLATTER O, 'Reversible phase transitions in emulsified nanostructured lipid systems', *Langmuir*, 2004 **20**(13) 5254–61.

111. GUSTAFSSON J, LJUSBERG-WAHREN H, ALMGREN M and LARSSON K, 'Cubic lipid-water phase dispersed into submicron particles', *Langmuir*, 1996 **12**(20) 4611–13.

112. GUSTAFSSON J, LJUSBERG-WAHREN H, ALMGREN M and LARSSON K, 'Submicron particles of reversed lipid phases in water stabilized by a nonionic amphiphilic polymer', *Langmuir*, 1997 **13**(26) 6964–71.

113. NAKANO M, SUGITA A, MATSUOKA H and HANDA T, 'Small-angle X-ray scattering and C-13 NMR investigation on the internal structure of "cubosomes"', *Langmuir*, 2001 **17**(13) 3917–22.

114. NAKANO M, TESHIGAWARA T, SUGITA A, LEESAJAKUL W, TANIGUCHI A, KAMO T, MATSUOKA H and HANDA T, 'Dispersions of liquid crystalline phases of the monoolein/oleic acid/Pluronic F127 system', *Langmuir*, 2002 **18**(24) 9283–8.

115. YAGHMUR A, DE CAMPO L, SAGALOWICZ L, LESER M E and GLATTER O, 'Emulsified microemulsions and oil-containing liquid crystalline phases', *Langmuir*, 2005 **21**(2) 569–77.

116. AMAR-YULI I, WACHTEL E, BEN SHOSHAN E, DANINO D, ASERIN A and GARTI N, 'Hexosome and hexagonal phases mediated by hydration and polymeric stabilizer', *Langmuir*, 2007 **23**(7) 3637–45.

117. EFRAT R, ASERIN A, DANINO D, WACHTEL E J and GARTI N, 'Novel discrete micellar cubic phase from a mixture of GMO/ethanol/water', *Australian Journal of Chemistry*, 2005 **58**(11) 762–6.

118. EFRAT R, ASERIN A, KASSELMAN E, DANINO D, WACHTEL E and GARTI N, 'Liquid micellar discontinuous cubic mesophase from ternary monoolein/ethanol/water mixtures', *Colloids and Surfaces A*, 2007 **299** 133–45.

119. SPICER P T, HAYDEN K L, LYNCH M L, OFORI-BOATENG A and BURNS J L, 'Novel process for producing cubic liquid crystalline nanoparticles (cubosomes)', *Langmuir*, 2001 **17**(19) 5748–56.

120. ENGSTROM S, ALFONS K, RASMUSSON M and LJUSBERG-WAHREN H, 'Solvent-induced sponge (L3) phases in the solvent-monoolein-water system', *Progr Colloid Polym Sci*, 1998 **108** 93–8.

121. WANG Z N, ZHENG L Q and INOUE T, 'Effect of sucrose on the structure of a cubic phase formed from a monoolein/water mixture', *Journal of Colloid and Interface Science*, 2005 **288**(2) 638–41.

122. RICHARDS C, TIDDY G J T and CASEY S, 'Liquid crystal and solution phases of sodium dodecyl-p-benzene sulfonate (LAS) and hexa-oxyethylene glycol dodecyl ether (C12E6); 1 : 1 mixtures in water', *Colloids and Surfaces A – Physicochemical and Engineering Aspects*, 2006 **288**(1–3) 103–12.

123. YANG B and WHITE J W, 'Isotope effects on the phase behaviour of cetyltrimethylammonium bromide in H-20 and D-20 studied by time resolved X-ray diffraction', *Colloids and Surfaces A – Physicochemical and Engineering Aspects*, 2006 **277**(1–3) 171–6.

124. BARAUSKAS J and LANDH T, 'Phase behavior of the phytantriol/water system', *Langmuir*, 2003 **19**(23) 9562–5.

125. JOHNSSON M, BARAUSKAS J and TIBERG F, 'Cubic phases and cubic phase dispersions in a phospholipid-based system', *J Am Chem Soc*, 2005 **127**(4) 1076–7.

126. TARAHOVSKY Y S, ARSENAULT A L, MACDONALD R C, MCINTOSH T J and EPAND R M, 'Electrostatic control of phospholipid polymorphism', *Biophys J*, 2000 **79**(6) 3193–200.
127. TENCHOV B G, MACDONALD R C and SIEGEL D P, 'Cubic phases in phosphatidylcholine-cholesterol mixtures: cholesterol as membrane 'fusogen'', *Biophys J*, 2006 **91**(7) 2508–16.
128. BARAUSKAS J, JOHNSSON M and TIBERG F, 'Self-assembled lipid superstructures: beyond vesicles and liposomes', *Nano Lett*, 2005 **5**(8) 1615–19.
129. SPICER P, 'Cubosomes: bicontinuous cubic liquid crystalline nanostructured particles', in *Encyclopedia of Nanoscience and Nanotechnology*, Schwarz JA, Contescu C and Putyera K, Eds, New York, Marcel Dekker, 2004.
130. LARSSON K, 'Cubic lipid-water phases – structures and biomembrane aspects', *Journal of Physical Chemistry*, 1989 **93**(21) 7304–14.
131. MONDUZZI M, LJUSBERG-WAHREN H and LARSSON K R, 'A C-13 NMR study of aqueous dispersions of reversed lipid phases', *Langmuir*, 2000 **16**(19) 7355–8.
132. NETO C, ALOISI G, BAGLIONI P and LARSSON K, 'Imaging soft matter with the atomic force microscope: cubosomes and hexosomes', *Journal of Physical Chemistry B*, 1999 **103**(19) 3896–9.
133. BUCHHEIM W and LARSSON K, 'Cubic lipid-protein-water phases', *Journal of Colloid and Interface Science*, 1987 **117**(2) 582–3.
134. LARSSON K, 'Aqueous dispersions of cubic lipid-water phases', *Current Opinion in Colloid and Interface Science*, 2000 **5**(1–2) 64–9.
135. SAGALOWICZ L, LESER M E, WATZKE H J and MICHEL M, 'Monoglyceride self-assembly structures as delivery vehicles', *Trends in Food Science & Technology*, 2006 **17**(5) 204–14.
136. DONG Y D, LARSON I, HANLEY T and BOYD B J, 'Bulk and dispersed aqueous phase behavior of phytantriol: effect of vitamin E acetate and F127 polymer on liquid crystal nanostructure', *Langmuir*, 2006 **22**(23) 9512–18.
137. PEVZNER S and REGEV O, 'The *in situ* phase transitions occurring during bicontinuous cubic phase formation', *Microporous and Mesoporous Materials*, 2000 **38**(2–3) 413–21.
138. BORNE J, NYLANDER T and KHAN A, 'Vesicle formation and other structures in the aqueous dispersions of monolein and sodium oleate', *Journal of Colloid and Interface Science*, 2001 **257**(2) 310–20.
139. RAEMY A, NOUZILLE C A, FROSSARD P, SAGALOWICZ L and LESER M E, 'Thermal behaviour of emulsifier-water systems studied by micro-DSC', *Journal of Thermal Analysis and Calorimetry*, 2005 **80**(2) 439–43.
140. VAN DIJCK P W M, DE KRUJFF B, VAN DEENEN L L, DE GIER J and DEMEL R A, 'The preference of cholesterol for phosphatidylcholine in mixed phosphatidylcholine – phosphatidylethanolamine bilayers', *Biochimica et Biophysica Acta*, 1976 **455** 576–87.
141. MEZZENGA R, MEYER C, SERVAIS C, ROMOSCANU A I, SAGALOWICZ L and HAYWARD R C, 'Shear rheology of lyotropic liquid crystals: a case study', *Langmuir*, 2005 **21**(8) 3322–33.
142. ALMGREN M, 'Alexander Lecture 2003: Cubosomes, vesicles, and perforated bilayers in aqueous systems of lipids, polymers, and surfactants', *Australian Journal of Chemistry*, 2003 **56**(10) 959–70.
143. ICKENSTEIN L M, EDWARDS K, KARLSSON G and MAYER L D, 'Formation of bilayer discs in lysolipid-containing thermosensitive liposomes during phase transition: a new drug release mechanism', *Journal of Liposome Research*, 2003 **13**(1) 39–40.

144. DANINO D, BERNHEIM-GROSWASSER A and TALMON Y, 'Digital cryogenic transmission electron microscopy: an advanced tool for direct imaging of complex fluids', *Colloids and Surfaces A – Physicochemical and Engineering Aspects*, 2001 **183** 113–22.
145. DANINO D, GUPTA R, SATYAVOLU J and TALMON Y, 'Direct cryogenic-temperature transmission electron microscopy imaging of phospholipid aggregates in soybean oil', *Journal of Colloid and Interface Science*, 2002 **249**(1) 180–6.
146. DELACROIX H, 'Crystallographic analysis of freeze-fracture electron micrographs: application to the structure determination of cubic lipid-water phases', *J Microsc*, 1998 **192**(Pt 3) 280–92.
147. MEYER H W and RICHTER W, 'Freeze-fracture studies on lipids and membranes', *Micron*, 2001 **32**(6) 615–44.
148. ANGELOV B, ANGELOVA A, PAPAHADJOPOULOS-STERNBERG B, LESIEUR S, SADOC J F, OLLIVON M and COUVREUR P, 'Detailed structure of diamond-type lipid cubic nanoparticles', *J Am Chem Soc*, 2006 **128**(17) 5813–17.
149. ANGELOVA A, ANGELOV B, PAPAHADJOPOULOS-STERNBERG B, OLLIVON M and BOURGAUX C, 'Proteocubosomes: nanoporous vehicles with tertiary organized fluid interfaces', *Langmuir*, 2005 **21**(9) 4138–43.
150. ERICSSON B, ERIKSSON P O, LOFROTH J E and ENGSTROM S, 'Cubic phases as delivery systems for peptide drugs', *ACS Symposium Series*, 1991 **469** 251–65.
151. GARG G and SARAF S, 'Cubosomes: an overview', *Biol Pharm Bull*, 2007 **30**(2) 350–3.
152. AMAR-YULI I and GARTI N, 'Transitions induced by solubilized fat into reverse hexagonal mesophases', *Colloids Surf B – Biointerfaces*, 2005 **43**(2) 72–82.
153. MEZZENGA R, GRIGOROV M, ZHANG Z, SERVAIS C, SAGALOWICZ L, ROMOSCANU A I, KHANNA V and MEYER C, 'Polysaccharide-induced order-to-order transitions in lyotropic liquid crystals', *Langmuir*, 2005 **21**(14) 6165–9.
154. PITZALIS P, MONDUZZI M, KROG N, LARSSON H, LJUSBERG-WAHREN H and NYLANDER T, 'Characterization of the liquid-crystalline phases in the glycerol monooleate/diglycerol monooleate/water system', *Langmuir*, 2000 **16**(15) 6358–65.
155. EFRAT R, ASERIN A and GARTI N, 'Synergistic solubilization of mixed nutraceuticals in modified discontinuous micellar cubic structures', *Food Colloids: Self-Assembly and Material Science*, Dickinson E and Leser M E, Eds, Cambridge, Royal Society of Chemsitry, 2006, 87–102.
156. SPICER P, 'Progress in liquid crystalline dispersions: cubosomes', *Current Opinion in Colloid and Interface Science*, 2005 **10**(5–6) 274–9.
157. VAUTHEY S, VISANI P, FROSSARD P, GARTI N, LESER M E and WATZKE H J, 'Release of volatiles from cubic phases: monitoring by gas sensors', 2000 **21**(3) 263–78.
158. BOYD B J, 'Characterisation of drug release from cubosomes using the pressure ultrafiltration method', *International Journal of Pharmaceutics*, 2003 **260**(2) 239–47.
159. MORARU C I, PANCHAPAKESAN C P, HUANG Q R, TAKHISTOV P, LIU S and KOKINI J L, 'Nanotechnology: a new frontier in food science', *Food Technology*, 2003 **57**(12) 24–9.

9

Complexes and conjugates of biopolymers for delivery of bioactive ingredients via food

Y. D. Livney, Technion – Israel Institute of Technology, Israel

9.1 Introduction

Complexes and conjugates of biopolymers, like proteins and polysaccharides, which are the main structural and functional components of biological systems, are of great practical and theoretical importance. These hybrid entities, both natural and nature-inspired man-made, counterparts, synergistically combine properties and functionalities of the biopolymers comprising them. They often present new structural and functional capacities not possessed by their parent macromolecules. In particular, the combination of proteins having hydrophobic domains, with polysaccharides characterized by large hydrophilic structures, creates superb macromolecular surfactants, exhibiting excellent stabilization capability for water insoluble phases dispersed in aqueous media. Moreover, they often spontaneously self-assemble to form nanoparticles, or nanocoating layers which are useful for protection, delivery, and programmed or targeted release of bioactive ingredients.

Associative interactions of biopolymers, like proteins and polysaccharides, have a plethora of implications and applications in many fields of science and technology. However, this realization has developed only slowly over many years. The phenomenon was discovered in 1911 by Tiebackx[1] who observed the appearance of opacity or precipitation upon mixing of gelatin and gum arabic (GA) in an acetic acid solution, and was first systematically studied 18 years later on the same biopolymer system by Bungenberg de Jong and Kruit.[2] Research interest in these phenomena increased very gradually over the following decades, and only in the last 10–15 years has the volume of research and publications in this field grown dramatically, especially with respect to biological and related systems. Complex formation occurring mainly, but not solely, between oppositely charged biopolymers in food is of great importance to formation and stability of texture, appearance and other sensory properties of products, as well as to the development of important novel applications related to special functionalities including delivery, analysis, quality preservation, and packaging. This field has been reviewed from different viewpoints over the past few years.[3–9] This chapter is intended to review aspects of complexes and conjugates of biopolymers related to the delivery of functional components via foods.

The advantages of complexes and conjugates of proteins and polysaccharides used as building blocks for delivery systems are many. Most of the proteins and polysaccharides used to form the complexes originate from natural food raw materials, and are therefore considered GRAS (generally recognized as safe), natural, and label-friendly. Most food polysaccharides and proteins are widely available and, excluding a few exceptions, are relatively inexpensive. A great variety of types of biopolymers of various structures, properties, and functionalities are available, allowing the tailoring of specific combinations for unique applications. The ability of such biopolymeric carriers to bind or entrap both hydrophilic and hydrophobic compounds[10] offers an important advantage for delivery of combinations of synergistically active components, like oil and water soluble antioxidants. Conjugates and complexes of proteins and polysaccharides can often both adsorb to an oil-water surface to facilitate emulsion formation, and form a thick macromolecular layer around it, which may greatly facilitate stability by both steric and electrostatic repulsion mechanisms, and by increasing the viscosity of the continuous phase. If properly designed and prepared, these synergistically combined mechanisms can provide excellent stability for long shelf-life liquid systems, offering in some cases, better performance than low molecular weight surfactants. The design of multi-layer capsules based on the layer-by-layer deposition approach[11] offers controllable formation of microcapsules with improved stability, and protection of the core material. Moreover, similar multi-layered vehicles may enable the programmable release profile and targeting to specific zones in the digestive tract.

Based on knowledge gained over the past few decades, relying on theoretical, experimental and modeling studies, the most important parameters affecting the interactions between proteins and polysaccharides are now well documented,[5,8,9] facilitating the design, application, and control of new delivery systems based on complexes and conjugates of these biopolymers.

There are also certain limitations and challenges related to the use of complexes and conjugates of proteins and polysaccharides for delivery applications. Some natural conjugates of polysaccharides and proteins, such as gum arabic, are rather expensive,[12] which is a driving force for continued research for proper substitutes, some of which will be discussed herein. Electrostatic complexes are quite sensitive to pH and ionic strength, as well as to temperature changes, which may cause their dissociation or aggregation, moreover compared to nanocapsules of low molecular weight surfactants, it is much more difficult to create transparent systems based on biopolymeric nanoparticles, e.g. for delivery in clear drinks. The reason for this is their large molecular weight, which both scatters much more light, and limits the curvature of stabilized oil-water surfaces, resulting in larger droplets or particles. Continued research and development efforts are seeking new ways to overcome such limitations, and exploit the large potential of biopolymeric complexes and conjugates for delivery of functional components in food.

9.2 Complexes and coacervates

Complex formation occurs between biopolymers exhibiting attractive interactions, including mainly electrostatic attraction between oppositely charged fixed groups on the polymers. However, other attractive interactions may also play important roles, e.g. hydrogen bonds, Van der Waals and hydrophobic interactions.[5,6,8,9] Such interactions may lead to formation of soluble complexes or to associative phase separation and sometimes precipitation, conversely repulsive interactions may lead to segregative phase separation due to thermodynamic incompatibility, wherein each phase is rich in one of the biopolymers and poor in the other.[13,14] In the case of associative interactions, if not all of the macro ion charges are neutralized by complexation, complexes may remain soluble, but near charge neutrality, complexes typically aggregate to form dense phases.[15] The formation of two liquid phases, one concentrated and containing both oppositely charged macromolecules, and the other dilute in both, is termed complex coacervation, a term which was coined by Bungenberg de Jong and Kruyt,[2] as opposed to simple coacervation, which occurs when a single polymer phase-separates because of decreased solvent-quality conditions.[8]

9.2.1 Theoretical and experimental studies of complex coacervation mechanisms

Overbeek and Voorn[16] have proposed the first theoretical model to describe electrostatic complex formation, based on the experimental results of Bungenberg de Jong and Kruit[2] on the gelatin-GA system. They explained that complex coacervation is mainly driven by the decreasing electrostatic free energy, but is opposed by the loss in configurational and translational entropy of the poly-ions. The attracted chains 'entrap' water within the complex, to increase entropy. They based their theory on the Flory–Huggins lattice model, for the entropy calculation, and the electrostatic free energy calculations on the Debye–Huckel theory, which gives good description of low molecular weight ions in solution, but does not take into account spatial corrections for closely interacting macro-ions. Veis and Aranyi,[17,18] who studied complex coacervation of oppositely charged gelatins, found that the Overbeek and Voorn theory was unable to correctly describe their results. They therefore developed an improved model, by taking into account the non-zero heat of mixing, and calculating the Flory–Huggins interaction parameter, which was found to be a function of both the temperature and the initial mixing concentrations. They described the complex coacervation as a two-step process: first, oppositely charged macro-ions interact electrostatically, resulting in stable nearly neutral (ion-paired) aggregates of low configurational entropy. Second, these slowly rearrange to form a coacervates phase, driven by increasing configurational entropy through random mixing within the coacervates phase and by the dilution of the non-random aggregate phase. An equilibrium forms between the dilute solution of the poly-ion aggregates and the concentrated phase of intertwined, but not specifically aggregated random coils.

A more recent theory[19,20] offers a way to overcome a major approximation in the Overbeek and Voorn theory, i.e. the poor description of the spatial correlation between the oppositely charged flexible chains. This kind of weak spatial correlations has been described using the random phase approximation (RPA), resulting in an enhanced driving force for coacervation.[5,19,20] The RPA is predicted to be applicable to complex coacervation of weakly charged systems and high salt concentrations. Biesheuvel and Cohen-Stuart[21] recently proposed a model using a cylindrical diffuse double-layer model for free polymer in solution and electroneutrality within the complex. A theoretical stability diagram was constructed relating pH, salt concentration, and mixing ratio, in qualitative agreement with data in the literature for complex coacervation of arabic gum and gelatin. The theory further explained the increased tendency toward phase separation when the polymer becomes more strongly charged and suggests that complexation of polyacid or polybase with zwitterionic polymer (e.g. protein) of the same charge sign at the 'wrong' side of the isoelectric point may be due, in part, to an induced charge reversal of the protein. A further model developed by Biesheuvel and Cohen-Stuart[22] uses a novel approach assuming cylindrical cells around each polyelectrolyte chain in the complex. The electrostatic free energy is calculated using the Poisson–Boltzmann equation. They applied this model to describe multilayer film formation by alternately adding a polycation and a polyanion solution.

A more relevant model for complex formation between a protein and a polyelectrolyte has been presented recently by Allen and Warren[23] describing complexation and phase behavior of oppositely charged polyelectrolyte/macro-ions using a Poisson–Boltzmann cell model. The model agreed well with experiments for most features of associative phase separation of these systems. Sanchez *et al.*[24,25] studied complex coacervation between beta-lactoglobulin and GA using small angle static light scattering (SALS). Time evolution of turbidity during acidification revealed the existence of six pH-induced structural transitions. It was concluded that complex coacervation between beta lactoglobulin and GA was a nucleation and growth type process, where a diffusion-controlled growth mechanism is first observed during growth of soluble complexes, followed by initiation of phase separation of the coacervation process with an interfacially-controlled growth mechanism.

Biesheuvel *et al.*[26] studied phase behavior of oppositely charged nanoparticles using mixtures of lysozyme and chemically modified lysozyme whose charge was close in magnitude but opposite in sign. They observed a reversible phase separation which depended on protein concentration, ionic strength, but also on temperature. They introduced a heterogeneous Poisson–Boltzmann cell model that generally applies to mixtures of oppositely charged nanoparticles, and in addition to steric and electrostatic interactions they included a temperature-dependent short-ranged interaction between the lysozyme molecules, whose exact origin was unknown. The strength and temperature dependence of the short-ranged attraction was similar in magnitude to that between unmodified lysozyme molecules. That strong short-ranged attraction in their system precluded the formation of colloidal liquid phases (or complex coacervates).

Da Silva *et al.*[27] studied the peculiar phenomenon of association between proteins and polyelectrolytes carrying net charge of same sign, previously interpreted as an ion–dipole interaction overcoming the repulsive ion–ion interaction. Using Monte Carlo simulations and a perturbation theory, they proposed that the explanation for the association is 'charge regulation'. Studying three different protein-polymer complexes they found that the induced ionization of amino acid residues due to the polyelectrolyte led to unexpectedly strong attractive interaction between the two biopolymers. The extra attraction from this charge induced-charge interaction can reportedly be several kT, and for the three cases studied, lysozyme, alpha-lactalbumin, and beta-lactoglobulin, was of the same magnitude or stronger than the ion–dipole interaction.

9.2.2 Factors affecting formation and properties of the complexes

Cooper *et al.*[4] have summarized the main factors affecting adsorption of a polyelectrolyte to a spherical colloidal particle (1–10 nm) such as a protein or a micelle: binding of a polyelectrolyte onto an oppositely charged spherical particle with uniform charge distribution should increase with the colloid surface charge density, the charge per polymer repeat unit or the linear charge density, and decrease with increasing ionic strength. Experiment and theory both show that at any ionic strength and linear charge density, no interactions occur unless the colloid surface charge density exceeds some critical value. Furthermore, von Goeler and Muthukumar[4,28] described the effects of polyelectrolyte molecular weight, chain rigidity and the radius of the colloidal surface, as well as solution conditions like temperature and dielectric constant of the medium. At constant temperature, dielectric constant, ionic strength and linear charge density, the adsorption increases with colloid radius and decreases with chain rigidity.

When dealing with the adsorption of polyelectrolytes on oppositely and uniformly charged planar surfaces, polymer adsorption initially increases[4,29] with addition of salt and then decreases with further increase of the ionic strength. While the latter effect of a decreasing Debye screening length is obvious, the possibility of intrapolymer repulsion in the absorbed state has been suggested to explain the former effect. Some experimental evidence for this may exist below 20 mM salt.[4]

The effect of pH (2.0–7.0), on the formation of electrostatic complexes between sodium caseinate and GA was studied by Ye *et al.*, using absorbance measurements, and dynamic light scattering (DLS).[30] While pure sodium caseinate solutions showed a typical abrupt turbidity jump and precipitation below pH 5.4, the addition of GA at increasing concentrations to sodium caseinate solution shifted the abrupt increase in turbidity (designated $pH_{\theta 1}$) to decreasing pH values. A moderate rise in turbidity (to a level that increased with GA concentration) remained at pH 5.4 (designated pHc, and denoting the pH at which the absorbance of the sample increased, but phase separation did not occur). Dynamic light scattering (DLS) data showed the formation of 100–150 nm nanoparticles between pH_c and $pH_{\theta 1}$, below $pH_{\theta 1}$ the nanoparticles flocculated and the system phase-separated as a consequence of charge neutralization, as shown by zeta-potential measurements. Below about pH 2.5, denoted pH_d, the phase-separated system regained its solubility and transparency.[30]

Benichou *et al.* have studied the aqueous phase diagram of xanthan gum and whey protein.[31] At pH 7.0, and low total concentration of biopolymers, cosolubility was observed, and at high total concentrations, above a binodal line, incompatibility caused phase separation. However, an interesting zone was observed below the binodal line where solution was still transparent but evidence was obtained of associative interactions of the two negatively charged biopolymers. The authors used a spectrophotometric dye-binding method to study effects of pH and ratio on the electrostatic interactions of the two biopolymers. A positively charged dye, methylene blue (MB), shows a shift in absorbance upon binding to xanthan gum. A reversal of the shift, with addition of whey protein suggested, of MB release due to competition of the protein for binding to the gum. This was observed already at pH 7.0, and increased with decreasing pH, as associative interactions between the biopolymers strengthened.[31] Apparently at pH 7, there are still enough local patches of positive charge on whey proteins allowing these weak associations.

Based on a comprehensive literature review, de Kruif *et al.*[5] have drawn the following conclusions regarding protein-polysaccharide complexation:

- Phase separation is mainly entropically driven, most probably due to the delocalization of the counter ions of the protein and the polysaccharide.
- If one of the polyanions is a strong polyelectrolyte, a precipitate is formed and not a liquid coacervates phase.
- Complexes form in solution much before macroscopic phase separation.
- At a mixing ratio where coacervation is maximal, the complexes formed are neutral.
- Even at a large excess of one of the polyelectrolytes, the complexes formed are only moderately charged.
- Salt dissociates all polyelectrolyte complexes (while extremely low ionic strength also suppresses complexation).

- Salt concentration in the dilute and concentrated phases is nearly equal.
- In the liquid coacervates phase there is a clear mobility of both polymers.
- Temperature hardly affects phase boundary.

9.2.3 Utilization of biopolymer complexation for encapsulation and delivery of bioactives in food

The utilization of biopolymer complexation for encapsulation of nutraceuticals for food enrichment has been growing over the last two decades, thanks to its many advantages and numerous possible applications. While microencapsulation was the main theme in the delivery field towards the end of the 1990s, nano-encapsulation and nanocoating are currently at the forefront of research and development, and are gradually being applied industrially. In the following sub-sections, several recent examples are highlighted, with a focus on nanoparticles and nanocoating layers for microcapsules.

Protein-anionic polysaccharide complexes

Liu *et al.* prepared complex hydrogel beads of about 2.5 mm diameter from citrus pectin and zein, a protein from corn.[32] While pure pectin gel beads swelled more with increasing pH in the range 3.5–7.4, the pectin/zein complex hydrogels did not. The complex beads endured protease attack, but were hydrolyzed by pectinases. The system was designed as a colon-specific delivery vehicle for bioactive molecules.[32,33]

Benichou *et al.* used electrostatic complexes of whey protein and xanthan gum, instead of low molecular weight nonionic hydrophilic emulsifiers, to nanocoat and stabilize the outer interface of water-in-oil-in-water (W/O/W) double emulsions, with polyglycerol polyricinoleate (PGPR) stabilizing the inner interface. Binary aqueous mixtures of the biopolymers did not coacervate but formed stable complexes. Below the pI of the protein (pH 4.6), glucose encapsulation yield in the inner aqueous phase of the double emulsion was close to 95%. The complexes conferred stability to the droplets by both steric and electrostatic mechanisms (zeta potential was about +30 mV). When the pH was around or above the protein pI, the system exhibited reduced emulsification capabilities with droplets larger than 10 μm and encapsulation yield $< 10\%$. This was apparently because of partial complex dissociation. At pH $= 2$, however, the polymer adducts efficiently stabilized the double emulsions and formed a thick barrier slowing the release of vitamin B1 which was entrapped in the innermost phase of the W/O/W double emulsions. Vitamin release after three weeks of storage at room temperature was $< 18\%$ compared to 100% at neutral pH.[34]

Recently, an interesting application has been presented for electrostatic complexes in the formation of 'colloidosomes' which consist of large oil droplets with small oil droplets adsorbed to their surfaces. These 'colloidosomes' were formed by mixing two oil-in-water emulsions: one containing relatively large anionic droplets (d $\sim$ 0.6 μm) at pH $= 4$ stabilized by a double layer made of β-lactoglobulin (β-Lg) as the surfactant coated with pectin, with another oil-in-water emulsion containing relatively small cationic droplets (d $\sim$ 0.2 μm), coated at pH 4 with β-Lg only). As the concentration of small cationic droplets was increased, the measured z-potential increased from −30 mV to +25 mV, indicating that the small positively charged droplets adsorbed to the surface of the large negatively charged droplets until they eventually saturated the surface of the large droplets. Instability to droplet aggregation at intermediate small droplet concentrations was explained by bridging flocculation and at high small droplet concentrations by a possible depletion flocculation effect. Relatively stable colloidosomes were made over an intermediate range of small droplet concentrations, which consisted of large droplets surrounded by small droplets. Colloidosomes, especially ones made of food grade components only, may have a number of useful practical applications, in particular in controlled or triggered release of encapsulated components.[35]

Protein-cationic polysaccharide complexes

Chen and Subirade prepared core-shell nanoparticles (100–200 nm hydrodynamic diameter) for delivery of nutraceuticals by crosslinking chitosan (CS), a poly-cation, with sodium tripolyphosphate (TPP), followed by coating the nano-gel-beads with either native or denatured β-lactoglobulin (β-Lg).[36] Before β-Lg addition, particles had a zeta potential of 20–60 mV. Coating layer of β-Lg ranged from 1–60% of the nanoparticle weight, increasing with pH, particularly above the protein pI (~5), and peaking around 6.0, following a sharp decrease as the chitosan charge decreases. The adsorption was mainly attributed to electrostatic interactions, hydrophobic interactions and hydrogen bonding between β-Lg and CS.[36]

Microcapsules made of chitosan and gelatin, were developed for delivery of bioactive molecules (e.g. riboflavin) by ingestion. Chitosan, gelatin and the bioactive molecule solutions were mixed at a low pH (using acetic acid) and 37 °C, and then dropped into cold sesame-seed oil to induce gelation of the gelatin. Then, ionotropic crosslinking was done at pH 7.0 (a pH which apparently also induced electrostatic complexation of the gelatin and the chitosan) to improve the mechanical strength of the beads, using either sulfate, citrate or tripolyphosphate (TPP). TPP, which has the highest charge density, gave the best mechanical properties. However, the pH-responsive drug release performance was poor and in 24 h riboflavin release rates in simulated gastric fluid (SGF) and in simulated intestinal fluid (SIF) were 59% and 49%, respectively. Citrate and sulfate crosslinked beads, released about 60% and 70% of the riboflavin in 24 h in SIF, respectively, and completely released it in less than 5 h in SGF.[37] The studies discussed here also exemplify the important role that multivalent counterions may play in electrostatic interactions between biopolymers, in particular when encapsulation and controlled release is sought.

Protein-protein and protein-DNA complexes

These types of complexes are less common in delivery of bioactive molecules via foods, and are much more common in pharmaceutical applications. Gelatin can be modified during production from collagen to make it either negatively or positively charged at physiological pH, by changing its pI. This enables the versatility and control of electrostatic binding of either positively or negatively charged bioactive molecules including drugs, proteins and DNA. Various forms of gelatin carrier matrices were fabricated for this type of controlled-release applications.[38]

An innovative food application of DNA in a complex with gelatin and carrageenan has been presented by Jonganurakkun *et al.*[39] Single-stranded DNA, extracted from discarded industrial salmon milt, was used to prepare hydrogels and cocoa-butter emulsion gels, by crosslinking with gelatin and κ-carrageenan, for the protection of functional materials, proteins, microorganisms or drugs. The hydrogels provided protection in a simulated gastric juice against release of methyl orange, simulating a functional ingredient, and the emulsion gels improved survival of probiotic lactic bacteria in the acid conditions.[39,40]

Anionic polysaccharide-cationic polysaccharide complexes

The polycation chitosan and the polyanion alginate were used to form polyelectrolyte complexes for encapsulation of proteins, cells, and enzymes. Simsek-Ege *et al.* have used FTIR (Fourier transform infrared) spectroscopy to study the interaction between chitosan and alginate at different pH values. Using nitrogen analysis of the beads to measure the chitosan coating on the alginate surface, the authors demonstrated that the yield of the complex formation must be higher when the complex is prepared at pH 5 than at pH 2 since at the latter pH there are fewer available ionized carboxylic acid groups on the alginate surface.[41] In another study, sodium alginate was used to coat the surface of TPP/chitosan beads also containing gelatin, for the improvement of sustained release performance. The loading efficiency of model encapsulants (brilliant blue and FITC-dextran) in these beads was higher than 90%.[42]

An interesting concept in delivery was introduced by Liao *et al.*[43] They produced drug-loaded fibers, employing interfacial polyelectrolyte complexation, by pulling out complexes forming at the interface between solutions of, e.g., chitosan and alginate, which contained the encapsulant. Release time of the encapsulant ranged from hours to weeks depending on its properties, as well as on the properties of the biopolymers used.[43]

In another study sequentially coated multilayered beads were prepared by complex coacervation of alternating calcium alginate and chitosan layers. The beads containing vitamin B2 were studied for vitamin release at varying pH levels at 37 °C. Nearly 54 ± 4% of the vitamin were released in the first 3 h in a SGF (pH 1.0) while the remaining amount was released in the SIF (pH 7.4) within 6 h. The effect of composition and crosslinking degree of the outermost alginate layer on the stability of the beads and the release kinetics were also studied.[44]

Double-layered composite particles were made of pectin and chitosan which gelled in the presence of Ca^{+2}. Chitosan and Ca^{+2} concentrations significantly affected particle swelling at pH 1.4 and 7.4 after 4 h, whereas chitosan and pectin concentrations significantly influenced the swelling at pH 7.4 after 12 h. The results suggested that complex formation between chitosan and pectin occurred. Protein release from the chitosan-pectin particles at pH 1.4 was negligible. Chitosan-pectin particles could sustain the release of most loaded protein for 24 h at pH 7.4. Pectinase increased protein release from 39% to 63% within 12 h at pH 7.4. These characteristics of the chitosan-pectin particles were suggested to be applicable for colon-specific delivery of bioactive macromolecules.[45]

9.3 Conjugates – covalently bonded biopolymers

Covalently conjugated biopolymers not only combine the properties of the biopolymers comprising them, but synergistically exhibit superior properties. Some of these entities are especially useful for encapsulating and stabilizing dispersed phases. A conjugate of a protein that has hydrophobic domains and a polysaccharide is a form of a block copolymer, which can adsorb to an oil–water interface as a surfactant, and simultaneously form a stabilizing 'hairy layer' with its polysaccharide block. This 'hairy layer' stabilizes the colloidal emulsion system by steric exclusion and sometimes also electrostatic repulsion between the droplets, as well as by increasing the viscosity of the continuous phase around the droplet. Compared to non-covalent complexes of polysaccharides and proteins, covalent conjugates of proteins and polysaccharides are generally more effective as surfactants and as self-assemblers.

9.3.1 Naturally occurring conjugates of proteins and polysaccharides

Gum arabic

The classical example is gum arabic (gum acacia) which is an excellent emulsifier and a natural nano-encapsulator. It is extracted from exudates of acacia trees, mainly Acacia Senegal and also Acacia Seyal. Its structure and properties have been extensively studied and reviewed (see, e.g., references 46–49). It is a highly branched and somewhat anionic heteropolymer of a very complex structure.[46] GA contains a small protein fraction, apparently playing a fundamental role in its adsorption at oil–water interface, an uncharacteristic property for polysaccharides. The protein fraction is composed predominantly of hydroxyproline and serine and is covalently bound to the reducing end of the polysaccharide chains.[50,51] The monomeric composition of Acacia Senegal (on a moist base with 12.5–16% moisture) reportedly comprises the following: 39–42% galactose, 24–27% arabinose, 12–16% rhamnose and 15–16% glucuronic acid, and the protein fraction is 1.5–2.6%.[48] GA was separated into three fractions using hydrophobic affinity chromatography.[46,49] Most of the gum (88.4%) was referred to as an arabinogalactan (AG). It had only 0.35% protein, and a molecular mass of 3.8×10^5 Da. A second fraction (10.4%) referred to as an 'AG–protein complex' (AGP) had a molecular mass of 1.45×10^6 Da, and protein content of 11.8%. The smallest fraction (1.2% of the total gum) referred to as a low molecular weight glycoprotein (GP) had a molecular mass of 2.5×10^5 Da and a protein content of 47.3%. The more commonly accepted model for the structure of AGP is the wattle blossom model[52] describing this high molecular weight fraction of the gum as consisting of large carbohydrate blocks with a molecular mass of approximately 2.5×10^5 Da individually attached to a polypeptide chain.

GA has been extensively used as an emulsifier and stabilizer in beverages, mainly for emulsifying aroma oils in acid drinks.[47,53] Not only does it have good emulsifying and stabilizing capacities,[47,54] but it is an excellent natural encapsulator for water insoluble compounds, such as flavors[55] and nutraceuticals such as lycopene,[56] linoleic acid,[57] and conjugated linoleic acid.[58] Its main disadvantage is its relatively high price, which is driving research efforts in the quest for cheaper substitutes with a similar functional quality.[53]

Soy soluble polysaccharides

One such interesting potential substitute from a natural source is 'soybean water-soluble polysaccharide' (SSPS).[59–63] SSPS is extracted from a byproduct of soybean protein isolation and has been shown to be effective in stabilizing beverage emulsions[63] as well as acidic milk beverages.[64] SSPS is an acidic polysaccharide containing galacturonic acid. SSPS is composed of a main rhamnogalacturonan backbone branched by β-1,4-galactan and α-1,3- or α-1,5-arabinan chains.[59] SSPS may be used in flavor emulsions because of its high water solubility, pH stability, low bulk viscosity, emulsifying properties, and ability to form strong interfacial films.[59] SSPS may be separated into a high MW fraction (HMF; 310 kDa) and a low MW fraction (LMF; 20 kDa) by size exclusion chromatography. SSPS/HMF consisted of 91.6% carbohydrate and 2.2% protein and showed better emulsifying properties than those of the whole SSPS, whereas SSPS/LMF seemed to exert a negative effect on the adsorption behavior of SSPS. The protein fraction of SSPS/HMF had a molecular mass of 50 kDa, as determined by SDS-PAGE, and was predominantly composed of proline (23.1%) and glutamic acid (15.2%), as well as 8.8% of residual neutral sugar and 5.3% of uronic acid.[59] SSPS and its fractions, HMF and LMF with and without maltodextrins were used for microencapsulation of linoleic acid by emulsification and spray drying. Stability to oxidation of the encapsulant increased as the weight fraction of the LMF increased, which was explained by a high free radical scavenging effect of this fraction.[65]

9.3.2 Man-made covalent conjugates useful for encapsulation

The emulsifying properties of egg-protein were shown more than ten years ago to improve by covalent conjugation with a polysaccharide. Conjugates were formed in a controlled dry state (60 °C, 65% R.H.) through the Maillard reaction between the ϵ-amino groups in proteins and the reducing-end carbonyl residues in polysaccharides. Ovalbumin or lysozyme, and dextran or galactomannan were employed for protein-polysaccharide conjugation.[66] WPI-dextran conjugates were similarly coupled by dry heat treatment of the protein polysaccharide mixtures. The conjugates caused much better stability of the emulsions formed, compared to WP alone, or to GA. The improvement of WP functionality by the conjugation was explained by the enhanced steric stabilization provided by the bulky hydrophilic polysaccharide moiety.[67]

Sodium caseinate (SC)-dextran (Dex) conjugates were investigated for the possibility of stabilizing water-in-oil-in-water emulsions (W/O/W) to control the release of vitamin B12 from the inner water phase (W1) to the outer aqueous phase (W2). The conjugates were prepared by Maillard reaction of SC with Dex (250 000 or 500 000 Da) induced by incubating at 60°C and 79% R.H. for 8 h. The double emulsions, with encapsulated vitamin B12, were prepared by a two-step emulsification technique using polyglycerin polyricinoleate (PGPR, E476) as the hydrophobic emulsifier. Compared to the pure protein, SC-Dex conjugate yielded smaller and more narrowly distributed water-containing oil droplet emulsions. Under acidic conditions, the conjugate-containing emulsions were more stable to coalescence than the emulsions with SC, and the vitamin B12 release from the inner W1 phase was significantly decreased.[68]

9.4 Complexes of a protein with a polysaccharide-protein conjugate

Electrostatic complexes of GA, the natural polysaccharide-protein conjugate, and proteins were studied extensively with respect to encapsulation. Complex coacervation at pH 3.0–4.5 between whey protein and GA (WP/GA) was successfully used for microencapsulation of sunflower oil, lemon and orange oil flavor. The formation of a smooth biopolymer shell around the oil droplets was achieved close to pH 4.0 and the amount of loaded encapsulant was higher than 80%. Large droplets (>200 μm) had a typical shell/core structure, but small droplets (<50 μm) were easier to encapsulate within a coacervate matrix. The stability of the emulsion was strongly pH-dependent. Maximal creaming rate was observed at pH 4.0, where the zeta potential was close to zero. When various capsules of lemon oil were introduced into Gouda cheese, after one month the release of flavor from large capsules was higher than that from small ones, while the release from covalently crosslinked capsules was lower than that from uncrosslinked ones, probably because of a dense biopolymer wall which was difficult to break by chewing.[69]

Complexes of plant proteins with GA were studied through zeta potential and turbidity measurements. Two model proteins were observed: a cereal protein (alpha gliadin) and a leguminous protein (pea globulin). Optimum coacervation was obtained with GA at pH 2.75 for pea globulins with a protein-polysaccharide ratio of 30 : 70, and at pH 3 for alpha gliadins with a protein-polysaccharide ratio of 50 : 50. Both the charge density profile and the steric conformation of both macromolecules forming the complex determined the ability of the coacervates to encapsulate oil droplets.[70]

9.5 Future trends

The two most important current and future trends driving the encapsulation and delivery field are health and nanotechnology. There is a strong worldwide drive to develop methods and technologies which would enable nanometric manipulation of material properties and functionalities. In food science and technology, one of the major potential applications of nanotechnology is in the encapsulation, targeted delivery and controlled release of nutraceuticals. Potential applications of complexes and conjugates may include multi-nano-layer constructs which may enable a 'programmable' release profile during digestion. Nanoencapsulation promises new advantages in product enrichment without harming sensory properties, like transparency (e.g. enrichment of clear drinks with oils, soluble vitamins, and antioxidants), texture smoothness, flavor and smell (timing profile and burst effect). Nanoencapsulation also promises superior absorption of certain micro-nutrients, a trend which would necessitate reevaluation of recommended daily allowance values, and may lower the actual amounts supplemented. The related risks would need to be carefully studied and managed. However, the potential benefits are stimulating extensive investment in research and development. The role of biopolymers, and their complexes and conjugates in these developments is central, and therefore harnessing of basic physical chemistry, polymer science, and biological-material science becomes vital for success.

9.6 Sources of further information and advice

Reviews of protein-polysaccharide complexes can be found in references 71–73. Reviews of nanotechnology in food and delivery of nutraceuticals can be found in references 74–78.

9.7 References

1. TIEBACKX, F. W. Simultaneous coagulation of two colloids. *Zeitschrift für Chemie und Industrie der Kolloide* **1911**, *8*, 198–201.
2. BUNGENBERG DE JONG, H. G.; KRUYT, H. R. Coacervation (partial miscibility in colloid systems). *Proc. Acad. Sci Amsterdam* **1929**, *32*, 849–856.
3. BENICHOU, A.; ASERIN, A.; GARTI, N. Protein polysaccharide interactions for stabilization of food emulsions. *Journal of Dispersion Science and Technology* **2002**, *23*, 93–123.
4. COOPER, C. L.; DUBIN, P. L.; KAYITMAZER, A. B.; TURKSEN, S. Polyelectrolyte-protein complexes. *Current Opinion in Colloid & Interface Science* **2005**, *10*, 52–78.
5. DE KRUIF, C. G.; WEINBRECK, F.; DE VRIES, R. Complex coacervation of proteins and anionic polysaccharides. *Current Opinion in Colloid & Interface Science* **2004**, *9(5)*, 340–349.
6. DOUBLIER, J. L.; GARNIER, C.; RENARD, D.; SANCHEZ, C. Protein-polysaccharide interactions. *Current Opinion in Colloid & Interface Science* **2000**, *5(3,4)*, 202–214.
7. McCLEMENTS, D. J. Non-covalent interactions between proteins and polysaccharides.

Biotechnology Advances **2006**, *24*, 621–625.

8. SCHMITT, C.; SANCHEZ, C.; SOBRY-BANON, S.; HARDY, J. Structure and technofunctional properties of protein-polysaccharide complexes: a review. *Critical Reviews in Food Science and Nutrition* **1998**, *38[8]*, 689–753.
9. TURGEON, S. L.; BEAULIEU, M.; SCHMITT, C.; SANCHEZ, C. Protein-polysaccharide interactions: phase-ordering kinetics, thermodynamic and structural aspects. *Current Opinion in Colloid & Interface Science* **2003**, *8*, 401–414.
10. RENARD, D.; ROBERT, P.; LAVENANT, L.; MELCION, D.; POPINEAU, Y.; GUEGUEN, J.; DUCLAIROIR, C.; NAKACHE, E.; SANCHEZ, C.; SCHMITT, C. Biopolymeric colloidal carriers for encapsulation or controlled release applications. *International Journal of Pharmaceutics* **2002**, *242(1–2)*, 163–166.
11. AOKI, T.; DECKER, E. A.; McCLEMENTS, D. J. Influence of environmental stresses on stability of O/W emulsions containing droplets stabilized by multilayered membranes produced by a layer-by-layer electrostatic deposition technique. *Food Hydrocolloids* **2005**, *19*, 209–220.
12. MADENE, A.; JACQUOT, M.; SCHER, J.; DESOBRY, S. Flavour encapsulation and controlled release – a review. *International Journal of Food Science and Technology* **2006**, *41*, 1–21.
13. TOLSTOGUZOV, V. Compositions and phase diagrams for aqueous systems based on proteins and polysaccharides. *International Review of Cytology* **2000**, *192*, 3–31.
14. TOLSTOGUZOV, V. Phase behavior in mixed polysaccharide systems. *Food Science and Technology* **2006**, *160*, 589–627.
15. DE VRIES, R.; COHEN-STUART, M. A. Theory and simulations of macroion complexation. *Current Opinion in Colloid & Interface Science* **2006**, *11*, 295–301.
16. OVERBEEK, J. T.; VOORN, M. J. Phase separation in polyelectrolyte solutions; theory of complex coacervation. *J Cell Comp Physiol* **1957**, *49(Suppl 1)*, 7–22.
17. VEIS, A.; ARANYI, C. Phase separation in polyelectrolyte systems. I. Complex coacervates of gelatin. *Journal of Physical Chemistry* **1960**, *64*, 1203–1210.
18. VEIS, A. Phase separation in polyelectrolyte solutions. II. Interaction effects. *Journal of Physical Chemistry* **1961**, *65*, 1798–1803.
19. KUDLAY, A.; OLVERA DE LA CRUZ, M. Precipitation of oppositely charged polyelectrolytes in salt solutions. *Journal of Chemical Physics* **2004**, *120(1)*, 404–412.
20. CASTELNOVO, M.; JOANNY, J. F. Complexation between oppositely charged polyelectrolytes: beyond the Random Phase Approximation. *European Physical Journal E: Soft Matter* **2001**, *6(5)*, 377–386.
21. BIESHEUVEL, P. M.; COHEN-STUART, M. A. Electrostatic free energy of weakly charged macromolecules in solution and intermacromolecular complexes consisting of oppositely charged polymers. *Langmuir* **2004**, *20(7)*, 2785–2791.
22. BIESHEUVEL, P. M.; COHEN-STUART, M. A. Cylindrical cell model for the electrostatic free energy of polyelectrolyte complexes. *Langmuir* **2004**, *20(11)*, 4764–4770.
23. ALLEN, R. J.; WARREN, P. B. Complexation and phase behavior of oppositely charged polyelectrolyte/macroion systems. *Langmuir* **2004**, *20(5)*, 1997–2009.
24. SANCHEZ, C.; MEKHLOUFI, G.; SCHMITT, C.; RENARD, D.; ROBERT, P.; LEHR, C. M.; LAMPRECHT, A.; HARDY, J. Self-assembly of β-lactoglobulin and acacia gum in aqueous solvent: structure and phase-ordering kinetics. *Langmuir* **2002**, *18(26)*, 10323–10333.
25. SANCHEZ, C.; MEKHLOUFI, G.; RENARD, D. Complex coacervation between beta-lactoglobulin and Acacia gum: a nucleation and growth mechanism. *Journal of Colloid and Interface Science* **2006**, *299*, 867–873.

26. BIESHEUVEL, P. M.; LINDHOUD, S.; DE VRIES, R.; COHEN-STUART, M. A. Phase behavior of mixtures of oppositely charged nanoparticles: heterogeneous Poisson–Boltzmann cell model applied to lysozyme and succinylated lysozyme. *Langmuir* **2006**, *22(3)*, 1291–1300.
27. DA SILVA, F. L.; LUND, M.; JOENSSON, B.; AAKESSON, T. On the complexation of proteins and polyelectrolytes. *J. Phys. Chem. B* **2006**, *110(9)*, 4459–4464.
28. VON GOELER, F.; MUTHUKUMAR, M. Adsorption of polyelectrolytes onto curved surfaces. *The Journal of Chemical Physics* **1994**, *100*, 7796–7803.
29. DOBRYNIN, A. V.; RUBINSTEIN, M. Effect of short-range interactions on polyelectrolyte adsorption at charged surfaces. *J. Phys. Chem. B* **2003**, *107*, 8260–8269.
30. YE, A. Q.; FLANAGAN, J.; SINGH, H. Formation of stable nanoparticles via electrostatic complexation between sodium caseinate and gum arabic. *Biopolymers* **2006**, *82*, 121–133.
31. BENICHOU, A.; ASERIN, A.; LUTZ, R.; GARTI, N. Formation and characterization of amphiphilic conjugates of whey protein isolate (WPI)/xanthan to improve surface activity. *Food Hydrocolloids* **2007**, *21*, 379–391.
32. LIU, L.; FISHMAN, M. L.; HICKS, K. B.; KENDE, M.; RUTHEL, G. Pectin/zein beads for potential colon-specific drug delivery: synthesis and *in vitro* evaluation. *Drug Deliv.* **2006**, *13*, 417–423.
33. LIU, L.; FISHMAN, M. L.; HICKS, K. B. Pectin in controlled drug delivery – a review. *Cellulose* **2007**, *14(1)*, 15–24.
34. BENICHOU, A.; ASERIN, A.; GARTI, N. Double emulsions stabilized by new molecular recognition hybrids of natural polymers. *Polymers for Advanced Technologies* **2002**, *13(10–12)*, 1019–1031.
35. GU, Y. S.; DECKER, E. A.; McCLEMENTS, D. J. Formation of colloidosomes by adsorption of small charged oil droplets onto the surface of large oppositely charged oil droplets. *Food Hydrocolloids* **2007**, *21*, 516–526.
36. CHEN, L.; SUBIRADE, M. Chitosan/b-lactoglobulin core-shell nanoparticles as nutraceutical carriers. *Biomaterials* **2005**, *26(30)*, 6041–6053.
37. SHU, X. Z.; ZHU, K. J. Controlled drug release properties of ionically cross-linked chitosan beads: the influence of anion structure. *International Journal of Pharmaceutics* **2002**, *233*, 217–225.
38. YOUNG, S.; WONG, M.; TABATA, Y.; MIKOS, A. G. Gelatin as a delivery vehicle for the controlled release of bioactive molecules. *Journal of Controlled Release* **2005**, *109(1–3)*, 256–274.
39. JONGANURAKKUN, B.; NODASAKA, Y.; SAKAIRI, N.; NISHI, N. DNA-based gels for oral delivery of probiotic bacteria. *Macromolecular Bioscience* **2006**, *6(1)*, 99–103.
40. BENJAMAS, J.; XIANG, D. L.; YOSHINOBU, N.; MOTOYOSHI, N.; NORIO, N. Survival of lactic acid bacteria in simulated gastrointestinal juice protected by a DNA-based complex gel. *Journal of Biomaterials Science, Polymer Edition* **2003**, *14*, 1269–1281.
41. SIMSEK-EGE, F. A.; BOND, G. M.; STRINGER, J. Polyelectrolyte complex formation between alginate and chitosan as a function of pH. *Journal of Applied Polymer Science* **2003**, *88(2)*, 346–351.
42. SHU, X. Z.; ZHU, K. J. A novel approach to prepare tripolyphosphate/chitosan complex beads for controlled release drug delivery. *International Journal of Pharmaceutics* **2000**, *201(1)*, 51–58.
43. LIAO, I. C.; WAN, A. C. A.; YIM, E. K. F.; LEONG, K. W. Controlled release from fibers of polyelectrolyte complexes. *Journal of Controlled Release* **2005**, *104(2)*, 347–358.
44. BAJPAI, S. K.; TANKHIWALE, R. Investigation of dynamic release of vitamin B2 from

calcium alginate/chitosan multilayered beads: Part II. *Reactive & Functional Polymers* **2006**, *66(12)*, 1565–1574.

45. CHANG, K. L. B.; LIN, J. Swelling behavior and the release of protein from chitosan-pectin composite particles. *Carbohydrate Polymers* **2000**, *43(2)*, 163–169.
46. VERBEKEN, D.; DIERCKX, S.; DEWETTINCK, K. Exudate gums: occurrence, production, and applications. *Applied Microbiology and Biotechnology* **2003**, *63*, 10–21.
47. BUFFO, R. A.; REINECCIUS, G. A.; OEHLERT, G. W. Factors affecting the emulsifying and rheological properties of gum acacia in beverage emulsions. *Food Hydrocolloids* **2001**, *15*, 53–66.
48. IDRIS, O. H. M.; WILLIAMS, P. A.; PHILLIPS, G. O. Characterisation of gum from Acacia senegal trees of different age and location using multidetection gel permeation chromatography. *Food Hydrocolloids* **1998**, *12*, 379–388.
49. RAY, A. K.; BIRD, P. B.; IACOBUCCI, G. A.; CLARK, B. C. Functionality of gum-arabic – fractionation, characterization and evaluation of gum fractions in citrus oil-emulsions and model beverages. *Food Hydrocolloids* **1995**, *9*, 123–131.
50. CONNOLLY, S.; FENYO, J. C.; VANDEVELDE, M. C. Heterogeneity and homogeneity of an arabinogalactan-protein: Acacia senegal gum. *Food Hydrocolloids* **1987**, *1(5–6)*, 477–480.
51. RANDALL, R. C.; PHILLIPS, G. O.; WILLIAMS, P. A. The role of the proteinaceous component on the emulsifying properties of gum arabic. *Food Hydrocolloids* **1988**, *2(2)*, 131–140.
52. FINCHER, G. B.; STONE, B. A.; CLARKE, A. E. Arabinogalactan proteins: structure, biosynthesis and function. *Annu Rev Plant Physiol* **1983**, *34*, 47–70.
53. DICKINSON, E. Hydrocolloids at interfaces and the influence on the properties of dispersed systems. *Food Hydrocolloids* **2003**, *17*, 25–39.
54. YADAV, M. P.; IGARTUBURU, J. M.; YAN, Y.; NOTHNAGEL, E. A. Chemical investigation of the structural basis of the emulsifying activity of gum arabic. *Food Hydrocolloids* **2007**, *21(2)*, 297–308.
55. KRISHNAN, S.; KSHIRSAGAR, A. C.; SINGHAL, R. S. The use of gum arabic and modified starch in the microencapsulation of a food flavoring agent. *Carbohydrate Polymers* **2005**, *62(4)*, 309–315.
56. MONTENEGRO, M. A.; NUNES, I. L.; MERCADANTE, A. Z.; BORSARELLI, C. D. Photoprotection of vitamins in skimmed milk by an aqueous soluble lycopene-gum arabic microcapsule. *Journal of Agricultural and Food Chemistry* **2007**, *55(2)*, 323–329.
57. FANG, X.; KIKUCHI, S.; SHIMA, M.; KADATA, M.; TSUNO, T.; ADACHI, S. Suppressive effect of alkyl ferulate on the oxidation of microencapsulated linoleic acid. *European Journal of Lipid Science and Technology* **2006**, *108(2)*, 97–102.
58. JIMENEZ, M.; GARCIA, H. S.; BERISTAIN, C. I. Spray-dried encapsulation of conjugated linoleic acid (CLA) with polymeric matrices. *Journal of the Science of Food and Agriculture* **2006**, *86(14)*, 2431–2437.
59. NAKAMURA, A.; YOSHIDA, R.; MAEDA, H.; FURUTA, H.; CORREDIG, M. Study of the role of the carbohydrate and protein moieties of soy soluble polysaccharides in their emulsifying properties. *Journal of Agricultural and Food Chemistry* **2004**, *52(17)*, 5506–5512.
60. NAKAMURA, A.; YOSHIDA, R.; MAEDA, H.; CORREDIG, M. The stabilizing behaviour of soybean soluble polysaccharide and pectin in acidified milk beverages. *International Dairy Journal* **2006**, *16(4)*, 361–369.
61. NAKAMURA, A.; MAEDA, H.; CORREDIG, M. Emulsifying properties of enzyme-digested soybean soluble polysaccharide. *Food Hydrocolloids* **2006**, *20(7)*, 1029–1038.

62. NAKAMURA, A.; MAEDA, H.; CORREDIG, M. Influence of heating on oil-in-water emulsions prepared with soybean soluble polysaccharide. *Journal of Agricultural and Food Chemistry* **2007**, *55(2)*, 502–509.
63. MAEDA, H. Soluble soybean polysaccharide. In *Handbook of Hydrocolloids*, Phillips, G. O.; Williams, P. A., Eds.; Cambridge: CRC Press, 2000, 309–319.
64. NAKAMURA, A.; FURUTA, H.; KATO, M.; MAEDA, H.; NAGAMATSU, Y. Effect of soybean soluble polysaccharides on the stability of milk protein under acidic conditions. *Food Hydrocolloids* **2003**, *17*, 333–343.
65. XU, F.; WATANABE, Y.; ADACHI, S.; MATSUMURA, Y.; MORI, T.; MAEDA, H.; NAKAMURA, A.; MATSUNO, R. Microencapsulation of linoleic acid with low- and high-molecular-weight components of soluble soybean polysaccharide and its oxidation process. *Bioscience, Biotechnology, and Biochemistry* **2003**, *67(9)*, 1864–1869.
66. KATO, A.; IBRAHIM, H. R.; NAKAMURA, S.; KOBAYASHI, K. New methods for improving the functionality of egg white proteins. *Egg Uses and Processing Technologies* **1994**, 250–268.
67. AKHTAR, M.; DICKINSON, E. Emulsifying properties of whey protein-dextran conjugates at low pH and different salt concentrations. *Colloids and Surfaces, B: Biointerfaces* **2003**, *31(1–4)*, 125–132.
68. FECHNER, A.; KNOTH, A.; SCHERZE, I.; MUSCHIOLIK, G. Stability and release properties of double-emulsions stabilised by caseinate-dextran conjugates. *Food Hydrocolloids* **2007**, *21(5–6)*, 943–952.
69. WEINBRECK, F.; MINOR, M.; DE KRUIF, C. G. Microencapsulation of oils using whey protein/gum arabic coacervates. *Journal of Microencapsulation* **2004**, *21(6)*, 667–679.
70. DUCEL, V.; RICHARD, J.; SAULNIER, P.; POPINEAU, Y.; BOURY, F. Evidence and characterization of complex coacervates containing plant proteins: application to the microencapsulation of oil droplets. *Colloids and Surfaces, A: Physicochemical and Engineering Aspects* **2004**, *232(2–3)*, 239–247.
71. MAGNIN, D.; DUMITRIU, S. Interactions between polysaccharides and polypeptides. In *Polysaccharides* (2nd edn), Dumitriu S, Ed., 2005, 305–356.
72. GANZEVLES, R. A.; ZINOVIADOU, K.; VAN VLIET, T.; COHEN-STUART, M. A.; DE JONGH, H. H. J. Modulating surface rheology by electrostatic protein/polysaccharide interactions. *Langmuir* **2006**, *22(24)*, 10089–10096.
73. DICKINSON, E. Colloid science of mixed ingredients. *Soft Matter* **2006**, *2(8)*, 642–652.
74. CHEN, H.; WEISS, J.; SHAHIDI, F. Nanotechnology in nutraceuticals and functional foods. *Food Technology* **2006**, *60(3)*, 30–32, 34.
75. KIM, D. M.; CHO, G. S. Nanofood and its materials as nutrient delivery system (NDS). *Agricultural Chemistry and Biotechnology (English Edition)* **2006**, *49(2)*, 39–47.
76. WEISS, J.; TAKHISTOV, P.; McCLEMENTS, D. J. Functional materials in food nanotechnology. *Journal of Food Science* **2006**, *71*, R107–R116.
77. CHEN, L.; REMONDETTO, G. E.; SUBIRADE, M. Food protein-based materials as nutraceutical delivery systems. *Trends in Food Science & Technology* **2006**, *17(5)*, 272–283.
78. SHOJI, Y.; NAKASHIMA, H. Nutraceutics and delivery systems. *Journal of Drug Targeting* **2004**, *12(6)*, 385–391.

10

Food-protein-derived materials and their use as carriers and delivery systems for active food components

M. Subirade and L. Chen, Laval University, Canada

10.1 Introduction

The emergence of dietary compounds with health benefits offers an excellent opportunity to improve public health. Known as nutraceuticals, this category of compounds has received much attention in recent years from the scientific community, consumers and food manufacturers. The list of potential nutraceutical products (vitamins, probiotics, bioactive peptides, antioxidants, etc.) is endless, and scientific evidence to support the concept of health-promoting food ingredients is growing steadily. Even though the nature of the involvement of nutraceutical compounds in physiological functions is not yet fully understood, it is now well recognized that their addition to foods, as simple and direct means of decreasing the incidence of illnesses, holds much promise for benefit to society. The scientific community can therefore contribute to the development of innovative functional foods that bring physiological benefits or reduce long-term risks of developing diseases (Elliott and Ong, 2002).

The effectiveness of nutraceutical products in preventing diseases depends on preserving the bioavailability of the active ingredients. This represents a formidable challenge, given that only a small proportion of these remains bioavailable following oral administration, because of insufficient gastric residence time, low permeability and/or low solubility within the gastrointestinal (GI) tract, as well as instability under conditions encountered during food processing (temperature, oxygen, light) or in the gut (pH, enzymes, presence of other nutrients), all of which limit the activity and potential health benefits of nutraceutical molecules (Bell, 2001). The delivery of these molecules will therefore require food formulators and manufacturers to provide protective mechanisms that both maintain the active molecular form until the time of consumption, and deliver this form to the physiological target within the organism.

Polymer-based delivery systems that trap molecules of interest within networks have been developed extensively for the biomedical and pharmaceutical sectors to protect and transport bioactive compounds to target functions (Peppas *et al.*, 2000; Langer and Peppas, 2003). In spite of the success of carrier systems such as hydrogels, micro- and nanoparticles designed for enhanced delivery of medicinal drugs, drug targeting or time-controlled release, most of these polymer-based systems, cannot be used for food applications that require compounds generally recognized as safe (GRAS).

Protein is an essential component of the daily diet. Adult humans must eat dozens of grams of protein every day. It provides energy as well as materials for the synthesis of all bodily structures and enzymes, including all the hydrolases that are secreted daily into the GI tract to digest food. Most natural foods contain at least some protein and certain byproducts of the food-processing industry are rich in protein (whey from cheese production, press cake from vegetable oil production). Proteins are also widely used in formulated foods, partly because of their nutritional value, but especially for their functional properties, which include gelling, foaming and emulsification and underlie many food sensory attributes. Among these functional properties, the network forming ability of protein and the ability to adsorb spontaneously to interfaces to stabilize polyphasic systems offer the possibility of developing GRAS biocompatible carriers for oral administration of sensitive nutraceuticals in a wide variety of foods (Chen *et al.*, 2006). Moreover, research continues to confirm the association between consumption of food proteins and peptides and the health of populations, making them substrates of choice for nutraceutical and functional foods (Shahidi and Mine, 2005).

This chapter illustrates the potential of food protein-based matrices to serve as carriers for the controlled release of functional food components. Since the diversity of the systems and production processes under investigation is already vast, this chapter focuses on recent progress in the design, preparation and evaluation of food-protein-based delivery systems based on hydrogel or micro- and nanoparticles and their potential for the development of innovative functional foods.

10.2 Food protein hydrogels and related networks as nutraceutical carriers

Various strategies have been developed to protect and deliver bioactive molecules. One that has been the subject of numerous studies (Brannon-Peppas, 1993; Kamath and Park, 1993; Park and Park, 1996; Peppas, 1997) consists of trapping molecules of interest in hydrogels. A hydrogel is obtained by swelling an infinite network of hydrophilic polymers with a large amount of water without loss of network structure (Qiu and Park, 2001). A three-dimensional network is formed by crosslinking polymer chains through covalent bonds, hydrogen bonding, van der Waals interactions, or physical entanglements (Kamath and Park, 1993). Over the past decade, hydrogels have been studied extensively in biomedical and pharmaceutical applications, primarily because of their ability to protect drugs from hostile environments and to deliver them in response to environmental stimuli such as changes in pH and temperature (Qiu and Park, 2001; Park and Park 1996; Kashyap *et al.*, 2005). In the food industry, the development of food-protein-based environment-sensitive hydrogels for nutraceutical delivery is interesting in terms of both product performance and marketing. A fundamental advantage of this approach is that nutraceutical carrier gels are safe and can stabilize food texture, which is a highly desirable characteristic in the manufacturing of food products.

10.2.1 Environment-sensitive food protein hydrogels

Swelling properties

Food protein hydrogels are examples of polyelectrolyte networks swelling in water and retaining significant amounts of water within their structure (Peppas and Khare, 1993). Their ability to absorb water is due to the presence of ionizable groups such as carboxylic acid, sulfonic acid or amine groups that render the polymer more hydrophilic. It has been shown recently that β-lactoglobulin protein gels formed by the gelation/phase separation method may swell by absorption of water from three to 30 times by weight for hydrated and dehydrated gels, respectively (Reddy *et al.,* 2006). Among the factors that affect the swelling/de-swelling behavior of protein hydrogels are the degree of ionization, the ionization equilibrium and the nature of the counter-ions. As ionization increases, repulsive forces increase and the network becomes more hydrophilic. For example, pH variations cause the acidic (carboxylic) and basic (ammonium) groups in polypeptide chains to accept or release protons and thus determine swelling behavior, a well-researched mechanism for swelling-controlled product release systems (Qiu and Park, 2001).

Medium pH has a striking effect on the swelling of whey protein gels and the difference between the acidic pH in the stomach (1.9) and the neutral pH in the intestine (7.5) is large enough to allow timed release of molecules from protein gels given orally (Beaulieu *et al.*, 2002). Swelling is minimal at pH 4.5, near the isoelectric point or pI of whey protein (5.2) and greater at intestinal pH (7.5) or gastric pH (1.9). This suggests that swelling is governed mainly by the net charge of the protein molecules. Near pI, the net charge is zero for many of them, which minimizes electrostatic repulsions and hence gel swelling. Protein–protein interactions are favored over protein–solvent interactions. However, at pH other than pI, many whey protein molecules bear a net charge, increasing electrostatic repulsive forces and swelling. Whey protein gels are more swollen at intestinal pH than at gastric pH, since both the carboxyl and amino groups on all molecules are charged, maximizing electrostatic repulsive forces. At gastric pH, only the amine groups are charged. Hydrogel shrinkage due to attractions between opposite charges appears insignificant, suggesting that the various protein molecules tend not to bear equal numbers of carboxyl and amino groups. It can therefore, be concluded that the ionizable and/or ionized groups are the major factors that govern the pH-sensitive swelling mechanism of the gels. Not surprisingly, pH-responsive polymeric networks have been extensively studied and have been one of the most frequently used systems for the development of controlled release formulations for oral administration.

Biodegradation properties

Among the factors that modulate the release properties of protein hydrogels, those that promote gel breakdown play a major role. In orally administered products, the most important degradation mechanism is hydrolytic breakdown catalyzed by digestive enzymes. Food proteins break down to peptides, some of which may be bioactive and exert physiological effects *in vivo*, for example on the gastrointestinal, cardiovascular, endocrine, immune and nervous systems (Kilara and Panyam, 2003; Korhonen and Pihlanto, 2003). Since the formation of enzyme-substrate complexes is highly specific, enzymatic digestion is sensitive to changes that may affect substrate conformation or the active site of enzymes. For example, Fig. 10.1(a) shows the profile of breakdown by digestive enzymes (pepsin in the gastric step and pancreatin in the intestinal step) of two types of whey protein hydrogel that differ in structure (Remondetto *et al.*, 2004). The first one, called 'filamentous gel', is composed of flexible linear strands in a regular network created by linear aggregation maintained by hydrophobic interactions. The second one, called 'particulate gel', is composed of larger and almost spherical particles in random aggregation maintained essentially by van der Waals forces. This figure shows that the profile of breakdown is sensitive to gel structure, which may reflect the importance of enzyme diffusion into the network, since it is the surface area available for contact between the enzyme and the substrate that determines the rate of breakdown. This may be particularly important for highly crosslinked networks, which hinder enzyme penetration, thus making the system more resistant to the enzymatic attack. Parameters such as the crosslinking density of the hydrogel and/or the substrate conformation could be modified to obtain the system with the desired degradation profile (Kamath and Park, 1993).

Binding properties

Protein hydrogel is not an inert polymer delivery system since it has the ability to interact with a wide range of active compounds owing to the affinity of these compounds for gel-constituting proteins. The excellent review by De Wolf and Brett (2000) describes the ligand-binding properties of proteins and their potential applications in protein-mediated transport and controlled release of small molecules. Among these, bovine β-lactoglobulin (β-lg), which is the major whey protein in the milk of ruminants, has a high affinity for a wide range of compounds including retinol (Futterman and Heller, 1972), fatty acids (Spector

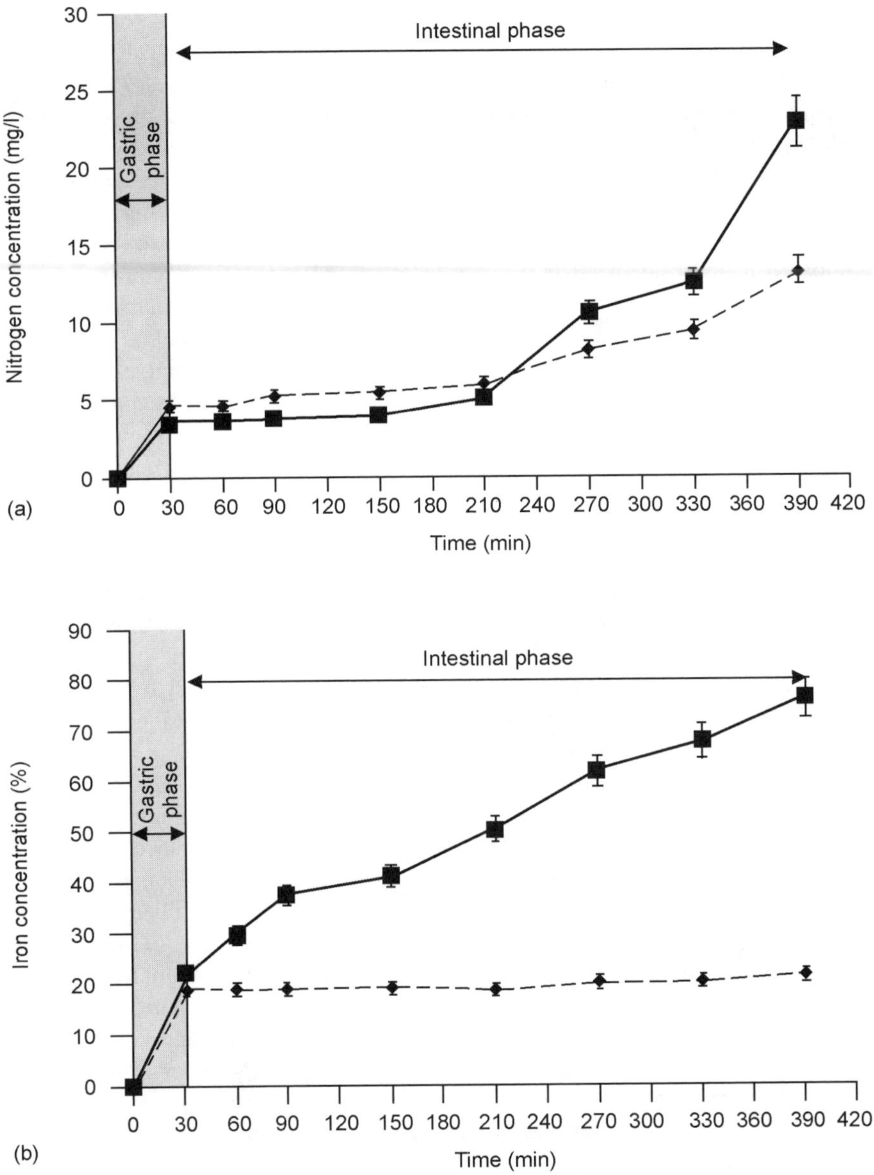

Fig. 10.1 Impact of simulated GI conditions on (a) soluble nitrogen release and (b) iron release from filamentous gels (solid line) and particulate gels (dashed line) (reproduced from Remondetto *et al.*, 2004).

and Fletcher, 1970) and phospholipids (Brown, 1984; Lefèvre and Subirade, 2001). After binding to β-lg, retinol and other hydrophobic molecules are protected from oxidative degradation (Hattori *et al.*, 1995). Thus, it has been proposed that β-lg be used as a versatile carrier of small hydrophobic molecules in controlled delivery applications (Batt *et al.*, 1994). In fact, many food proteins have high affinities for a wide range of molecules (Arts *et al.*, 2002; Jöbstl *et al.*, 2004; Jannin *et al.*, 2004; Rawel *et al.*, 2005; Papadopoulou *et al.*, 2005). These binding properties are dependant on the molecular structure of protein which under appropriate conditions (pH, temperature, ionic strength, etc.) can expose multiple functional groups modulating interactions with active compounds. Therefore, an active compound, having affinity for gel-constituting protein, is expected to be preferably loaded on the carrier and to be released slowly, as compared to inert gel where release rate of compounds typically relies on diffusion of the active through aqueous channels, the diffusion-controlled mechanisms reducing the potential for sustained release of active (Tada *et al.*, 2005).

Protein hydrogels thus have various properties, such as binding, swelling and breakdown, which can be modulated to control the release of encapsulated molecules.

10.2.2 Food protein hydrogel formation

The most popular hydrogel obtained from food proteins and used as a delivery carrier for active compounds is certainly gelatin-based hydrogel (Young *et al.*, 2005). Gelatin has desirable properties including biodegradability, low immunogenicity and low cytotoxicity. Gelatin alone forms physical hydrogels in response to temperature change (Einerson *et al.*, 2003). However, these have poor mechanical and chemical stability, necessitating crosslinking agents or mixing with other polymers (Yang and Kao, 2006).

Even though gelation of other food proteins, particularly of globular proteins (e.g. egg white, soybean and whey proteins), has attracted much attention over the years because of its physicochemical and industrial significance (Clark and Ross-Murphy, 1987; Ziegler and Foegeding 1990; Bryant and McClement 1998; Clark *et al.*, 2001), its application to active delivery systems is still limited. This may be due to the fact that gelation is traditionally achieved through heat treatment limiting their use as carriers to thermally stable compounds such as caffeine (Gunasekaran *et al.*, 2006). However, since many nutraceutical molecules would be irreversibly inactivated by exposure to the degree of heating involved, a method for producing food protein hydrogels under much milder conditions is required.

Recently, it has been reported that cold-induced gelation of globular whey proteins (Barbut and Foegeding, 1993) and more recently soybean proteins (Maltais *et al.*, 2005) can be achieved by adding Ca^{2+} ions to a predenatured protein suspension. This method requires a heating step during which proteins are denatured and polymerized into soluble aggregates, followed by a cooling step and subsequent salt addition, which results in the formation of a network via Ca^{2+}-mediated interactions of soluble aggregates. A mechanism of crosslinking carboxyl groups with Ca^{2+} has been suggested for the gelation of pre-denatured whey proteins at ambient temperature (Roff and Foegeding, 1996) with ionic strength playing a major role (Remondetto and Subirade, 2003). The formation of cold-set gels opens interesting opportunities for food proteins as carriers of sensitive nutraceutical compounds and in the development of innovative functional food ingredients. One advantage of this approach is exposure of multiple functional groups within the protein upon denaturing, which could be exploited to create different interactions between nutraceutical compounds and polypeptide chains such as (i) hydrogen-bonding, (ii) hydrophobic interactions and (iii) electrostatic interactions, which may in turn be applied to target nutraceutical delivery.

Based on the same process, cold-set gels were obtained by adding a ferrous salt to solutions of denatured β-lactoglobulin, addressing a mineral deficiency that concerns a large number of people all over the world (Remondetto *et al.*, 2002). Gel mechanical properties and microstructural analyses showed that two types of gel, 'filamentous' and 'particulate', could thus be obtained, depending on the iron/protein ratio. These different microstructural properties are strongly linked to aggregate molecular structure, as revealed by Fourier transform infrared spectroscopy and rheological data (Remondetto and Subirade, 2003). Release data presented in Fig. 10.1(b) suggest that these microstructures would have a major impact on iron delivery, due to their different sensitivities to environmental conditions such as pH and digestive enzymes and that filamentous gels show promise as matrices for transporting iron and promoting its absorption (Remondetto *et al.*, 2004). Modulation of gel microstructure and functional properties by cold gelation should allow tailoring of water-soluble delivery devices for nutraceutical and functional food system development.

Moreover, in the past few years, there has been growing interest in the application of composite systems such as emulsion gels to food formulation (Dickinson and Chen, 1999). Because of their flexibility and amphiphilic nature, globular proteins such as whey proteins rapidly adsorb to the emulsion interface, where they self-aggregate and form continuous and homogeneous membranes around oil droplets through intermolecular β-sheet interactions (Lefèvre and Subirade, 2003). By coating oil droplets with charged layers, protein films provide an electrostatic barrier against flocculation and coalescence (Phillips *et al.*, 1994). Numerous works have demonstrated the ability of oil-in-water emulsions stabilized by whey protein isolates (WPI) to form gels by heat treatment, extending the possibilities for creating foods with new and improved organoleptic properties (Jost *et al.*, 1986, 1989; McClements *et al.*, 1993; Dickinson *et al.*, 1996; Chen and Dickinson, 1998). Recently, it has been shown that β-lactoglobulin emulsions can be gelled at room temperature and that gel microstructure, final physical characteristics and functional properties can be modulated by varying the oil and calcium concentrations (Leung Sok Line *et al.*, 2005a). The suitability of such matrices as carriers for heat-sensitive and fat-soluble antioxidants has been demonstrated using α-tocopherol, as bioactive compound model (Leung Sok Line *et al.*, 2005b). It has been shown that cold-set β-lactoglobulin emulsion gels protect α-tocopherol under GI conditions and that α-tocopherol release is correlated with matrix degradation. Furthermore, degradation of these gels during the gastric step was only 20%, indicating that the matrices are gastro-resistant. Their gastro-resistance could be related to the interaction of the dispersed oil droplets with the gel matrix, which strengthens the three-dimensional network (Leung Sok Line *et al.*, 2005a).

Gelation of globular proteins has been the subject of numerous reviews, indicating both its scientific and commercial relevance (Totosaus *et al.*, 2002). The ability of hydrogels made from such proteins to entrap both fat-soluble and water-soluble food components of high health impact and to maintain the active forms of these components should contribute to the development of innovative functional solid foods. Nevertheless, understanding the role of nutraceutical ingredients in gel formation and studying their interactions within the protein network remain essential to anticipating their release profiles. Moreover, depending on the production method, variations in the electrochemical and physical properties of protein-based controlled release systems could be achieved.

10.3 Food protein micro- and nanoparticles

Micro and sub-micro particles are defined generally as particles less than about 1000 μm in size and which may be substantially less than 1 μm in diameter. ('Sub-micro' refers to nanoparticles in this chapter.) Advantages of micro and sub-micro particles are their abilities to control the release rate of the incorporated materials and deliver them to the right place at the right time (Schäfer *et al.*, 1992; Brannon-Peppas, 1995). This can be achieved by careful formulation of the desired internal composition and surface coating to allow matrix degradation and consequent contents release to occur at different times, for example to mimic the administration schedules of different nutraceutical compounds (Pothakamury and Barbosa-Gnovas, 1995), or by precise control of the particle size (Torrado *et al.*, 1989; Berkland *et al.*, 2002; Amidon *et al.*, 2003; Hori *et al.*, 2005). Larger particles generally release encapsulated compounds more slowly and over longer time periods, while particle size reduction introduces several bio-adhesive improvement factors, including increased adhesive force and prolonged GI transit time, leading to a higher drug bioavailability. Also, by simply altering the size, it is possible to achieve localization of microspheres in a particular organ or tissue. For example, Desai *et al.* (1997) showed that 100 nm nanoparticles had 2.5-fold greater uptake compared to 1 μm and 6-fold higher uptake compared to 10 μm microparticles in Caco-2 cells.

The functional properties of proteins make them very good coating material for the encapsulation of bioactive compounds (Gennadios, 2002; Radwick and Burgess, 2002). During the past two decades, interest in developing protein micro (sub-micro) particles as delivery systems has grown, and various kinds of animal proteins have been investigated, including gelatin (Franz *et al.*, 1998; Paynea *et al.*, 2002), collagen (Alex and Bodmeier, 1990; Rössler e*t al.*, 1995; Swatschek *et al.*, 2002), casein (Latha *et al.*, 1995; Latha *et al.*, 2000; Sahin *et al.*, 2002), albumin (Sokoloski and Royer, 1984; Tomlinson and Burger, 1985; Arnedo *et al.*, 2002; Yeo *et al.*, 2003) and whey protein (Rosenberg and Young,

Table 10.1 Manufacturing techniques used for preparing protein nanoparticles

Technique	Protein	Particle size	Reference
Extrusion	Gelatin	1–2.5 mm	Di Silvio *et al.* (1994)
			Franz *et al.* (1998)
	Whey	1–2 mm	Beaulieu *et al.* (2002)
Emulsification-stabilization	BSA	22–30 nm	Sahin *et al.* (2002)
	Casein	75–100 μm	Latha *et al.* (2000)
	Collagen	120–300 nm	Swatschek *et al.* (2002)
	Gelatin	6–400 μm	Esposito *et al.* (1996)
		20–160 μm	Dinarvand *et al.* (2004)
		1–7 μm	Mladenovska *et al.* (2002)
	Ovalbumin, HSA	1–32 μm	Torrado *et al.* (1989)
	Soy glycinin	100 μm	Lazko *et al.* (2004)
	Whey protein	10–70 μm	Bhattacharjee and Das (2001)
		10–50 μm	Picot and Lacroix (2004)
		3–10 μm	Chen *et al.* (2006)
Spray-drying	Albumin	1–5 μm	Li *et al.* (2001)
	Gelatin	2.5 μm	Bruschi *et al.* (2003)
	Insulin	88–110 nm	Gomez *et al.* (1998)
	Silk protein	5 μm	Hino *et al.* (2003)
Coacervation	BSA	250 nm	Arnedo *et al.* (2002)
	Gliadin	500 nm	Mauguet *et al.* (2002)
	HAS, gelatin	200–400 nm	Weber *et al.* (2000)
	Zein	300–1200	Liu *et al.* (2005)
	Gelatin/gum arabic, Xanthan	50–200 μm	De Kruif *et al.* (2004)
	Albumin/gum arabic		
	β-lactoglobulin/gum arabic		

1993; Beaulieu *et al.*, 2002; Picot and Lacroix, 2004) in addition to plant proteins such as soy glycinin (Lazko *et al.*, 2004), zein (Liu *et al.*, 2005) and wheat gliadin (Ezpeleta *et al.*, 1996). Multiple component matrices such as protein-polysaccharide (Guérin *et al.*, 2003; Chen and Subirade, 2006) and protein-synthetic polymer (Kasper *et al.*, 2005; Arbós *et al.*, 2002) micro (sub-micro) particles have also been designed.

So far, a great variety of processes have been developed to prepare protein-based micro (sub-micro) particles, including extrusion, emulsification-stabilization, spray drying and coacervation (Table 10.1). Since particle size plays a very important role in regulating polymer matrix degradation rate as well as nutraceutical compound release rate, uptake and distribution in the body, we will address these techniques in increasing order of final product particle size.

10.3.1 Techniques of micro (sub-micro) particle preparation

Extrusion

Extrusion is achieved by forcing protein droplets containing the fixing agent into a gel-forming solution and solidifying the falling droplets in microbeads or capsules. The smallest particles so produced are about 1 mm in diameter. Extrusion is a convenient process for encapsulating bioactive compounds in food protein matrices and these provide good barrier properties. For example, gelatin microspheres may be prepared by forcing the heated gelatin solution into chilled liquid paraffin (Di Silvio *et al.*, 1994; Franz *et al.*, 1998). Cooling of the dispersion results in the formation of beads of 1.6–2 mm, which are subsequently crosslinked to form microspheres, that are insoluble at 37 °C. Aldehyde derivatives such as formaldehyde, glutaraldehyde or other bifunctional reactants have been used as crosslinking agents. Recently, a protein-based 'all aqueous' encapsulation method has been developed as a two-stage process, which consists of first forming 'liquid core' microspheres with droplets of whey protein encapsulated in alginate (Al) 'skin' by dropping whey protein/alginate mixed solution into Ca^{2+} solution and then crosslinking the microsphere matrix either by heating or chemically by glutaraldehyde (Rosenberg and Lee, 2004). The formation of stable microspheres is critically dependent on rapid and effective formation of Ca-Al 'skin' around the dispersed whey droplets, which could be expected to depend on rapid diffusion of Ca^{2+} ions into the surface of the whey protein droplets to provide effective crosslinking of the Na-Al. Whey protein micro-beads (1–2 mm) have also been formed by an extrusion method consisting of dropping the whey protein/oil emulsion into a calcium chloride ($CaCl_2$) solution (Beaulieu *et al.*, 2002). The principle of this method is based on a Ca^{2+}-induced gelation of pre-denatured whey protein. Preparation conditions influence micro-bead sphericity and homogeneity, increased $CaCl_2$ favoring a more regular shape. In fact, high $CaCl_2$ concentration results in accelerated gelation kinetics, which leads to rapid entrapment and stabilization of emulsion droplets by the protein network. Properties such as swelling, elasticity, deformability and resistance at fracture depend on pH and preparation conditions. FTIR spectroscopic analysis shows that bead formation has a pronounced effect on the secondary structure of whey protein, leading to the formation of intermolecular hydrogen-bonded β-sheet structures. Originally, this process allowed the formation of beads of about 1 or 2 mm. However, it has been shown recently that calcium concentration can be judiciously controlled to obtain desired particle sizes in the μm to mm range (Lavoie *et al.*, 2005).

Emulsification/stabilization process

The emulsification-stabilization technique is a two-step process comprising the emulsification of a protein solution containing the encapsulated substance, followed by particle hardening by protein precipitation using a crosslinking agent or by changing the pH and temperature. Protein microsphere diameter can be set anywhere over a wide range by formulation and controlling the emulsifying conditions. For example, by varying the gelatin concentration from 5% to

65%, microspheres ranging from 10 μm to 1500 μm have been produced (Dinarvand *et al.*, 2004). Higher protein concentrations result in larger droplets due to the effect of increased viscosity on mixer efficiency. Moreover, both the stirring rate and the phase volume ratio strongly influence microsphere size distribution. The degree of crosslinkage does not significantly influence size distribution, although it can be adjusted to improve microsphere smoothness and more importantly to control microsphere swelling properties (Dinarvand *et al.*, 2004). Since crosslinkers are usually toxic or may engage in unwanted reactions with encapsulated cores, thermal hardening has been applied to stabilize BSA, gelatin and β-lactoglobulin droplets, by maintaining the w/o emulsion above 90 °C (Bhattacharjee and Das, 2001). Compared to chemical crosslinking microspheres, spheres formed by heat crosslinking are soft and spongy and have smoother surfaces with very smaller pores.

Microencapsulation by forming a surface layer of protein-surfactant insoluble complex on oil droplets has also been developed, using gelatin type A and sodium dodecyl sulphate (SDS), an anionic surfactant (Magdassi and Vinetsky, 1995). Binding isotherms and the specific molar ratios of SDS to gelatin at which the insoluble complex is formed have been determined. Encapsulability of oil by using soy glycinin, as wall forming material, was also realized by changing pH and temperature of the protein solution. Mild acid (pH 2.0) and heating (55 °C) the reaction medium during the emulsification step significantly enhance surface deposition of glycinin. Slow readjustment of the pH limits aggregation and leads to homogeneous precipitation of proteins. The homogeneity of microcapsule size could be improved by shortening the time required for the precipitation step (Lazko *et al.*, 2004).

Recently, alginate-whey protein microspheres (100 μm) have been developed by emulsification/internal cold gelation (Chen and Subirade, 2006) without organic crosslinking agent or heating. The method is based on the release of calcium ions from an acid-soluble calcium salt in emulsified pre-denatured protein solution. This is achieved by acidification with an oil-soluble acid, which partitions to the dispersed aqueous phase, releasing soluble calcium and initiating gelation. Microscopy, presented in Fig. 10.2(a), revealed a special matrix/granular structure for these microspheres with whey protein granules (3–10 μm) homogeneously distributed inside an alginate spherical matrix. By controlling emulsification conditions, the particle size can be controlled from 30 to 800 μm in diameter (Chen and Subirade, 2007). This method opens interesting opportunities for the use of food proteins as carriers of sensitive nutraceutical compounds and in the development of innovative functional food ingredients.

Spray drying

Spray drying is a method of preparing protein microparticles free of organic solvent residues. Protein solutions are typically pumped into a drying chamber and pneumatically atomized through a two-fluids external mixing nozzle using compressed air. The inlet temperature is generally set at about 100 °C or higher. The atomized droplets dry rapidly because of their high surface area and

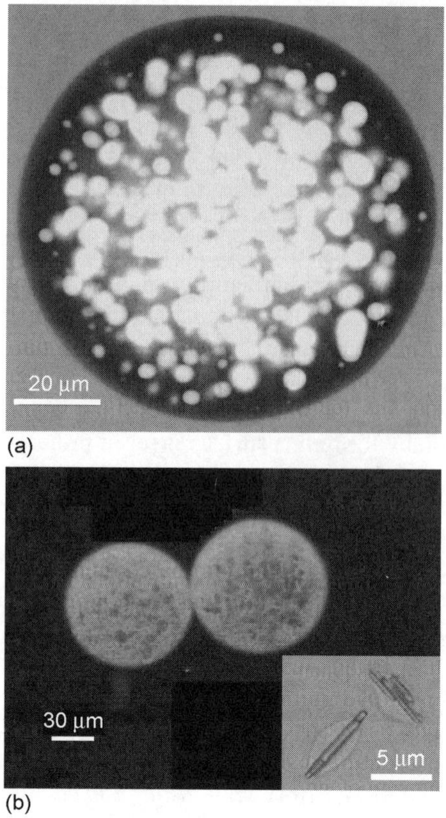

Fig. 10.2 Micrographs of AL–WPI microspheres: (a) labeled with phen green (b) loaded with riboflavin (WPI/AL ratio 8:2; particle size 100 μm and CAL+WPI 5%). Insert shows microsphere internal structure with loaded riboflavin inside granules about 3–10 μm in diameter (reproduced from Chen and Subirade, 2006).

intimate contact with the drying gas. Spray drying can reduce particle diameter to less than 5 μm. Albumin microparticles of spherical shape and indented surface have been obtained by spraying through the two-flow pressure nozzle (Li *et al.*, 2001). Particle diameters in the 1–5 μm range depend largely on spray dryer settings (feed flow per nozzle, compressed gas flow rate and pressure). Protein microparticle agglomeration can be overcome by addition of mannitol to the gelatin solution (Bruschi *et al.*, 2003). SEM photomicrographs of spray-dried gelation microparticles obtained with mannitol show a very smooth and uniform surface and spherical shape. A low number of coalesced microparticles and less agglomeration are observed, suggesting that by interacting with gelatin, mannitol probably reduces polymer chain contact. Moreover, mannitol molecules bind water molecules and could fill the empty space within the microparticles, preserving hydration, reducing surface depressions and ensuring a more uniform wall. Spray drying can alter the secondary structure (α-helix, β-sheet, and random coil) of some proteins. For example, it has been shown by powder X-ray diffractometry, FTIR and Raman spectroscopy that the conformation of silk proteins used to prepare spray-dried microspheres (5 μm) changed from anti-parallel β-sheet to amorphous (Hino *et al.*, 2003). These microspheres formed gels immediately after soaking in purified water and solutions buffered at pH 1.2 and 6.8, followed by slight gel dissolving.

Cone-jet-mode electrosprays offer the appealing feature of droplet mono-dispersion. In this system, liquid is dispersed primarily by electrical forces, which uncouples atomization from gas flow. Droplets/particles of a wide range of sizes can be produced, from molecular dimensions to hundreds of microns, depending on liquid flow rate, applied voltage and liquid electrical conductivity. The feasibility of producing relatively monodisperse and biologically active insulin nanoparticles by electrospray drying has been demonstrated (Gomez *et al.*, 1998). The process comprised dissolving dry insulin in an acidic ethanol/ water solution, electrospraying the solution and, after solvent evaporation, collecting dry residues on suitable deposition substrates. Particle diameter ranged from about 88 to 110 nm and the distributions were quasi-monodisperse with relative standard deviation estimated at approximately 10%.

Coacervation

Coacervation is suited to the formation of particles of sub-micron size in the range of 20 nm to 300 μm. The process essentially involves changing the physicochemical environment of a protein in aqueous solution to cause varying degrees of phase separation of the protein depending on its molecular weight, isoelectric point and compatibility with solvents. A coacervation process can be 'simple' or 'complex'.

'Simple' coacervation can result from the removal of the solvent from a single colloid dispersed in solution by addition of chemical compounds which have a greater affinity for solvent than the colloid. This causes the colloid chains to come closer together and form the coacervate. Gelatin, albumin, gliadin and zein sub-microparticles (100 nm–1.5 μm) have been prepared by coacervation using water, alcohol, or acetone as the desolving agent (Weber *et al.*, 2000; Arnedo *et al.*, 2002; Mauguet *et al.*, 2002; Liu *et al.*, 2005). It has been shown that the nanoparticle size obtained by the HSA desolvation process depended mainly on the amount of added desolvating agent (Weber *et al.*, 2000). Addition of up to 1.5 volumes of ethanol per volume of initial HSA solution increased particle size significantly, while over 1.5 volumes of ethanol maintained constant particle size but increased particle concentration. Nanoparticles prepared with acetone were generally smaller and had a lower polydispersity compared to those prepared with ethyl alcohol (Azarmi *et al.*, 2006). Sub-microparticles have also been prepared by salting-out coacervation by adding a salt solution to protein solutions (Mauguet *et al.*, 2002). The study of gliadin coacervation has shown that the richer in protein the continuous phase, the smaller the quantity of salt required. The main problem of salting-out coacervation is control of particle size and agglomeration. This can be achieved by adjusting the kinetics of the salt addition. Aggregation is decreased considerably at very low salt solution addition rates.

'Complex' coacervation is caused by the interaction of two oppositely charged colloids. Complexes of food protein (gelatin, albumin, β-lactoglobulin, casein, legumin) with polysaccharide (acacia gum, pectin, gellan gum, xanthan gum, alginate, carrageenan and carboxymethyl cellulose) have been considered as prospective sources of material for microcapsule formation (de Kruif *et al.*, 2004). The gelatin-gum arabic system is the most studied complex coacervation system. Complex coacervation is possible only at pH values below the iso-electric point of gelatin. It is at these pH values that gelatin becomes positively charged, while gum arabic remains negatively charged. The phase separation is mainly entropically driven, and may most probably be attributed to the delocalization of the counter ions of the protein and gum arabic. If one of the poly-ions is a strong poly-electrolyte, a precipitate is formed rather than a liquid coacervate phase. Complex coacervates formed are usually neutral or modestly charged. Salt has a dissociating effect on all poly-electrolyte complexes and increasing ionic strength diminishes the protein-polysaccharide interaction.

The techniques presented above have their specific advantages and dis-advantages and their use will depend on the intended product use. Extrusion allows narrower size distributions in the final product and is thought to be less detrimental to the bioactive compounds encapsulated. However, extrusion of protein solutions through nozzles to produce either beads or capsules is mainly a laboratory technique and is difficult to scale up. The main advantages of emulsification are its feasibility at production scale and its ability to produce microparticles with a wide range of diameters. However, large amounts of oil are needed for extraction and for washing. Spray drying produces microparticles in continuous operation and with a narrow size distribution. Its inconveniences are related to the fine control of processing variables that is required in order to avoid difficulties such as low yield, sticking or high moisture content. Although free of oil residue or organic solvent problems, thermal treatment has limited applicability to formulations containing heat-sensitive ingredients. Coacervation methods are simpler to perform and the solvents used are less toxic but the particles so produced are not very stable and aggregate easily into larger microspheres. Finally, the choice of cross-linking agents for protein particle stabilization has a considerable influence on final microsphere characteristics and should be carefully considered from both a technological and toxicological perspective. There is thus a continuing need for improved methods of preparation of protein micro (sub-micro) particles that do not suffer from the disadvantages summarized above.

10.3.2 Selected examples of food protein micro (sub-micro) particle controlled release systems

Protein-based micro (sub-micro) particles have found wide and rapidly increasing applications in the food industry because they can be precisely designed for use in many food formulations and virtually any ingredient can be encapsulated, whether hydrophobic, hydrophilic, or even bacterial. The following section discusses a few typical protein-based microparticles as delivery systems for functional food ingredients.

Vitamins

Vitamins are naturally occurring compounds present in most foods. They are essential in small quantities for all body functions including growth, repair of tissues, and the maintenance of health. However, they are generally sensitive to light, pH and oxidation.

Figure 10.2 shows whey protein microparticles of diameters ranging from 3 to 10 μm homogeneously distributed within an alginate matrix (Chen and Subirade, 2006). Riboflavin, visible as yellow crystals by light microscopy (Fig. 10.2(b)), was chosen as a typical water-soluble vitamin for incorporation into these microparticles. High encapsulation efficiency (above 80%) was obtained with loading efficiency reaching 20–30%. Degradation assays and release experiments were performed in a simulated GI tract in the presence of digestive enzymes. In simulated gastric fluid (SGF), the alginate matrix remained in a shrunken state due to conversion of sodium alginate to insoluble alginic acid, which acted as a barrier to whey protein microparticle degradation by pepsin, thereby retarding riboflavin release. When transferred to simulated intestinal fluid (SIF), the alginate matrix eroded, liberating the whey protein microparticles, which were then digested by pancreatin, followed by complete release of the riboflavin. Release data were analyzed according to zero- and first-order kinetics as well as a diffusion-controlled mechanism using linear regression analysis. The release of riboflavin in SGF and in SIF without pancreatin followed the Higuchi diffusion model, while release in SIF in the presence of pancreatin was attributed to complete degradation of the whey protein particles.

For encapsulation of fat-soluble molecules, whey protein (WPI) beads were successfully produced using a new emulsification/cold gelation process (Beaulieu *et al.*, 2002). Uniform (homogeneous) oil globules are distributed in a gel protein network. Retinol, a model fat-soluble vitamin sensitive to environmental conditions, particularly to stomach acid, is dissolved in the oil phase. Pre-denatured proteins, acting as an emulsifier, rapidly adsorb to the surface of the oil droplets, which creates a stabilizing layer that protects the fine droplets against coalescence and provides physical stability to the emulsion. Addition of Ca^{2+} causes initial gelation, leading to the rapid entrapment and stabilization of the droplets by the protein network. Release experiments in the simulated GI tract demonstrated a matrix biodegradation release mechanism. Little retinol was released in SGF due to gastric digestion. These beads notably exhibit resistance to pepsin, since its preferred sites of activity, hydrophobic amino acid residues, are trapped within the protein network upon addition of Ca^{2+}. These same protein microparticles were completely degraded by pancreatin, leading to total release of the bioactive compounds. HPLC shows no sign of oxidation of the retinol after its release. It can be suggested that bioavailability of retinol can be improved when the fat globules are stabilized by protein molecular chains.

This research suggests the use of food protein microparticles as a matrix to protect gastric-pH-sensitive bioactive molecules for delivery in the small intestine, where they can be absorbed through the brush border walls.

Microorganisms

The use of probiotic bacterial cultures to maintain healthy gut microflora may provide protection against gastrointestinal disorders including gastrointestinal infections, inflammatory bowel disease and even cancers. It has been recommended that foods that are marketed as probiotic should contain at least 10^6 *live* microorganisms per g or mL of product at the time of consumption in order to produce therapeutic benefits. However, the microorganisms tend to be sensitive to many environmental conditions including acidic pH.

To overcome this problem, immobilization of bifidobacteria in food-grade water-insoluble microcapsules has been successfully achieved (Picot and Lacroix, 2004). Microcapsules of diameters ranging from 3 to 75 μm were prepared using whey protein and spray-dried in a single continuous process. Even though losses in cell viability occur during spray-drying, it has been shown that encapsulation of bifidobacteria in these microcapsules can increase their tolerance to highly acidic environments, making this approach potentially useful for delivery of probiotic cultures to the gastrointestinal tract of humans.

Protein-polysaccharide gel beads have also been used to increase the survival of bifidobacteria during transit through the GI tract (Guérin *et al.*, 2003). A transacylation reaction resulting in the formation of amide bonds between protein and alginate has been used to produce a crosslinked membrane on the surface of mm-sized alginate beads, which resists gastric pH and pepsin activity (Levy and Edwards-Levy, 1996). It was shown that bifidobacteria immobilized in these beads were more resistant to simulated human GI tract conditions. Their mortality in an acidic solution (pH 2.5) containing pepsin was less than 2.5 log after two hours of incubation, while free cells did not survive. In treatments with bile salts, membrane-coated beads performed better than membrane-free beads, in which cells were less resistant than free cells. In fact, this matrix could protect bacteria in both the carrier food and in the human stomach. An initial immobilized cell population of 10^{10} cell/g could thus reach the intestine at a viable concentration of $10^{7.5}$ cell/g, which would be sufficient to produce the alleged beneficial health effects (Marteau *et al.*, 1997).

Macromolecules

Many proteins and peptides are therapeutic compounds with highly specific and potent biological functions. Fuelled by recent advances in biotechnology, the introduction of these compounds is growing rapidly (Shahidi and Mine, 2005). However, they are easily degradable both chemically and enzymatically and may also lose their biological activity through conformational changes and aggregation. In order to maintain their biological activity, they must be administered with carrier systems.

Protein-loaded gelatin microspheres were prepared by emulsification of an aqueous solution of gelatin and BSA in oil and stabilized by crosslinking with glutaraldehyde (Mladenovska *et al.*, 2002). By varying process conditions, microspheres were prepared with high loading efficiency (80–95%) and an optimal size (1–10 μm) for uptake into mucosal subepithelia. The BSA model protein provided understanding of the release kinetics. Degradation data showed that crosslinking prolongs degradation of these gelatin microspheres by collagenase. BSA *in vitro* release profiles were biphasic, characterized by an initial relatively rapid release period influenced by degradation and diffusive processes, followed by a slower release phase influenced by diffusive processes only. Strong negative correlations between particle size and degradation rate and between particle size and BSA release rate were noted. Particles smaller than 4 μm demonstrated a higher initial release rate, which is probably a result of greater water penetration into the spheres, due to faster enzymatic degradation. Biodistribution studies of the same particles were carried out on BALB/c mice after oral administration (Mladenovska *et al.*, 2003) and was followed periodically over 15 days as a percentage of total BSA present in stomach, small intestine with Peyer's patches and mesentery, colon with Peyer's patches, appendix and mesentery, liver, spleen, blood, kidney, lungs and heart. The biodistribution data confirmed that uptake in mice into Peyer's patches and passage to the liver and spleen via the mesentery lymph supply and nodes, increased with decreasing particle size.

A tetra-peptide, FITC-labelled lysine–arginine–phenylalanine–lysine, was encapsulated in 50 μm alginate microspheres coated with a serum albumin-alginate membrane using an impregnation technique comprising dropping the peptide onto freeze-dried microparticles (Hurteaux *et al.*, 2005). This procedure avoids peptide exposure to the preparation treatments and losses during the loading step. Although partial adsorption on the particle surface cannot be excluded, encapsulation equivalent to 100% of the microsphere capacity was achieved, based on confocal microscopy images of uniformly fluorescent microcapsules, confirming that the peptide was evenly distributed throughout the entire microsphere volume. The peptide was released more slowly from microspheres surrounded by membrane than from pure alginate microspheres. The filtering effect of the membrane, combined with the binding effect of HSA, potentiated by the presence of calcium ions, contributed to slowing down peptide release. It should be emphasized that the coated microspheres resisted stirring at 37 °C for over eight days, while uncoated microspheres were more or less degraded at the end of the 24 h test. This degradation also contributed to faster release from the control batch, and confirms again the advantage of a membrane at the periphery of the microspheres.

Unsaturated fatty acids

Unsaturated fatty acids are important nutrients involved in many body functions, including neuro-protective, antioxidant, anti-inflammatory effects and cardiovascular health. However, their development as nutritional supplements is limited by their high susceptibility to oxidative rancidity, which leads to the formation of off-flavors and potentially toxic compounds. Food proteins are very effective at stabilizing oil-in-water emulsions by adsorbing to the oil droplet surface, lowering interfacial tension, forming protective membranes around the lipid and imparting an electrical charge to the droplet surface when the pH is not near the pI of the interfacial protein. Unsaturated fatty acid encapsulated in protein microparticles can be converted to a stable powder form that is more resistant to oxidation.

Omega-3 fatty acids have recently been encapsulated in an emulsified pre-denatured whey protein matrix using an extrusion method comprising dropping the emulsion into calcium chloride solution (Lavoie *et al.*, 2005). Calcium concentration was found to be very important for regulating particle size and particles ranging from μm to mm sizes could be obtained by varying the calcium concentration from 0.01 M to 5 M. This process of extrusion and aggregation of a pre-formed emulsion is very efficient and allowed high encapsulation efficiency at all calcium concentrations used. Omega-3 fatty acid was encapsulated at 77% with 0.1 M calcium and at 98% with 3 M calcium, which can be explained by differences in resulting matrix structure. At low calcium concentration, whey proteins were solidified on the oil droplet surface, while at high calcium concentration, protein forms a three-dimensional network that entraps additional oil droplets. The oxidative stability of omega-3 fatty acid encapsulated in 1 mm beads was investigated during storage at 60 °C. Linolenic acid (C18:3n-3) content in free unsaturated fatty acid remained constant for the first six days, followed by a rapid decrease to 4% of its initial level. Unsaturated fatty acid encapsulated in whey protein beads lost some C18:3n-3 during the early stage because of some oxidation during the particle formation process. However, C18:3n-3 content decreased slowly during storage, by only 15% after 15 days. This study demonstrated the ability of whey protein matrix to prevent linseed oil oxidation.

Transition metals are found in all foods, since they are common constituents of raw food materials, water, ingredients and packaging materials. The formation of free radicals by interactions between lipid hydroperoxides and transition metals is important in the promotion of lipid oxidation in both model system studies and foods. The impact of chelators like ethylenediamine tetraacetic acid (EDTA), citrate and polyphosphates on the physical and oxidative stability of omega-3 fatty acid oil-in-water emulsions, stabilized by whey protein isolate, has been investigated (Hu *et al.*, 2004). EDTA improved the oxidative stability of emulsions stabilized by whey protein isolate, at both pH 3.0 and 7.0, at concentrations $\geq 1\ \mu$M without any impact on the physical stability of the emulsions. These results indicate that the oxidative stability of whey-protein-isolate-stabilized oil-in-water emulsions can be increased by EDTA.

As can be inferred from the above examples, microparticles based on food proteins allow effective protection of bioactive compounds until the time of consumption and controlled, sustainable release of the compound to the physiological target within the organism.

Nanoparticles

Although the applications of nano-scale particles in therapeutic systems have been well documented (Langer and Peppas, 2003), their use in the food industry is still limited, even though nanoparticles offer promising means of improving the bioavailability of nutraceutical compounds, especially poorly soluble substances such as functional lipids (e.g. carotenoids, phytosterols, ω-3 fatty acids), natural antioxidants and numerous other compounds that are widely used as active ingredients in various food products. Among the various polymer-based nanoparticulate delivery systems potentially available to the food industry, those based on proteins have gained the most interest. The proteins investigated include gelatin (Balthasar *et al.*, 2005; Dinauer *et al.*, 2005; Leo *et al.*, 1997) gliadin (Duclairoir *et al.*, 1999, 2003; Ezpeleta *et al.*, 1996), sericin (Cho *et al.*, 2003), serum albumin (Weber *et al.*, 2000; Lin *et al.*, 1997, 2001; Merodio *et al.*, 2001) or vicilin (Ezpeleta *et al.*, 1996). Protein nanoparticles are particularly interesting because they can incorporate a wide variety of nutrients in a relatively non-specific fashion or conjugate nutrients via either primary amino groups or sulfhydryl groups (Weyermann *et al.*, 2005). Furthermore, proteins offer functional surface coatings for polymer nanoparticles, since they can complex with polysaccharides, lipids or other polymers to increase particle circulation time in the GI tract or enhance their ability to penetrate the mucosa.

Recently, it has been shown that β-lactoglobulin-coated chitosan nanoparticles (around 100 nm), obtained by cold ionic gelation of chitosan and β-lactoglobulin mixtures with sodium tripolyphosphate, provide protection of sensitive nutraceutical compounds in the GI tract (Chen and Subirade, 2005). Release profiles for these nanoparticles were investigated in a simulated GI tract, using brilliant blue as the model molecule. Native β-lactoglobulin-coated nanoparticles exhibited the desirable property of retarding molecule release into the gastric environment. A pepsin degradation assay revealed that the coating resists acid and pepsin, which may be attributed to the hydrophobic amino acids of native β-lactoglobulin being hidden inside its calyx-like globular structure, given that pepsin preferentially attacks peptide bonds involving hydrophobic aromatic amino acids. Under simulated intestinal conditions, the coatings were degraded by pancreatin. Due to its muco-adhesive nature, the chitosan core should then adhere to the intestinal wall to facilitate the absorption of the nutraceutical compound. This study suggests that β-lactoglobulin, especially in its native configuration, is an interesting candidate for coating nanoparticles in order to protect carried nutrients from exposure to gastric fluids.

Research has revealed that covalent binding of tomato lectin (TL), a bio-adhesive glycoprotein, to the surface of nanoparticles increases carrier residence time in the gut, due mainly to a significant decrease in the terminal elimination rate of the fraction adhering to the gastrointestinal mucosa (Arbós *et al.*, 2002). When bound to 500 nm nanoparticles, intestinal uptake occurs not only through the gut associated lymphoid tissue but also through normal enterocytes. It has been shown that the uptake rate of TL coupled to nanoparticles was 41.5 ng/mg h^{-1},

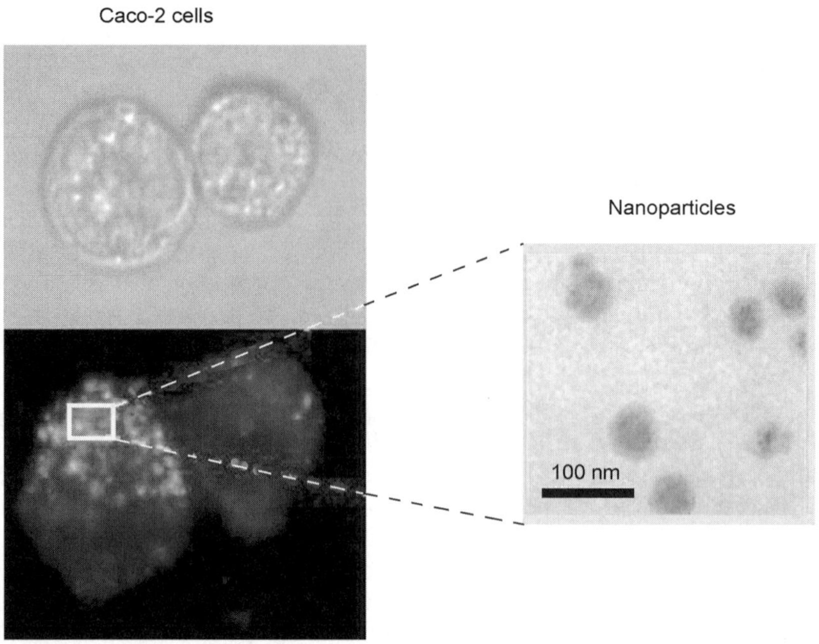

Fig. 10.3 Images of Caco-2 cells after incubation alone (top) or in the presence of FITC-labeled nanospheres (bottom). Picture on the right shows transmission electron micrographs of nanoparticles.

which was 4-fold higher than BSA coupled to nanoparticles (11.8 ng/mg h^{-1}) and 11-fold higher than free TL (3.9 ng/mg h^{-1}) (Carreno-Gómez *et al.*, 1999).

Recently, whey protein nanoparticles (40 nm), prepared by coacervation with calcium, have shown potential as carriers for orally administered nutraceutical agents (Leclerc *et al.*, 2005). This was evaluated *in vitro* by monitoring the interaction between fluorochrome-labeled nanospheres and cells of intestinal origin (Caco-2 cells). Figure 10.3 shows Caco-2 cells incubated without (top) and with (bottom) FITC-labeled nanospheres. FITC-labeled nanospheres are clearly seen on the top half of the cells (i.e. the exposed surface area) and are absent from the bottom half (i.e. the surface area against the flask). Interestingly, these nanospheres are not only able to adhere to the cell surface but are also visible within the cells. These experiments suggest that whey protein nanoparticles could be internalized by cells and degraded therein to release the nutraceutical compounds, significantly improving nutraceutical compound bioavailability.

10.4 Future trends

Food proteins show great promise for developing and engineering a range of new GRAS matrices with the potential to incorporate nutraceutical compounds and provide controlled release via the oral route. Clear advantages of food protein matrices include their high nutritional value, the abundance and wide variety of their sources and their acceptability as naturally occurring food components degradable by digestive enzymes. As discussed in this chapter, food proteins can be used to prepare a wide range of matrices and multi-component matrices in the form of hydrogel, micro- or nanoparticles, all of which can be tailored for specific applications in the development of innovative functional food products. The ability to control the particle size of proteinaceous materials is of primary importance, not only for food product properties such as taste, aroma, texture and appearance, but also for determining the release rates of the carried bioactive compounds and ultimately how much is absorbed into the body and hence overall compound efficacy. Although protein hydrogels, and to a lesser extent protein microparticles, have been the subject of intensive research over the years, nanoparticles are relatively new in the food industry and poorly documented in spite of their potential to improve the bioavailability of nutraceutical compounds, especially of poorly soluble substances. However, the past decade has seen great advances in understanding of protein structure and function and it can now be envisaged that knowledge of protein self-assembly into nanostructures with controlled properties and high surface area will rapidly improve.

10.5 Sources of further information and advice

There are a number of excellent books that the reader may consult for further information on controlled release of active components in general and in the specific context of biopolymer delivery systems. The interested reader is advised to proceed to the original papers and references cited therein for more detailed information.

10.6 Acknowledgements

This work was financially supported by the Natural Sciences and Engineering Council of Canada and by the Fonds Québécois de la Recherche sur la Nature et les Technologies (Action Concertée FQRNT-AEE-MAPAQ-MIC)

10.7 References and further reading

ALEX R and BODMEIER R (1990), 'Encapsulation of water-soluble drugs by a modified solvent evaporation method. I. Effect of process and formulation variables on drug entrapment', *J Microencapsul*, 7, 347–355.

AMIDON G L, LEVY R J and LABHASETWAR V (2003), 'Polymer degradation and *in vitro* release of a model protein from poly(D,L-lactide-co-glycolide) nano- and microparticles', *J Control Release*, 92, 173–187.

ARBÓS P, ARANGOA M A, CAMPANERO M A and IRACHE J M (2002), 'Quantification of the bioadhesive properties of protein-coated PVM/MA nanoparticles', *Int J Pharm*, 242, 129–136.

ARNEDO A, ESPUELAS S and IRACHE J M (2002), 'Albumin nanoparticles as carriers for a phosphodiester oligonucleotide', *Int J Pharm*, 244, 59–72.

ARTS M J T J, HAENEN G R M M, WILMS L C, BEETSTRA S A J N, HEIJNEN C G M, VOSS H P and BAST A (2002), 'Interactions between flavonoids and proteins: effect on the total antioxidant capacity', *J Agric Food Chem*, 50, 1184–1187.

AZARMI S, HUANGY, CHEN H, McQUARRIE S, ABRAMS D, ROA W, FINLAY W H, MILLER G G and LÖBENBERG R (2006), 'Optimization of a two-step desolvation method for preparing gelatin nanoparticles and cell uptake studies in 143B osteosarcoma cancer cells', *J Pharm Pharmaceut Sci*, 9, 124–132.

BALTHASAR S, MICHAELIS K, DINAUER N, VON BRIESEN H, KREUTER J and LANGER K (2005), 'Preparation and characterisation of antibody modified gelatin nanoparticles as drug carrier system for uptake in lymphocytes', *Biomaterials* 26(15), 2723–2732.

BARBUT S and FOEGEDING EA (1993), 'Ca^{2+}-induced gelation of pre-heated whey protein isolate', *Journal of Food Science*, 58(4), 867–871.

BATT CA, BRADY J and SAWYER L (1994) 'Design improvements of β-lactoglobulin', *Trends Food Sci Technol*, 5, 261–265.

BEAULIEU L, SAVOIE L, PAQUIN P and SUBIRADE M (2002), 'Elaboration and characterization of whey protein beads by an emulsification/cold gelation process: application for the protection of retinol', *Biomacromolecules*, 3, 2239–2248.

BELL L N (2001), 'Stability testing of nutraceuticals and functional foods', in Wildman R E C, *Handbook of Nutraceuticals and Functional Foods*, New York: CRC Press, 501–516

BERKLAND C, KING M, COX A, KIM K and PACK D W (2002), 'Precise control of PLG microsphere size provides enhanced control of drug release rate', *J Control Release*, 82, 137–147.

BHATTACHARJEE C and DAS K P (2001), 'Characterization of microcapsules of β-lactoglobulin formed by chemical cross-linking and heat setting', *J Dispersion Sci Tech*, 22, 71–78.

BRANNON-PEPPAS L (1993), 'Controlled release in the food and cosmetics industries', in El-Nokaky M A, Piatt D M and Charpentier B A, *Polymeric Delivery Systems*, Washington, DC: American Chemical Society, 42–52.

BRANNON-PEPPAS L (1995), 'Recent advances on the use of biodegradabble microparticles and nanoparticles in controlled drug delivery', *Int J Pharm*, 116, 1–9.

BROWN E M (1984), 'Interactions of β-lactoglobulin and α-lactalbumin with lipids: a review', *J Dairy Sci*, 67, 713–722.

BRUSCHI M L, CARDOSO M L C, LUCCHESI M B and GREMIAO M P D (2003), 'Gelatin microparticles containing propolis obtained by spray-drying technique: preparation and characterization', *Int J Pharm*, 264, 45–55.

BRYANT C M and McCLEMENT D J (1998), 'Molecular basis of protein functionality with special consideration of cold-set gels derived from heat-denatured whey', *Trends Food Sci Technol*, 9, 143–151.

CARRENO-GÓMEZ B, WOODLEY J F and FLORENCE A T (1999), 'Studies on the uptake of tomato lectin nanoparticles in everted gut sacs', *Int J Pharm*, 183, 7–11.

CHEN J and DICKINSON E (1998), 'Viscoelastic properties of heat-set whey protein emulsion gels', *Journal of Texture Studies*, 29, 285–304.

CHEN L and SUBIRADE M (2005), 'Chitosan/β-lactoglobulin coreshell nanoparticles as nutraceutical carriers', *Biomaterials*, 26, 6041–6953.

CHEN L and SUBIRADE M (2006), 'Alginate-whey protein granular microspheres as oral delivery vehicles for bioactive compounds', *Biomaterials*, 27, 4646–4654.

CHEN L and SUBIRADE M (2007), 'Effect of preparation conditions on the nutrient release properties of alginate/whey protein granular microspheres', *Eur J Pharm Biopharm*, 65, 354–362.

CHEN L, REMONDETTO G E and SUBIRADE M (2006), 'Food protein-based materials as nutraceutical delivery systems', *Trends Food Sci Technol*, 17, 272–283.

CHEN Y and HAGERMAN A E (2004), 'Characterization of soluble non-covalent complexes between bovine serum albumin and 1,2,3,4,6-penta-o-galloyl-β-D-glucopyranose by MALDI-TOF MS', *J Agric Food Chem*, 52, 4008–4011.

CHO K Y, MOON J Y, LEE Y W, LEE K G, YEO J H, KWEON H Y, KIM K H and CHO C S (2003), 'Preparation of self-assembled silk sericin nanoparticles', *Int J Biol Macromol*, 32(1–2), 36–42.

CLARK A H and ROSS-MURPHY S B (1987), 'Structural and mechanical properties of biopolymer gels', *Advances in Polymer Science*, 83, 57–192.

CLARK A H, KAVANAGH G M and ROSS-MURPHY S B (2001), 'Globular protein gelation: theory and experiments', *Food Hydrocolloids*, 15, 383–400.

DE KRUIF C G, WEINBRECK F and VRIES R (2004), 'Complex coacervation of proteins and anionic polysaccharides', Lecture notes, Utrecht, October.

DE WOLF F A and BRETT G M (2000), 'Ligand-binding proteins: their potential for applications in systems for controlled delivery and uptake of ligands', *Pharmacological Review*, 52, 207–232.

DESAI M P, LABHASETWAR V, WALTER E, LEVY R J and AMIDON G L (1997), 'The mechanism of uptake of biodegradable microparticles in Caco-2 cells is size dependent', *Pharm Res*, 14, 1568–1573.

DI SILVIO L, GURAV N, KAYSER M V, BRADEN M and DOWNES S (1994), 'Biodegradable microspheres: a new delivery system for growth hormone', *Biomaterials*, 15, 931–936.

DICKINSON E and CHEN J (1999), 'Heat-set whey protein emulsion gels: role of active and inactive filler particles', *J Dispersion Sci Tech*, 20, 197–213.

DICKINSON E, HONG S T and YAMAMOTO Y (1996), 'Rheology of heat-set emulsion gels containing β-lactoglobulin and small-molecule surfactants', *Netherlands Milk and Dairy Journal*, 50, 199–207.

DINARVAND R, MAHMOODI S and FARBOUD E (2004), 'Effect of process variables on particle size of gelatin microspheres containing lactic acid', *Pharm Dev Technol*, 9, 291–299.

DINAUER N, BALTHASAR S, WEBER C, KREUTER J, LANGER K and VON BRIESEN H (2005), Selective targeting of antibody-conjugated nanoparticles to leukemic cells and primary T-lymphocytes. *Biomaterials*, 26(29), 5898–5906.

DUCLAIROIR C, IRACHE J M, NAKACHE E, ORECCHIONI A M, CHABENAT C and POPINEAU Y (1999), 'Gliadin nanoparticles: formation, all-*trans*-retinoic acid entrapment and release, size optimization', *Polymers International*, 79, 327–333.

DUCLAIROIR C, ORECCHIONI A M, DEPRAETERE P, OSTERSTOCK F and NAKACHE E (2003), 'Evaluation of gliadin nanoparticles as drug delivery systems: a study of three different drugs', *Int J Pharm*, 253(1–2), 133–144.

EINERSON N J, STEVENS K R and KAO W J (2003), 'Synthesis and physicochemical analysis of gelatin-based hydrogels for drug carrier matrices', *Biomaterials*, 24(3), 509–523.

ELLIOTT R and ONG T J (2002), 'Science, medicine, and the future of nutritional genomics', *Br Med J*, 324, 1438–1442.

ESPOSITO E, CORTESI R and NASTRUZZI C (1996), 'Gelatin microspheres: influence of preparation parameters and thermal treatment on chemico-physical and biopharmaceutical properties', *Biomaterials*, 17, 2009–2020.

EZPELETA I, IRACHE J M, STAINMESSE S, CHABENAT C, GUEGUEN J and POPINEAU Y *ET AL.* (1996), 'Gliadin nanoparticles for the controlled release of all-*trans*-retinoic acid', *Int J Pharm*, 131, 191—200.

FRANZ J, POKOROVA D, HAMPL J and DITTRICH M (1998), 'Adjuvant efficacy of gelatin particles and microparticles', *Int J Pharm*, 168, 153–161.

FUTTERMAN S and HELLER J (1972), 'The enhancement of fluorescence and the decreased susceptibility to enzymatic oxidation of retinol complexed with bovine serum albumin, β-lactoglobulin and the retinol-binding protein of human plasma', *J Biol Chem*, 247(16), 5168–5172.

GENNADIOS A (Ed.) (2002), *Protein-based Films and Coatings*, New York: CRC Press.

GOMEZ A, BINGHAM D, JUAN L and TANGT K (1998), 'Production of protein nanoparticles by electrospray drying', *J Aerosol Sci*, 29, 561–574.

GUÉRIN D, VUILLEMARD J C and SUBIRADE M (2003), 'Protection of bifidobacteria encapsulated in polysaccharide–protein gel beads against gastric juice and bile', *J Food Protec*, 66(11), 2076–2084.

GUNASEKARAN S, XIAO L and OULD ELEYA M M (2006), 'Whey protein concentrate hydrogels as bioactive carriers', *J Appl Poly Sci*, 99, 2470–2476.

HATTORI M, WATABE A and TAKAHASHI K (1995), 'β-Lactoglobulin protects β-ionone related compounds from degradation by heating, oxidation and irradiation', *Biosci Biotechnol Biochem*, 59, 2295–2297.

HINO T, TANIMOTO M and SHIMABAYASHI S (2003), 'Change in secondary structure of silk fibroin during preparation of its microspheres by spray-drying and exposure to humid atmosphere', *J Colloid Interface Sci*, 266, 68–73.

HORI M, ONISHI H and MACHIDA Y (2005), 'Evaluation of Eudragit-coated chitosan microparticles as an oral immune delivery system', *Int J Pharm*, 97, 223–234.

HU M, McCLEMENTS D J and DECKER E A (2004), 'Impact of chelators on the oxidative stability of whey protein isolate-stabilized oil-in-water emulsions containing ω-3 fatty acids', *Food Chem*, 88, 57–62.

HURTEAUX R, EDWARDS-LEVY F, LAURENT-MAQUIN D and LEVY M C (2005), 'Coating alginate microspheres with a serum albumin-alginate membrane: application to the encapsulation of a peptide', *Eur J Pharm Sci*, 24, 187–197.

JANNIN B, MENZEL M, BERLOT J P, DELMAS D, LANÇON A and LATRUFFE N (2004), 'Transport of resveratrol, a cancer chemopreventive agent to cellular targets: plasmatic protein binding and cell uptake', *Biochem Pharmacol*, 68, 1113–1118.

JÖBSTL E, O'CONNELL J, FAIRCLOUGH J P A and WILLIAMSON M P (2004), 'Molecular model for astringency produced by polyphenol/protein interactions', *Biomacromolecules*, 5, 942–949.

JOST R, BAECHLER R and MASSON G (1986), 'Heat gelation of oil-in-water emulsions stabilized by whey protein', *J Food Sci*, 51, 440–444, 449.

JOST R, DANNENBERG F and ROSSET J (1989), 'Heat-set gels based on oil/water emulsions: an application of whey protein functionality', *Food Microstructure*, 8, 23–28.

KAMATH K R and PARK K (1993), 'Biodegradable hydrogels in drug delivery', *Adv Drug Deliv Rev*, 11, 59–84.

KASHYAP N, KUMAR N and RAVI KUMAR M N (2005), 'Hydrogels for pharmaceutical and biomedical applications', *Crit Rev Therapeut Drug Carrier Sys*, 22 (2), 107–150.

KASPER F K, KUSHIBIKI T, KIMURA Y, MIKOS A G and TABATA Y (2005), '*In vivo* release of plasmid DNA from composites of oligo(poly(ethylene glycol)fumarate) and cationized gelatin microspheres', *J Controlled Release*, 107, 547–61.

KILARA A and PANYAM D (2003), 'Peptides from milk proteins and their properties', *Crit Rev Food Sci Nutr*, 43(6), 607–633.

KORHONEN H and PIHLANTO A (2003), 'Bioactive peptides: new challenges and opportunities for the dairy industry', *Australian J Dairy Technol*, 58 (2), 129–134.

LANGER R and PEPPAS N A (2003), 'Advances in biomaterials, drug delivery, and bionanotechnology', *AIChE Journal*, 49, 2990–3006.

LATHA M S, RATHINAM K, MOHANAN P V and JAYAKRISHNAN A (1995), 'Bioavailability of theophylline from glutaraldehyde cross-linked casein microspheres in rabbits following oral administration', *J Control Release*, 34, 1–7.

LATHA M S, LAL A V, KUMARY T V, SREEKUMAR R and JAYAKRISHNAN A (2000), 'Progesterone release from glutaraldehyde cross-linked casein microspheres: *in vitro* studies and *in vivo* response in rabbits', *Contraception*, 61, 329–334.

LAVOIE C, REMONDETTO G E, ANGERS P and SUBIRADE M (2005), 'Encapsulation of omega-3 fatty acids in a protein matrix: characterization and protective effect study', in *XIIth International Workshop on Bioencapsulation*, Kingston, 24–26 June 2005.

LAZKO J, POPINEAU Y and LEGRAND J (2004), 'Soy glycinin microcapsules by simple coacervation method', *Colloids Surf B Biointerfaces*, 37, 1–8.

LECLERC P L, REMONDETTO G E, RAMASSAMY C and SUBIRADE M (2005), 'Whey protein nanospheres as drug carriers for oral administration', in *XIIth International Workshop on Bioencapsulation*, Kingston, 24–26 June 2005.

LEFÈVRE T and SUBIRADE M (2001), 'Conformational rearrangement of β-lactoglobulin upon interaction with anionic membrane', *Biochim Biophys Acta* 15, 37–50.

LEFÈVRE T and SUBIRADE M (2003), Formation of intermolecular β-sheet structures: a phenomenon relevant to protein film structure at oil–water interfaces of emulsions, *Journal of Colloid and Interface Science*, 263, 59–67.

LEO E, VANDELLI M A, CAMERONI R and FORNI F (1997), 'Doxorubicin-loaded gelatin nanoparticles stabilized by glutaraldehyde: involvement of the drug in the cross-linking process', *Int J Pharm*, 12, 75–82.

LEUNG SOK LINE V, REMONDETTO G E and SUBIRADE M (2005a), 'Cold gelation of β-lactoglobulin oil-in-water emulsions', *Food Hydrocolloids*, 19, 269–278.

LEUNG SOK LINE V, REMONDETTO G E and SUBIRADE M. (2005b) Controlled release of α-tocopherol from cold-set β-lactoglobulin emulsion gels', in *XIIth International Workshop on Bioencapsulation*, Kingston, 24–26 June 2005, 87–89.

LEVY M C and EDWARDS-LEVY F (1996), 'Coating alginate beads with cross-linked biopolymers: a novel method based on a transacylation reaction', *J Microencapsul*, 13, 169–183.

LI F Q, HU J H, LU B, YAO H and ZHANG W G (2001), 'Ciprofloxacin-loaded bovine serum albumin microspheres: preparation and drug-release *in vitro*', *J Microencapsul*, 18, 825–829.

LIN W, GARNETT M C, DAVIES M C, BIGNOTTI F, FERRUTI P, DAVIS S S and ILLUM L (1997), 'Preparation of surface-modified albumin nanospheres', *Biomaterials*, 18(7), 559–565.

LIN W, GARNETT M C, DAVIS S S, SCHACHT E, FERRUTI P and ILLUM L (2001), 'Preparation and characterisation of rose Bengal-loaded surface-modified albumin nanoparticles', *J Control Release*, 71(1), 117–126.

LIU X, SUN Q, WANG H, ZHANG L and WANG J (2005), 'Microspheres of corn protein, zein, for an ivermectin drug delivery system', *Biomaterials*, 26, 109–115.

McCLEMENTS D J, MONAHAN F J and KINSELLA J E (1993), 'Effect of emulsion droplets on the rheology of whey protein isolate gels', *J Texture Stud*, 24, 411–422.

MAGDASSI S and VINETSKY Y (1995), 'Microencapsulation of O/W emulsions by formation of a protein-surfactant insoluble complex', *J Microencapsul*, 2, 537–545.

MALTAIS A, REMONDETTO G E, GONZALES R and SUBIRADE M (2005), 'Formation of soy protein isolate cold-SET gels: protein and salt effects', *J Food Sci*, 70, 67–73.

MARTEAU P, MINEKUS M, HAVENAAR R and HUIS IN'T VELD J H J (1997), 'Survival of lactic acid bacteria in a dynamic model of the stomach and small intestine: validation and the effects of bile', *J Dairy Sci*, 80, 1031–1037.

MAUGUET M C, LEGRAND J, BRUJES L, CARNELLE G, LARRE C and POPINEAU Y (2002), 'Gliadin matrices for microencapsulation processes by simple coacervation method', *J Microencapsul*, 19, 377–384.

MERODIO M, ARNEDO A, RENEDO M J and IRACHE J M (2001), 'Ganciclovir-loaded albumin nanoparticles: characterization and *in vitro* release properties', *Eur J Pharm Sci*, 12(3), 251–259.

MLADENOVSKA K, KLISAROVA LJ, JANEVIK E I and GORACINOVA K (2002), 'BSA-loaded gelatin microspheres: preparation and drug release rate in the presence of collagenase', *Acta Pharm*, 52, 91–100.

MLADENOVSKA K, JANEVIK E I, GLAVAS, M D, KUMBARADZI E F and GORACINOVA K (2003), 'Biodistribution of 131I-BSA loaded gelatin microspheres after peroral application to BALB/c mice – particle size study', *Acta Pharm*, 53, 187–197.

PAPADOPOULOU A, GREEN R J and FRAZIER R A (2005), 'Interaction of flavonoids with bovine serum albumin: a fluorescence quenching study', *J Agric Food Chem*, 53, 158–163.

PARK H and PARK K (1996), 'Hydrogels in bioapplications', in Ottenbrite R M, Huang S J and Park K, *Hydrogels and Biodegradable Polymers for Bioapplications*, Washington, DC: American Chemical Society, 2–9.

PAYNEA R G, YASZEMSKIB M J, YASKOC A W and MIKOS A G (2002), 'Development of an injectable, *in situ* crosslinkable, degradable polymeric carrier for osteogenic cell populations. Part 1. Encapsulation of marrow stromal osteoblasts in surface crosslinked gelatin microparticles', *Biomaterials*, 23, 4359–4371.

PEPPAS N A (1997), 'Hydrogels and drug delivery', *Curr Opin Coll Int Sci*, 2, 531–537.

PEPPAS N A and KHARE A R (1993), 'Preparation, structure and diffusional behavior of hydrogels in controlled release', *Adv Drug Deliv Rev*, 11, 1–35.

PEPPAS N A, BURES P, LEOBANDUNG W and ICHIKAWA H (2000), 'Hydrogels in pharmaceutical formulations', *Eur J Pharm Biopharm*, 50 (1), 27–46.

PHILLIPS L G, WHITEHEAD D M and KINSELLA J E (1994), *Structure–Function Properties of Food Proteins*, San Diego, CA: Academic Press.

PICOT A and LACROIX C (2004), 'Encapsulation of bifidobacteria in whey protein-based microcapsules and survival in simulated gastrointestinal conditions and in yoghurt', *Int Dairy J*, 14, 505–515.

POTHAKAMURY U R and BARBOSA-GNOVAS G V (1995), 'Fundamental aspects of controlled release in foods', *Trends Food Sci Tech*, 61, 397–406.

QIU Y and PARK K (2001), 'Environment-sensitive hydrogels for drug delivery', *Adv Drug Deliv Rev*, 53, 321–339.

RADWICK, A E and BURGESS, D J (2002), 'Proteins as microencapsulating agents and microencapsulated active compounds in pharmaceutical applications', in

Gennadios A (ed.), *Protein-based Films and Coating*, New York: CRC Press, 341–366.

RAWEL H M, MEIDTENER K and KROLL J (2005), 'Binding of selected compounds to proteins', *J Agric Food Chem*, 53, 4228–4235.

REDDY T T, LAVENANT L, LEFEBVRE J and RENARD D (2006), 'Swelling behaviour and controlled release of theophylline and sulfamethoxazole drugs in β-lactoglobulin protein gels obtained by phase separation in water/ethanol mixture', *Biomacromolecules*, 7, 323–330.

REMONDETTO G E and SUBIRADE M (2003), 'Molecular mechanisms of Fe^{2+}-induced β-lactoglobulin cold gelation: an interactions story', *Biopolymers*, 69, 461–469.

REMONDETTO G E, PAQUIN P and SUBIRADE M (2002), 'Cold gelation of β-lactoglobulin in the presence of iron', *J Food Sci*, 67, 586–595.

REMONDETTO G E, BEYSSAC E and SUBIRADE M (2004), 'Iron availability from whey protein hydrogels: an *in vitro* study', *J Agric Food Chem*, 52 (26), 8137–8143.

ROFF C F and FOEGEDING E A (1996), 'Dicationic-induced gelation of pre-denatured whey protein isolate', *Food Hydrocolloids*, 10(2), 193–198.

ROSENBERG M and LEE S J (2004), 'Calcium-alginate coated, whey protein-based microspheres: preparation, some properties and opportunities', *J Microencapsul*, 21, 263–81.

ROSENBERG M and YOUNG S L (1993), 'Whey proteins as microencapsulating agents. Microencapsulation of anhydrous milkfat – structure evaluation', *Food Structure*, 12, 31–41.

RÖSSLER B, KREUTER J and SCHERER D (1995), 'Collagen microparticles: preparation and properties', *J Microencapsul*, 12, 49–57.

SAHIN S, SELEK H, PONCHEL G, ERCAN M T, SARGON M, HINCAL A A and KAS H S (2002), 'Preparation, characterization and *in vivo* distribution of terbutaline sulfate loaded albumin microspheres', *J Control Rel*, 82, 345–358.

SCHÄFER V, VON BRIESEN H, ANDREESEN R, STEFFAN A M, ROYER C and TRÖSTER S, *et al.* (1992), 'Phagocytosis of nanoparticles by human immunodeficiency virus (HIV)-infected macrophages: a possibility for antiviral drug targeting', *Pharm Res*, 9, 541–546.

SHAHIDI F and MINE Y (2005), *Nutraceutical Proteins and Peptides in Health and Disease*, Boca Raton, FL: CRC Press/Taylor and Francis.

SOKOLOSKI T D and ROYER G P (1984), 'Drug entrapment with innovative albumin beads', in Davis S S, Illum L, McVie J G and Tomlinson E (eds), *Microspheres and Drug Therapy, Pharmaceutical, Immunological and Medical Aspects*, Amsterdam: Elsevier, pp. 295–307.

SPECTOR A A and FLETCHER J E (1970), 'Binding of long-chain fatty acids to β-lactoglobulin', *Lipids*, **5**, 403–411.

SWATSCHEK D, SCHATTON W, MüLLER W E G and KREUTER J (2002), 'Microparticles derived from marine sponge collagen (SCMPs): preparation, characterization and suitability for dermal delivery of all-*trans*-retinol', *Eur J Pharm Biopharm*, 54, 125–133.

TADA D, TANABE T, TACHIBANA A and YAMAUCHI K (2005), 'Drug release from hydrogel containing albumin as cross-linker', *J Biosc Bioengin*, 100(5), 551–555.

TOMLINSON E and BURGER J J (1985), 'Incorporation of water soluble drugs in albumin microspheres', in Widder J and Green R, *Methods in Enzymology*, Vol. 112, New York: Academic Press, pp. 27–43.

TORRADO J J, ILLURN L and DAVIS S S (1989), 'Particle size and size distribution of albumin microspheres produced by heat and chemical stabilization', *Int J Pharm*, 51, 85–93.

TOTOSAUS A, MONTEJANO J, SALAZAR J and GUERRERO I (2002), 'A review of physical and chemical protein gel induction', *Int J Food Sci Technol*, 37, 589–601.

VAZ C M, DE GRAAF L A, REIS R L and CUNHA A M (2004), 'pH-sensitive soy protein films for the controlled release of an anti-inflammatory drug', *Mater Res Innovations*, 8(3), 149–150.

WEBER C, COESTER C, KREUTER J and LANGER K (2000), 'Desolvation process and surface characterisation of protein nanoparticles', *Int J Pharm*, 194, 91–102.

WEYERMANN J, LOCHMANN D, GEORGENS C and ZIMMER A (2005), 'Albumin–protamine–oligonucleotide-nanoparticles as a new antisense delivery system. Part 2: Cellular uptake and effect', *Eur J Pharm Biopharm*, 59, 431–438.

YANG H and KAO W J (2006), 'Thermoresponsive gelatin/monomethoxy poly(ethylene glycol)–poly(D,L-lactide) hydrogels: formulation, characterization, and anti-bacterial drug delivery', *Pharm Res*, 23(1), 205–214.

YEO J H, LEE K G, LEE Y W and KIM S Y (2003), 'Simple preparation and characteristics of silk fibroin microspheres', *Eur Poly J*, 39, 1195–1199.

YOUNG S, WONG M, TABATA Y and MIKOS A G (2005), 'Gelatin as a delivery vehicle for the controlled release of bioactive molecules', *J Control Rel*, 109 (1–3), 256–274.

ZIEGLER G R and FOEGEDING EA (1990), 'The gelation of proteins', *Adv Food Nutr Res*, 34, 203–298.

11

Starch as an encapsulation material to control digestion rate in the delivery of active food components

E. Shimoni, Technion – Israel Institute of Technology, Israel

11.1 Introduction

Starch and its derivatives are widely used in the encapsulation of various food components. Indeed the use of starch in many encapsulation processes has provided solutions to problems such as thermal stabilization, process induced controlled release, and extended shelf-life of sensitive compounds. While in most cases this encapsulation is aimed at protecting the ingredient from such stressors, consideration such as the loci and mode of release are often not the main aim of these encapsulation systems. The aim of this chapter is to look at the benefits of starch encapsulation to control digestion rate, in particular, the use of starch as a controlled delivery system for functional food ingredients. It will review the principles that may serve for encapsulation technologies used to control the digestion rate of starch. Existing technologies as well as those under development will be reviewed with emphasis on the relation of molecular architecture and the coating properties. The use of starch for nano-scale encapsulation of bioactive nutraceuticals will also be discussed. Where applicable, the effect of the specific techniques on food quality will be reviewed as well as the commercial potential.

Starch is one of the most abundant carbohydrates in plants. Starch in plants is packaged in granules, and can be fractionated into two glucose homo-polysaccharide macromolecules: amylose and amylopectin. Amylose is essentially a linear molecule in which 500–600 D-glucose units are linked by α-(1-4) glycoside links. Amylopectin contains both α-(1-4) and a number of α-(1-6) linkages. The amylopectin branched molecule consists of short α-(1-4)-D-glucan chains linked by occasional α-(1-6) bonds. Starch composition varies from almost pure amylopectin in waxy barleys and rice, to high levels of amylose in amylomaize.

The crystallinity of native starch changes from 15 to 45% (Zobel, 1988). For most native starches, X-ray diffraction (XRD) yields two types of spectral patterns, A type and B type. Cereal starches yield A-type pattern, while tuber starches and amylase-rich starches yield the B-type pattern. An intermediate pattern (C-type) found in legumes appears to be a mixture of A and B types rather than a distinct structure (Gernat *et al.*, 1990). The A and B patterns are a result of the α-glucan chains forming left-handed, parallel-stranded double helices (Imberty and Perez, 1988; Imberty *et al.*, 1988). In native starch, these crystalline forms are mostly amylopectin (Zobel, 1992). When heated in the presence or absence (annealing) of water, the A and B diffraction patterns are lost, and a new V-type pattern may develop as a result of the formation of amylose-lipid complexes.

Generally, heat treatment processing of starch increases the resistant starch (RS) content in foods, as was found in grain products (Kishida *et al.*, 2001; Marlett and Longacre, 1996; Parchure and Kulkarni, 1997; Tovar and Melito, 1996). Upon heating, the crystalline structures within the starch granule lose order. This disordering, referred to as melting, occurs at temperatures that vary with moisture content, and with the origin of the starch. Melting temperature varies from 60 to 70 °C in an excess of water to over 150 °C in the absence of water. In most cases, when moisture is present and gelatinization of starch has occurred, the amylose and amylopectin molecules form a gel (Miles *et al.*, 1985). This gel network is formed as the glucans re-adopt the helical structure, and form a B-type pattern (Colonna *et al.*, 1992). The increase in B-crystallinity, referred to as retrogradation, takes several hours (in high amylose starch) to several days (in high amylopectin starches) (Eerlingen *et al.*, 1994). It was shown that decrease in branching (Zhang and Jackson, 1992) and repeated heating and cooling cycles (Sievert *et al.*, 1991) increase the retrogradation. RS type III is the fraction of the retrograded starch that resists α-amylase digestion. Thus, this property of the starch can be used to control its digestion rate, and thus the release of the encapsulated material.

Starch has been used in a variety of encapsulation processes, which have been reviewed in detail elsewhere (Forssell, 2004). As noted before, the use of starch in these encapsulation processes has provided solutions to problems such as thermal stabilization, process induced controlled release, and extended shelf-life of sensitive compounds. However, the scope of this chapter is the use of starch to improve the delivery and controlled release properties of the encapsulation system.

11.2 Uses of starch to control delivery in microencapsulation

Delivery of active sensitive ingredients to specific sites of the gastrointestinal (GI) tract is important in drug therapies for colonic diseases, as well as for oral administration of bioactive proteins, and delivery of vitamins and sensitive food ingredients. This approach is attractive for the food industry, when trying to protect sensitive functional compounds, and deliver them safely to the intestine and colon. However, one should keep in mind that in the pharma and cosmetics industries, the design of such systems could rely on materials that were tailored, for example, for specific dissolution properties, and are allowed for use in that specific system, however, are not food grade. This is often a limitation in the design of delivery systems for food, thus making their design a challenging task, which relies on a limited arsenal of building blocks (e.g. only these approved for use in food).

Starch and its derivatives are used in a wide variety of encapsulation procedures for the purpose of protection and controlled release of nutrients and bioactive chemicals. Some examples are the use of modified starches in suspension to form microspheres with fish oil (Tan *et al.*, 2005). Others showed the potency of starch and its combination with cyclodextrin as encapsulation matrix for ascorbic acid (Trindade and Grosso, 2000; Uddin *et al.*, 2001). The interesting concept of encapsulating lipids into starch granules by their short-term swelling using microwave heating was demonstrated by Kapušniak and Tomasik (2006). As noted before, a variety of applications of starch in encapsulation are reviewed elsewhere (Forssell, 2004). The key point is that most of these studies were not aimed at, nor revealed the mode of, release of the core material. In addition, many do not question the stability of the nutrient during storage and processing.

Microencapsulation is often the method of choice when the need arises to apply and protect a bioactive substance in a food product or formulation. To control the release of the bioactive in these microparticles, capsules can be coated with one or several layers to obtain several degrees of protection, for example, against oxidation, thermal stress, or dissolution. An efficient way to control the release of an active compound into the GI tract is the use of enteric coating materials. In the pharmaceuticals industry, the most widely used enteric coating materials are synthetic polymers, such as polymetacrylates, that are practically insoluble in acid (as in the stomach) and readily dissolve in mild acid to neutral solutions (as in the small intestine). These polymers are not approved for use in foods, and thus pose a challenge for the food encapsulation developers.

A number of solutions have been suggested to provide a similar solution based on food grade materials. These include the development of various enteric coatings to protect sensitive compounds from dissolution and decomposition in the GI tract. Common enteric coating materials are natural, semi synthetic and synthetic polymers, such as zein and shellac (Mueller and Yunis-Specht, 1994); amylose acetate phthalate and cellulose derivatives (CAP, HPMCP) (Schmidt and Niemann, 1992; Bechard *et al.*, 1995; Williams and Liu, 2000); and various ethylacrylate and methylmethacrylate polymers (Plaizier-Vercammen and Van Molle, 1991; Felton *et al.*, 1995; Semde *et al.*, 2000). The common mechanism of function of these materials is their reaction to pH changes. Thus, the release of the active core is determined by the local pH in the intestine (Dew *et al.*, 1982; Schroeder *et al.*, 1987; Ashford *et al.*, 1993; Gazzaniga *et al.*, 1994, 1995; Makino, 1994; Wilding *et al.*, 1994; Ishibashi *et al.*, 1998; Marvola *et al.*, 1999; Pillay and Fassihi, 1999; Krogars *et al.*, 2000). Therefore, these materials protect the functional materials from the acidic environment of the stomach, but do not protect the active ingredients from dissolution and enzymatic digestion in the small intestine. In order to reach the colon, the materials that are insoluble in both acidic and neutral environments and not digestible by pancreatic enzymes should be used for enteric coating.

One attractive alternative to the commonly used coatings is ingested starch that escapes digestion in the small intestine and reaches the large intestine. Starch is an excellent candidate for this purpose; if one can control its digestion rate by pancreatic amylases, the release of the microcapsule core can be tuned to specific regions in the GI. Russel (1983) showed that the biodegradation of starch microspheres is highly dependent on their size, but can also be controlled by the degree of starch crosslinking. These microspheres were degraded with half-life time of over an hour. This time should be long enough to impose release in the lower part of the small intestine. Cummings *et al.* (1996) used amylose and ethyl-cellulose films as coating for a colon-targeted system. The release of the active core materials was delayed at pH of 1–9 at 37 °C. In these systems, enzymatic digestion was the trigger for the release. One should keep in mind, however, the high cost of pure amylose when considering it for food use. An interesting proof of the efficacy of starch as a delivery vehicle comes from the work of Heritage *et al.* (1996), showing that starch microspheres loaded with an antigen induced an immune response in mice after oral administration.

Another approach in the use of starch as an encapsulation material is the use of retrograded indigestible starch, named resistant starch (RS) (Englyst *et al.*, 1983). While not digested in the small intestine, it is fermented and disintegrated by the microflora in the large intestine (Ring *et al.*, 1988; Englyst *et al.*, 1992; Kritchevsky, 1995; Muir *et al.*, 1995; Phillips *et al.*, 1995; Silvester *et al.*, 1995). RS type III that is formed by a hydrothermal treatment followed by retrogradation is of particular interest as a potential coating material, due to its poor solubility, resistance to enzymatic digestion in the small intestine, and its decomposition by the colon microflora. Some success in applying starch as a coating in the pharmaceuticals industry suggests that the combination of some biopolymers with starch is a potential formula for enteric coating. However, being a critical factor in their functionality, the homogeneity of films made of polymer mixtures should be closely monitored. One of the ways this can be done is by imaging coating surfaces using atomic force microscopy (AFM).

The above-mentioned approach was used in our laboratory in the development of coatings based on RS enriched HACS and pectin, designed to protect active core materials from dissolution and destruction in both the stomach and small intestine (Dimantov *et al.*, 2004). The first step towards the development of such coatings was to coat glass slabs with mixtures of treated HACS and pectin, and their surface was characterized by AFM. The surface of selected coatings based on mixtures of HACS and HM pectin was characterized by AFM using the tapping mode. These mixtures were chosen for their higher stability in preliminary dissolution tests as compared to HACS–LM pectin mixtures. Scanning of coated slabs was performed in a liquid cell using n-butanol to prevent water interference. The topographic images in 2- and 3-D of surfaces of different coatings at 5 mm × 5 mm scan are shown in Fig. 11.1. The surface of 100% HM pectin coating (Fig. 11.1(a)) revealed homogeneous topography with a z-range of 111.3 nm. The z-range of coatings containing 10%, 50%, and 75% HACS was 80.25, 192.6, and 360.7, respectively (Fig. 11.1(b)–(d)). Beyond the increasing z-range, an elevated heterogeneity in the topography of coating surfaces is evident, as the HACS content increases.

The stability of the coatings for dissolution in the stomach and intestine was tested in model buffers and enzyme solutions. Dissolution of coatings decreased with the increase in HACS in both stomach and intestine pH conditions. HACS coatings also showed higher resistance to pancreatic amylolysis. The potential of these coatings for targeted release in the colon was therefore demonstrated.

In the second phase of this study, we tested the use of HACS coatings on glass beads as a model for encapsulated core. These coatings were applied using the fluidized bed coating technique. Coated beads were characterized by SEM, dissolution and enzymatic digestion tests (Dimantov *et al.*, 2004). The first coatings produced by this method were stable in dissolution tests, simulating stomach and small intestine pHs, but completely digested in enzymatic digestion tests. However, by increasing the thickness of the coating improvements of the coating stability to digestion by pancreatic amylases was accomplished. The addition of resistant starch (RS) resulted in considerable improvement of the coating stability to enzymatic digestion, but also in the appearance of cracks on the coating's surface. A closer look at the kinetics of the degradation of the coatings revealed that cracking occurs during the first 3 h of digestion by amylases. Interestingly, the critical cracking and 'peeling' of the coatings occurs after about 2 h. This time is about the residence time of the 'head of the meal' in the stomach and intestine before it reaches the colon. Apparently, by optimization of their processing and application, these coatings may prove useful for delivery of functional ingredients to the colon.

Some of the more interesting uses of starch in microencapsulation processes are in the encapsulation of probiotic bacteria. High amylose starch was used in its granular form to protect bifidobacterium adhered to the granule (Wang *et al.*, 1999). This adhesion was shown to improve the survival of the bacteria six-fold. Other extensive work performed by Crittenden *et al.* (2001) showed that this adhesion is common to many, but not all, bifidobacteria. Regardless, for adhering species, this phenomenon can be used as the first step in their encapsulation. The group of Mattila-Sandholm of VTT (Mattila-Sandholm *et al.* 2002; Myllarinen *et al.* 1999) used this phenomenon to produce a starch-based microcapsule for the protection and delivery of probiotic bifidobacteria to the colon. Using amylases, growing the bacteria in the presence of granules, they obtained a freeze-dried powder that was then coated with high amylose starch.

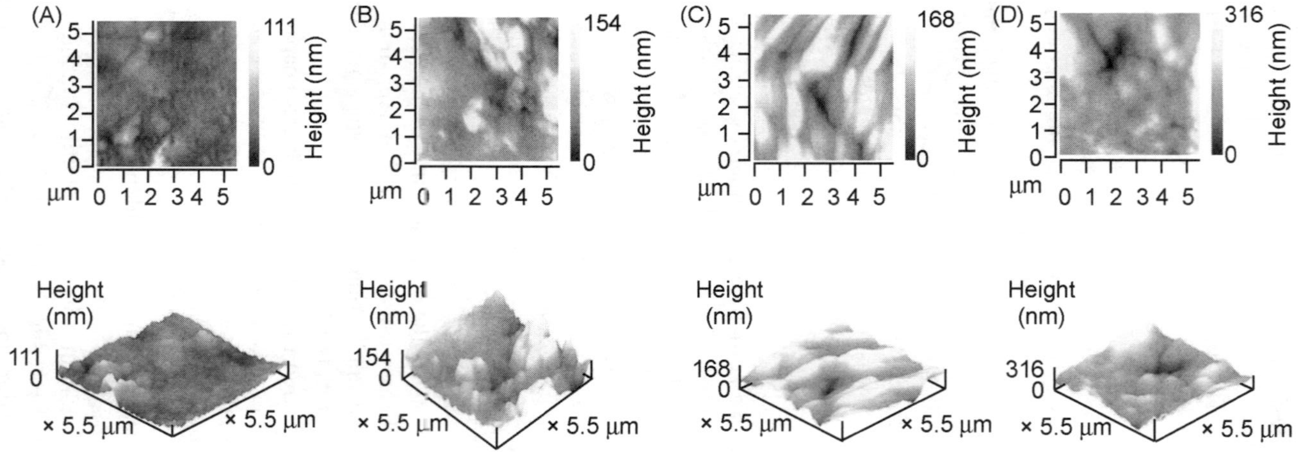

Fig. 11.1 Two- and three-dimensional imaging of coating surfaces composed of different starch–pectin mixtures: (A) 100% pectin, (B) 90% pectin/10% HACS, (C) 50% pectin/50% HACS, (D) 25% pectin/75% HACS. AFM scanning was performed under butanol at taping mode.

This system was shown to stabilize the bacteria at ambient temperatures, during their passage through stimulated upper GI conditions, and to release the bacteria in simulated colon fermentation system. As the main encapsulation material in microspheres produced by spray drying, starch was also used as a pre-biotic companion to probiotics, used mostly to increase their viability. Sultana *et al.* (2000) showed that the addition of cornstarch to alginate-based microcapsules improved the survival of the encapsulated culture. In a recent study by Crittenden *et al.* (2006), high amylose cornstarch was used as part of a formulation based on emulsion encapsulation and other polysaccharides. While over 95% of the bacteria were lost during the first few weeks of storage, the formulation seems to stabilize the remaining bacteria in simulated GI conditions. While the various systems provide solutions to different performance parameters required from an efficient probiotic food-grade delivery system, a system that can provide high survival rate during microcapsule formation, storage, thermal treatment, and the GI has not yet been developed.

11.3 Starch as nanoencapsulation material

For many technological challenges, microencapsulation provides adequate solutions. However, microcapsules are usually dry products, or beads, that are suitable for cloudy and dry applications. Regardless of their nature, dry or hydrated, microcapsules are also in many cases large enough to affect the textural perception of the product. In nano-encapsulation, products can be reduced to a size of tens of nanometers, and thus one can avoid such visual and textural effects. Unlike in microencapsulation, where a core is being protected by a coating or a matrix formed by a technological procedure (e.g. fluidized bed technique), in nano-encapsulation the process is based on self-assembly phenomenon, controlled at the molecular level. Nano-size encapsulation systems may therefore provide improved activity level and solubilized ingredients for clear beverages that previously could not be used. It may further enhance the stability of the products; generally to a pH of 2, allow pasteurization (thereby achieving thermodynamic stability), and provide better control of reaction rate.

The majority of the previously described encapsulation concepts use a general concept of a bioactive core, coated with a protective layer. This is also the case for some starch based nano-sized encapsulation systems such as vitamin E nanoparticles (Chen and Wagner, 2004). While these methods provide adequate solutions for the delivery of bioactive substances in foods, the thermal stability of these compositions may be limited due to the liquid or non-glassy state of the encapsulated core. Tackling this problem, we suggested using a different approach, based on molecular inclusion complexes that encapsulate the active ingredient on a molecular basis. Furthermore, the release of the active compound in this case is based on the loci, in which the encapsulating molecule is degraded in the digestive tract. Our model for this concept was amylose as the encapsulating material.

The use of amylose as encapsulating material for molecular entrapment is based on the interaction between amylose and lipids, which is characterized by amylose chains forming semi crystalline V forms. The V form is an amylose chain that forms a helix with a large cavity in which low molecular weight chemicals can be situated. The size of the ligand determines the number of glucosyl residues per turn (6, 7, or 8) (Snape *et al.*, 1998). It was suggested that the crystalline state of amylose–fatty acid complex involves the V-amylose six-fold single chain left-handed helix, well known among starch polymorphs (Buleon *et al.*, 1998). The common perception is that the fatty acid is a 'stem' (planar zigzag) inside the helix, whose inner surface is hydrophobic because of the carbon–hydrogen of the six-fold helix (Godet, 1993). The V-helical complex segments are interrupted by short sections of uncomplexed amylose that permit random orientation of the helical segments in the type I complexes, and folding into parallel and anti-parallel arrays in the crystalline type II complexes (Biliaderis and Galloway, 1989; Karkalas and Raphaelides, 1986; Karkalas *et al.*, 1995).

Amylose-lipid complexes are of interest mostly because of their technological importance in starchy food systems, since they modify the texture and structural stability of starch-based products (e.g. reduction in stickiness, improved freeze-thaw stability and retardation of retrogradation) (Tufvesson *et al.*, 2003; Eliasson and Krog, 1985; Godet *et al.*, 1996). The complexes were also studied in view of their contribution to the bioavailability of starch, in terms of its enzymatic digestion (Holm *et al.*, 1984; Crowe *et al.*, 2000). The various studies on complexes produced from mono- and diglycerides (Eliasson and Krog, 1985; Biliaderis and Galloway, 1989; Seneviratne and Biliaderis, 1991; Tufvesson and Eliasson, 2000; Tufvesson *et al.*, 2001), saturated fatty acids (Karkalas *et al.*, 1995; Tufvesson *et al.*, 2003; Lebail *et al.*, 2000; Godet *et al.*, 1993, 1995), as well as unsaturated fatty acids (Snape *et al.*, 1998; Karkalas *et al.*, 1995; Tufvesson *et al.*, 2003; Eliasson and Krog, 1985; Szejtli and Banky-Elod, 1975), have shown that they have high melting temperatures; that the complexed fatty acid is efficiently protected from oxidation; and that the digestibility of starch is influenced by complex formation, which decreases the digestibility of starch.

In light of these studies, we suggested that amylose-lipid complexes could be used as a delivery system for polyunsaturated fatty acids (PUFA). Our concept was to use these complexes to provide protection during processing and storage, and release the PUFA in the intestine following enzymatic hydrolysis of the amylose. The feasibility of this concept was examined using conjugated linoleic acid (CLA) as a model. Numerous physiological properties have been attributed to CLA including action as an antiadipogenic, antidiabetogenic, anticarcinogenic, and antiatherosclerotic agent. In addition, CLA has effects on bone formation and the immune system as well as fatty acid and lipid metabolism and gene expression in numerous tissues. For these reasons CLA was chosen as the ligand in the development of amylose complexes. Amylose-CLA complexes were produced with the intention to provide optimal stability to oxidation and thermal treatments, to dissolution in the stomach, and efficient release by mammalian amylases.

Amylose-CLA complexes were produced using the method of Eliasson and Krog (1985) in a water/DMSO mixture. Amylose-CLA complexes were also produced by the method of Karkalas *et al.* (1995). Amylose and CLA were dissolved in an alkaline solution at 90 °C and then mixed; acidification of the mixture caused the formation of the complexes. The formation of a V type complex was verified by X-ray diffraction (XRD). The X-ray diffraction diagram of the complexes obtained was typical of the V form of amylose with the three main reflections corresponding to the Bragg angles $2\theta = 7.73°$, $2\theta = 12.87°$, $2\theta = 20°$. To test the thermal stability of the complexes they were also analyzed by differential scanning calorimetry (DSC). DSC analysis was performed by a Perkin Elmer DSC-7 system. To a complex sample of 10 mg, 80 mg of water was added and the reference sample was 50 mg of water. Sample and reference were placed in high-pressure DSC pans. Temperature scans ranged from 20 to 200 °C with a 3 °C/min ramping. The DSC analysis of the amylose-CLA complex created by KOH/HCl precipitation showed a transition temperature of the complex ranging from 88 to 95 °C, suggesting stability of the complexes during food processing.

To correlate their structure with their physical properties, the complexes were scanned by AFM. The complexes created in water/DMSO and KOH/HCl solutions at different crystallization temperatures were characterized by AFM using the tapping mode, and the images of complexes created at 90 °C, at 5 μm × 5 μm scan, revealed globular structures of heterogeneous nature with an average z-range ($n = 15$) of 71.6 nm ± 59 nm, and diameter of 152 nm ± 39 nm. Larger structures are seen, but closer inspection indicates that these are clusters of the previously described spheres. Complexes created by KOH/HCl solution at 90 °C exhibited mainly rod-like structures of heterogeneous nature with an average z-range ($n = 15$) of 37.8 nm ± 18 nm, width of 43–160 nm, and length, which ranged from 0.35 μm to 3.2 μm. Globular structures with average diameter of 60 nm ± 8 nm were also present. As previously described, the amylose-CLA complexes produced by the water/DMSO complexation system yield nanoparticulated molecular capsules.

It is our hypothesis that CLA complexation with amylose protects it from oxidation. To test this hypothesis amylose-CLA complexes were placed in airtight serum bottles (2 ml) under atmospheric oxygen pressure, and incubated for 70 h at 37 °C. The headspace oxygen content was determined by GC. In all tested samples, CLA oxidation was lower in complex with amylose. An example is given in Fig. 11.2, describing the results for complexes produced by the water/DMSO method. The protective effect of the complexes against CLA oxidation was higher for complexes created in water/DMSO solution than that of complexes created by KOH/HCl solution. This difference was more pronounced for complexes created at 90 °C and 30 °C. For complexes made by water/DMSO solution, 60 °C was the temperature that supplied the lower protection amongst complexes made by this method. Regardless of the complexation method and

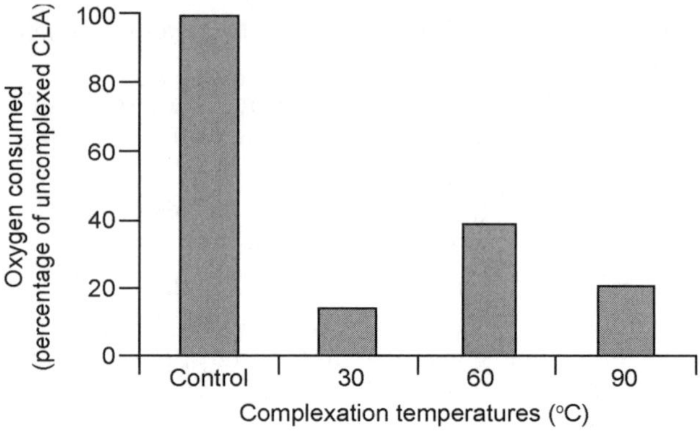

Fig. 11.2 Oxidative stability of encapsulated CLA.

temperature, it is evident that the complexes protect and inhibit the oxidation of the ligand.

Enzymatic digestion of amylose-FA complexes was carried out to examine our hypothesis that the CLA, molecularly encapsulated in helical amylose cavity, can be released by enzymatic activity in the intestine. Mammalian pancreatic amylases, and various microbial enzymes were used to measure the susceptibility of the complexes to enzymatic degradation. The degradation of the complexes was measured at 24 h by measuring the reducing sugar content in the solution (calibrated for maltose or glucose). Our results showed that amylolytic enzymes can digest the complex, and the susceptibility to amylolytic hydrolysis was much higher for the mammalian pancreatin. Interestingly, extremely low digestion was observed for β-amylase and amyloglucosidase, two exo-amylases, possibly due to steric interference of the fatty acid, which is partly located outside the helix. Following the digestion experiments, the amount of CLA released to the medium was measured. CLA was extracted from the reaction media (including the undigested complex residue) by hexane, and their concentration was determined by GC. The results show that indeed CLA was released only when complexes were digested by amylases (Fig. 11.3). These results demonstrate the potential of the proposed system for molecular encapsulation of hydrophobic functional materials.

11.4 Conclusion

As noted before, we have recently demonstrated the potential of amylose-lipid complexes to become a molecular encapsulation system, providing thermal and oxidative stability, and serving as a controlled release system, activated by pancreatic amylases. Recently, a number of studies suggesting additional uses of these complexes have been published. These focused mostly on encapsulation of flavor and aroma compounds (Wulff *et al.*, 2005; Jouquand *et al.*, 2006). Other studies have demonstrated synthetic pathways for the formation of such

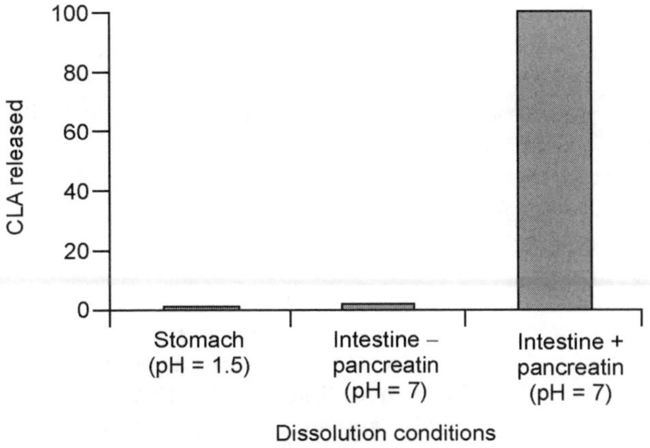

Fig. 11.3 Release of CLA under conditions simulating the stomach (pH = 1.5), amylase-free intestine (pH = 7), and the small intestine enzymatic attack (pH = 7 + pancreatin).

complexes by enzymatic polymerization (Gelders *et al.*, 2005). All these developments suggest that the full potential of this system is yet to be revealed. One can therefore envision the potential of these systems for the oral delivery of functional ingredients. The precise mode in which these low molecular weight substances are entrapped in the helical amylose it not fully elucidated yet, nor its relation to the thermal stability of the complexes and the rate at which they are released by enzymatic digestion. These issues are critical for the intelligent design of tailored delivery systems based on these particles. This task will also require the development of methodologies for pilot and commercial scale production of such molecular inclusion bodies.

11.5 References

ASHFORD, M., FELL, J. T., ATTWOOD, D., SHARMA, H., WOODHEAD, P. J. An *in vivo* investigation into the suitability of pH-dependent polymers for colonic targeting. *International Journal of Pharmaceutics*, 95(1–3), 193–199, 1993.

BECHARD, S., LEVY, L., CLAS, S. Thermal, mechanical and functional properties of cellulose acetate phthalate (CAP) coatings obtained from neutralized aqueous solutions. *International Journal of Pharmaceutics*, 114(2), 205–213, 1995.

BILIADERIS, C. G., GALLOWAY, G. Crystallization behavior of amylose-V complexes: structure-property relationships. *Carbohydrate Research*, 189, 31–48, 1989.

BULEON, A., COLONNA, P., PLANCHOT, V., BALL, S. Starch granules: structure and biosynthesis. *International Journal of Biological Macromolecules*, 23(2), 85–112, 1998.

CHEN, C. C., WAGNER, G. Vitamin E nanoparticle for beverage applications, *Chemical Engineering Research and Design*, 82(A11), 1432–1437, 2004.

COLONNA, P., LELOUP, V., BULEON, A. Limiting factors of starch hydrolysis. *Eur. J. Clin. Nutr.*, 46(Suppl), S17–S32, 1992.

CRITTENDEN, R., LAITILA, A., FORSSELL, P., MATTO, J., SAARELA, M., MATTILA-SANDHOLM, T.,

MYLLARINEN, P. Adhesion of bifidobacteria to granular starch and its implications in probiotic technologies. *Applied and Environmental Microbiology*, 67(8), 3469–3475, 2001.

CRITTENDEN, R., WEERAKKODY, R., SANGUANSRI, L., AUGUSTIN, M. Synbiotic microcapsules that enhance microbial viability during nonrefrigerated storage and gastrointestinal transit. *Applied and Environmental Microbiology*, 72(3), 2280–2282, 2006.

CROWE, T. C., SELIGMAN, S. A., COPELAND L. Inhibition of enzymic digestion of amylose by free fatty acids *in vitro* contributes to resistant starch formation. *J. Nutr.*, 130, 2006–2008, 2000.

CUMMINGS, J. H., MILOJEVIC, S., HARDING, M., COWARD, W. A., GIBSON, G. R., BOTHAM, RING, S. G., WREIGHT, E. P., STOCKHAM, M., ALLWOOD, M. C., NEWTON, J. M. *In vivo* studies of amylose and ethylcellulose coated [13C] glucose microspheres as a model for drug delivery to the colon. *J. Controlled Release*, 40, 123–131, 1996.

DEW, M., HUGHES, P., LEE, M., EVANS, B., RHODES, J. An oral preparation to released drug in the human colon. *British Journal of Clinical Pharmacology*, 14, 405–408, 1982.

DIMANTOV, A., GREENBERG, M., KESSELMAN, E., SHIMONI, E. Study of high amylose corn starch as food grade enteric coating in a microcapsule model system. *Innovative Food Science and Emerging Technologies*, 5, 93–100, 2004.

EERLINGEN, R. C., JACOBS, H., DELCOUR, J. A. Enzyme resistant starch. V. Effect of retrogradation of waxy maize starch on enzyme susceptibility. *Cereal Chemistry*, 71, 351–355, 1994.

ELIASSON, A.C., KROG, N. Physical properties of amylose-monoglyceride complexes. *J. Cereal Sci.*, 3, 239–248, 1985.

ENGLYST, H. N., ANDERSON, V., CUMMINGS, J. H. Starch and non-starch polysaccharides in some cereal foods. *Journal of Science and Food Agriculture*, 34(12), 1434–1440, 1983.

ENGLYST, H. N., KINGMAN, S. M., CUMMINGS, J. H. Classification and measurement of nutritionally important starch fractions. *European Journal of Clinical Nutrition*, 46 (Suppl 2), 33–50, 1992.

FELTON, L. A., HAASE, M. M., SHAH, N. H., ZHANG, G., INFELD, M. H., MALICK, A.W., MCGINITY, J. W. Physical and enteric properties of soft gelatin capsules coated with Eudragit L 30 D-55. *International Journal of Pharmaceutics*, 113(1), 17–24, 1995.

FORSSELL, P. Starch-based microencapsulation. In: *Starch in Food: Structure, Function and Applications*, A. C. Elliasson (ed.), Woodhead Publishing, 2004, pp. 461–474.

GAZZANIGA, A., IAMARTINO, P., MAFFIONE, G., SANGALLI, M. Oral delayed-released system for colonic specific delivery. *International Journal of Pharmaceutics*, 108, 77–83, 1994.

GAZZANIGA, A., BUSETTI, C., SANGALLI, M., GIORDANO, F. Time-dependent oral delivery system for the colon targeting. *S.T.P. Pharma Sciences*, 5, 83–88, 1995.

GELDERS, G. G., GOESAERT, H., DELCOUR, J. A. Potato phosphorylase catalyzed synthesis of amylose-lipid complexes. *Biomacromolecules*, 6(5), 2622–2629, 2005.

GERNAT, C., RODOSTA, S., DAMASCHYUN, G., SCHIERBAUM, F. Supermolecular structure of legume starches revealed by X-ray scattering. *Starch/Starke*, 42, 175–178, 1990.

GODET, M. C., BULEON, A., TRAN, V., COLONNA, P. Structural features of fatty acid-amylose complexes. *Carbohydrate Polymers*, 21, 91–95, 1993.

GODET, M. C., TRAN, V., COLONNA, P., BULEON, A. Inclusion/exclusion of fatty acids in amylose complexes as a function of the fatty acid chain length. *Int. J. Biol. Macromol.*, 17(6), 405–408, 1995.

GODET, M. C., BOUCHET, B., COLONNA, P., GALLANT, D. J., BULEON, A. Crystalline amylose-fatty acid complexes: morphology and crystal thickness. *Journal of Food Science*, 61(6), 1196–1201, 1996.

HERITAGE, P. L., LOOMES, L. M., JIANXIONG, J., BROOK, M. A., UNDERDOWN, B. J., MCDERMOTT, M. R. Novel polymer grafted starch microparticles for mucodal delivery of vaccines. *Immunology,* 88, 162–168, 1996.

HOLM, J., BJORCK, I., OSTROWSKA, S., ELIASSON, A. C., ASP, N. G., LARSSON, K., LUNDQUIST, I. Fats (lipids) baking extrusion. *LIPIDFORUM Symposium*, 1984, 52–59.

IMBERTY, A., PEREZ, S. A revisit to the three dimensional structure of B-type starch. *Biopolymers*, 27, 1205–1221, 1988.

IMBERTY, A., CHANZY, H., PEREZ, S., BULEON, A., TRAN, V. The double helical nature of crystalline part of A-starch. *J. Mol. Biol.* 201, 365–378, 1988.

ISHIBASHI, T., HATANO, H., KOBAYASHI, M., MIZOBE, M., YOSHINO, H. Design and evaluation of a new capsule-type dosage form for colon- targeted delivery of drug. *International Journal of Pharmaceutics*, 168, 31–40, 1998.

JOUQUAND, C., DUCRUETA, V., LE BAIL, P. Formation of amylose complexes with C6-aroma compounds in starch dispersions and its impact on retention. *Food Chemistry*, 96(3), 461–470, 2006.

KAPUSNIAK, J., TOMASIK, P. Lipid microencapsulation in starch. *Journal of Microencapsulatiuon*, 23(3), 341–348, 2006.

KARKALAS, J., RAPHAELIDES, S. Quantitative aspects of amylose-lipid interactions. *Carbohydrate Research*, 157, 215–234, 1986.

KARKALAS, J., MA, S., MORRISON, W., PETHRICK, R. A. Some factors determining the thermal properties of amylose inclusion complexes with fatty acids. *Carbohydrate Research*, 268, 233–247, 1995.

KISHIDA, T., NOGAMI, H., HIMENO, S., EBIHARA, K. Heat moisture treatment of high amylose cornstarch increases resistant starch content but not its physiologic effects in rats. *J. Nutrition,* 131, 2716–2721, 2001.

KRITCHEVSKY, D. Epidemiology of fiber, resistant starch and colorectal cancer. *European Journal of Cancer Prevention*, 4, 345–352, 1995.

KROGARS, K., HEINAMAKI, J., VESALAHTI, J., MARVOLA, M., ANTIKAINEN, O., YLIRUUSI, J. Extrusion-spheronization of pH-sensitive polymeric matrix pellets for possible colonic drug delivery. *International Journal of Pharmaceutics*, 199(2), 187–194, 2000.

LEBAIL, P., BULEON, A., SHIFTAN, D., MARCHESSAULT, R. H. Mobility of lipid in complexes of amylose-fatty acids by deuterium and carbon-13 solid-state NMR. *Carbohydrate Polymers*, 43(4), 317–326, 2000.

MAKINO, K. Stimuli-sensitive microcapsules. *Pharm Technology of Japan*, 10(8), 913–920, 1994.

MARLETT, J. A. AND LONGACRE, M. J. Comparison of *in vitro* and *in vivo* measures of resistant starch in selected grain products. *Cereal Chemistry*, 73, 63–68, 1996.

MARVOLA, M., NYKVNEN, P., RAUTIO, S., ISONEN, N., AUTERE, A. Enteric polymers as binders and coating materials in multiple-unit site-specific drug delivery systems. *European Journal of Pharmaceutical Sciences*, 7, 259–267, 1999.

MATTILA-SANDHOLM, T. MYLLARINEN, P. CRITTENDEN, R. MOGENSEN, G. FONDEN, R., SAARELA, M. Technological challenges for future probiotic foods. *International Dairy Journal*, 12(2–3), 173–182, 2002.

MILES, M. J., MORRIS, V. J., ORFORD, P. D. AND RING, S. G. The roles of amylose and amylopectin in the gelation and retrogradation of starch. *Carbohydrate Research*, 135, 271–281, 1985.

MUELLER, B.W., YUNIS-SPECHT, F. Ammonium shellac – a new formulation of shellac for the application of a formerly used natural polymer for enteric coatings. *Protocol of 21st International Symposium – Controlled Release of Bioactive Materials*, 760–761, 1994.

MUIR, J. G., LU, Z. X., YOUNG, G. P., CAMERON-SMITH, D., COLLIER, G.R., O'DEA, K. Resistant starch in the diet increases breath hydrogen and serum acetate in human subjects. *American Journal of Clinical Nutrition*, 61, 792–799, 1995.

MYLLARINEN, P., FORSSELL, P., VON WRIGHT, A., ALANDER, M., MATTILA-SANDHOLM, T., POUTANEN, K. Starch capsules containing microorganisms and/or polypeptides or proteins. PCT Int. Appl. (1999), WO 9952511.

PARCHURE, A. A., KULKARNI, P. R. Effect of food processing treatments on generation of resistant starch. *International Journal of Food Sciences and Nutrition*, 48, 257–260, 1997.

PHILLIPS, J., MUIR, J. G., BIRKETT, A., LU, Z. X., JONES, G. P., O'DEA, K., YOUNG, G. P. Effects of resistant starch on fecal bulk and fermentation-dependent events in humans. *American Journal of Clinical Nutrition*, 62, 121–130, 1995.

PILLAY, V., FASSIHI R. *In vitro* release modulation from crosslinked pellets for site-specific drug delivery to the gastrointestinal tract. I. Comparison of pH- responsive drug release and associated kinetics. *Journal of Controlled Release*, 59, 229–242, 1999.

PLAIZIER-VERCAMMEN, J., VAN MOLLE, M. Evaluation of aqueous dispersions of Eudragit L 100-55 for their enteric coating properties. *S.T.P. Pharma Sciences*, 1(4), 267–271, 1991.

RING, S., GEE, J. M., WHITTAN, P., OXFORD, I., JOHNSON, I. Resistant starch: its chemical forms in foodstuffs and effect on digestibility *in vitro*. *Food Chemistry*, 28, 97–109, 1988.

RUSSEL, G. F. J. Starch microspheres as drug delivery systems. *Pharm. Internat.*, 4, 260–262, 1983.

SCHMIDT, P. C., NIEMANN, F. The MiniWiD-coater: II. Comparison of acid resistance of enteric-coated bisacodyl pellets coated with different polymers. *Drug Development and Industrial Pharmacy*, 18(18), 1969–1979, 1992.

SCHROEDER, K., TREMAINE, W., ILSTRUP, D. Coated oral 5-aminosalicylic acid therapy for mildly to moderately active ulcerative colitis. *New England Journal of Medicine*, 317, 1625–1629, 1987.

SEMDE, R., AMIGHI, K., DEVLEESCHOUWER, M. J., MOES, A. J. Studies of pectin HM/Eudragit RL/Eudragit NE film-coating formulations intended for colonic drug delivery. *International Journal of Pharmaceutics*, 197, 181–192, 2000.

SENEVIRATNE, H. D., BILIADERIS, C. G. Action of α-amylases on amylose-lipid complex superstructures. *Journal of Cereal Science*, 13(2), 129–143, 1991.

SIEVERT, D., CZUCHAGOWSKA, A., POMERANTZ, Y. Enzyme resistant starch, III. X-ray diffraction of autoclaved amylomaize. *Cereal Chemistry*, 68, 86–91, 1991.

SILVESTER, K. R., ENGLYST, H. N., CUMMINGS, J. H. Ileal recovery of starch from whole diets containing resistant starch measured in vitro and fermentation of ileal effluent. *American Journal of Clinical Nutrition*, 62, 403–411, 1995.

SNAPE, C. E., MORRISON, W. R., MAROTO-VALER, M. M., KARKALAS, J., PETHRICK, R. A. Solid state ^{13}C NMR investigation of lipid ligands in V-amylose inclusion complexes. *Carbohydrates Polymers*, 36, 225–237, 1998.

SULTANA, K., GODWARDA, G., REYNOLDS, N., ARUMUGASWAMYC, R., PEIRISC, P., KAILASAPATHY, K. Encapsulation of probiotic bacteria with alginate–starch and evaluation of survival in simulated gastrointestinal conditions and in yoghurt. *International Journal of Food Microbiology*, 62(1–2), 47–55, 2000.

SZEJTLI, J., BANKY-ELOD, E. Inclusion complexes of unsaturated fatty acids with amylose and cyclodextrin. *Starch*, 27, 368–376, 1975.

TAN, L. H., CHAN, L. W., HENG, P. W. S. Effect of oil loading on microspheres produced by spray drying. *J. Microencapsulation*, 22(3), 253–259, 2005.

TOVAR, J., MELITO, C. Steam-cooking and dry heating produce resistant starch in legumes. *Journal of Agricultural and Food Chemistry*, 44, 2642–2645, 1996.

TRINDADE, M. A., GROSSO, C. R. F. The stability of ascorbic acid microencapsulated in granules of rice starch and in gum arabic. *J. Microencapsulation*, 17(20), 169–176, 2000.

TUFVESSON, F., ELIASSON, A. C. Formation and crystallization of amylose-monoglyceride complex in a starch matrix. *Carbohydrate Polymers*, 43(4), 359–365, 2000.

TUFVESSON, F., SKRABANJA, V., BJORCK, I., ELMSTAHL, H. L., ELIASSON, A. C. Digestibility of starch systems containing amylose-glycerol monopalmitin complexes. *Lebensmittel-Wissenschaft und -Technologie*, 34(3), 131–139, 2001.

TUFVESSON, F., WAHLGREN, M., ELIASSON, A. C. Formation of amylose-lipid complexes and effects of temperature treatment. Part 2. Fatty acids. *Starch*, 55, 138–149, 2003.

UDDIN, M. N., HAWLANDER, M. N. A., ZHU, H. J. Microencapsulation of ascorbic acid: effect of process variables on product characteristics. *J. Microencapsulation*, 18(2), 199–209, 2001.

WANG, X., CONWAY, P. L., BROWN, I. L., EVANS, A. J. *In vitro* utilization of amylopectin and high-amylose maize (Amylomaize) starch granules by human colonic bacteria. *Applied and Environmental Microbiology*, 65, 4848–4854, 1999.

WILDING, I. R., DAVIS, S. S., O'HAGAN, D. T. Targeting of drug and vaccine to the gut. *Pharmacology and Therapeutics*, 62, 97–124, 1994.

WILLIAMS, R.O. III, LIU, J. Influence of processing and curing conditions on beads coated with an aqueous dispersion of cellulose acetate phthalate. *European Journal of Pharmaceutics and Biopharmaceutics*, 49(3), 243–252, 2000.

WULFF, G., AVGENAKI, G., GUZMANN, M. S. P. Molecular encapsulation of flavours as helical inclusion complexes of amylase. *Journal of Cereal Science*, 41(3), 239–249, 2005.

ZHANG, W., JACKSON, D. S. Retrogradation behaviour of wheat starch gels with differing molecular profiles. *J. Food Sci.*, 57, 1428–1432, 1992.

ZOBEL, H. F. Starch crystal transformation and their industrial importance. *Starch/Starke*, 40, 1–7, 1988.

ZOBEL, H. F. Starch granule structure. In: *Developments in Carbohydrates Chemistry*, Alexander, R. J., Zobel, H. F. (eds), St. Paul, MN: American Assoc. Cereal Chem., 1992, pp. 1–36.

Part III

Delivery and controlled release of particular nutraceuticals

Part III

[illegible]
[illegible]

12

Encapsulation and controlled release of antioxidants and vitamins

C. M. Sabliov and C. E. Astete, Louisiana State University Agricultural Center, USA

12.1 Introduction

Antioxidants and vitamins are fundamental for human health. Natural antioxidants such as vitamin E (α-tocopherol), vitamin C and flavonoids (e.g. quercitin, catequin, among others), are compounds which have the potential to modulate oxidative stress. Many of the principal diseases affecting human health are related with the oxidative process, among which cardiovascular diseases, atherosclerosis, and cancer are the most prevalent. Antioxidants can prevent these diseases by reducing the reactive oxygen species (ROS) and thus limiting cell damage by ROS.

Natural antioxidants are prone to degradation and their bioavailability is limited by low absorption and degradation during delivery. Encapsulation of antioxidants and vitamins in polymeric nanoparticles is a promising way of improving the bioavailability of these components. Nanoparticles can be designed to deliver the antioxidant/vitamin to the specific site of action with minimal degradation of the bioactive component during delivery. The antioxidant/vitamin is then potentially released in a controlled manner by controlling the degradation profile of the polymer, or by using 'smart' polymers which interact with the cell environment (e.g. temperature, pH, enzymes, ionic strength) in a predictable way.

In this chapter, the importance of antioxidants to human health will be addressed, advantages of nanoencapsulation of these components over traditional delivery methods will be discussed, top-down techniques available to entrap antioxidants and vitamins in biodegradable and biocompatible polymeric nanoparticles will be discussed, methods available for polymeric nanoparticle characterization will be briefly mentioned, and some insights on the release profile of antioxidants and vitamins from polymeric nanoparticles will be presented. The chapter will close with conclusions and future research directions.

12.2 Antioxidants and vitamins in protecting human health

12.2.1 Reactive oxygen species and oxidation process

Halliwell and Gutteridge[1] defined antioxidants as 'any substance that, when present in low concentrations compared to that of an oxidisable substrate, significantly delays or inhibits the oxidation of that substrate'. Enzymatic and nonenzymatic antioxidants prevent cellular damage surged from chemical reactions that involve reactive oxygen species (ROS).[2–4] The reactive oxygen species play beneficial roles in the cell metabolism influencing cell growth, energy production, and intracellular signaling, but ROS can be damaging to different tissues due to their involvement in lipid peroxidation, DNA modification, carbohydrates oxidation, and/or protein alteration. Damage caused by ROS has been related to cancer, diabetes, cardiovascular disease, inflammatory responses, degenerative diseases, aging, liver injury, cataract, and others.[5–25]

Intracellular ROS are mainly generated by autooxidation of small molecules and from the activity of certain enzymes like oxidases, peroxidases, lipoxygenases, dehydrogenases, and cyclooxygenases; they can also be generated by exogenous sources (i.e. pollutants, cigarettes). The cellular sites of reactive oxygen species generation are mitochondria, peroxisomes, lysosomes, endoplasmic reticular membrane, nuclear membrane, and plasma membrane.[26–29]

The state of a biological system exposed to excessive formation of ROS and limited antioxidant defense is defined as oxidative stress.[30–32] The human body has developed an internal defense against oxidative stress (Fig. 12.1). Metal binding proteins control the concentration of free metal ions and reduce the potential risk of antioxidant molecules to act as pro-oxidants, whereas enzymatic and natural antioxidants act mainly as suppressors of radicals, protecting the cellular environment from oxidative damage by ROS which could otherwise result in oxidative stress-related diseases. Enzymatic antioxidants and proteins are mainly hydrophilic and therefore act in a hydrophilic environment. Natural antioxidants acquired from food act in a hydrophilic or hydrophobic environment, based on their chemical make-up, and complement the enzymatic defensive system, enhancing the protection of intracellular hydrophobic molecules. If oxidative stress occurs, the repair system, consisting of different enzymes (i.e. lipases, DNA repair enzymes, proteases, and others) can repair the damage originated from the oxidation process.[27]

A balanced food intake which includes vegetables, legumes, fruits and nuts has been suggested to ensure optimal tissue levels of antioxidants[33] necessary to maintain a balance between ROS formation and antioxidant defense. The distribution of natural antioxidant molecules in the cell is depicted in Fig. 12.2. The different sites of action of natural antioxidants suggest that it is important to

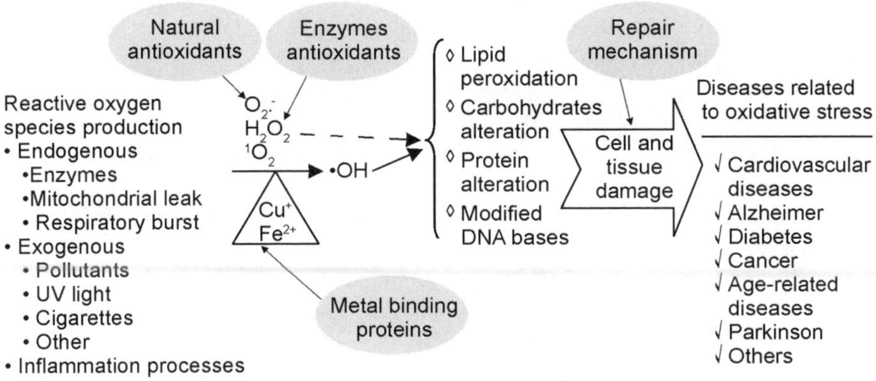

Fig. 12.1 Schematic representation of oxidative damage and defense systems. The oval shapes represent the defense system against reactive oxygen species. Adapted from Ref. 3.

maintain adequate concentrations of both hydrophobic and hydrophilic antioxidant molecules to reduce the risk of oxidative stress in all cell compartments.

The understanding of the antioxidant characteristics is crucial for designing an adequate delivery and targeting system for the specific antioxidants of interest. In the next section, several chemical components of demonstrated

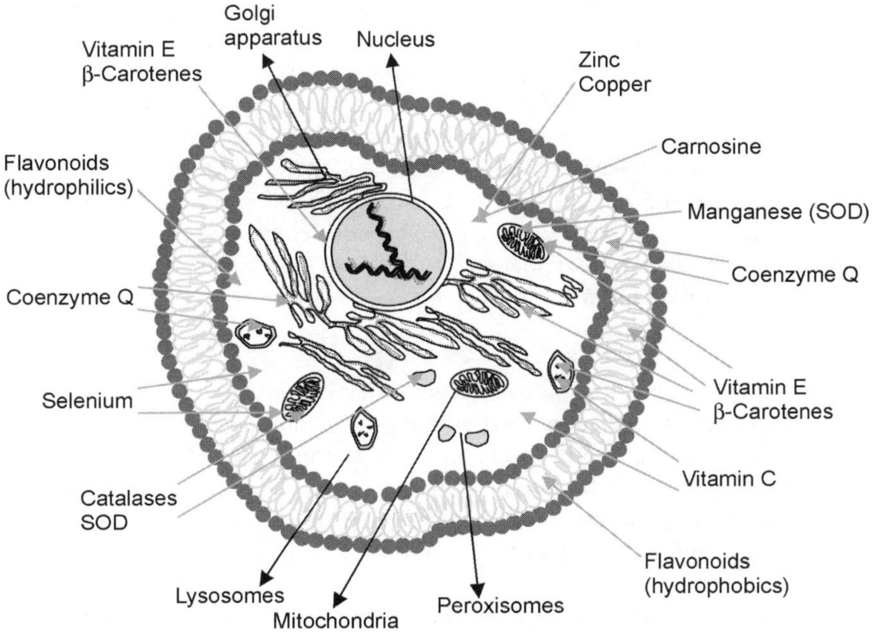

Fig. 12.2 Localization of different antioxidant molecules in the cell. The distribution of natural antioxidants in the cell components is dictated by their solubility in a hydrophilic (water) or a hydrophobic medium. Adapted from Ref. 26.

antioxidant activity are described with emphasis on how nanoencapsulation can improve their delivery and function.

12.2.2 Natural antioxidants in biological systems: characteristics and human health implications

Natural antioxidants and vitamins can be classified based on their solubility in two classes, hydrophobic (water insoluble) and hydrophilic (water soluble). Examples of hydrophilic antioxidants are vitamin C and carnosine, whereas vitamin E, Coenzyme Q_{10}, and carotenoids are examples of hydrophobic antioxidants.

Vitamin E

Vitamin E is a hydrophobic vitamin which has been intensively studied due to its antioxidant properties and non-antioxidant actions.[34–38] The vitamin E family is composed of four tocopherols (α, β, γ, δ) and four tocotrienols (α, β, γ, δ) (Fig. 12.3). Sources of vitamin E are wheat-germ oil, sunflower seed, almond, cereals, and others. The daily recommended intake of α-tocopherol is 22.7 mg for adults and 18.7 mg for lactating women.[39]

Vitamin E is absorbed in the intestine[40] to form chylomicron particles. Circulation of chylomicron particles (a structure formed with fatty acids, phospholipids, cholesterol, vitamin E and other hydrophobic molecules) in plasma is one mechanism for vitamin E transfer to muscles, adipose tissue and brain. Once delivered to the tissue, tocopherols are localized in the cellular membranes with the chromanol nucleus located in the phospholipid bilayer.[41] The role of vitamin E supplementation in prevention of several diseases such as cardiovascular diseases, atherosclerosis, cancer, arthritis, Alzheimer's, and Parkinson's, among others has been documented and linked to its ability to trap peroxyl radicals and to break the chain reaction of lipid peroxidation.[34,42–49] *Targeted delivery of vitamin E to the cytoplasmatic compartments achieved with polymeric nanoparticles may increase its efficiency as an antioxidant and may aid in prevention and treatment of oxidative stress-related diseases.*

Coenzyme Q_{10}

Coenzyme Q_{10} is a hydrophobic molecule which is synthesized by humans (Fig. 12.3). It can act as an electron carrier and proton translocator. Coenzyme Q is also found in plants, animals, and in most microorganisms. The amount of Coenzyme Q_{10} that can be provided by the diet is variable; for examples the Coenzyme Q_{10} content of pork heart is 203 μg/g, 41 μg/g is found in beef heart, 17 μg/g in chicken leg, 27 μg/g in herring fish, 2.2 μg/g in spinach, and 2.2 μg/g in oranges.[50]

Coenzyme Q plays fundamental roles in energy conversion, regeneration of other antioxidant molecules, and cell death inhibition and cell growth regulation.[50–53] The intestinal absorption of Coenzyme Q_{10} seems to be similar to the vitamin E absorption mainly due to its lipophilic characteristics.[54] The intestinal absorption is enhanced in the presence of lipids. In the small intestine, Coenzyme Q_{10} is assimilated into chylomicrons to reach the plasma via the lymphatic system. The plasma concentration of Coenzyme Q_{10} in healthy people can range from 0.4–0.6 μmol/l and can be improved with supplementation.[54–56] The

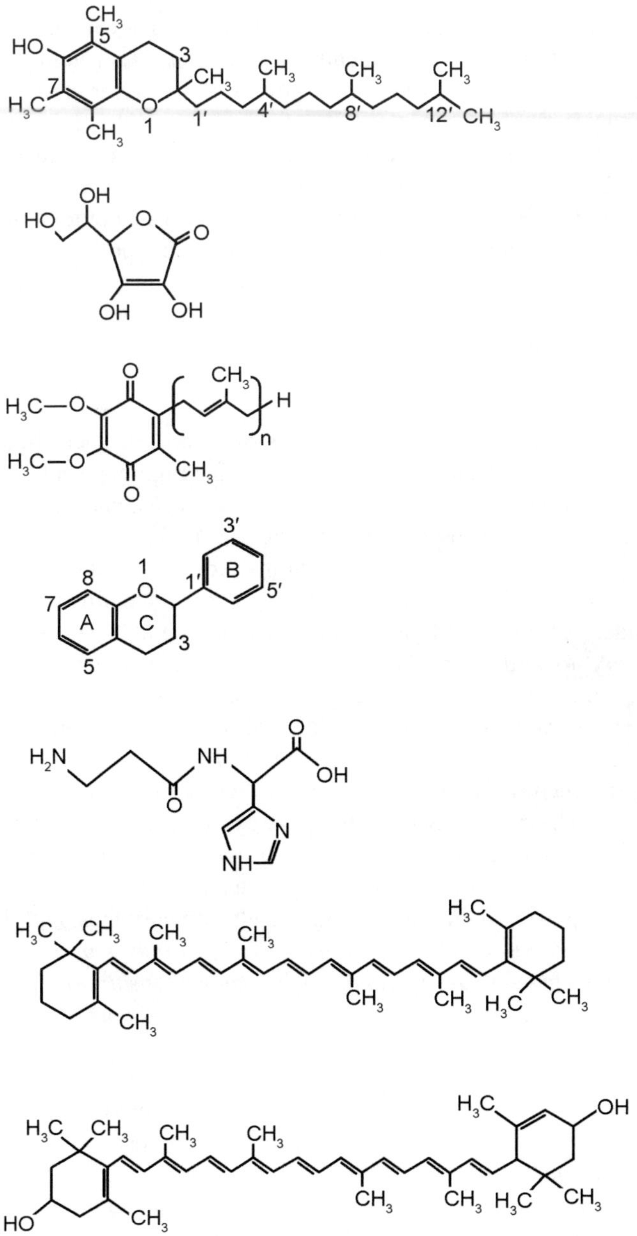

α-Tocopherol

Dehydroascorbic acid

Coenzyme Q (Ubiquinone)

General flavonoid structure

Carnosine

β-Carotene

Lutein

Fig. 12.3 Chemical structure of selected natural antioxidants and vitamins.

toxicity of Coenzyme Q_{10} is extremely low. Supplementation of Coenzyme Q_{10} up to 1200 mg/day has been used in trials without any noted adverse effects.[54]

The plasma level of Coenzyme Q_{10} decreases with aging[57] and supplementation or stimulation of Q_{10} biosynthesis are needed to increase the cellular levels of Coenzyme Q_{10}. Supplements of Q_{10} have been used for their potential benefits against cardiovascular and neurodegenerative diseases, cancer, aging, migraine, and others.[58–67] However, supplements of Coenzyme Q_{10} administered orally have been shown to have a low absorption efficiency (around 3–5%), mainly due to the hydrophobic characteristics of the bioactive component. Oil, emulsion, and powder formulations have been used in an attempt to increase the Coenzyme Q_{10} absorption, with the solubilized forms being most effective.[54,68,69] Tikhaze *et al.*[70] found that the solubilized form of ubiquinone form of emulsion increased the antioxidant activity of this component by more than 36 times as opposed to its insoluble form. *Nanoparticle formulation of Coenzyme Q_{10} may yet show improved bioavailability characteristics as compared to the emulsion formulations.*

Carotenoids

Carotenoids are a class of antioxidants composed of more than 500 identified components (Fig. 12.3). Carotenoids are abundant in vegetable pigments and are found mainly in carrots, broccoli, spinach, apricots, tomatoes, peas, and orange peppers among others.[71] The carotenoids level in the US population ranges between 0.05 and 0.5 μg/ml.[72] The characteristic of carotenoids is the unsaturated 40 carbons skeleton which confers the hydrophobicity specific to this compound. The most abundant carotenoids found in humans are α- and β-carotene, lycopene, zeaxanthin, cryptoxanthin and lutein.[73] Absorption of carotenoids is passive by intestinal absorption and it is similar to the absorption of tocopherol and Coenzyme Q_{10}. The carotenoids plasma concentration should be around 0.4–0.5 μM for good health, as suggested by Albertini *et al.*[24]

The lipophilicity and structure of carotenoids affect their distribution in the cellular systems *in vivo*. These components are poor water soluble and are usually associated with other components for improved absorption (i.e. lipoproteins). The lipophilic cell components are the main locations where carotenoids are found (i.e. plasma membranes, mitochondria membranes, Golgi membranes). *Improved distribution of carotenoids in the cytoplasm or other cellular hydrophilic compartments can be achieved by nanoencapsulation and delivery with polymeric nanoparticles dispersable in aqueous mediums.*

Polyphenols

Polyphenols (i.e. flavonoids) are natural compounds found in many vegetables, fruits, nuts, cereals, and herbs.[74–76] Polyphenols is a family composed of more than 4000 identified molecules with the general structure depicted in Fig. 12.3. The daily intake of polyphenols can reach 800 mg/day,[77] with a mean intake of 23 mg/day in Western countries.[78] In a normal diet, the plasma concentration of flavonoids is less than 1 μM.[79] The importance of flavonoids is represented by three characteristics: inhibitions of certain enzymes, potent antioxidant properties, and iron chelator ability.[78,80–83]

The potential health benefits of flavonoids are related to anti-inflammatory, anti-hepatotoxic, antiviral, antimicrobial, anti-allergic and anti-tumor activities, and to vascular protection, regulation and control of diabetic cataracts, among others.[75,84–88] *For those individuals less inclined to consume fruits and vegetables, sources of polyphenols, supplementation of the diet with nano-encapsulated flavonoids may be the path to a healthier life.*

Ascorbic acid

Ascorbic acid is a water soluble vitamin fundamental in metabolic reactions, and as a cofactor of various enzymes involved in the hydroxylation process; it cannot be synthesized by humans (Fig. 12.3). Different natural sources provide the necessary amounts of ascorbic acid, such as fruits (e.g. orange, papaya, strawberries) and vegetables (e.g. broccoli, red or green pepper, brussel sprouts).[32,89,90] Albertini *et al.*[24] suggested that the optimal concentration of plasma vitamin C should be at least 40–50 μM for antioxidant purposes.

Vitamin C has been widely reported as a potent co-antioxidant with vitamin E, due to the synergism between the two antioxidants.[91] The process involves α-tocopherol regeneration from α-tocopheryl radicals in lipoproteins and membranes in the presence of vitamin C. A synergetic action has also been suggested between vitamin C, vitamin E, and β-carotene against the oxidative damage generated by peroxynitrite anion and nitrogen dioxide.[92] *Efficient co-delivery of synergistic antioxidants and vitamins can be achieved by entrapping and delivering these components in a controlled manner with polymeric nanoparticles.*

Carnosine

Carnosine is a hydrophilic dipeptide (β-alanyl-L-histidine) which is found exclusively in animal tissue and in the brain (Fig. 12.3). This histidine containing dipeptide is part of the ω-aminoacylamino acids family. Other compounds are anserine, homocarnosine, and homoanserine.[93] It has been shown by different studies that carnosine acts as an antioxidant, carbonyl scavenger (anti-glycating agent), and metal ion chelator.[94–97] Due to its hydrophilic nature, carnosine protects proteins from oxidation.[96,98–102] It has also been pointed out that carnosine could be beneficial for people with diabetes, especially to control the secondary effects of this disease.[101,103] Hipkiss[104] found that carnosine has a potential therapeutic effect on neurodegenerative diseases related with high levels of zinc such as Alzheimer's disease. *Protection of carnosine against degradation during delivery to the site of action by entrapment in a polymeric matrix is a viable method to increase its bioavailability and efficiency in vivo.*

To conclude this section, there is a wide variety of natural antioxidants and vitamins which have been shown to possess beneficial properties and could potentially improve human health. The antioxidant efficiency of these components *in vivo* is related not only to their chemical reactivity, but also to their localization and mobility. Bioavailability, synergetic inhibition of the oxidation process, and pro-oxidant effects of natural antioxidants are other important factors to consider in the evaluation of potential antioxidant activity of different compounds *in vivo*.[105] Scientific proof of the advantages offered by encapsulation and delivery of antioxidants/vitamins with polymeric nanoparticles are presented below.

12.3 Advantages of encapsulation over traditional delivery methods

Nanoparticles can offer several advantages over traditional delivery methods for antioxidants and vitamins, which include protection of the bioactive component, an increase in its bioavailability, delivery of the component to the site of action in a controlled manner, as well as the possibility to integrate the bioactive component in various foods for enhanced food quality (Fig. 12.4).

12.3.1 Protection and bioavailability

Bioavailability of antioxidants and vitamins can be enhancecd by delivering the bioactive components via polymeric nanoparticles for several reasons:

- the antioxidant is protected against harsh conditions (i.e. low stomach pH, enzymatic and microflora intestinal degradation),
- the extracellular interaction between the components of interest and other molecules is minimized,
- low absorption of some antioxidants and vitamins (mainly hydrophobic) is improved.

Several studies available in the literature are focused on protection of the bioactive component against natural degradation by nanoencapsulation. For example, the stability against UV and thermal degradation of nanoencapsulated Q_{10} was studied by Kwon *et al.*[106] The polymer used to entrap Q_{10} was poly(methyl methacrylate) (PMMA), and the nanoparticles were prepared with different surfactants (i.e. poly(vinyl alcohol), sodium dodecyl sulfate). The entrapped Coenzyme Q_{10} proved to be structurally more stable against UV and high temperature-induced inactivation as compared with an oil-based Q_{10} formulation.

Chen and Subirade[107] tested biodegradable nanoparticles made from chitosan and β-lactoglobulin designed for oral delivery of nutraceuticals. The protein, β-lactoglobulin protected the active component against pepsin and acid degradation in the gastric environment. The study suggested that these nanoparticles could protect the nutraceuticals entrapped against degradation by adverse conditions during food processing, or in the gastrointestinal tract, and the absorption of these components could therefore be improved.

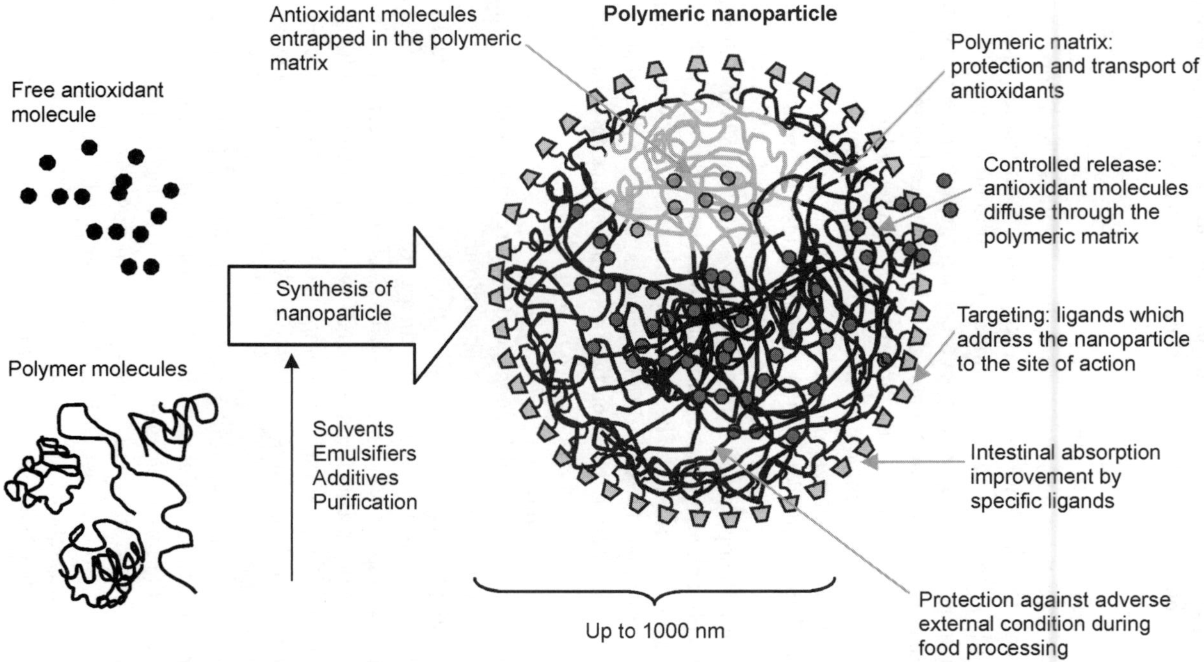

Fig. 12.4 Description of the nanoparticle antioxidant system with emphasis on the advantages offered by nanoparticle delivery systems.

An improvement in the intestinal absorption of antioxidants is a means to improve their bioavailability. The attachment of ligands like lectins, invasins, folic acid, vitamin B12, biotin, glycoproteins (i.e. transferrin), membrane transduction sequences[108] was shown to improve intestinal uptake of nanoparticles with entrapped poor absorbable antioxidant molecules.

12.3.2 Targeting and cellular uptake

Polymeric nanoparticles can be surface functionalized to improve half life circulation and to minimize opsonins recognition of the nanoparticles (i.e. by addition of PEG), to improve the intestinal absorption (i.e. by use of a cationic polymer such as chitosan), to improve cellular uptake (i.e. by addition of a membrane disruption agent or agents which promote endosomal release), or to ensure that the antioxidant/vitamin reaches intracellular targets such as mitochondria (i.e. by attachment of lipophilic triphenylphosphonium).

Nanoparticle functionalization with vitamins is an interesting means to improve the cellular uptake of the nanoparticles. Vitamin H (biotin) was covalently attached to pullulan acetate, a hydrophobized polysaccharide by Na *et al.*[109] to form nanoparticles with a mean size of 99 ± 52 nm (20 vitamin H groups per 100 anhydroglucose units of pullulan acetate), with entrapped adriamycin as a drug model. The release profile of adriamycin loaded nanoparticle was controlled by vitamin H content attached to pullulan; a specific interaction of the functionalized particle with hepatoma cell line (HepG2) by ligand-receptor interaction was proven and shown to be responsible for increased nanoparticle uptake.

The antioxidant action can be enhanced by targeting, when the antioxidant molecule is addressed to the specific site into the cell, such as mitochondria, nucleus, or endoplasmic reticulum, where oxidative stress occurs and causes damage. A significant decrease in heart dysfunction, cell death, and mitochondrial damage was shown after ischemia reperfusion with oral administration of a target functionalized system (ubiquinol plus lipophilic triphenylphosphonium) in rats. In comparison, ubiquinol delivered without targeted functionalization was ineffective for cardiac ischemia-reperfusion injury.[13]

Muro *et al.*[110] entrapped an antioxidant molecule (catalase) in a functionalized nanoparticle, called anti-ICAM1. This system targeted endothelial cells and entered the cell via an amiloride-sensitive endocytic pathway. The anti-ICAM nanoparticle system was effective in offering protection to endothelial cells against oxidation for over 2 h after cellular uptake. It was suggested that due to the targeting ability of the nanoparticulate system, the oxidants were intercepted near their sites of generation or at sites where sensitive damage occurred and therefore a better protection against oxidative damage was possible.

12.3.3 Controlled release

Degradation of the polymeric matrix as a result of interaction with different environmental parameters (temperature, enzymes, pH, ionic strength, and other) allows control of the amount of antioxidant released over time. Several studies published in the literature focus on the release of vitamin E, a model component for hydrophobic vitamins and antioxidants.

Shea *et al.*[111] entrapped vitamin E into a PEG-decanyl polymeric nanoparticle. The antioxidant property of the released α-tocopherol was tested in human neuroblastoma cells (SH-SY-5Y). The vitamin E added in the preparation was 20 mg, equivalent to 20% w/w (there is no information about the entrapment efficiency and size of the nanoparticle synthesized). The results showed that encapsulated vitamin E was more effective against ROS as compared to free vitamin E. The free vitamin E was effective only if administered before or during treatment with amyloid-β (a neurotoxic peptide which accumulates in Alzheimer's disease), but was not effective in protection against oxidation if administered 1 h after the addition of amyloid-β to the cell culture. The encapsulated form of vitamin E was effective even when administered 1 h after the amyloid-β treatment against ROS.

Duclairoir *et al.*[112] used wheat gliadin nanoparticles for vitamin E encapsulation by nanoprecipitation method. The mean size of the nanoparticle obtained was 900 nm. The entrapment efficiency ranged from 77% to 95%. The entrapment efficiency of vitamin E decreased as more vitamin E was added relative to the gliadin protein weight. The release profile showed an initial fast release (burst effect) and a further slower release.

Zigoneanu *et al.*[113] entrapped vitamin E in PLGA nanoparticles at two initial loadings of 8% and 16%. The nanoparticles synthesized by emulsion evaporation method measured 100 nm for nanoparticles prepared with SDS as a surfactant, and about 200 nm for nanoparticles prepared with PVA. The entrapment efficiency of α-tocopherol in the polymeric matrix was approximately 89% and 95% for nanoparticles with 8% and 16% α-tocopherol theoretical loading, respectively. The release profile showed an initial burst followed by a slower release of the α-tocopherol entrapped inside the PLGA matrix, similar to the findings of Shea *et al.*[111] and Duclairoir *et al.*[112]

12.3.4 Integration and safety

The nanoencapsulated antioxidants and vitamins can be integrated into foods and beverages to enhance the quality (sensorial, chemical, and microbiological properties) of the food. Chen and Wagner[114] entrapped vitamin E in starch sodium octenyl succinate nanoparticles designed for beverage applications. The mean size of the nanoparticles was 112 nm; the size varied as a function of the number of homogenization cycles used in the preparation. The vitamin E nanoparticles were more stable in terms of turbidity and aggregation over time as compared to the non-nanoparticle systems. The suspension with nanoparticles did not change in turbidity over 6 months and no perceptible creaming and/or ringing were observed over the same period of time. The stability of the nanoparticle suspension in apple juice also presented low turbidity without change over time (6 months). The vitamin E used was vitamin E acetate, and the stability studies were performed after pasteurization at 75 °C for 10 min. There was no information provided about the entrapment efficiency and release profiles of the vitamin.

It should be noted that prior to integration of nanoparticles in food systems, it is necessary to develop a thorough understanding of the nanoparticle physicochemical characteristics, and very importantly of their safety for human consumption. The safety of the nanoparticles is dictated by their components (polymer, surfactant, antioxidant, additives), by the properties of the nanoparticle (size, size distribution, surface characteristics), and by the nanoparticle bio-interaction (bioavailability, cellular uptake, targeting). Limited information is available in the literature regarding the safety of nanoparticles for human health, whether used in pharmaceutical or food applications.

Several steps can be taken to insure nanoparticle safety for human consumption. Biocompatible, biodegradable components should be selected for nanoparticle synthesis. Synthetic polymers and natural macromolecules (e.g. albumin, chitosan) have been extensively researched as materials for nanoparticle production designed for drug delivery. Synthetic polymers have the advantage of high purity and reproducibility over natural polymers. Among the synthetic polymers, the polyesters family (i.e. poly(lactic acid) (PLA), poly(ϵ-caprolactone) (PCL), poly(glycolic acid) (PGA)) are of interest in the biomedical area because of their biocompatibility and biodegradability properties. In particular, poly(lactide-co-glycolide) (PLGA) has been approved by the US Food and Drug Administration (FDA) for human therapy.[115] Food biopolymers (i.e. albumen, chitosan, dextran, alginate, etc.), on the other hand, are the alternative option when considering the safety of the nanoparticles for human food consumption.

The size and size distribution of the polymeric nanospheres used for delivery of antioxidants and vitamins, among other physical characteristics, are affected by the technique used for the nanoparticle production and the pertinent synthesis parameters, i.e. polymer used, polymer molecular weight and concentration, the addition of active components, surfactants, solvents, and other additives, shear stress, evaporation rate, and other physical parameters.[115–123] These characteristics, such as size, size distribution and surface characteristics, affect the behavior of the nanoparticles and their interaction with a biological system and ultimately their safety. It is important to understand how the physical characteristics relate to the efficiency of the active component delivery at the site of action. Entrapment and delivery of antioxidants via polymeric nanoparticles may be very efficient as compared to the delivery of the free component (i.e. non-nanoencapsulated), in which case the safety of the antioxidant delivered in high amounts becomes an issue which must be addressed.

Synthesis of polymeric nanoparticles and the parameters that affect the nanoparticle physical characteristics have been intensively investigated during

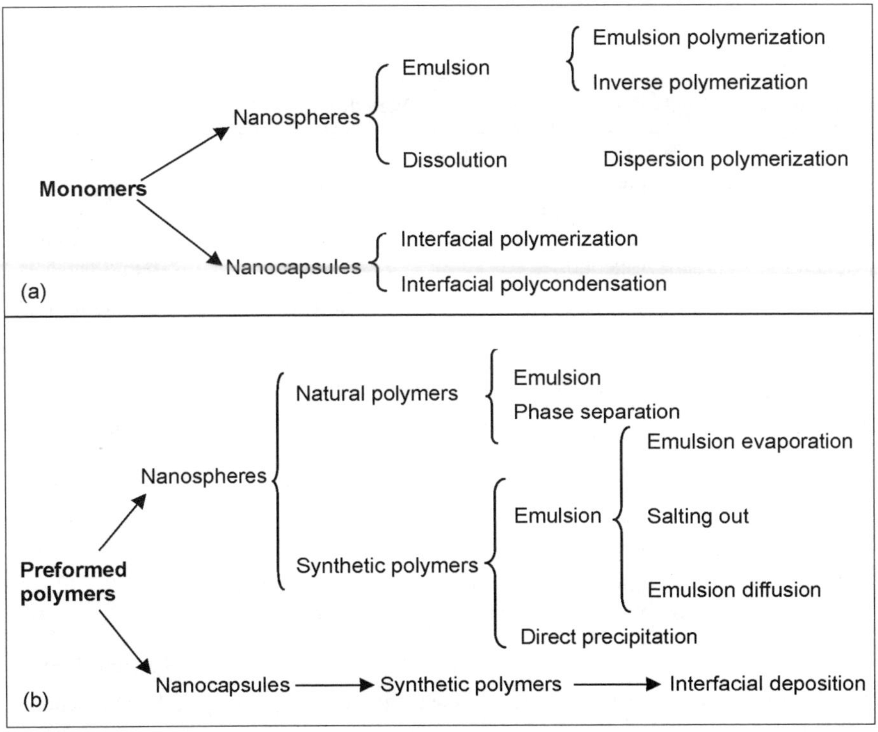

Fig. 12.5 Polymeric nanoparticle synthesis techniques for encapsulation of antioxidants: (a) bottom-up techniques, (b) top-down techniques.

the last decade.[121,124–131] Methods available for nanoparticle synthesis can be divided into two classes: bottom-up and top-down techniques. The bottom-up techniques, such as emulsion or microemulsion polymerization, interfacial polymerization, and precipitation polymerization, employ a monomer as a starting point (Fig. 12.5). Polymerization methods require the addition of the antioxidant during polymerization, and crosslinking of the antioxidant with the polymer synthesized during the nanoparticle formation is possible. The crosslink modifies the antioxidant molecules with further loss of its antioxidant activity. The top-down techniques minimize the damage to the active components entrapped (vitamin E, flavonoids, carotenoids, etc.) as compared to the polymerization methods. Some details on the top-down techniques are presented in the next section.

12.4 Top-down techniques used for encapsulation of antioxidants and vitamins in polymeric nanoparticles

Emulsion evaporation, emulsion diffusion, nanoprecipitation (or solvent displacement), and salting out are top-down techniques in which the nanoparticles

Table 12.1 Characteristics of top-down techniques used for antioxidant entrapment in polymeric nanoparticles

Characteristics	Methods			
	Emulsion diffusion	Salting out	Nano-precipitation	Emulsion evaporation
Solvent availability	Low toxicity (benzyl alcohol)	Low toxicity	Low toxicity (acetone)	Low toxicity (ethyl acetate)
Energy consumption	Low	Low	Low	Medium
Water requirements	High	Medium	High	Low
Polymer concentration	Linear relation with Np size	Medium–high	Low to no effect on Np size	Medium–high without increase Np size
Hydrophobic antioxidants	High entrapment	High entrapment	High entrapment	High entrapment
Hydrophilic antioxidants	Low entrapment	Low entrapment	Medium entrapment	Medium entrapment (by double emulsion)
Additives for size control	Recommended	Recommended	Recommended	Recommended
Process time	Medium–high	Medium	High	Reduced
Purification	Medium	High	Medium	Medium
Observations	Solvent limitation	Presence of salts	Polar solvents	Single emulsion or double emulsion

are synthesized from pre-formed polymers. The characteristics of these methods are presented in Table 12.1.

12.4.1 Emulsion diffusion method

In this synthetic scheme, the polymer is dissolved in a suitable organic phase (e.g. benzyl alcohol, propylene carbonate, ethyl acetate), which must be partially miscible in water. The organic phase is then emulsified with an aqueous solution of a suitable surfactant (i.e. anionic sodium dodecyl sulfate (SDS), non-ionic polyvinyl alcohol (PVA), or cationic didodecyl dimethyl ammonium bromide (DMAB)), under stirring. The diffusion of the organic solvent and the counter diffusion of water into the emulsion droplets induce polymer nanoparticle formation.[132] Important parameters that affect the nanoparticle size synthesized by emulsion evaporation are: copolymer ratio, polymer concentration, solvent nature, surfactant polymer molecular weight, viscosity, phase ratios, stirring rate, solvent nature, temperature and flow of water added.[133–138]

12.4.2 Salting out method

In this method, the polymer is dissolved in the organic phase, which should be water-miscible, such as acetone or tetrahydrofuran ethanol. The organic phase is emulsified in an aqueous phase under strong mechanical shear stress. The aqueous phase contains the emulsifier and a high concentration of salts which are not soluble in the organic phase. Contrary to the emulsion diffusion method, there is no diffusion of the solvent due to the presence of salts. The fast addition of pure water to the o/w emulsion, under mild stirring, reduces the ionic strength and leads to the migration of the water-soluble organic solvent to the aqueous phase inducing nanosphere formation.[120] The final step is purification by cross flow filtration or centrifugation required to remove the salting out agent. Common salting out agents are electrolytes (sodium chloride, magnesium acetate, or magnesium chloride) or non-electrolytes, such as sucrose.[129,139–141] Important parameters to be considered to control the physical characteristics of the synthesized nanoparticles by this method are: polymer concentration and molecular weight, stirring rate and time, nature and concentration of surfactant and solvent, and cryoprotectans.

12.4.3 Nanoprecipitation (solvent diffusion, or solvent displacement) method

Polymer and drug are dissolved in a polar, water-miscible solvent such as acetone, acetonitrile, ethanol, or methanol. The solution is then poured in a controlled manner (i.e. drop-wise addition) into an aqueous solution with surfactant. Nanoparticles are formed instantaneously by rapid solvent diffusion. Finally, the solvent is removed under reduced pressure by evaporation. Important parameters to be considered in controlling the nanoparticle size, among other physical characteristics, are: polymer/surfactant ratio, polymer concentration, surfactant nature and concentration, solvent nature, viscosity, additives, active component, and phase injection rate.[142–151]

12.4.4 Emulsion evaporation method

Emulsion evaporation is the oldest method used to form polymeric nanoparticles from preformed polymers. The method is based on the emulsification of an organic solution of the polymer in an aqueous phase followed by the evaporation of the organic solvent. Normal emulsions of oil-in-water (o/w) or water-in-oil (w/o) and double emulsions (w/o/w) can be used to accommodate the entrapment of active components with different properties. The o/w emulsion can be used for entrapment of hydrophobic compounds, whereas w/o/w double emulsion is used for the entrapment of hydrophilic compounds. The method is widely used for microencapsulation because it is easy to scale up, it does not require high shear stress, and it can be adjusted (by use of the double emulsion method) to encapsulate water soluble drugs.[152–160] More details of single and double emulsion evaporation methods are presented below.

Single emulsion method (o/w)
This method is based on the emulsification of an organic solution which contains the polymer and the active component in an aqueous phase, followed by the evaporation of the organic solvent. Different surfactants such as poly(vinyl alcohol), sodium dodecyl sulfate, pluronic F68 can be dissolved in the aqueous phase. The size reduction of the emulsion droplet is accomplished by sonication or microfluidization for mini-emulsion formation. The evaporation step is required to eliminate the organic solvent present in the organic phase. This leads to the precipitation of the polymer as nanoparticles with a diameter in the nanometers range. Important parameters which affect the nanoparticle size are: polymer molecular weight and concentration, copolymer ratio and end groups, surfactant nature, phase ratio, solvent nature, evaporation rate, drug entrapment, additives, shear stress, and sterilization.[160–165]

Double emulsion method (w/o/w)
The first step of the double emulsion method is the formation of a water-in-oil (w/o) emulsion where the aqueous solution contains the hydrophilic active component and the organic phase contains the polymer and a suitable surfactant (Span 80, pluronic F68, and others) of a low HLB. The mini-emulsion is formed under strong shear stress (i.e. sonication, microfluidization, high speed homogenization). Next, the (water-in-oil) in water (w/o/w) emulsion is formed and sonicated or homogenized for droplet size reduction. This second size reduction should be controlled to minimize the hydrophilic active component diffusion to the external aqueous phase. Evaporation, the final step, is used to remove the organic solvent. Evaporation is done under vacuum to avoid polymer and active component damage, and to promote final nanoparticle size reduction.

The main drawback of the double emulsion method is the large size of the nanoparticles formed and the leakage of the hydrophilic active component, responsible for low entrapment efficiencies. The coalescence and Ostwald ripening[166–168] are the two important mechanisms that destabilize the double emulsion droplets, and the diffusion through the organic phase of the hydrophilic active component is the main mechanism responsible for low levels of entrapped active component by this method.[168] The leakage effect can be reduced by using a high polymer concentration and a high polymer molecular weight, accompanied by an increase in the viscosity of the inner water phase, and an increase in the surfactant molecular weight.[169] Important parameters which affect nanoparticle size are: polymer/surfactants ratio, polymer concentration, surfactant nature, viscosity, solvent nature, shear stress, evaporation, additives, and first/second phase ratios.[161,162,165,170–178]

Following the synthesis, the polymeric nanoparticles specifically designed for antioxidant and vitamin delivery are characterized in terms of morphology, size, size distribution, surface charge, and surface characteristics amongst others.

12.5 Characterization methods

The methods most broadly used to characterize nanoparticle morphology are transmission electron microscopy (TEM), scanning electron microscopy (SEM), cryogenic transmission electro microscopy (cryo-TEM) and atomic force microscopy (AFM). Dynamic light scattering (DLS) is the most widely used technique to determine size and size distribution of the polymeric nanoparticles. Laser doppler anemometry is used to measure zeta potential, an important parameter when considering the stability of the nanoparticles *in vitro*. For chemical characterization, Fourier transform infrared spectroscopy (FTIR) is used to analyze surface modification of the nanoparticles by the attachment of specific components. The quantitative determination of the active component entrapped in the nanoparticles is done by extraction of the drug and measurement of the drug concentration by spectroscopy at defined wavelengths (related to the active component) or by high pressure liquid chromatography (HPLC).

12.6 Controlled release of antioxidants and vitamins

The release of the active components from the nanoparticles is critical in understanding the effects of the active components at the site of action. The release profiles of the antioxidant/vitamin will impact the availability, interactions, pharmacokinetics, and the formation of metabolites at the site of action. Different parameters affect the release profile of the antioxidant, which include the chemical structure of the active component, amount of drug entrapped, polymer hydrophobic/hydrophilic behavior, temperature, pH, presence of enzymes, sterilization, polymer structure (i.e. copolymer ratio, crystallinity), polymer molecular weight, and nanoparticle size.

The common mechanisms which explain the drug release from polymeric nanoparticles are the natural degradation of the polymeric matrix (passive release), and diffusion of the drug through the matrix (active release). The diffusion process is dominant at the beginning of the active component release when minimum degradation of the particle is observed. During this process, the active component diffuses from the polymeric matrix to the hydrophilic external environment and water molecules diffuse into the nanoparticle polymeric matrix. The diffusion rate is controlled by the hydrophilicity of the polymeric matrix and that of the active component. Diffusion is followed by a constant release of the drug over time which is controlled by the erosion process of the polymer (degradation of the whole matrix with final mass loss). These mechanisms result in a typical release profile for active components entrapped in polymeric nanoparticles consisting of an initial fast release (burst effect), followed by a more constant release over time (Fig. 12.6). The burst effect can be beneficial when an initial high release improves the action of the active component or detrimental when a constant release of the antioxidant over time is required.

The release studies are usually performed *in vitro* with buffer solutions incubated at 37 °C. The main methods used to quantify the release are depicted

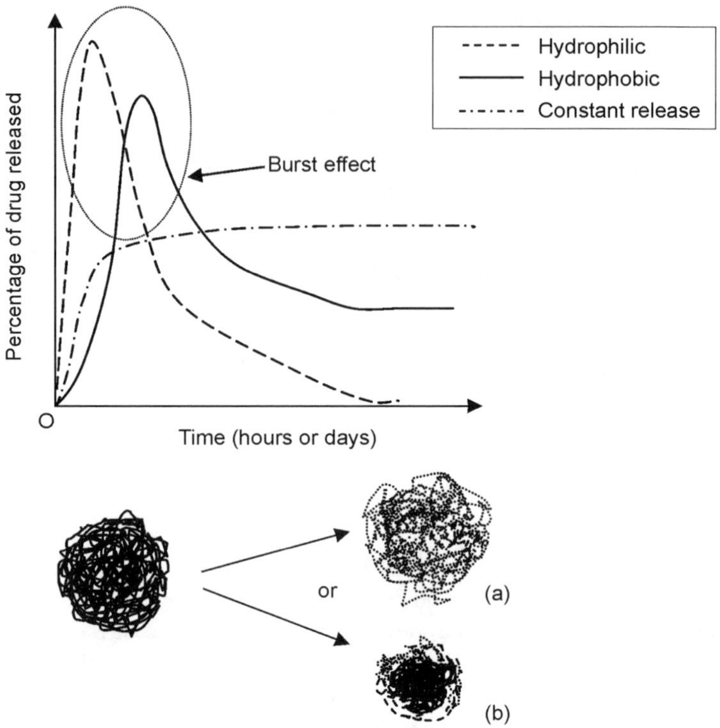

Fig. 12.6 Typical release profile. In general, hydrophilic drugs release faster as compared to hydrophobic drugs. Diffusion forces are predominant initially, after which the degradation process is more predominant as the time elapses. The nanoparticle degradation can occur by bulk erosion (a), or surface erosion (b).

in Fig. 12.7. The main parameters that affect the release profile are summarized below.

12.6.1 Active component (type and amount)

The entrapped active component interacts with the polymer matrix by hydrophobic–hydrophilic interaction, ionic interactions, van der Waals forces, and steric effect, amongst others. Chemical and physical interactions between the active component and polymeric matrix affects the diffusion of the entrapped antioxidant.[143,172]

The amount of entrapped active component (i.e. antioxidant or vitamin) affects the release, in that the higher the initial amount of active component, the faster the release is. The amount of entrapped active component is expressed by two parameters, entrapment efficiency and drug content. Drug content is defined as mass of drug in the nanoparticles divided by mass of nanoparticles, usually expressed in % w/w. A normal value for the drug content ranges from 0.5 to 4% w/w for hydrophilic drugs, and 10 to 15% w/w for hydrophobic drugs. Entrapment efficiency is calculated by dividing the amount of drug entrapped by the

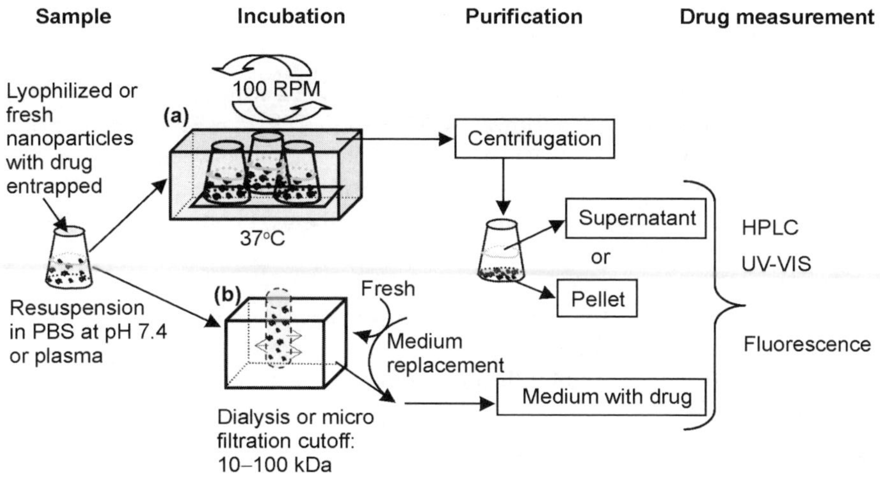

Fig. 12.7 Summary of the common procedures used to quantify *in vitro* drug release profiles. The first method (a) uses centrifugation to separate the drug released from the nanoparticle suspension, the second method (b) uses dialysis or filtration for separation. The suspension medium is usually PBS (phosphate bovine serum). The temperature is maintained at 37 °C with gentle agitation. The volume is maintained constant, in the first method, by addition of PBS in the same amount as that extracted for analysis. The other approach is to divide the sample in as many sub-samples as needed to study the release profile for the desired period of time. The second approach needs to maintain the concentration gradient necessary for the diffusion process to occur.

theoretical amount of drug (the initial amount used in the nanoparticle formation). Entrapment efficiency for hydrophobic drugs varies from 80% to 95%, and for hydrophilic drugs it varies from 30% to 70%.

12.6.2 Polymer (type, copolymer ratio, MW)

Different polymers have been used for nanoparticle synthesis. Chitosan, dextran, albumin, pullulan, poly(lactic acid) (PLA), poly(lactide-*co*-glycolide acid), poly(ethyle oxide) (PEO), poly(caprolactone), poly(3-hydroxybutyrate) and others are natural and synthetic polymers used for nanoparticle synthesis. The degradation profiles of these components affect the drug release profile. For example, poly(lactic acid), poly(lactide-*co*-glycolide acid) (PLGA) and PEO-PLGA nanoparticles appear to degrade homogeneously in time without autocatalysis.[140,179,180] The degradation rate of PEO-PLGA nanoparticles is slightly higher than that of PLGA particles.

The hydrophilicity (hydrophilic–hydrophobic ratio) of the polymer impacts the release profile of the active component entrapped. An example is PLGA, a hydrophobic copolymer formed with lactide and glycolide monomers. Changing the copolymer ratio, the PLGA copolymer can be more or less hydrophilic (50:50, 75:25, and 85:15 PLGA molar ratios are the most common copolymer ratios available commercially) which affects the polymer degradation profile.

PLGA copolymers formed from D,L-PLA and PGA are amorphous in nature while L-PLA and PGA are crystalline.[130,181] The copolymer 75/25 showed faster drug release as compared with 85/15.[178] Nanoparticles with a higher amount of lactide monomers showed a slower release rate. The more hydrophobic the polymer, the more intense the hydrophobic interaction between polymer and drug entrapped is, and thus a slower release profile is observed.[148,182–187]

The molecular weight of the polymer impacts the degradation process which in turn affects the release profile, mainly during the degradation step. Polymer molecular weight ranges from a few thousands Da to over 100 000 Da. Polymers with higher molecular weight present a slower degradation associated with a prolonged constant drug release over time. Diffusion processes are affected by the polymer molecular weight if the drug is mainly entrapped in the nanoparticle core.[115,183,184,188]

12.6.3 Nanoparticle size

Particle size is dictated by the polymer used, the method used, the drug entrapped, and the physical and chemical interactions between the drug and the polymer. The nanoparticle size, in turn can affect the release profile. Bigger nanoparticles degrade more slowly, and present slower active component diffusion through the polymeric matrix which affects the release profiles. The initial burst effect is reduced with a further prolonged slow release due to slow nanoparticle degradation.[189,190] The drug released from microparticles is relatively slower than release from nanoparticles because of the higher surface area and smaller size of nanoparticles.[130,178,191]

12.6.4 Environmental conditions (pH, temperature, release medium)

Temperature, pH, ionic strength, among other environmental conditions affect the bioactive component release profile due to an increase or decrease in polymeric degradation rates and changes in the diffusion process. The polymer behavior changes as a function of pH, temperature or other parameters. For example, poly(ortho esters) are stable at alkaline pH with increased degradation at acidic pH.[192] It would be ideal to use the internal cellular characteristics to release the drug in a controlled time-dependent manner. Pullulan, hystidine base peptides, amine base polymers, poly(vinylpyrrolidone-*co*-dimethylmaleic anhydride) and others are examples of pH sensitive polymers that release the drug entrapped differently at different pHs. It is important to note that physiological pH is around 7, but organelles have a different pH, endosomes are more acidic, and lysosomes have a pH as low as 5. At this pH, nanoparticles suffer chemical alterations that activate degradation mechanisms or polymer reconfiguration which affect the release of the drug entrapped. Temperature can affect the stability of the nanoparticles, which in turn affects the release of the active component. The polymers poly(butyl methacrylate) and poly(*N*-isopropylacrylamide) are examples of temperature dependent drug release polymers.[193–198]

The affinity between the bioactive component and the release medium affects the release. Redhead *et al.*[199] worked with release of a hydrophilic component in different mediums. The release profiles performed in PBS showed an initial fast release, between 40% and 60% of the hydrophilic active component was released in 24 h. When serum was used, the release rates were faster with almost 50% of drug released in 1 h. At 24 h, almost 100% was released. This difference was attributed to the presence of serum proteins and the affinity between the proteins and the drug.[199] The configuration that polymers acquire (i.e. random coil, oval, rod, and other) in presence of good, ideal, or bad solvents could also affect the drug release profile which adds more complexity to the solvent election as the organic phase in the synthesis process.

12.6.5 Complexation

Conjugation of the polymer with the active component causes a decrease in the diffusion process of the drug, and the degradation process becomes the dominant mechanism for the release of the active component. Yoo *et al.*[200] studied the anti-tumor activity of entrapped doxorubicin into nanoparticles. The non-conjugated drug entrapped in the matrix showed a fast release with almost 80% in the first day, and 100% of release after 5 days. The sample with drug conjugated to PLGA (PLGA MW of 5 kDa) showed slower release with 40% after one day which continued up to 24 days, when 100% of the drug was released.

12.7 Conclusions

Natural antioxidants and vitamins have been shown to possess beneficial properties that could potentially improve human health. Targeted delivery of vitamins and antioxidants via polymeric nanoparticles can offer several advantages over traditional delivery methods, which include protection of the bioactive component from degradation, an increase in the bioavailability of the vitamin, delivery of the component to the site of action in a controlled manner, as well as the possibility to integrate the bioactive component in various foods for enhanced food quality. Safety of the polymeric nanoparticles for antioxidant and vitamin delivery is a very important factor which must be considered in designing and building polymeric nanoparticles for targeted controlled delivery of antioxidants and vitamins.

Several methods are available to synthesize polymeric nanoparticles with entrapped antioxidants/vitamins, by starting with a preformed polymer. Each method has its advantages and disadvantages; the selection of the most suitable method should be based on the chemical characteristics of the active component and that of the polymer, as well as on the final use of the nanoparticles. Methods which are based on the diffusion of the organic solvent to form nanoparticles are limited to low polymer concentration for nanoparticle mean sizes of less than

200 nm. Methods that involve solvent evaporation are more time consuming and more expensive, but are less sensitive to changes in the polymer concentration. Emulsion evaporation, in particular, can be used to entrap hydrophilic (w/o/w emulsion) or hydrophobic (w/o emulsion) antioxidants, which is an advantage. The salting out method is suitable for formation of nanoparticles at higher polymer concentrations, but the involved purification process is a limitation of this synthesis method. Surfactant concentration, polymer concentration, polymer molecular weight, solvents, surfactant concentrations, and phase ratios play an important role in controlling the size of the nanoparticles in all methods available for nanoparticle formation.

The methods most broadly used to characterize nanoparticles in terms of morphology, size and size distribution, surface chatacteristics, entrapment efficiency, and release of the entrapped component, are the following: transmission electron microscopy (TEM), scanning electron microscopy (SEM), cryogenic transmission electro microscopy (cryo-TEM), atomic force microscopy (AFM), dynamic light scattering (DLS), laser doppler anemometry, Fourier transform infrared spectroscopy (FTIR), spectroscopy, and high pressure liquid chromatography (HPLC).

The release of the active component consists of two steps, diffusion and polymer degradation. The diffusion process is predominant at the initial states of the release, followed by the release of the active component due to the polymer degradation. Different parameters affect the release profile, which include chemical structure of the active component, amount of drug entrapped, polymer hydrophobic–hydrophilic behavior, temperature, pH, presence of enzymes, method of sterilization, polymer structure (i.e. copolymer ratio, crystallinity), polymer molecular weight, and nanoparticle size, among others.

12.8 Future trends

Entrapment and delivery of antioxidants and vitamins with polymeric nanoparticles is anticipated to become a significant research field in the near future, due to the extensive benefits that nanoencapsulation can offer. Single antioxidant entrapment approach will be enhanced by including multiple action antioxidants with potential synergistic effects in a single nanoparticle delivery system.

Nanoparticle size, among other physical characteristics, affects the release profiles of the bioactive component, and many researchers are continually attempting to decrease the average nanoparticle size with better control. There are important advances in the literature on understanding the mechanisms involved in manipulation and control of the nanoparticle characteristics and the improvement in the drug entrapment efficiency by carefully controlling these parameters.

Antioxidants entrapped into polymeric nanoparticles will present release profiles strictly modulated by carefully designing the particle to interact with the environment (i.e. pH, temperature) in a predictable way. Formation of nanoparticles that can interact with the human body and can modify their responses based on changes in the environment is an important research area in the field. Several questions will be addressed to reach this goal, such as tailoring synthesis of new polymers for modulated responses and a better performance.

Very importantly, understanding of the safety of the nanoparticle delivery system is critical for the final use of the system designed in food. It involves analysis of the complex interactions between the nanoparticles and the cellular environment, and the mechanism involved in the delivery of the antioxidant/vitamin at the site of action.

12.9 References

1. HALLIWELL B and GUTTERIDGE J N C (1995), 'The definition and measurement of antioxidants in biological systems', *Free Radical Bio Med* 18 (1), 125–126.
2. HALLIWELL B (1999), 'Antioxidant defence mechanisms: from the beginning to the end (of the beginning)', *Free Radical Research*, 31 (4), 261–272.
3. YOUNG I S and WOODSIDE J V (2001), 'Antioxidants in health and disease', *Journal of Clinical Pathology*, 54 (3), 176–186.
4. KNIGHT J A (2000), 'Review: Free radicals, antioxidants, and the immune system', *Annals of Clinical and Laboratory Science*, 30 (2), 145–158.
5. DRAPER H H, DRAPER H H and BIRD R P (1984), 'Antioxidants and cancer', *Journal of Agricultural and Food Chemistry*, 32, 433–435.
6. ZHANG Y H, KRAMER T R, TAYLOR P R, LI J Y, BLOT W J, BROWN C C, GUO W, DAWSEY S M and LI B (1995), 'Possible immunologic involvement of antioxidants in cancer prevention', *The American Journal of Clinical Nutrition*, 62 (6 Suppl), 1477S–1482S.
7. VINA J, LLORET A, ORTI R and ALONSO D (2004), 'Molecular bases of the treatment of Alzheimer's disease with antioxidants: prevention of oxidative stress', *Molecular Aspects of Medicine*, 25 (1–2), 117–123.
8. PANI G, COLAVITTI R, BEDOGNI B, FUSCO S, FERRARO D, BORRELLO S and GALEOTTI T (2004), 'Mitochondrial superoxide dismutase: a promising target for new anticancer therapies', *Current Medicinal Chemistry – Anti-Inflammatory and Anti-Allergy Agents*, 11 (10), 1299–1308.
9. HENSLEY K, ROBINSON K A, GABBITA S P, SALSMAN S and FLOYD R A (2000), 'Reactive oxygen species, cell signaling, and cell injury', *Free Radical Bio Med*, 28 (10), 1456–1462.
10. TAYLOR A, JACQUES P F and EPSTEIN E M (1995), 'Relations among aging, antioxidant status, and cataract', *The American Journal of Clinical Nutrition*, 62 (6 Suppl), 1439S–1447S.
11. HALLIWELL B (2000), 'Lipid peroxidation, antioxidants and cardiovascular disease: how should we move forward?' *Cardiovascular Research*, 47 (3), 410–418.
12. HAZEN S L (2000), 'Oxidation and atherosclerosis', *Free Radical Bio Med*, 28 (12), 1683–1684.
13. ADLAM V J, HARRISON J C, PORTEOUS C M, JAMES A M, SMITH R A J, MURPHY M P and SAMMUT I A (2005), 'Targeting an antioxidant to mitochondria decreases cardiac ischemia-reperfusion injury', *FASEB J*, 19 (9), 1088–1095.

14. BERGENDI L, BENES L, DURACKOVA Z and FERENCIK M (1999), 'Chemistry, physiology and pathology of free radicals', *Life Sciences*, 65 (18–19), 1865–1874.
15. AZZI A, DAVIES K J and KELLY F (2004), 'Free radical biology – terminology and critical thinking', *FEBS Letters*, 558 (1–3), 3–6.
16. KOJDA G and HARRISON D (1999), 'Interactions between NO and reactive oxygen species: pathophysiological importance in atherosclerosis, hypertension, diabetes and heart failure', *Cardiovascular Research*, 43 (3), 562–571.
17. NISHIKAWA T, EDELSTEIN D, DU X L, YAMAGISHI S, MATSUMURA T, KANEDA Y, YOREK M A, BEEBE D, OATES P J, HAMMES H P, GIARDINO I and BROWNLEE M (2000), 'Normalizing mitochondrial superoxide production blocks three pathways of hyperglycaemic damage', *Nature*, 404 (6779), 787–790.
18. FERRARI C K (2004), 'Functional foods, herbs and nutraceuticals: towards biochemical mechanisms of healthy aging', *Biogerontology*, 5 (5), 275–289.
19. VITAGLIONE P, MORISCO F, CAPORASO N and FOGLIANO V (2004), 'Dietary antioxidant compounds and liver health', *Critical Reviews in Food Science and Nutrition*, 44 (7–8), 575–586.
20. HIGUERA-CIAPARA I, FELIX-VALENZUELA L and GOYCOOLEA F M (2006), 'Astaxanthin: a review of its chemistry and applications', *Critical Reviews in Food Science and Nutrition*, 46 (2), 185–196.
21. CUTLER R G (2005), 'Oxidative stress profiling: Part I. Its potential importance in the optimization of human health', *Annals of the New York Academy of Sciences*, 1055, 93–135.
22. CHISOLM G M and STEINBERG D (2000), 'The oxidative modification hypothesis of atherogenesis: an overview', *Free Radical Biol Med*, 28 (12), 1815–1826.
23. YANG T T C, DEVARAJ S and JIALAL I (2001), 'Oxidative stress and atherosclerosis', *Journal of Clinical Ligand Assay*, 24 (1), 13–24.
24. ALBERTINI R, MORATTI R and DE LUCA G (2002), 'Oxidation of low-density lipoprotein in atherosclerosis from basic biochemistry to clinical studies', *Current Molecular Medicine*, 2 (6), 579–592.
25. ALLEN R G and TRESINI M (2000), 'Oxidative stress and gene regulation', *Free Radical Biol Med*, 28 (3), 463–499.
26. MACHLIN L J and BENDICH A (1987), 'Free radical tissue damage: protective role of antioxidant nutrients', *FASEB J*, 1 (6), 441–445.
27. WILLCOX J K, ASH S L and CATIGNANI G L (2004), 'Antioxidants and prevention of chronic disease', *Critical Reviews in Food Science and Nutrition*, 44 (4), 275–295.
28. NOHL H and GILLE L (2005), 'Lysosomal ROS formation', *Redox Report*, 10 (4), 199–205.
29. BABIOR B M (1997), 'Superoxide: a two-edged sword', *Brazilian Journal of Medical and Biological Research*, 30 (2), 141–155.
30. TURRENS J F (2003), 'Mitochondrial formation of reactive oxygen species', *The Journal of Physiology*, 552 (Pt 2), 335–344.
31. DOTAN Y, LICHTENBERG D and PINCHUK I (2004), 'Lipid peroxidation cannot be used as a universal criterion of oxidative stress', *Progress in Lipid Research*, 43 (3), 200–227.
32. SAGIN F G and SOZMEN E Y (2004), 'Anti-inflammatory effects of dietary antioxidants', *Current Medicinal Chemistry – Anti-Inflammatory and Anti-Allergy Agents*, 3 (1), 19–30.
33. LACHANCE P A, NAKAT Z and JEONG W S (2001), 'Antioxidants: an integrative approach', *Nutr*, 17 (10), 835–838.

34. BRIGELIUS-FLOHE R and TRABER M G (1999), 'Vitamin E: function and metabolism', *FASEB J*, 13 (10), 1145–1155.
35. WAGNER K H, ISNARDY B and ELMADFA I (2004), 'Gamma- and delta-tocopherols are more effective than alpha-tocopherol on the autoxidation of a 10% rapeseed oil triacylglycerol-in-water emulsion with and without a radical initiator', *European Journal of Lipid Science and Technology*, 106 (1), 44–51.
36. WAGNER K H, KAMAL-ELDIN A and ELMADFA I (2004), 'Gamma-tocopherol – an underestimated vitamin?' *Annals of Nutrition & Metabolism*, 48 (3), 169–188.
37. JIANG Q, CHRISTEN S, SHIGENAGA M K and AMES B N (2001), 'Gamma-tocopherol, the major form of vitamin E in the US diet, deserves more attention', *The American Journal of Clinical Nutrition*, 74 (6), 714–722.
38. LEONARD S W, PATERSON E, ATKINSON J K, RAMAKRISHNAN R, CROSS C E and TRABER M G (2005), 'Studies in humans using deuterium-labeled alpha- and gamma-tocopherols demonstrate faster plasma gamma-tocopherol disappearance and greater gamma-metabolite production', *Free Radical Biol Med*, 38 (7), 857–866.
39. BERMAN K and BRODATY H (2004), 'Tocopherol (vitamin E) in Alzheimer's disease and other neurodegenerative disorders', *CNS Drugs*, 18 (12), 807–825.
40. COHN W (1997), 'Bioavailability of vitamin E', *European Journal of Clinical Nutrition*, 51, S80–S85.
41. WANG X and QUINN P J (1999), 'Vitamin E and its function in membranes', *Progress in Lipid Research*, 38 (4), 309–336.
42. KAUL N, DEVARAJ S and JIALAL I (2001), 'Alpha-tocopherol and atherosclerosis', *Experimental Biology and Medicine*, 226 (1), 5–12.
43. MEYDANI S N, WU D, SANTOS M S and HAYEK M G (1995), 'Antioxidants and immune response in aged persons: overview of present evidence', *The American Journal of Clinical Nutrition*, 62 (6 Suppl), 1462S–1476S.
44. HARRIS A, DEVARAJ S and JIALAL I (2002), 'Oxidative stress, alpha-tocopherol therapy, and atherosclerosis', *Current Atherosclerosis Reports*, 4 (5), 373–380.
45. UPSTON J M, KRITHARIDES L and STOCKER R (2003), 'The role of vitamin E in atherosclerosis', *Progress in Lipid Research*, 42 (5), 405–422.
46. BRAMLEY P M, ELMADFA A, KAFATOS F J, KELLY Y, MANIOS H E, ROXBOROUGH W, SCHUCH P J, SHEEHY A and WAGNER K H (2000), 'Review vitamin E', *Journal of the Science of Food and Agricultural*, 80, 913–938.
47. WAYNER D D M, BURTON G W, INGOLD K U, BARCLAY L R C and LOCKE S J (1987), 'The relative contribution of vitamin E, urate, ascorbate and proteins to the total peroxyl radical-trapping antioxidant activity of human blood plasma', *Biochimica et Biophysica Acta (BBA) – General Subjects*, 924 (3), 408–419.
48. THOMAS S R and STOCKER R (2000), 'Molecular action of vitamin E in lipoprotein oxidation: implications for atherosclerosis', *Free Radical Biol Med*, 28 (12), 1795–1805.
49. JIALAL I and DEVARAJ S (2005), 'Scientific evidence to support a vitamin E and heart disease health claim: research needs', *The Journal of Nutrition*, 135 (2), 348–353.
50. CRANE F L (2001), 'Biochemical functions of coenzyme Q10', *Journal of the American College of Nutrition*, 20 (6), 591–598.
51. QUINN P J, FABISIAK J P and KAGAN V E (1999), 'Expansion of antioxidant function of vitamin E by coenzyme Q', *Biofactors*, 9 (2–4), 149–154.
52. KAIKKONEN J, TUOMAINEN T P, NYYSSONEN K and SALONEN J T (2002), 'Coenzyme Q10: absorption, antioxidative properties, determinants, and plasma levels', *Free Radical Research*, 36 (4), 389–397.

53. NOHL H, STANIEK K, KOZLOV A V and GILLE L (2003), 'The biomolecule ubiquinone exerts a variety of biological functions', *Biofactors*, 18 (1–4), 23–31.
54. BHAGAVAN H N and CHOPRA R K (2006), 'Coenzyme Q10: absorption, tissue uptake, metabolism and pharmacokinetics', *Free Radical Research*, 40 (5), 445–453.
55. LAAKSONEN R, RIIHIMAKI A, LAITILA J, MARTENSSON K, TIKKANEN M J and HIMBERG J J (1995), 'Serum and muscle tissue ubiquinone levels in healthy subjects', *The Journal of Laboratory and Clinical Medicine*, 125 (4), 517–521.
56. ZITA C, OVERVAD K, MORTENSEN S A, SINDBERG C D, MOESGAARD S and HUNTER D A (2003), 'Serum coenzyme Q10 concentrations in healthy men supplemented with 30 mg or 100 mg coenzyme Q10 for two months in a randomised controlled study', *Biofactors*, 18 (1–4), 185–193.
57. TURUNEN M, SINDELAR P and DALLNER G (1999), 'Induction of endogenous coenzyme Q biosynthesis by administration of peroxisomal inducers', *Biofactors*, 9 (2–4), 131–139.
58. OVERVAD K, DIAMANT B, HOLM L, HOLMER G, MORTENSEN S A and STENDER S (1999), 'Coenzyme Q10 in health and disease', *European Journal of Clinical Nutrition*, 53 (10), 764–770.
59. SARTER B (2002), 'Coenzyme Q10 and cardiovascular disease: a review', *The Journal of Cardiovascular Nursing*, 16 (4), 9–20.
60. LANGSJOEN P H and LANGSJOEN A M (1999), 'Overview of the use of CoQ10 in cardiovascular disease', *Biofactors*, 9 (2–4), 273–284.
61. BEAL M F (2002), 'Coenzyme Q10 as a possible treatment for neurodegenerative diseases', *Free Radical Research*, 36 (4), 455–460.
62. GALATI G and O'BRIEN P J (2003), 'Cytoprotective and anticancer properties of coenzyme Q versus capsaicin', *Biofactors*, 18 (1–4), 195–205.
63. CONKLIN K A (2005), 'Coenzyme Q10 for prevention of anthracycline-induced cardiotoxicity', *Integrative Cancer Therapies*, 4 (2), 110–130.
64. MORTENSEN S A (2003), 'Overview on coenzyme Q10 as adjunctive therapy in chronic heart failure. Rationale, design and end-points of "Q-symbio" – a multinational trial', *Biofactors*, 18 (1–4), 79–89.
65. GENOVA M L, PICH M M, BIONDI A, BERNACCHIA A, FALASCA A, BOVINA C, FORMIGGINI G, PARENTI CASTELLI G and LENAZ G (2003), 'Mitochondrial production of oxygen radical species and the role of Coenzyme Q as an antioxidant', *Experimental Biology and Medicine*, 228 (5), 506–513.
66. PICH M M, CASTAGNOLI A, BIONDI A, BERNACCHIA A, TAZZARI P L, D'AURELIO M, CASTELLI G P, FORMIGGINI G, CONTE R, BOVINA C and LENAZ G (2002), 'Ubiquinol and a Coenzyme Q reducing system protect platelet mitochondrial function of transfusional buffy coats from oxidative stress', *Free Radical Res*, 36 (4), 429–436.
67. SANDOR P S, DI CLEMENTE L, COPPOLA G, SAENGER U, FUMAL A, MAGIS D, SEIDEL L, AGOSTI R M and SCHOENEN J (2005), 'Efficacy of coenzyme Q10 in migraine prophylaxis: a randomized controlled trial', *Neurology*, 64 (4), 713–715.
68. MILES M V, HORN P, MILES L, TANG P, STEELE P and DEGRAUW T (2002), 'Bio-equivalence of coenzyme Q(10) from over-the-counter supplements', *Nutrition Research*, 22 (8), 919–929.
69. ZAGHLOUL A A, GURLEY B, KHAN M, BHAGAVAN H, CHOPRA R and REDDY I (2002), 'Bioavailability assessment of oral coenzyme Q10 formulations in dogs', *Drug Development and Industrial Pharmacy*, 28 (10), 1195–1200.
70. TIKHAZE A K, KONOVALOVA G G, LANKIN V Z, KAMINNYI A I, KAMINNAJA V I, RUUGE E K and KUKHARCHUK V V (2005), 'Effect of ubiquinone Q(10) and antioxidant vitamins

on free radical oxidation of phospholipids in biological membranes of rat liver', *Bulletin of Experimental Biology and Medicine*, 140 (2), 181–183.
71. HUGHES D A (2001), 'Dietary carotenoids and human immune function', *Nutrition*, 17 (10), 823–827.
72. YOUNG A J and LOWE G M (2001), 'Antioxidant and prooxidant properties of carotenoids', *Archives of Biochemistry and Biophysics*, 385 (1), 20–27.
73. COOKE M S, EVANS M D, MISTRY N and LUNEC J (2002), 'Role of dietary antioxidants in the prevention of *in vivo* oxidative DNA damage', *Nutrition Research Reviews*, 15 (1), 19–41.
74. ZHENG W and WANG S Y (2001), 'Antioxidant activity and phenolic compounds in selected herbs', *J Agric Food Chem*, 49 (11), 5165–5170.
75. DI CARLO G, MASCOLO N, IZZO A A and CAPASSO F (1999), 'Flavonoids: old and new aspects of a class of natural therapeutic drugs', *Life Sciences*, 65 (4), 337–353.
76. TAPIERO H, TEW K D, BA G N and MATHE G (2002), 'Polyphenols: do they play a role in the prevention of human pathologies?' *Biomedicine & Pharmacotherapy*, 56 (4), 200–207.
77. PIETTA P G (2000), 'Flavonoids as antioxidants', *Journal of Natural Products*, 63 (7), 1035–1042.
78. VAN ACKER S A, VAN DEN BERG D J, TROMP M N, GRIFFIOEN D H, VAN BENNEKOM W P, VAN DER VIJGH W J and BAST A (1996), 'Structural aspects of antioxidant activity of flavonoids', *Free Radical Biol Med*, 20 (3), 331–342.
79. HOLLMAN P C H (2004), 'Absorption, bioavailability, and metabolism of flavonoids', *Pharmaceutical Biology*, 42, 74–83.
80. COTELLE N, BERNIER J L, CATTEAU J P, POMMERY J, WALLET J C and GAYDOU E M (1996), 'Antioxidant properties of hydroxy-flavones', *Free Radical Biol Med*, 20 (1), 35–43.
81. CROFT K D (1998), 'The chemistry and biological effects of flavonoids and phenolic acids', *Annals of the New York Academy of Sciences*, 854, 435–442.
82. KARAKAYA S (2004), 'Bioavailability of phenolic compounds', *Critical Reviews in Food Science and Nutrition*, 44 (6), 453–464.
83. SILVA M M, SANTOS M R, CAROCO G, ROCHA R, JUSTINO G and MIRA L (2002), 'Structure-antioxidant activity relationships of flavonoids: a re-examination', *Free Radical Res*, 36 (11), 1219–1227.
84. GALATI G and O'BRIEN P J (2004), 'Potential toxicity of flavonoids and other dietary phenolics: significance for their chemopreventive and anticancer properties', *Free Radical Biol Med*, 37 (3), 287–303.
85. NEUHOUSER M L (2004), 'Flavonoids and cancer prevention: what is the evidence in humans?' *Pharmaceutical Biology*, 42, 36–45.
86. KAUR C and KAPOOR H C (2001), 'Antioxidants in fruits and vegetables – the millennium's health', *International Journal of Food Science and Technology*, 36 (7), 703–725.
87. KONDRATYUK T P and PEZZUTO J M (2004), 'Natural product polyphenols of relevance to human health', *Pharmaceutical Biology*, 42, 46–63.
88. JEONG W S and KONG A N T (2004), 'Biological properties of monomeric and polymeric catechins: green tea catechins and procyanidins', *Pharmaceutical Biology*, 42, 84–93.
89. CARR A C and FREI B (1999), 'Toward a new recommended dietary allowance for vitamin C based on antioxidant and health effects in humans', *The American Journal of Clinical Nutrition*, 69 (6), 1086–1107.

90. PADAYATTY S J, KATZ A, WANG Y, ECK P, KWON O, LEE J H, CHEN S, CORPE C, DUTTA A, DUTTA S K and LEVINE M (2003), 'Vitamin C as an antioxidant: evaluation of its role in disease prevention', *Journal of the American College of Nutrition*, 22 (1), 18–35.
91. COULTER I D, HARDY M L, MORTON S C, HILTON L G, TU W, VALENTINE D and SHEKELLE P G (2006), 'Antioxidants vitamin C and vitamin E for the prevention and treatment of cancer', *Journal of General Internal Medicine*, 21 (7), 735–744.
92. BOHM F, EDGE R, McGARVEY D J and TRUSCOTT T G (1998), 'Beta-carotene with vitamins E and C offers synergistic cell protection against NOx', *FEBS Letters*, 436 (3), 387–389.
93. BAUER K (2005), 'Carnosine and homocarnosine, the forgotten, enigmatic peptides of the brain', *Neurochemical Research*, 30 (10), 1339–1345.
94. FONTANA M, PINNEN F, LUCENTE G and PECCI L (2002), 'Prevention of peroxynitrite-dependent damage by carnosine and related sulphonamido pseudodipeptides', *Cellular and Molecular Life Sciences*, 59 (3), 546–551.
95. HIPKISS A R (1998), 'Carnosine, a protective, anti-ageing peptide?' *The International Journal of Biochemistry & Cell Biology*, 30 (8), 863–868.
96. HIPKISS A R, BROWNSON C and CARRIER M J (2001), 'Carnosine, the anti-ageing, anti-oxidant dipeptide, may react with protein carbonyl groups', *Mechanisms of Ageing and Development*, 122 (13), 1431–1445.
97. BARAN E J (2000), 'Metal complexes of carnosine', *Biochemistry (Moscow)*, 65 (7), 789–797.
98. BOLDYREV A A (2005), 'Protection of proteins from oxidative stress: a new illusion or a novel strategy?' *Annals of the New York Academy of Sciences*, 1057, 193–205.
99. HIPKISS A R and BROWNSON C (2000), 'Carnosine reacts with protein carbonyl groups: another possible role for the anti-ageing peptide?' *Biogerontology*, 1 (3), 217–223.
100. HIPKISS A R, PRESTON J E, HIMSWORTH D T, WORTHINGTON V C, KEOWN M, MICHAELIS J, LAWRENCE J, MATEEN A, ALLENDE L, EAGLES P A and ABBOTT N J (1998), 'Pluripotent protective effects of carnosine, a naturally occurring dipeptide', *Annals of the New York Academy of Sciences*, 854, 37–53.
101. HIPKISS A R (2005), 'Glycation, ageing and carnosine: are carnivorous diets beneficial?' *Mechanisms of Ageing and Development*, 126 (10), 1034–1039.
102. GUIOTTO A, CALDERAN A, RUZZA P and BORIN G (2005), 'Carnosine and carnosine-related antioxidants: a review', *Current Medicinal Chemistry – Anti-Inflammatory and Anti-Allergy Agents*, 12 (20), 2293–2315.
103. NAGAI K, NIIJIMA A, YAMANO T, OTANI H, OKUMRA N, TSURUOKA N, NAKAI M and KISO Y (2003), 'Possible role of L-carnosine in the regulation of blood glucose through controlling autonomic nerves', *Experimental Biology and Medicine*, 228 (10), 1138–1145.
104. HIPKISS A R (2005), 'Could carnosine suppress zinc-mediated proteasome inhibition and neurodegeneration? Therapeutic potential of a non-toxic but non-patentable dipeptide', *Biogerontology*, 6 (2), 147–149.
105. NIKI E and NOGUCHI N (2004), 'Dynamics of antioxidant action of vitamin E', *Accounts of Chemical Research*, 37 (1), 45–51.
106. KWON S S, NAM Y S, LEE J S, KU B S, HAN S H, LEE J Y and CHANG I S (2002), 'Preparation and characterization of coenzyme Q(10)-loaded PMMA nanoparticles by a new emulsification process based on microfluidization', *Colloids and Surfaces A – Physicochemical and Engineering Aspects*, 210 (1), 95–104.
107. CHEN L and SUBIRADE M (2005), 'Chitosan/beta-lactoglobulin core-shell nano-

particles as nutraceutical carriers', *Biomaterials*, 26 (30), 6041–6053.
108. RUSSELL-JONES G J (2004), 'Use of targeting agents to increase uptake and localization of drugs to the intestinal epithelium', *Journal of Drug Targeting*, 12 (2), 113–123.
109. NA K, BUM LEE T, PARK K H, SHIN E K, LEE Y B and CHOI H K (2003), 'Self-assembled nanoparticles of hydrophobically-modified polysaccharide bearing vitamin H as a targeted anti-cancer drug delivery system', *Euopean Journal of Pharmaceutical Sciences*, 18 (2), 165–173.
110. MURO S, CUI X, GAJEWSKI C, MURCIANO J C, MUZYKANTOV V R and KOVAL M (2003), 'Slow intracellular trafficking of catalase nanoparticles targeted to ICAM-1 protects endothelial cells from oxidative stress', *American Journal of Physiology – Cell Physiology*, 285 (5), C1339–C1347.
111. SHEA T B, ORTIZ D, NICOLOSI R J, KUMAR R and WATTERSON A C (2005), 'Nanosphere-mediated delivery of vitamin E increases its efficacy against oxidative stress resulting from exposure to amyloid beta', *Journal of Alzheimer's Disease*, 7 (4), 297–301.
112. DUCLAIROIR C, ORECCHIONI A M, DEPRAETERE P and NAKACHE E (2002), 'Alpha-tocopherol encapsulation and *in vitro* release from wheat gliadin nanoparticles', *J Microencapsul*, 19 (1), 53–60.
113. ZIGONEANU I G, ASTETE C E, MONROE W T and SABLIOV C M (2006), 'Sustained release and antioxidant activity of alpha-tocopherol entrapped in Poly(D,L-lactide-*co*-glycolide) nanoparticles', IFT Annual Meeting. Orlando, FL.
114. CHEN C C and WAGNER G (2004), 'Vitamin E nanoparticle for beverage applications', 82 (A11), 1432–1437.
115. ANDERSON J M and SHIVE M S (1997), 'Biodegradation and biocompatibility of PLA and PLGA microspheres', *Advanced Drug Delivery Reviews*, 28 (1), 5–24.
116. WESTESEN K, BUNJES H, HAMMER G and SIEKMANN B (2001), 'Novel colloidal drug delivery systems', *PDA Journal of Pharmaceutical Science and Technology/PDA*, 55 (4), 240–247.
117. BRIGGER I, DUBERNET C and COUVREUR P (2002), 'Nanoparticles in cancer therapy and diagnosis', *Advanced Drug Delivery Reviews*, 54 (5), 631–651.
118. CHORNY M, FISHBEIN I, DANENBERG H D and GOLOMB G (2002), 'Lipophilic drug loaded nanospheres prepared by nanoprecipitation: effect of formulation variables on size, drug recovery and release kinetics', *Journal Controlled Release*, 83 (3), 389–400.
119. BARRAT G, COURRAZE G, COUVREUR C, FATTAL E, GREF R, D. L, LEGRAND P, PONCHEL G and VAUTHIER C (2000), 'Polymeric micro- and nanoparticles as drug carriers', in Dumitriu S, *Polymeric Biomaterials*, New York, Marcel Dekker, 753–781.
120. NAKACHE E, POULAIN N, CANDAU F, ORECCHIONI A M and IRACHE J M (2000), 'Biopolymer and polymer nanoparticles and their biomedical applications', in Nalwa H S, *Handbook of Nanostructured Materials and Nanotechnology*, New York, Academic Press, 577–635.
121. HANS M L and LOWMAN A M (2002), 'Biodegradable nanoparticles for drug delivery and targeting', *Current Opinion in Solid State and Materials Science*, 6 (4), 319–327.
122. SEKHARA RAO G C, KUMAR M S, MATHIVANAN N and BHANOJI RAO M E (2004), 'Nano-suspensions as the most promising approach in nanoparticulate drug delivery systems', *Phamazie*, 59 (1), 5–9.
123. ASTETE C E and SABLIOV C M (2006), 'Synthesis and characterization of PLGA nanoparticles', *Journal of Biomaterial Sciences – Polymer Edition*, 17 (3), 247–289.

124. ALLEMANN E, GURNY R and DOELKER E (1993), 'Drug-loaded nanoparticles – preparation methods and drug targeting issues', *European Journal of Pharmceutics and Biopharmaceutics*, 39 (5), 173–191.
125. HIREMATH S R R and HOTA A (1999), 'Nanoparticles as drug delivery systems', *Indian Journal of Pharmaceutical Sciences*, 61 (2), 69–75.
126. QUINTANAR-GUERRERO D, ALLEMANN E, FESSI H and DOELKER E (1998), 'Preparation techniques and mechanisms of formation of biodegradable nanoparticles from pre-formed polymers', *Drug Development and Industrial Pharmacy*, 24 (12), 1113–1128.
127. MOGHIMI S M, HUNTER A C and MURRAY J C (2001), 'Long-circulating and target-specific nanoparticles: theory to practice', *Pharmacological Reviews*, 53 (2), 283–318.
128. PANYAM J and LABHASETWAR V (2003), 'Biodegradable nanoparticles for drug and gene delivery to cells and tissue', *Advanced Drug Delivery Reviews*, 55 (3), 329–347.
129. AVGOUSTAKIS K (2004), 'Pegylated poly(lactide) and poly(lactide-*co*-glycolide) nanoparticles: preparation, properties and possible applications in drug delivery', *Curr Drug Deliv*, 1 (4), 321–333.
130. BALA I, HARIHARAN S and KUMAR M N (2004), 'PLGA nanoparticles in drug delivery: the state of the art', *Critical Reviews in Therapeutic Drug Carrier Systems*, 21 (5), 387–422.
131. ALEXIS F (2005), 'Factors affecting the degradation and drug-release mechanism of poly(lactic acid) and poly[(lactic acid)-*co*-(glycolic acid)]', *Polym Int*, 54 (1), 36–46.
132. QUINTANAR-GUERRERO D, ALLEMANN E, DOELKER E and FESSI H (1997), 'A mechanistic study of the formation of polymer nanoparticles by the emulsification-diffusion technique', *Colloid Polym Sci*, 275 (7), 640–647.
133. KONAN Y N, CERNY R, FAVET J, BERTON M, GURNY R and ALLEMANN E (2003), 'Preparation and characterization of sterile sub-200 nm meso-tetra(4-hydroxylphenyl)porphyrin-loaded nanoparticles for photodynamic therapy', *European Journal of Pharmceutics and Biopharmaceutics*, 55 (1), 115–124.
134. LEE S J, JEONG J R, SHIN S C, KIM J C, CHANG Y H, LEE K H and KIM J D (2005), 'Magnetic enhancement of iron oxide nanoparticles encapsulated with poly(D,L-latide-*co*-glycolide)', *Colloids and Surfaces A – Physicochemical and Engineering Aspects*, 255 (1–3), 19–25.
135. KUMAR R M N, BAKOWSKY U and LEHR C M (2004), 'Preparation and characterization of cationic PLGA nanospheres as DNA carriers', *Biomaterials*, 25 (10), 1771–1777.
136. KUMAR R M N, MOHAPATRA S S, KONG X, JENA P K, BAKOWSKY U and LEHR C M (2004), 'Cationic poly(lactide-*co*-glycolide) nanoparticles as efficient *in vivo* gene transfection agents', *Journal of Nanoscience and Nanotechnology*, 4 (8), 990–994.
137. AHLIN P, KRISTL J, KRISTL A and VRECER F (2002), 'Investigation of polymeric nanoparticles as carriers of enalaprilat for oral administration', *International Journal of Pharmaceutics*, 239 (1–2), 113–120.
138. KWON H Y, LEE J Y, CHOI S W, JANG Y S and KIM J H (2001), 'Preparation of PLGA nanoparticles containing estrogen by emulsification-diffusion method', *Colloids and Surfaces A – Physicochemical and Engineering Aspects*, 182 (1–3), 123–130.
139. KONAN Y N, GURNY R and ALLEMANN E (2002), 'Preparation and characterization of sterile and freeze-dried sub-200 nm nanoparticles', *International Journal of Pharmaceutics*, 233 (1–2), 239–252.
140. ZWEERS M L, ENGBERS G H, GRIJPMA D W and FEIJEN J (2004), '*In vitro* degradation of nanoparticles prepared from polymers based on DL-lactide, glycolide and poly(ethylene oxide)', *Journal Controlled Release*, 100 (3), 347–356.

141. ELEY J G, PUJARI V D and McLANE J (2004), 'Poly(lactide-*co*-glycolide) nanoparticles containing coumarin-6 for suppository delivery: *in vitro* release profile and *in vivo* tissue distribution', *Drug Delivery*, 11 (4), 255–261.
142. PRAKOBVAITAYAKIT M and NIMMANNIT U (2003), 'Optimization of polylactic-*co*-glycolic acid nanoparticles containing itraconazole using 2(3) factorial design', *AAPS PharmSciTech*, 4 (4), 1–9.
143. GOVENDER T, STOLNIK S, GARNETT M C, ILLUM L and DAVIS S S (1999), 'PLGA nanoparticles prepared by nanoprecipitation: drug loading and release studies of a water soluble drug', *Journal Controlled Release*, 57 (2), 171–185.
144. CSABA N, CAAMANO P, SANCHEZ A, DOMINGUEZ F and ALONSO M J (2005), 'PLGA: poloxamer and PLGA:poloxamine blend nanoparticles: new carriers for gene delivery', *Biomacromolecules*, 6 (1), 271–278.
145. NIWA T, TAKEUCHI H, HINO T, KUNOU N and KAWASHIMA Y (1993), 'Preparations of biodegradable nanospheres of water-soluble and insoluble drugs with D,L-lactide glycolide copolymer by a novel spontaneous emulsification solvent diffusion method, and the drug release behavior', *Journal Controlled Release*, 25 (1–2), 89–98.
146. AMELLER T, MARSAUD V, LEGRAND P, GREF R and RENOIR J M (2004), 'Pure anti-estrogen RU 58668-loaded nanospheres: morphology, cell activity and toxicity studies', *Euopean Journal of Pharmaceutical Sciences*, 21 (2–3), 361–370.
147. OSTER C G, WITTMAR M, UNGER F, BARBU-TUDORAN L, SCHAPER A K and KISSEL T (2004), 'Design of amine-modified graft polyesters for effective gene delivery using DNA-loaded nanoparticles', *Pharm Res*, 21 (6), 927–931.
148. SAXENA V, SADOQI M and SHAO J (2004), 'Indocyanine green-loaded biodegradable nanoparticles: preparation, physicochemical characterization and *in vitro* release', *International Journal of Pharmaceutics*, 278 (2), 293–301.
149. CSABA N, GONZALEZ L, SANCHEZ A and ALONSO M J (2004), 'Design and characterisation of new nanoparticulate polymer blends for drug delivery', *Journal of Biomaterials Science – Polymer Edition*, 15 (9), 1137–1351.
150. PANAGI Z, BELETSI A, EVANGELATOS G, LIVANIOU E, ITHAKISSIOS D S and AVGOUSTAKIS K (2001), 'Effect of dose on the biodistribution and pharmacokinetics of PLGA and PLGA-mPEG nanoparticles', *International Journal of Pharmaceutics*, 221 (1–2), 143–152.
151. DE JAEGHERE F, DOELKER E and GURNY R (1999), 'Nanoparticles', in Mathiowitz E, *Encyclopedia of Control Drug Delivery*, New York, John Wiley & Sons, 641–664.
152. ALEX R and BODMEIER R (1990), 'Encapsulation of water-soluble drugs by a modified solvent evaporation method. I. Effect of process and formulation variables on drug entrapment', *J Microencapsul* 7(3), 347–355.
153. TAKADA S, YAMAGATA Y, MISAKI M, TAIRA K and KUROKAWA T (2003), 'Sustained release of human growth hormone from microcapsules prepared by a solvent evaporation technique', *Journal Controlled Release*, 88 (2), 229–242.
154. OBEIDAT W M and PRICE J C (2003), 'Viscosity of polymer solution phase and other factors controlling the dissolution of theophylline microspheres prepared by the emulsion solvent evaporation method', *J Microencapsul*, 20 (1), 57–65.
155. PARK S J and KIM S H (2004), 'Preparation and characterization of biodegradable poly(L-lactide)/poly(ethylene glycol) microcapsules containing erythromycin by emulsion solvent evaporation technique', *J Colloid Interf Sci*, 271 (2), 336–341.
156. MOHAMMED F A and HASSAN M A (2003), 'Formulation and evaluation of ciprofloxacin hydrochloride and norfloxacin microspheres prepared by an enhanced emulsion-solvent evaporation process', *Stp Pharma Sci*, 13 (5), 319–327.

157. LOPEZ-QUINTELA M A (2003), 'Synthesis of nanomaterials in microemulsions: formation mechanisms and growth control', *Curr Opin Colloid In*, 8 (2), 137–144.
158. BOUCHEMAL K, BRIANCON S, PERRIER E and FESSI H (2004), 'Nano-emulsion formulation using spontaneous emulsification: solvent, oil and surfactant optimisation', *International Journal of Pharmaceutics*, 280 (1–2), 241–251.
159. SCHMITT V, LEAL-CALDERON F and BIBETTE J (2003), 'Preparation of monodisperse particles and emulsions by controlled shear', *Topics in Current Chemistry*, 227, 195–215.
160. ZHAO X and GOVEAS J L (2001), 'Size selection in viscoelastic emulsions under shear', *Langmuir*, 17 (13), 3788–3791.
161. PANYAM J, WILLIAMS D, DASH A, LESLIE-PELECKY D and LABHASETWAR V (2004), 'Solid-state solubility influences encapsulation and release of hydrophobic drugs from PLGA/PLA nanoparticles', *Journal of Pharmaceutical Sciences*, 93 (7), 1804–1814.
162. PIETZONKA P, WALTER E, DUDA-JOHNER S, LANGGUTH P and MERKLE H P (2002), 'Compromised integrity of excised porcine intestinal epithelium obtained from the abattoir affects the outcome of *in vitro* particle uptake studies', *European Journal of Pharmaceutical Sciences*, 15 (1), 39–47.
163. JULIENNE M C, ALONSO M J, AMOZA J L G and BENOIT J P (1992), 'Preparation of poly(D,L-lactide glycolide) nanoparticles of controlled particle-size distribution – application of experimental designs', *Drug Development and Industrial Pharmacy*, 18 (10), 1063–1077.
164. CHUNG T W, HUANG Y Y, TSAI Y L and LIU Y Z (2002), 'Effects of solvent evaporation rate on the properties of protein-loaded PLLA and PDLLA microspheres fabricated by emulsion-solvent evaporation process', *J Microencapsul*, 19 (4), 463–471.
165. SONG C X, LABHASETWAR V, MURPHY H, QU X, HUMPHREY W R, SHEBUSKI R J and LEVY R J (1997), 'Formulation and characterization of biodegradable nanoparticles for intravascular local drug delivery', *Journal Controlled Release*, 43 (2–3), 197–212.
166. PAYS K, GIERMANSKA-KAHN J, POULIGNY B, BIBETTE J and LEAL-CALDERON F (2001), 'Coalescence in surfactant-stabilized double emulsions', *Langmuir*, 17 (25), 7758–7769.
167. PAYS K, GIERMANSKA-KAHN J, POULIGNY B, BIBETTE J and LEAL-CALDERON F (2002), 'Double emulsions: how does release occur?' *Journal Controlled Release*, 79 (1–3), 193–205.
168. FICHEUX M F, BONAKDAR L, LEAL-CALDERON F and BIBETTE J (1998), 'Some stability criteria for double emulsions', *Langmuir*, 14 (10), 2702–2706.
169. DOMB A J, BERGELSON L and AMSELEM S (1996), 'Lipospheres for controlled delivery of substances', in Benita S, *Microencapsulation: Methods and Industrial Applications*, New York, Marcel Dekker, 377–410.
170. AUKUNURU J V, AYALASOMAYAJULA S P and KOMPELLA U B (2003), 'Nanoparticle formulation enhances the delivery and activity of a vascular endothelial growth factor antisense oligonucleotide in human retinal pigment epithelial cells', *The Journal of Pharmacy and Pharmacology*, 55 (9), 1199–1206.
171. VANDERVOORT J and LUDWIG A (2002), 'Biocompatible stabilizers in the preparation of PLGA nanoparticles: a factorial design study', *International Journal of Pharmaceutics*, 238 (1–2), 77–92.
172. DILLEN K, WEYENBERG W, VANDERVOORT J and LUDWIG A (2004), 'The influence of the use of viscosifying agents as dispersion media on the drug release properties from PLGA nanoparticles', *European Journal of Pharmceutics and Biopharma-*

ceutics, 58 (3), 539–549.

173. PRABHA S and LABHASETWAR V (2004), 'Critical determinants in PLGA/PLA nanoparticle-mediated gene expression', *Pharm Res*, 21 (2), 354–364.
174. VANDERVOORT J, YONCHEVA K and LUDWIG A (2004), 'Influence of the homogenisation procedure on the physicochemical properties of PLGA nanoparticles', *Chemical & Pharmaceutical Bulletin*, 52 (11), 1273–1279.
175. SAHOO S K, PANYAM J, PRABHA S and LABHASETWAR V (2002), 'Residual polyvinyl alcohol associated with poly (D,L-lactide-*co*-glycolide) nanoparticles affects their physical properties and cellular uptake', *Journal Controlled Release*, 82 (1), 105–114.
176. DILLEN K, VANDERVOORT J, VAN DEN MOOTER G, VERHEYDEN L and LUDWIG A (2004), 'Factorial design, physicochemical characterisation and activity of ciprofloxacin-PLGA nanoparticles', *International Journal of Pharmaceutics*, 275 (1–2), 171–187.
177. MU L and FENG S S (2003), 'A novel controlled release formulation for the anticancer drug paclitaxel (Taxol): PLGA nanoparticles containing vitamin E TPGS', *Journal Controlled Release*, 86 (1), 33–48.
178. JEON H J, JEONG Y I, JANG M K, PARK Y H and NAH J W (2000), 'Effect of solvent on the preparation of surfactant-free poly(DL-lactide-*co*-glycolide) nanoparticles and norfloxacin release characteristics', *International Journal of Pharmaceutics*, 207 (1–2), 99–108.
179. LEMARCHAND C, COUVREUR P, BESNARD M, COSTANTINI D and GREF R (2003), 'Novel polyester-polysaccharide nanoparticles', *Pharm Res*, 20 (8), 1284–1292.
180. LI Y, PEI Y, ZHANG X, GU Z, ZHOU Z, YUAN W, ZHOU J, ZHU J and GAO X (2001), 'PEGylated PLGA nanoparticles as protein carriers: synthesis, preparation and biodistribution in rats', *Journal Controlled Release*, 71 (2), 203–211.
181. BELETSI A, LEONTIADIS L, KLEPETSANIS P, ITHAKISSIOS D S and AVGOUSTAKIS K (1999), 'Effect of preparative variables on the properties of poly(DL-lactide-*co*-glycolide)-methoxypoly(ethyleneglycol) copolymers related to their application in controlled drug delivery', *International Journal of Pharmaceutics*, 182 (2), 187–197.
182. FONSECA C, SIMOES S and GASPAR R (2002), 'Paclitaxel-loaded PLGA nanoparticles: preparation, physicochemical characterization and *in vitro* anti-tumoral activity', *Journal Controlled Release*, 83 (2), 273–286.
183. LAMPRECHT A, UBRICH N, YAMAMOTO H, SCHAFER U, TAKEUCHI H, MAINCENT P, KAWASHIMA Y and LEHR C M (2001), 'Biodegradable nanoparticles for targeted drug delivery in treatment of inflammatory bowel disease', *The Journal of Pharmacology and Experimental Therapeutics*, 299 (2), 775–781.
184. LEE W K, PARK J Y, YANG E H, SUH H, KIM S H, CHUNG D S, CHOI K, YANG C W and PARK J S (2002), 'Investigation of the factors influencing the release rates of cyclosporin A-loaded micro- and nanoparticles prepared by high-pressure homogenizer', *Journal Controlled Release*, 84 (3), 115–123.
185. MAINARDES R M and EVANGELISTA R C (2005), 'PLGA nanoparticles containing praziquantel: effect of formulation variables on size distribution', *International Journal of Pharmaceutics*, 290 (1–2), 137–144.
186. WIN K Y and FENG S S (2005), 'Effects of particle size and surface coating on cellular uptake of polymeric nanoparticles for oral delivery of anticancer drugs', *Biomaterials*, 26 (15), 2713–2722.
187. ZHANG Y and ZHUO R X (2005), 'Synthesis and drug release behavior of poly (trimethylene carbonate)-poly (ethylene glycol)-poly (trimethylene carbonate) nanoparticles', *Biomaterials*, 26 (14), 2089–2094.

188. AVGOUSTAKIS K, BELETSI A, PANAGI Z, KLEPETSANIS P, KARYDAS A G and ITHAKISSIOS D S (2002), 'PLGA-mPEG nanoparticles of cisplatin: *in vitro* nanoparticle degradation, *in vitro* drug release and *in vivo* drug residence in blood properties', *Journal Controlled Release*, 79 (1–3), 123–135.
189. ALLEMANN E, LEROUX J C, GURNY R and DOELKER E (1993), '*In vitro* extended-release properties of drug-loaded poly(DL-lactic acid) nanoparticles produced by a salting-out procedure', *Pharm Res*, 10 (12), 1732–1737.
190. PRABHA S, ZHOU W Z, PANYAM J and LABHASETWAR V (2002), 'Size-dependency of nanoparticle-mediated gene transfection: studies with fractionated nanoparticles', *International Journal of Pharmaceutics*, 244 (1–2), 105–115.
191. PANYAM J, DALI M M, SAHOO S K, MA W, CHAKRAVARTHI S S, AMIDON G L, LEVY R J and LABHASETWAR V (2003), 'Polymer degradation and *in vitro* release of a model protein from poly(D,L-lactide-*co*-glycolide) nano- and microparticles', *Journal Controlled Release*, 92 (1–2), 173–187.
192. GOPFERICH A (1996), 'Polymer degradation and erosion: mechanisms and applications', *European Journal of Pharmceutics and Biopharmaceutics*, 42 (1), 1–11.
193. CHUNG J E, YOKOYAMA M and OKANO T (2000), 'Inner core segment design for drug delivery control of thermo-responsive polymeric micelles', *Journal Controlled Release*, 65 (1–2), 93–103.
194. CHUNG J E, YOKOYAMA M, YAMATO M, AOYAGI T, SAKURAI Y and OKANO T (1999), 'Thermo-responsive drug delivery from polymeric micelles constructed using block copolymers of poly(*N*-isopropylacrylamide) and poly(butylmethacrylate)', *Journal Controlled Release*, 62 (1–2), 115–127.
195. GIACOMELLI C, LE MEN L, BORSALI R, LAI-KEE-HIM J, BRISSON A, ARMES S P and LEWIS A L (2006), 'Phosphorylcholine-based pH-responsive diblock copolymer micelles as drug delivery vehicles: light scattering, electron microscopy, and fluorescence experiments', *Biomacromolecules*, 7 (3), 817–828.
196. GILLIES E R and FRECHET J M (2005), 'pH-Responsive copolymer assemblies for controlled release of doxorubicin', *Bioconjugate Chemistry*, 16 (2), 361–368.
197. KAMADA H, TSUTSUMI Y, YOSHIOKA Y, YAMAMOTO Y, KODAIRA H, TSUNODA S, OKAMOTO T, MUKAI Y, SHIBATA H, NAKAGAWA S and MAYUMI T (2004), 'Design of a pH-sensitive polymeric carrier for drug release and its application in cancer therapy', *Clinical Cancer Research*, 10 (7), 2545–2550.
198. XU P, VAN KIRK E A, MURDOCH W J, ZHAN Y, ISAAK D D, RADOSZ M and SHEN Y (2006), 'Anticancer efficacies of cisplatin-releasing pH-responsive nanoparticles', *Biomacromolecules*, 7 (3), 829–835.
199. REDHEAD H M, DAVIS S S and ILLUM L (2001), 'Drug delivery in poly(lactide-*co*-glycolide) nanoparticles surface modified with poloxamer 407 and poloxamine 908: *in vitro* characterisation and *in vivo* evaluation', *Journal Controlled Release*, 70 (3), 353–363.
200. YOO H S, LEE K H, OH J E and PARK T G (2000), '*In vitro* and *in vivo* anti-tumor activities of nanoparticles based on doxorubicin-PLGA conjugates', *Journal Controlled Release*, 68 (3), 419–431.

13

Encapsulation and controlled release of folic acid

K. Kailasapathy, University of Western Sydney, Australia

13.1 Introduction

Microencapsulation (ME) is an inclusion technique for confining a bioactive substance into a polymeric matrix coated by one or more semi-permeable polymers, by virtue of which the encapsulated compound becomes more stable and protected than its isolated or free form. Encapsulation of food bioactives can provide the final product with better technological and nutritional properties and, in addition, give controlled release of encapsulated microcomponents under specific conditions.

Encapsulation technology in food processing includes coating of minute particles or ingredients (e.g. flavours, colours, fats, acidulants) and other nutrients (vitamins, minerals and antioxidants) which may be accomplished by microencapsulation and macrocoating techniques (Kirby, 1991; Schrooyen *et al.*, 2001).

Folic acid is a member of the vitamin family and essential for the healthy functioning of a variety of physiological processes in humans. Chemically, the folates are a group of heterocyclic compounds and have a common basic structure, N-4[{2-amino-1, 4-dihydro 4-oxo-6 pteridinyl}-methyl}amino] benzoyl] glutamic acid, with or without additional L-glutamic acid residues conjugated via peptide linkages (Hawkes and Villota, 1989) (Fig. 13.1). There are many chemical derivatives of folic acid that exhibit a common vitamin activity as folates, and these have been well described (Scott, 1989; Wagner, 1985).

Interest in the health benefits of folic acid has increased considerably over the last 15 years. It has been known for some time that optimising blood folate levels around the time of conception and in the early weeks of pregnancy can

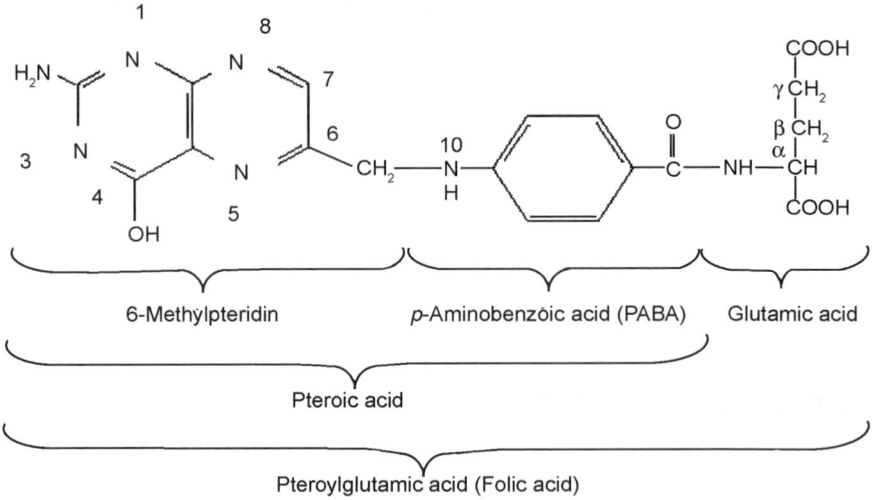

Fig. 13.1 Chemical structure of folic acid (Hawkes and Villota, 1989).

significantly reduce the chance of neural tube defects (NTD) during pregnancy (MRC Vitamin Study Group, 1991). Folates provide protection against coronary heart disease (Brouwer *et al.*, 1999), cognitive functions (Seshadri *et al.*, 2002) and prevention of certain forms of cancer (Giovannucci *et al.*, 1995). Elevated plasma homocysteine, a consequence of marginal folate deficiency, is emerging as an independent risk factor for several types of vascular stroke (Quere *et al.*, 2002; Vollset *et al.*, 2001) and neuro-psychiatric disturbances, including depression and dementia (Bottiglieri, 1996). A number of countries have set up nutritional intervention programmes aimed at increasing the folate status of women. Some have recommended folic acid supplementation (typically 400 μg/day) for all women of child bearing age. In some countries, there is a mandatory folic acid food fortification programme focusing on flour and many have stressed the importance of folate-rich foods (green vegetables, liver, oranges, and whole grain foods).

Most naturally occurring folate derivatives in foods are highly sensitive to oxygen, temperature, pH and light, and thus their stability is affected during harvest, processing or storage of food sources of this vitamin (Hawkes and Villota, 1989). With an increasing proportion of the world food supply being subjected to some sort of processing and storage, most studies demonstrate negative effects on folate stability from both industrial processing and household preparation causing increased losses with increasing severity of heating, temperature and time (Vahteristo *et al.*, 1998; Wigertz *et al.*, 1997; Williams *et al.*, 1995) (Table 13.1). Compared to natural folates, synthetic forms of folic acid are stable (Blakely, 1969) and more bioavailable (Gregory, 1997; Cuskelly *et al.*, 2001) as the bioavailability of natural folates is affected by incomplete conjugation on the brush border of the mucosal cells (Gregory, 1997). The low bioavailability and, more importantly, the poor chemical stability of the natural

Table 13.1 Effect of thermal processing on the loss of folates in foods

Processing techniques	Time exposed	Food sample	Folate losses (%)	Cause of loss	Reference
Water blanching, steam blanching	3–6 min	Spinach Broccoli	42–83 70–91	Leakage, oxidation	DeSouza and Eitenmiller (1990)
Steaming (pressure)	20–40 min	Broccoli	24–41	Leakage, oxidation	Petersen (1993)
Sous-vide processing in vacuum bags	40 min	Broccoli	11	Oxidation	Petersen (1993)
Tinning		Spinach, Broccoli,	50	Leakage, oxidation	DeSouza and Eitenmiller (1990)
Tinning		Brussel sprouts	30	Leakage, oxidation	Malin (1977)
Boiling or pressure cooking	10–30/5 min	Brussel sprouts	0	Leakage, oxidation	Malin (1977)
Infrared or convection heating		Cooked vegetables	4–24	Oxidation	Williams *et al.* (1995)
Warm holding	1 h (72 °C)	Cooked vegetables	14	Oxidation	Williams *et al.* (1995)
Autoclaving	20 min	Folate standards in food model	0[a] 75[b]	Oxidation	Ristow *et al.* (1982)
Oven baking	25–35 min (200 °C)	Rainbow trout Pollack Chicken breast fillet	30[b] 46[b] 34[b]	Oxidation	Vahteristo *et al.* (1998)
Pasteurisation UHT	15 sec (74 °C) 5 sec (140 °C)	Milk Milk	8[b] 19[b]	Oxidation	Wigertz *et al.* (1997)

[a] losses of folic acid; [b] losses of 5-methyltetrahydrofolate
Adapted from Madziva (2006)

folates have a profound influence on the development of folate-based food nutraceuticals. Microencapsulation could provide the necessary protection needed for the folates. Microencapsulation technology could be used to encapsulate synthetic forms of folates, incorporate them into a food vehicle and deliver them to consumers in greater stable concentrations. The selection of appropriate food vehicle for enrichment is an important consideration. A number of organisations (Codex Alimentarius Commission, 1991; Darnton-Hill, 1998) have published criteria for selecting an appropriate food vehicle including the need for the selected vehicle to be regularly consumed by the population at risk in stable, predictable amounts. The consumption of cheese in the world has grown by 15% since 1997 with the forecast set to increase further during the coming decades (US FDA, 1998). The nutritional importance and the popularity of cheese in providing the required bioavailable calcium in preventing osteoporosis make it an ideal vehicle to enrich with folates.

The focus of this chapter will be to deliver microencapsulated folates through food such as cheese and its controlled release.

13.2 Encapsulation technologies for folates

A number of microencapsulation techniques have been reported for food and nutraceuticals. Readers are referred to other chapters in this book. Technologies applied to folates are not widely reported and utilise gel particles by extrusion and emulsions. The focus of microencapsulation of folates, therefore, will cover techniques using extrusion and hydrogelling.

13.2.1 Microentrapment in hydrogel beads

Microentrapment in hydrogel capsules is probably the only reported ME technology for folic acid. The technology involves creating solid hydrogel beads which contain the folic acid in its matrix. During the formation of folate hydrogels, folates are immobilised and the gel particle size ranges from 450 to 1000 μm in diameter depending on the type of ME equipment used. It could be said that this process may not be true ME as it does not produce a capsule with a membrane coating. The terms immobilisation and encapsulation are used interchangeably in most of the reported literature. While encapsulation is the process of forming a continuous coating around an inner matrix that is wholly contained within the capsule wall as a core of encapsulated material, immobilisation refers to the trapping of the material within or throughout the matrix. Thus bioencapsulation (when referring to folic acid), matrix encapsulation or micro-entrapment could also be used to describe this process. The hydrogel beads containing folic acid can be coated further with polymers and they can be used to delay the release of folic acid and provide protection and enhance the stability of folic acid.

13.2.2 Extrusion method for folic acid encapsulation

Alginate beads have been used in microencapsulation of folates because they are easy to prepare on a laboratory scale, it is a mild hydrogelling process and comparatively cheap and safe for use in incorporation into food formulations. The major drawbacks of alginate microcapsules are, however, scaling up difficulties and the porous nature of microcapsules that will leak folic acid, hence, resulting in low encapsulation efficiency. This section will, however, focus on the development of a pilot scale prototype for scaling up of hydrogelling encapsulation of folic acid (Nguyen *et al.*, 2005) and use of mixed polymers to prevent leakage of folic acid from the hydrogel beads into the environment (Madziva *et al.*, 2005).

Sodium alginate (viscosity 2%, (w/v) solution at 25 °C, 250 cps), potassium salt of pectin (28% degree of esterification, low methoxy and 20% amidated) were used as hydrogelling polymers to microencapsulate folic acid. A bench-top encapsulator fitted with a 300 μm nozzle was used to prepare folic acid microcapsules in small quantities. Aqueous polymer solutions were first prepared (alginate, alginate and pectin in various combinations) and the solutions were left overnight to fully hydrate at room temperature. The folic acid was dispersed in the polymer solutions, the mixture was pumped through the encapsulation nozzle with a continuous flow of nitrogen into a gently agitated aqueous solution of 0.1 M calcium chloride solution at room temperature when discrete hydrogel beads entrapping folic acid was formed upon contact with the polymerising calcium chloride solution. The polymerising vessel with calcium chloride as the polymerising solution was constantly stirred to keep the polymer from clumping (Fig. 13.2). A dropping height of about 10 cm between the nozzle and the reaction vessel was maintained to ensure that spherical droplets were formed. The crosslinking process between alginate–pectin and calcium ions began almost instantaneously; however, the formed capsules were kept in the polymerising solution to harden and form a capsular wall at the surface of the beads. Blending of alginates and pectin polymers and microencapsulating folic acid produced the highest encapsulation efficiency (EE) compared to a number of single and mixed polymers including carrageenan, gelatin and xanthan gum (Madziva *et al.*, 2005).

A University of Western Sydney prototype (Fig. 13.3) was developed to scale up the hydrogel capsule formation. This prototype was based on a falling film polymerisation technique. The prototype consists of a tank (alginate tank) where the polymer solution is mixed with folic acid and made into a suspension. This suspension is allowed to hydrate and then pumped through a feed tube onto the central area of a rotating disk which spreads it outwards by centrifugal force. The disk is rotated by a variable speed drive. The disk is specially designed so that the alginate suspension is thrown away as microdroplets. These microdroplets strike against a falling film of the polymerising solution (calcium chloride) and instantly form micro hydrogel particles entrapping folic acid. These beads then fall on to the polymerising solution (calcium chloride) at the bottom of the encapsulation tank. The polymerising solution (calcium chloride)

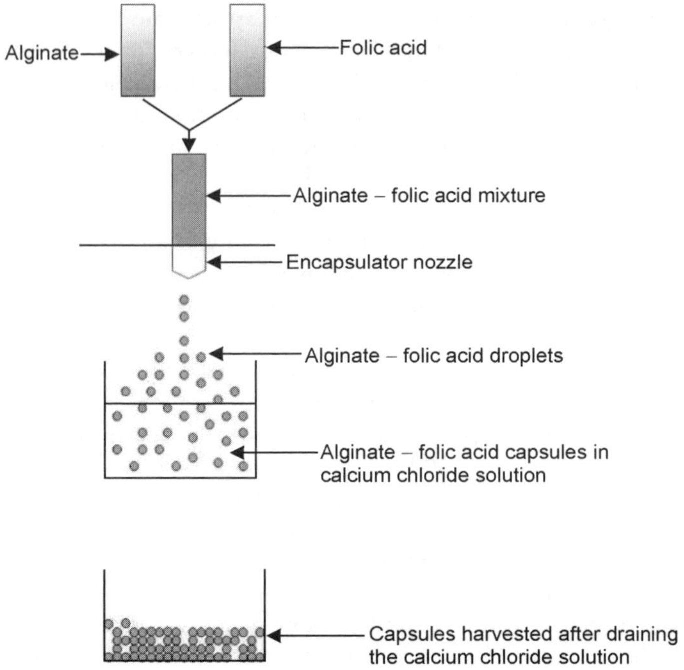

Fig. 13.2 A simplified schematic illustration of microencapsulation of folates.

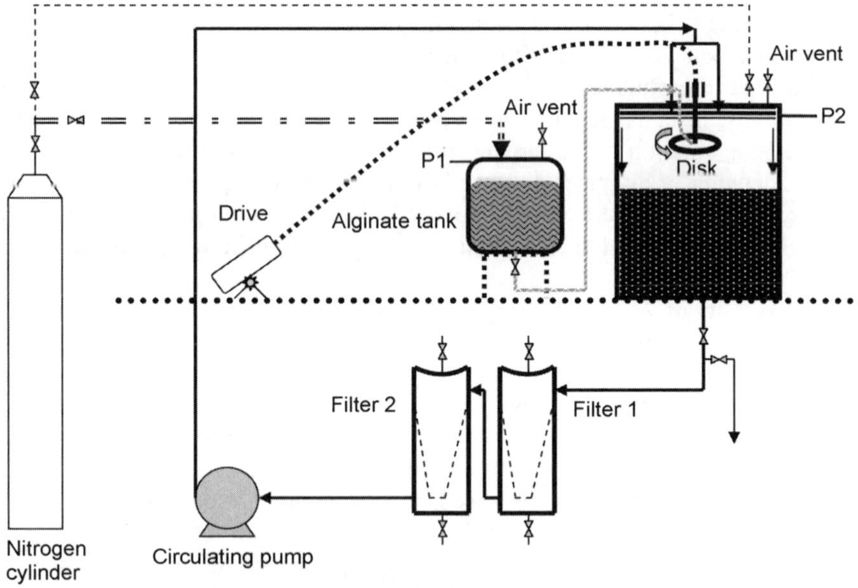

Fig. 13.3 The UWS Prototype Encapsulator. P1 – feed tank pressure; P2 – encapsulation tank pressure; Drive – spinning disk motor (adapted from Nguyen *et al.*, 2005).

is circulated through a specially designed pumping system to recirculate calcium chloride solution. The formed and hardened hydrogel beads are then harvested using cloth filters 1 and 2. These filters of 400 μm (Filter 1) and 55 μm (Filter 2) capture the gelled beads and the calcium chloride solution is recirculated by a pump. The polymerising solution then enters the encapsulation vessel and falls along the walls as a falling film. A nitrogen gas cylinder produces a pressure difference between the alginate feed tank and the encapsulation tank (P1 = driving pressure for alginate suspension; P2 = pressure inside encapsulation tank) and aid in the flow of alginate suspension with folic acid on to the rotating disk and also to maintain a nitrogen blanket above the solutions during the operation, which helps to prevent any oxidation of the folic acid.

13.3 Controlled release of folate in the gastrointestinal system

It is beneficial for the encapsulated folic acid to be released in the small intestine where most of the absorption of vitamins takes place. Therefore the polymers used for microencapsulation should be able to protect the folate in the acidic stomach conditions and release the folate in the alkaline conditions in the small intestine.

Many studies show that ME in alginate or pectin-based beads can be used for controlled release of bioactive products. The alginate–pectin hydrogels showed remarkable stability in acidic conditions, while they completely released folic acid in alkaline conditions (Madziva *et al.*, 2005). This *in vitro* release step was designed to stimulate gastric and intestinal conditions in the human gut. Pectin gels are stronger and more stable than alginate gels in acidic and in alkaline conditions. It was observed that folic acid loss was greater in alginate gel than in alginate–pectin gel at pH 1.2 (Madziva *et al.*, 2005). Data from *ex vivo* conditions in porcine GI contents using alginate-pectin beads are in agreement with these findings. Thus *ex vivo* porcine studies showed an important phenomenon in targeted delivery. The alginate–pectin hydrogels containing folate remained largely intact in gastric contents but released folic acid in both small intestine and colon. The folic acid release in the small intestinal contents as well as that of the colonic contents follows a very similar pattern except the former is greater (Fig. 13.4). These two very different environments showed different effects on the release of capsules. The presence of bacterial enzymes localised in the colonic region may well have contributed to the disintegration of the gel capsules leading to the release of folic acid. The least amount of folic acid released was in the stomach contents (Fig. 13.4). Thus it could be said that encapsulated folic acid was protected from the deteriorative gastric conditions (Madziva, 2006).

The jejunum is the site of maximum absorption of free folates, where absorption occurs by a pH dependent, carrier-mediated system (Mason, 1990; Strum, 1981). It is demonstrable from the *in vitro* and *ex vivo* porcine intestinal content studies that the alginate–pectin hydrogels released folic acid in alkaline

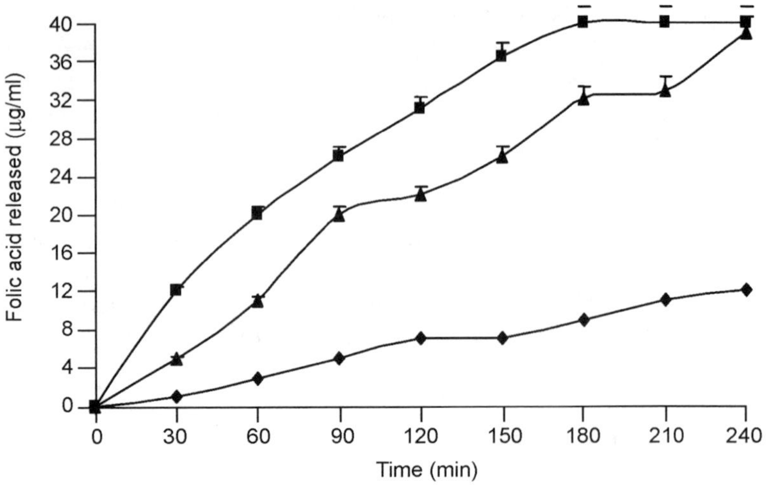

Fig. 13.4 Release of encapsulated folic acid in *ex vivo* porcine gastrointestinal contents. The error bars represent ± s.d ($n = 3$). Symbols: ■ small intestine, ▲ colon, ◆ stomach (Madziva, 2006).

conditions (Figs 13.4 and 13.5). This suggests that the hydrogels have the capability to protect folic acid during their transit through the upper gastro-intestinal tract and release them in the alkaline conditions of the small intestine, where the folic acid is absorbed.

13.4 Applications in functional foods

Fortification with folic acid in one or more of the commonly consumed dietary items is now regarded as the best nutritional intervention method to ensure increased folate intake and the reduction of the risks associated with folate deficiency.

Although increasing folate levels in fermented dairy products through 'natural' synthesis using bacterial cultures has been reported (Hoppner and Lampi, 1990), Crittenden *et al.* (2002) reported only a sixfold increase in folate (72 mg per g) with careful selection of microorganisms. This means that daily consumption of 100 g of such a dairy product would contribute approximately 15% of the minimum required folate intake, but only 2% of recommended consumption for women of child bearing age. Thus it is clear that biofermentation using cultures may not be able to produce sufficient quantities of folates in recommended doses for consumption, hence the need for fortification.

Preliminary studies for the suitability of folic acid fortification in cheddar cheese showed satisfactory results (Madziva *et al.*, 2006). Folic acid micro-encapsulated with combined alginate–pectin polymers showed the highest encapsulation efficiency and increased stability and retention when tested in a milk system where pH was adjusted from 6.7 to 4.5 over a 4 h period. The

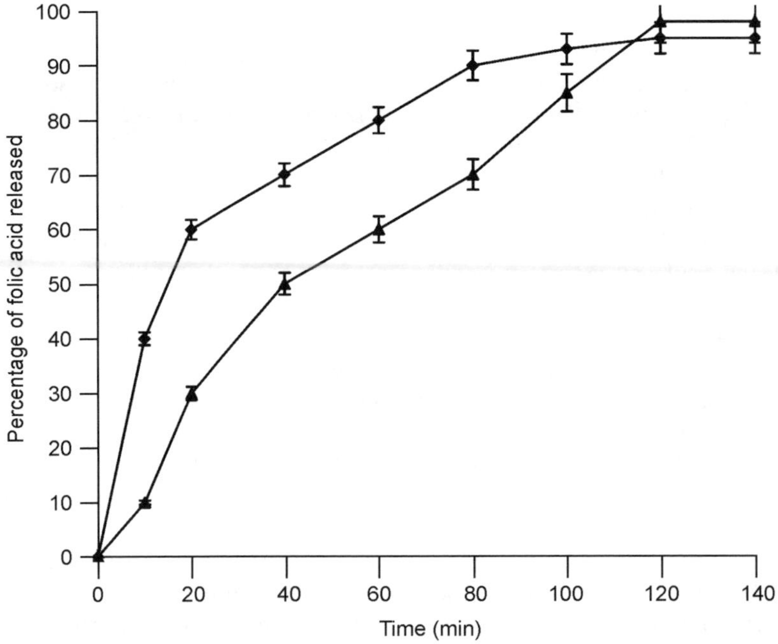

Fig. 13.5 Effect of pectin ratio on the release of alg–pect hydrogels in 0.1 M phosphate buffer pH 8.2 alginate: pectin: ■ A100:P0; ◆ A70:P30. Values are mean ± s.d. ($n = 9$) (Madziva, 2006).

retention of folic acid under the test conditions was 360 μg/100 g within the alginate–pectin capsules indicating their ability to remain intact in the acidified milk system. Stress tolerance of the folic acid capsules tested under simulated cheese press pressures for 4 h showed folic acid retention of up to 288 μg/100 g. Three stages in Cheddar cheese manufacturing, namely, addition of folate capsules in the milk, incorporation of capsules after milling the curd and injection into the pressed block of raw cheese when studied, showed that addition of capsules in the milk showed the most even distribution of capsules while the latter showed poor and irregular distribution. Encapsulated folic acid produced greater stability in Cheddar cheese over a three-month ripening period compared to free folic acid. Thus this study showed that folic acid could be successfully microencapsulated in alginate–pectin polymers and incorporated into milk to make Cheddar cheese with little loss of folates in whey and excellent stability during cheese ripening compared to free folates. These results showed clearly that Cheddar cheese would be an effective food vehicle for folic acid delivery.

13.4.1 Controlled delivery and bioactivity of encapsulated folic acid in Cheddar cheese in reducing hyperhomocysteinemia

The efficiency of the microencapsulated folic acid in Cheddar cheese was studied using a methionine-induced hyperhomocysteinemia mice model

(Madziva *et al.*, 2006). Reported studies based on sound epidemiological evidence have shown a strong association of elevated homocysteine with incidence of vascular disease (Bautista *et al.*, 2002; Homocysteine Studies Collaboration, 2002; Wald *et al.*, 2002). Elevation of plasma total homocysteine may occur due to genetic defects and/or an inadequate status of folate and co-factors. Hyperhomocysteinaemia is regarded as an independent risk factor for cardiovascular diseases.

To study the maintenance of bioactivity of encapsulated folic acid, the microencapsulated capsules were incorporated into a diet with excess dietary methionine using Cheddar cheese as the food carrier and fed to mice. It was found that hyperhomocysteinemia caused by the dietary addition of 10 g/kg methionine was completely counteracted by the encapsulated folic acid compared to free folic acid. Encapsulated folic acid caused substantial reduction in plasma homocysteine and arterial lesions in mice fed the diet supplemented with 20 g/kg methionine compared with free folic acid. This study showed that folic acid in its encapsulated form in Cheddar cheese was released in the intestine of mice and reduced homocysteine levels and aortic pathological events to a greater degree than free folic acid. Alginate–pectin capsules enhanced folic acid stability and bioactivity in sufficient quantities to normalise plasma total homocysteine levels when Cheddar cheese was used as the food carrier.

This study therefore confirmed the preventative therapeutic effect of folic acid more so in its encapsulated form than its free form as seen by a reduction in homocysteine levels and aortic pathological events. Alginate–pectin encapsulation of folic acid enhanced its stability and was bioavailable in sufficient quantities to normalise plasma homocysteine levels when Cheddar cheese was used as the food carrier. This makes a strong case for possible application of this technique in food systems for folic acid delivery.

13.5 Future trends

The work on folic acid encapsulation reported by Madziva *et al.* (2005, 2006) has demonstrated that ME using food grade polymers confers stability to otherwise unstable folic acid. With increasing reports that this vitamin is essential for cell replication and has an important role in the prevention of neural tube defects (Botto *et al.*, 1999; Cragan *et al.*, 1995), possibly coronary heart disease (Brouwer *et al.*, 1999) and indications for positive effects on good folate status for cognitive functions (Seshadri *et al.*, 2002) and on prevention of certain forms of cancer (Giovannucci *et al.*, 1995), fortification with folic acid is the best nutritional intervention method to ensure increased folate intake and reduce risk associated with folate deficiency. The United States Food and Drug Administration (US FDA) made the fortification of grain products with folic acid mandatory from 1 January 1998. Most industrialised countries practise voluntary fortification with folic acid. Folic acid fortification of other food products such as juices and flours are being carried out elsewhere. However, the stability of free folic acid during processing and storage and the efficiency of its bioactivity may be questionable, hence, microencapsulation techniques with tailor-made polymers will need to be developed to suit different processing conditions, storage conditions and release conditions.

Research on food grade polymers that provide higher encapsulation efficiency and greater stability need to be done in order to increase the application base and variety of food delivery vehicles. Coating of gel capsules with polymers such as poly-L-lysine or chitosan during crosslinking should be investigated.

Other ME techniques such as spray drying, spray chilling, spray coating or a combination of these techniques should be developed for better product integrity and folate stability and preserving their therapeutic bioactivity during processing and storage in different food products.

There is good indication that the gel capsules can be further modified to achieve targeted release not just along the whole length of the small intestine, but, specifically the jejunum of the small intestine where folic acid absorption is maximal. This can be studied by monitoring transient time of the capsules in the upper gastrointestinal tract until such time when folic acid release starts. Such a study could be carried out using synthetic folic acid and suitable dye binding techniques.

Sensory aspects of foods are critical in acceptance of new food products. The new food product development with microencapsulated folates will need to be evaluated for any impact on subtle changes in texture or flavour. As the benefit of folate becomes increasingly known, it is to be expected that the product be labelled with a specific health claim. It can be expected that in future microencapsulation will not only be used for the controlled delivery of folates in food nutraceuticals but also to differentiate folate enriched foods from others on the supermarket shelf. Thus, ME could be a future marketing tool for the functional foods and nutraceuticals industry.

13.6 References

BAUTISTA LE, ARENAS IA, PENUELA A, MARTINEZ LX (2002) 'Total plasma homocysteine level and risk of cardiovascular disease: a meta-analysis of prospective cohort studies' *J Clin Epidemiol* 55, 882–887.

BLAKELY RL (1969) *The Biochemistry of Folic Acid and Related Pteridines*. Amsterdam: North Holland.

BOTTIGLIERI T (1996) 'Folate, vitamin B12, and neuropsychiatric disorders' *Nutr Rev* 54, 382–390.

BOTTO LD, MOORE CA, KHOURY MJ, ERICKSON JD (1999) 'Medical progress: neural-tube defects' *N Eng J Med* 341, 1512.

BROUWER IA, VAN DUSSELDORP M, THOMAS CM, DURAN M, HAUTAVAST JG, ESKES TK, STEEGERS-THEUNISSEN RP (1999) 'Low dose folic acid supplementation decreases plasma homocysteine concentrations: a randomised trial' *Am J Clin Nutr* 69, 99–104.

CODEX ALIMENTARIUS COMMISSION (1991) *General Principles for the Addition of Essential Nutrients to Foods*. CAC/GL 09-1987 (Amended 1989, 1991).

CRAGAN JD, ROBERTZ HE, EDMONDS LD, KHOURY MJ, KIRBY RS, SHAW GM, VELIE EM, MERZ RD, FORRESTER MB, WILLIAMSON RA, KRISHNAMURTI DS, STEVENSON RE, DEAN JH (1995) 'Surveillance for anencephaly and spina bifida and the impact of prenatal diagnosis – United States, 1985–1994', *MMWR CDC Surveill Summ* 44, 1–13.

CRITTENDEN RG, MARTINEZ NR, PLAYNE MJ (2002) 'Synthesis and utilisation of folates by yoghurt starter cultures and probiotic bacteria' *Int J Food Microbiol* 80, 217–222.

CUSKELLEY GJ, STACPOOLE, PW, WILLIAMSON J, BAUMGARTNER TG, GREGORY JF (2001) 'Deficiencies of folate and vitamin B6 exert distinct effects on homocysteine, serine and methionine kinetics' *Am J Physiol Endocrinol Metab* 281, E1182–E1190.

DARNTON-HILL I (1998) 'Rationale and elements of a successful food-fortification program'. Food and Nutrition Bulletin 19(2).

DESOUZA S, EITENMILLER R (1990) 'Effects of different enzyme treatments on extraction of total folate from various foods prior to microbiological and radio assay' *J. Micronutri. Anal* 7, 37–57.

GIOVANNUCCI E, RIMM EB, ASCHERIO A, STAMPFER ME, COLDITZ GA, WILLET WC (1995) 'Alcohol, low-methionine-low-folate diets, and risk of colon cancer in man' *JNCI* 87, 265–273.

GREGORY JF (1997) 'Bioavailability of folate' *Eur J Clin Nutr* 51, S54–59.

HAWKES JG, VILLOTA R (1989) 'Folates in foods: reactivity, stability during processing and nutritional implications' *Crit Rev Food Sci Nutr* 28, 439–539.

HOMOCYSTEINE STUDIES COLLABORATION (2002) 'Homocysteine and risk of ischemic heart disease and stroke: a meta-analysis' *J Am Med Assoc* 288, 2015–2022.

HOPPNER K, LAMPI B (1990) 'Total folate, pantothenic acid and biotin content of yoghurt products' *Canadian Inst Food Sci Technol J* 23, 223–225.

KIRBY CJ (1991) 'Microencapsulation and controlled delivery of food ingredients' *Food Sci Technol Today* 5, 74–80.

MADZIVA H (2006) 'Studies on stability and efficiency of microencapsulated folic acid in Cheddar cheese and in methionine induced hyperhomocysteinemia in mice'. PhD Thesis, University of Western Sydney, Australia.

MADZIVA H, KAILASAPATHY K, PHILLIPS M (2005) 'Alginate-pectin microcapsules as a potential for folic acid delivery in foods' *J Microencap* 22, 343–351.

MADZIVA H, KAILASAPATHY K, PHILLIPS M (2006) 'Evaluation of alginate–pectin capsules in Cheddar cheese as a food carrier for the delivery of folic acid' *LWT-Food Sci Technol* 39, 146–151.

MALIN JD (1977) 'Total folate activity in Brussels sprouts: the effects of storage, processing, cooking and ascorbic acid content' *J Food Technol.* 12, 623–632.

MASON JB (1990) 'Intestinal transport of monoglutamyl folates in mammalian systems' in: Picciano MF, Stokstad ELR, Gregory JF (eds), *Evaluation of Folic Acid Metabolism in Health and Disease*, New York: Alan R. Liss, pp. 47–63.

MRC VITAMIN STUDY GROUP (1991) 'Prevention of neural tube defects: results of the Medical Research Group vitamin study' *Lancet* 338, 131–137.

NGUYEN MH, KAILASAPATHY K, BUI V (2005) 'The performance of a prototype Encapsulator for food applications' *Proc. XIIIth Int Workshop on Bioencapsulation,* Kingston, Ontario, Canada.

PETERSON MA (1993) 'Influence of *sous vide* processing, steaming and boiling on vitamin retention and sensory quality in broccoli florets. *Eur Food Res Technol* 197(4), 375–380.

QUERE I, PERNEGER T, ZITTOUN J, BELLET H, GRIS J, DAURES J, SCHVED J, MERCIER E, LAROCHE J, DAUZAT M (2002) 'Red blood cell methyl folate and plasma homocysteine as risk factors for venous thromboembolism: a matched case-control study' *Lancet* 359, 747–752.

RISTOW KA, GREGORY JF, DAMRON BL (1982) 'Effect of dietary fibre on the bio-availability of folic acid monoglutamate' *J Nutr* 112, 750–758.

SCHROOYEN PMM, VAN DER MEER R, DE KRUIF CG (2001) 'Microencapsulation: its application in nutrition' *Proc Nutr Soc* 475–479.

SCOTT KJ (1989) 'Micronutrients in dairy products'. In: Renner E (ed.) *Micronutrients in Milk and Milk-based Products*, London: Elsevier Applied Science, 71–123.

SESHADRI S, BEISER A, SELHUB J, JACQUES PF, ROSENBERG IH, D'AGOSTINO RB, WILSON PW, WOLF PA (2002) 'Plasma homocysteine as a risk factor for dementia and Alzheimers disease *New Engl J Med* 346, 476–483.

STRUM WB (1981) 'Characteristics of the transport of pteroylglutamate and amethopterin in rat jejunum' *Pharmacol* 216(2), 329–333.

US FDA (1998) Food standards: amendment of standards of identity for enriched cereal grain products to require the addition of folic acid (final rule). *Fed Register* 61(44), 8781–8797.

VAHTERISTO L, LEHIKOINEN K, OLLILAINEN V, KOIVISTOINEN PE, VARO P (1998) 'Oven-baking and frozen storage affect folatc vitamer retention' *Lebensm-Wiss U-Technol* 31, 329–333.

VOLLSET EM, REFSUM H, TVERDAL A, NYGARD O, NORDREHAUG JE, TELL GS, UELAND PM (2001) 'Plasma total homocysteine and cardiovascular and non-cardiovascular mortality: the Hordaland Homocysteine Study' *Am J Clin Nutr* 74, 130–136.

WAGNER C (1985) 'Folate-binding proteins' *Nutr Rev* 43, 293–299.

WALD DS, LAW M, MORRIS JK (2002) 'Homocysteine and cardiovascular disease: evidence on causality from a meta-analysis' *Br Med J* 325, 1202–1208.

WIGERTZ K, HANSEN SI, HOIER-MADSEN L, WITTHOFT CM, HOLM J, JAGERSTAD M (1997) 'Effect of milk processing on the concentration of folate-binding protein (FBP), the folate-binding capacity and the retention of 5-methyltetrahydrofolate' *Int J Food Sci Nutr* 47, 315–322.

WILLIAMS PG, ROSS H, MILLER B (1995) 'Ascorbic acid and 5-methyltetrahydrofolate losses in vegetables with cook/chill or cook/hot-hold food service systems' *J Food Sci* 60, 541–546.

14

Encapsulation of probiotics

C. P. Champagne, Agriculture and Agri-Food Canada, Canada and K. Kailasapathy, University of Western Sydney, Australia

14.1 Introduction

Microencapsulation (ME) is a technology used to 'package' various materials, principally solids or liquids, in miniature capsules which release their contents at controlled rates. There are many instances where it is useful to encapsulate ingredients for functional foods or nutraceuticals (FFN). Controlled release of compounds is a critical benefit of ME. However, there are many other advantages of ME for bioactive ingredients in the food matrices: taste or colour masking, increased stability of the bioactive ingredient during processing and storage, and reduced interactions between the ingredient and the environment. For these reasons, numerous vitamins, minerals or antioxidants destined for the FFN markets are encapsulated (Schrooyen *et al.*, 2001), but this chapter will focus on probiotics.

Probiotics are considered as being 'Defined, live microorganisms which, administered in adequate amounts, confer a beneficial physiological effect on the host' (Araya *et al.*, 2002). The potential health benefits of probiotic bacteria (*Lactobacillus acidophilus* and bifidobacteria) are presented in Table 14.1. Increasing the intestinal population in these beneficial bacteria can be accomplished in two ways: directly consuming probiotic bacteria or consuming products that will selectively improve their growth in the gastrointestinal (GI) system. Such products, which are called prebiotics, are not digested by humans and are typically composed of oligosaccharides (inulin, fructooligosaccharides, lactulose, galactooligosaccharides).

Although there are instances where non-viable probiotics can still exert their biological effects (Ouwehand and Salminen, 1998; Tabrizi *et al.*, 2004; Stanton *et al.*, 2005) it is widely considered that products which contain viable probiotics

Table 14.1 Potential health benefits of probiotic bacteria (from Ouwehand *et al.*, 2003 and Kailasapathy and Chin, 2000)

Prevention/reduction of symptoms of:	Improvement of:
• hypercholesterolaemia • diarrhoea • inflammatory bowel disease (ulcerative colitis – Crohn) • necrotising enterocolitis • irritable bowel syndrome • constipation • *Helicobacter pylori* infection (ulcers) • small bowel bacterial overgrowth • colorectal cancer • superficial bladder cancer • cervical cancer • breast cancer • allergy • coronary heart disease • urinary tract infection • upper respiratory tract infections	• lactose assimilation • food digestibility • immune response • blood pressure • oral health

are the most desirable. Viability of probiotic cultures in FFN is therefore a critical issue of their functionality. Viability and activity of the probiotic bacteria are important considerations, because the bacteria must survive in the food or nutraceutical during shelf-life and during transit through the acidic conditions of the stomach, and resist any degradation by hydrolytic enzymes and bile salts in the small intestine. It is essential that products sold with any health claims meet the population levels stated on the label, presumably those found effective in clinical trials, and the minimum therapeutic dose per day is often found to be 10^8–10^9 cells (Kurmann and Rasic, 1991; Kailasapathy and Rybka, 1997; Lourens-Hattingh and Viljoen, 2001; Shah, 2000).

Various studies have shown that probiotic organisms survive poorly in foods like yoghurt and fermented milks with cell numbers being much lower than the recommended levels at the expiry date (Iwana *et al.*, 1993; Klaver *et al.*, 1993; Hughes and Hoover, 1995; Dave and Shah, 1997; Kailasapathy and Rybka, 1997; Rybka and Fleet, 1997). Therefore there is a need to protect the probiotic bacteria against adverse processing and storage conditions as well as during transit in the GI tract. Microencapsulation has recently gained considerable attention as the likely means to provide the required protection to probiotic bacteria.

When considering controlled delivery of bioactive compounds to humans, one usually thinks of caplets and controlled delivery in the GI system. With respect to probiotic-containing caplets (nutraceuticals), encapsulation has some definite advantages in controlling the release in the GI system, and some data will be presented in this respect. However, it must be stressed that ME is as extensively used for the *delivery* of probiotics into foods. This aspect will thus be extensively covered.

14.2 Encapsulation technologies for probiotics

There are numerous microencapsulation techniques. However, technologies applied to probiotics are generally limited to gel particles, spray-coating, spray-drying, extrusion and emulsions. Therefore, the focus will be on these approaches. Other valuable techniques for ME such as liposomes, coacervation, co-crystallisation or molecular inclusion (cyclodextrins) do not yet appear to have extensive use in the preparation of probiotics for FFN either because of cost, of the large size of bacteria or of the requirement of food-grade compounds in the process.

14.2.1 Microentrapment in gel particles

Microentrapment in gel beads is probably the most widely studied ME technology for probiotics. It basically consists in the production of solid matrices which contain the cells. It could be argued that this process is not true microencapsulation because some particle sizes can reach 2 mm in diameter, and because we do not obtain a capsule with a membrane coating. Thus, bio-encapsulation (when applied to probiotics), matrix encapsulation or micro-entrapment could also be used to describe this process. Nevertheless, the particles obtained can often achieve the delayed release and protection properties required. Probiotic bacteria (1–3 μm in size) are well retained in the alginate hydrogel matrix which is estimated to have a pore size of less than 17 nm (Klein *et al.*, 1983).

Extrusion and emulsion methods for probiotics

Various polymers can be used to prepare ionic gel particles (Doleyres and Lacroix, 2005), but alginate, carrageenan, gellan and pectin are the most frequently used. The technique relies on the ability of these polymers to form gels in the presence of minerals, principally calcium and potassium.

Three methods of producing alginate beads which carry probiotics are most commonly used (Fig. 14.1). The most popular is the alginate extrusion process in a calcium chloride solution. In some cases, starch (Sultana *et al.*, 2000; Dembczynski and Jankowski, 2002; Muthukumarasamy *et al.*, 2006), pectin and whey proteins (Guérin *et al.*, 2003) have been blended with alginate in order to improve the matrix for subsequent applications. There seems to be little effect of $CaCl_2$ on cell viability during microentrapment, but high alginate content, high cell loading and large particles appear to be preferable when the products are subsequently challenged in simulated gastric environments (Chandramouli *et al.*, 2004). There are various scale-up methods as well as industrial suppliers of extrusion bead-forming equipment (Heinzen, 2002; Champagne, 2006), and the production can be contracted out.

The oil emulsion processes (Fig. 14.1) are less simple but easier to scale up (Krasaekoopt *et al.*, 2003). They have the advantage of enabling the production of smaller beads (Gouin, 2004), which can be critical in preventing sensory changes in the functional foods enriched with the probiotic-containing beads.

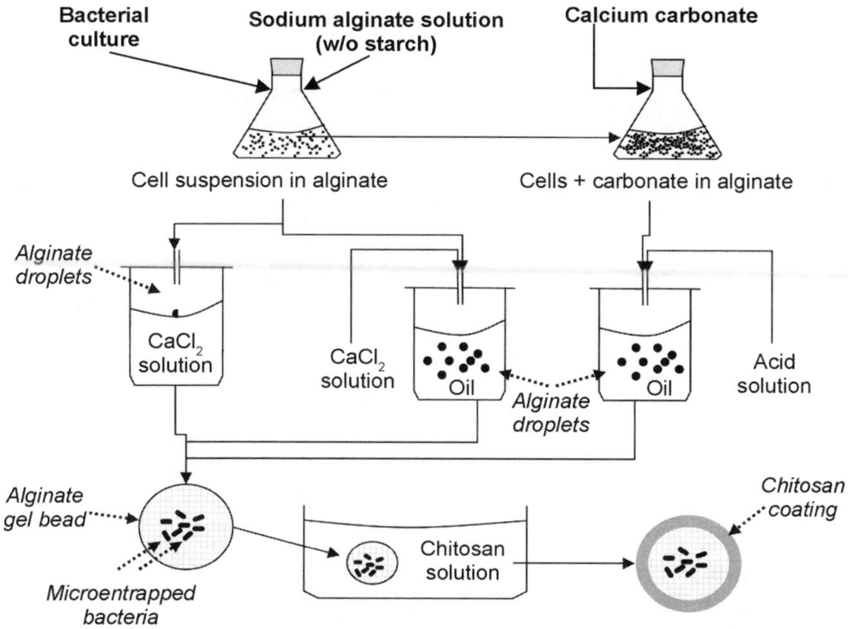

Fig. 14.1 Several techniques used to microentrap probiotic bacteria in alginate gel beads.

Coating of the alginate beads can also be carried out to improve their protective properties. The beads are simply dipped in a solution containing a cationic polymer, such as chitosan (Fig. 14.1), gelatin or poly-L-lysine (Groboillot *et al.*, 1993; Krasaekoopt *et al.*, 2004). Other coating or bead modification methods for alginate have been attempted (Krasaekoopt *et al.*, 2003; Le-Tien *et al.*, 2004), but important losses in viability have resulted in some instances (Larish *et al.*, 1994).

Some protein-based ME techniques are also carried out by ionic gelification. Limited data suggest that alginate gels offer better protection to probiotics than other carbohydrate-based or protein-based gels (Kailasapathy and Sureeta, 2004; Muthukumarasamy *et al.*, 2006) but more data are needed on the effect of the gelling matrix. The reader is directed to Chapter 10 for more details on the use of protein matrices for ME technologies.

Spray-chilling technology

The production of gel beads can also be achieved by spray-chilling, which is considered the least expensive encapsulation technology (Gouin, 2004). In spray-chilling, a molten matrix of hydrogenated or fractionated vegetable oils with low melting point (32 to 42 °C) containing the bioactive compound is atomized through a nozzle into a vessel. This process is similar to spray-drying with respect to the production of fine droplets. However, it is based on the injection of cold air in the vessel to enable solidification of the gel particle, rather that hot air which dries the droplet into a fine powder particle. The liquid droplet, which is rather small, thus solidifies and entraps the bioactive product. The matrix is consequently a thermogelling compound.

Starch granules

It has long been recognized that bacteria adhere to starch particles (Anderson and Salyers, 1989), and this property has been specifically used for ME of probiotics in native maize (corn) potato, oat or barley (Myllarinen *et al.*, 2000; Crittenden *et al.*, 2001; Myllarinen 2002). The cells can adhere to the granules or grow inside the matrix. Thus, probiotic-containing starch granules could constitute carriers for probiotics in the development of functional foods (Fig. 14.2). Resistant starch can be used to ensure the viability of probiotic organisms in food and in the gastrointestinal tract. Resistant starch is also a suitable surface for adherence of probiotics to starch granule during processing, storage and transit through the upper gastrointestinal tract, providing protection against environmental stresses. However, as for other ME techniques based on particles, a disadvantage is the potential effect of texture (Table 14.2). Bacterial populations in the particles are lower that the other gel particle systems; this system might need higher addition levels in foods to provide the required number of probiotic cells.

Nevertheless, the system could potentially be used for controlled release of the probiotics in the GI system. In an *in vitro* model that simulated the upper gastrointestinal tract, adhesion was inhibited both by the action of proteinase, by pH values less than 3 and by exposure to pancreatin (Crittenden *et al.*, 2001). Thus, it could be expected that some cells would be released under these conditions. Although the properties of the starch-adhering cells were not compared

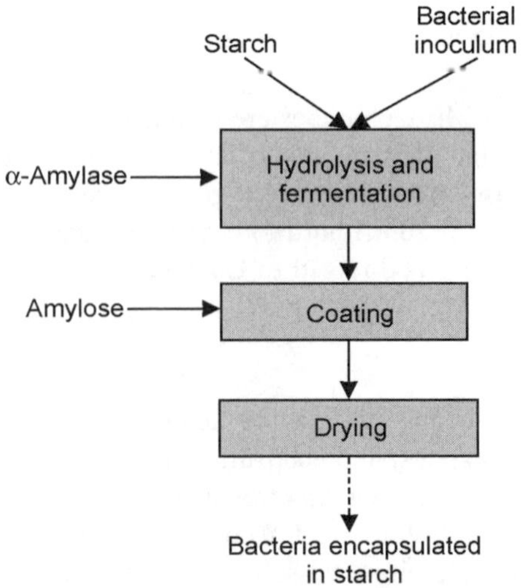

Fig. 14.2 VTT process for ME in starch granules (Myllarinen *et al.*, 2000; Myllarinen, 2002).

Table 14.2 Some properties of microencapsulation (ME) techniques for probiotics

ME technique	Characteristics	Benefits	Problems
Gel particles	• Cells entrapped in a gel matrix: polymer-based system • Gel particle coatings can be applied (chitosan) • High cell load (> 10^{11} cfu/g) in pectin or alginate beads, but lower for starch granules	• Can be used to produce biomass itself • Particles can be air-dried • Many techniques can be used • Spray-chilling is the lowest-costing ME technology • Demonstrated protection against freezing, oxygen and GI[a] environment • Cell release in intestines • Carrier can have bioactive properties	• Lower yields in biomass production • Particles can affect food texture • Particles may dissolve in environments having chelating compounds (phosphates, citrates)
Spray-coating	• True coating; cells in core powder • Coat generally lipid-based • High cell load (> 10^{11} cfu/g). • Most widely used commercially	• High protection to short exposure periods (a few days) to acids, humidity and oxygen • Multiple layers can be added for controlled release or density adjustments	• Phase separation in beverages if coating is lipid-based • Slow cell release at low temperatures • Particles can affect food texture
Spray-drying	• Cells 'coated' individually with the ingredients in the drying medium • Medium cell load (10^{10}–10^{11} cfu/g)	• Many possibilities for coating materials • Rapid cell release if ingredients dissolve rapidly • Less effects on food/beverage texture if they dissolve	• High temperatures in the process kill many strains; may require pre-heating adaptation step • Short-term protection (a few hours) linked to particle dissolution

Table 14.2 *Continued*

ME technique	Characteristics	Benefits	Problems
Extrusion	• Cells blended with various powders and then extruded • Low cell load (10^9–10^{10} cfu/g)	• Many possibilities for coating materials • Particles can be air-dried • Particles can dissolve, as a function of carrier • Carrier can be prebiotic	• Low populations require large inoculation rates
Emulsion	• Homogenization of aqueous and lipid phases	• Can be combined with spray-drying • Little data on applications; lipid-based systems theoretically good for protection against acids and oxygen	• Difficult to apply to probiotics • Emulsifiers can be detrimental to viability • High losses in liquid phase • Little data on stability during storage; liquid core may be unfavourable to long-term stability

[a] GI = *in-vitro* gastrointestinal conditions.

to those of free cells, some data suggest that cultures which grow in biofilms could be physiologically different than free cells (Doleyres and Lacroix, 2005), and have improved abilities to survive stressful conditions of the GI system.

14.2.2 Encapsulation by spray-coating

Although there is much less scientific data on the effect of spray-coating for probiotics than with microentrapment in gels, this technology has been the most extensively applied commercially. Many of the process details are proprietary, but the extensive industrial interest in this technology is reflected by the patent activity in the field (Durand and Panes, 2001; Shin *et al.*, 2002; Ubbink *et al.*, 2003).

In spray-coating, the core material needs to be in a solid form, although not necessarily completely dried. The core material is kept in motion in a specially designed vessel, either by injection of air at the bottom or by a rotary action of some sort (vessel or pale) (Fig. 14.3). A liquid coating material is sprayed over the core material and solidifies to form a layer at the surface. The coating material can be injected from many angles (Fig. 14.3), and this influences the properties of the coating. In food applications the coating is mostly lipid-based, but a wide variety of compounds can be used (Ubbink *et al.*, 2003):

- Lipids:
  - waxes (paraffin, beeswax, carnauba, candellila, shellac)
  - fatty acids (oleic, stearic, palmitic) and their salts

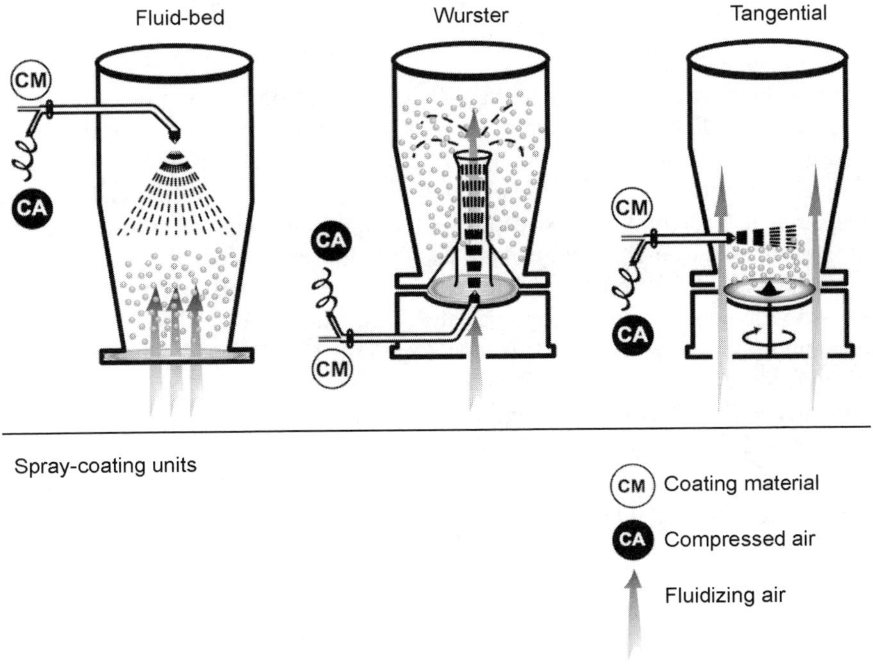

Fig. 14.3 Various designs of spray-coating processes.

  - hydrogenated or hardened oils
  - mono- or diglycerides, and their derivatives
  - fats of animal origin (milkfat)
  - phospholipids
- Proteins: gluten, casein
- Carbohydrates: cellulose and its derivatives, carrageenan, alginate.

The temperature of this coating material is typically kept between 60 and 120 °C during spraying while the temperature of the fluidizing air can vary between 10 and 50 °C (Durand and Panes, 2001). The process conditions must be adjusted to prevent a strong increase in the temperature of the core material. Initial particle size ranges between 100 and 1000 μm in diameter, and the coating will ultimately represent between 30% and 80% of the weight of the product.

Depending on the particle size to be coated, various pieces of equipment are preferred. For instance the pellet-coating process (Ubbink *et al.*, 2003) is recommended for particles having a volume of at least 0.02 cm^3.

Spray-coating is particularly well suited for coats having multiple layers. For example, a first layer of lipids can be applied in order to confer a protective property for the probiotic bacteria in the core, while a second layer, protein or other, can be used in order to adjust the specific gravity or stabilizers for the coated probiotics to remain in suspension in beverages (Shin *et al.*, 2002). There are also other potential advantages of spray-coating and a few are listed in Table 14.2.

14.2.3 Spray-drying

Spray-drying is extensively used in the production of microencapsulated ingredients. A liquid suspension of the bioactive product is sprayed at the top of a large vessel, creating a mist of fine droplets, in conjunction with hot air. The droplets dry into solid particles. Spray-drying has traditionally had limited industrial use for probiotics because of the high temperatures used during the process which would generate high viability losses. Indeed, the inlet air can typically be at 120–160 °C, and is still at 50–80 °C at the outlet. The use of protective compounds and adapting cells (Desmond *et al.*, 2002) are critical in preventing viability losses. It is considered that when cells and protective compounds are emulsified with oils prior to spray-drying (Ross *et al.*, 2005; Crittenden *et al.*, 2006), or that when probiotics are emulsified in an oil phase (see Section 14.2.5), they are encapsulated. No data are available on cell release of these systems. However, it can be expected that the benefits/problems of the spray-dried ME products will be linked to their dissolution properties (Table 14.2).

14.2.4 Extrusion

Extrusion was proposed almost 20 years ago as a method of stabilizing lactic cultures (Kim *et al.*, 1988). In this process, dried cultures are mixed with bulking agents (cellulose, glycerol) and small amounts of water to form wet granule.

This mixture is processed in an extruder with screens having pores about 1 mm in diameter. The wet particles can subsequently be air-dried or spray-coated. Refinements of this method have appeared where ingredients designed for enteric protection have been added in the bulking agents blends (Simmons *et al.*, 2004). The disadvantage of this method is the low ratio of probiotic bacteria to protective compounds. In extrusion, the initial probiotic culture represents less than 10% of the final product, while in spray-coating, for example, it can represent 50–80% of the mass. Thus, products obtained from extrusion have rather low populations, typically in the 10^9–10^{10} cfu/g range. This is acceptable for the nutraceutical market, but might be a bit low as a supplement to foods. Indeed, many uncoated freeze-dried probiotics are marketed with 10^{11} cfu/g populations.

A variation of extrusion technology was developed by Chan and Zhang (2002) with a direct compression approach. In this technique, a probiotic-containing powder is placed between two layers of a powder which will act as a protective coating. The blend is compressed in a dye to obtain a tablet. The powder containing the probiotics is located in the centre of the tablet. This enables the production of chewable tablets, particularly enjoyed by children.

14.2.5 Emulsions

Oil-in-water and water-in-oil emulsions are prepared by homogenizing an oil phase and an aqueous phase together in the presence of an emulsifier. The type and concentrations of the various ingredients used to formulate these emulsions, the homogenization and other processing conditions employed determine their physicochemical properties and stability (McClements, 2005). However, stabilization for food systems remains an issue due to the droplets coalescence or due to the diffusion of the water molecules from the internal aqueous phase to the bulk aqueous phase (Garti and Benichou, 2004).

Emulsions are mostly applied to chemical bioactives, but a few processes have involved probiotics. One method involves mixing a probiotic cell suspension with a mixture of sesame oil bodies in sesame oil (Hou *et al.*, 2003).

Combining some processes can generate interesting approaches. Picot and Lacroix (2003a,b) have combined emulsification and spray-drying to develop a low-cost ME method adaptable to large-scale production. In this method, a micronized powder, containing the probiotic bacteria, is added to a hydrophobic phase, which can be anhydrous milk fat. The hydrophobic phase containing the dry material to be encapsulated is added to a hydrophilic phase, in this case a 10% whey protein isolate solution. An oil-in-water emulsion is prepared using a dynamic loop mixer, and the emulsion is then spray-dried. The product contains a core of probiotic powder having a first coating layer of milk fat and a second coating layer of whey proteins, generated during spray-drying. A similar approach has been proposed by Crittenden *et al.* (2005) using different homogenization systems and different emulsifiers.

14.3 Applications in nutraceuticals

A nutraceutical can be defined as a product isolated or purified from foods that is generally sold in medicinal forms not usually associated with food. Probiotic-containing nutraceuticals are produced by specialized manufacturers who typically carry out the biomass production step under pH control in vats of 2000 to 10 000 L, recover and concentrate the cells be centrifugation or filtration, dry the concentrate and market it in the form of a powder, caplets or chewable tablets. An increasing number of products are encapsulated. In current commercial practices, the encapsulation process is carried out by spray-coating (Section 14.2.2), either on the small dried particles of probiotics (prior to insertion in a caplet) or on the larger caplet itself. Examples include the STAR and Probiocap technologies (Goulet and Wozniak, 2002). However, some strains cannot survive this traditional process and it will be examined how ME technologies can offer novel production strategies for the manufacturers of nutraceuticals.

14.3.1 Encapsulation of powders or caplets

Early on, it was shown that coating of dried probiotic cultures (Rao *et al.*, 1989) would increase survival of the cells in a simulated gastric fluid (SGF). When a cellulose acetate phthalate cover was applied, complete survival of the culture was noted following exposure to pH 1.33 SGF (Rao *et al.*, 1989). However, such phthalate polymers may not be of food grade, and manufacturers have shifted to fat-based compounds, at least with respect to spray-coating of fine particles of probiotic cultures (Fig. 14.3). Fat coating of small particles generates extremely good survival rates to short exposures to SGF environments (Goulet and Wozniak, 2002).

A coating which is not fat based has also been developed for large caplets (Goulet and Wozniak, 2002). In this instance, the coating prevents the disintegration of the caplet in the stomach thus preventing the loss of viability due to exposure to the SGF.

In addition to highly improved survival to acid environments, small lipid-coated particles have shown a fourfold increase in stability during a 4-month storage period at 30 °C (Siuta-Cruce and Goulet, 2001). Therefore, spray-coating encapsulation acts in at least two ways to improve viability of probiotics delivered as nutraceuticals: increased viability during storage and enhanced survival to SGF. Although no comparative study has been carried out, data from the literature suggest that the lipid-based spray-coating method generates better protection against SGF that does ME in alginate gels, and appears to be the most recommended ME technology for nutraceuticals products (and possibly functional foods as well).

Although data are available on the survival of probiotics in lipid-coated particles to gastric environments, little is known about the effect of the nature of the lipid envelope on release of the probiotics in the intestine or the colon. Fats used in the coating process typically have melting points above 40 °C and should not melt in the GI system. Therefore, it is assumed that the action of bile salts and pancreatic lipases would serve to remove the lipid coating and gradually allow release of the cells.

14.3.2 Advantages of encapsulation in tablet manufacture

Compression of powders into tablets induces some mortality. When spray-coated particles were exposed to the process, the survival rate was two times higher than with uncoated particles (Goulet and Wozniak, 2002). Furthermore, when free-cell and compression-generated tablets (Section 14.2.4) were exposed to the SGF, much higher survival was recorded with the cells in the tablet, but only in very acidic environments (between pH 1.2 and 2.0) (Chan and Zhang, 2005). In the compression-ME system, the formation of a hydrogel around the cell was thought to be the basis for cell protection (Chan and Zhang, 2005), but the fact that only 50% of the tablet was hydrated after 120 min of exposure to the SGF suggests that mass transfer limitation of the acidic solution inside the tablet could also be involved. These data are useful to ascertain the value of encapsulation when the tablet is taken whole. However, further studies are needed to examine the value of the technique when the tablet is chewed and broken down to small particles. It can be expected that the 'limitation of mass transfer of SGF' type of protection offered by encapsulation would be lower in that instance. This would suggest that when spray-coated particles are compressed into tablets, such products would have two methods of protection against exposure to the SGF. The first is the lipid coating of each small particle. The second would be through the tablet itself, which limits mass transfer of the SGF.

The encapsulated powders with enteric coating may protect the probiotic bacteria from gastric acidity and deliver large numbers of bacteria to the small intestine.

14.3.3 Production of ME cultures in gel beads

The industrial process used to manufacture a probiotic nutraceutical (or any powder or frozen concentrate for direct inoculation of foods as well) generates a certain number of stresses to the cells which produce sub-lethal or lethal cell damage (Goulet and Wozniak, 2002). Examples of stressful conditions include:

- oxygen stress due to agitation during pH control fermentations
- oxygen and pressure stresses during centrifugation or filtration
- membrane damage during freezing and/or drying
- pressure and heat stresses during the production of the tablets.

Producing concentrated cultures in alginate or carrageenan gel beads is an alternative to the traditional process, if viability drops are unacceptably high during processing (Champagne, 2006). In this technology, bacteria are microentrapped in gel particles of 1–2 mm which are then added to the growth

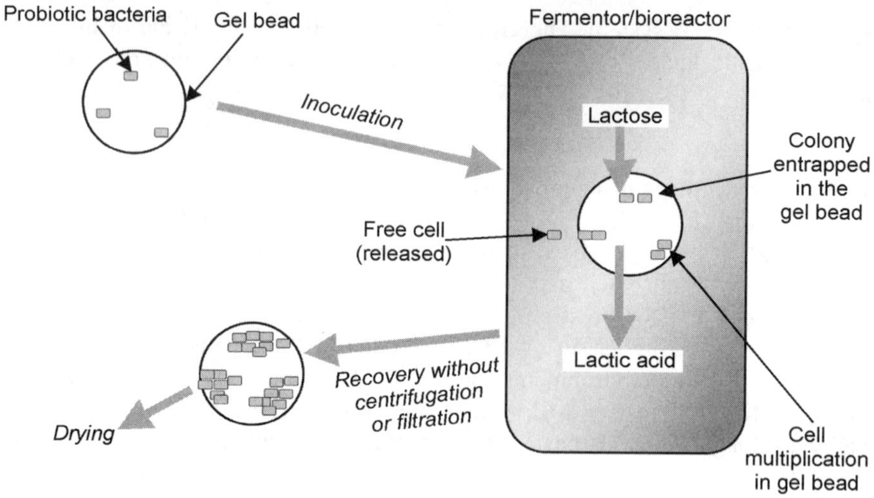

Fig. 14.4 Production of concentrated probiotics in gel particles.

medium (Fig. 14.4). Bacterial growth occurs in the gel beads and the population can reach levels well above 10^{10} cfu/g of gel. In the central portion of the bead, the dissolved oxygen level is lower than that in the liquid medium, which is critical to oxygen-sensitive cultures. The concentrated cultures are then recovered without requiring centrifugation or filtration, which solves the problem of strains which are sensitive to these processing stresses. The approach also enables an alternate method of drying, since lower-costing air-drying (fluidized bed or other) can be used instead of freeze-drying. The alginate beads can also be stored in low water activity solutions and sold fresh (Champagne *et al.*, 1994). However, this immobilized cell technology (ICT) suffers from a number of drawbacks:

- the need for a bead production step in the process
- lower overall biomass yields for most cultures
- solid state of the culture which can affect texture in foods and beverages.

In the past, cultures which did not survive well the technological stresses of the traditional process were simply discarded from the product line. However, in the future, culture producers will have to examine alternatives such as ICT, if they must market a very sensitive strain having valuable and demonstrable health effects.

Survival in the gastrointestinal system

In addition to solving technological problems in the production of probiotics, many studies have shown that cultures which are microencapsulated in alginate beads show improved survival to acid in the gastric environment or to bile solutions (Fig. 14.5). It is important to mention that even though there are numerous examples where ME in alginate is beneficial (Chandramouli *et al.*,

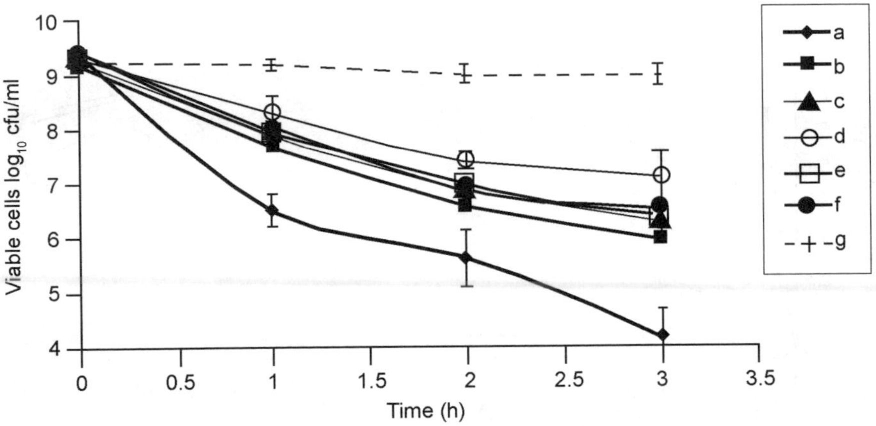

Fig. 14.5 Effect of different coating materials on survival of *Lactobacillus acidophilus* CSCC 2400 under *in vitro* acidic conditions. The error bars represent mean ± SD of three replicates for all treatments. a = free cells; b = cells alginate-encapsulated; c = cells co-encapsulated in alginate and starch; d = cells co-encapsulated in alginate and starch and further coated with alginate; e = cells co-encapsulated in alginate and starch and further coated with poly-L-lysine; f = cells co-encapsulated in alginate and starch and further coated with chitosan; g = free cells pH 6.5 (control). (From Iyer and Kailasapathy, 2005.)

2003, 2004; Guérin *et al.*, 2003; Le Tien *et al.*, 2004; Iyer and Kailasapathy, 2005; Mandal *et al.*, 2006) there are also reports with negative data (Sultana *et al.*, 2000; Truelstrup-Hansen *et al.*, 2002). Possible reasons for these discrepancies could be:

- method of bead production or coating method (Lee at al., 2004; Krasaekoopt *et al.*, 2004; Mandal *et al.*, 2006) (Fig. 14.5)
- size of the bead particle (Lee and Heo, 2000; Chandramouli *et al.*, 2003)
- cell load (Chandramouli *et al.*, 2004).

Controlled release of the probiotics in the GI system

It is beneficial for the encapsulated probiotic bacteria to be released in the small intestine where the Peyers patches exist to activate the immune system. Therefore the polymers used for microencapsulation should be able to protect the bacteria in acidic stomach conditions and release the bacteria in the alkaline conditions in the small intestine.

Many data show that ME in alginate or pectin-based beads can be used for controlled release of bioactive products. *In vitro* studies on calcium alginate or pectinate beads showed that they maintain their integrity in the simulated upper gastrointestinal tract, but subsequently release the agents (Sriamornsak, 1999, Mandal *et al.*, 2006). Data from *ex vivo* conditions in porcine GI contents using alginate beads are in agreement with these findings. Thus, Iyer *et al.* (2004) showed that little bacterial release occurred in the gastric environment, but that bacteria are subsequently released in the intestinal environment. Furthermore, Iyer *et al.* (2005) reported that *Lactobacillus casei* strain Shirota micro-

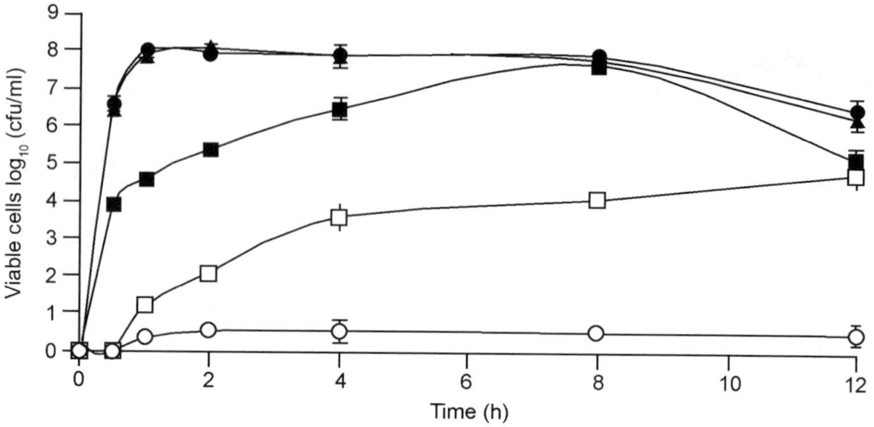

Fig. 14.6 Release of encapsulated bacteria in porcine intestinal contents. The error bars represent mean ± SD ($n = 3$). Symbols: ○ gastric, □ duodenum, ● small intestine (anterior), ▲ small intestine (posterior), ■ colon (Iyer *et al.*, 2004).

encapsulated with chitosan-coated alginate-starch was released in *ex vivo* porcine gastrointestinal contents. They showed that there was almost complete release of the bacteria from capsules within 8 h of incubation in small intestinal contents (Fig. 14.6). With alginate, cell release is basically the result of dissolution of the gel particle due to calcium binding by phosphates or other anions in the medium.

14.4 Applications in functional foods

During processing and storage of foods, many situations can be detrimental to the viability of probiotics:

- inhibitory compounds produced by starter cultures (fermented milks)
- heating – pasteurization (most products)
- freezing (ice cream, frozen products)
- oxygen (during blending in processing or during storage in many packaged products)
- ingredients (salt, spices, flavours)
- drying (powders).

In the past, companies which added a probiotic culture to a food product basically selected the strain as a function of its ability to survive during processing and storage. As scientific evidence of the health effects of probiotics accumulates, it appears that the strain itself may be a critical component of the functionality of the product. Therefore, strains are now selected not so much on their technological properties but rather on their ability to confer a health effect, and companies are now faced with the challenge of finding ways to ensure viability. Encapsulation offers the opportunity to enhance the survival of probiotics in some situations and this section will examine these conditions. Thus, with respect to functional foods, ME is used not so much for controlled release into foods but for incorporation and protection of viable cells into the products.

14.4.1 Dairy products

Yoghurt

Although there are discrepancies, most teams report increased viability of probiotics during storage of yoghurt when microentrapped in alginate, carrageenan or gellan/xanthan gel beads (Adhikari *et al.*, 2000, 2003; Anjani *et al.*, 2004). Talwalkar and Kailasapathy (2003) observed that the benefits of microentrapment occurred when there is oxygen in the medium. Microentrapment thus appears to provide a microenvironment having reduced oxygen levels, which prevent viability losses to oxygen-sensitive strains (Talwalkar and Kailasapathy, 2004) rather than protect the cells against the detrimental effect of the acid environment in yoghurt. Data from McMaster *et al.* (2005) also show an enhanced degree of oxygen tolerance by bifidobacteria in gel beads. Thus, it can be hypothesized that the discrepancies in the literature could be due to various oxygen sensitivities between the strains (microentrapment being ineffective when the strains are not sensitive to oxygen) or to variable dissolved oxygen levels in the products due to differences in experimental procedures.

A second reason for the discrepancies might be improvements in the alginate matrix. Thus, addition of starch in the alginate core or coating the bead with chitosan (Figs 14.1 and 14.6) appear to more systematically improve the viability of the cultures (Sultana *et al.*, 2000; Krasaekoopt *et al.*, 2006).

Cheese

In cheesemaking, a fraction of the bacterial cells are lost in the whey. There appears to be a higher retention of the microencapsulated probiotics in the cheese matrix, as evidenced by higher counts of microentrapped cells in the first weeks following production. Thus, ME in gel particles could be considered for controlled release of the probiotic in the cheese matrix. However, contrary to yoghurt, ME of probiotics in alginate beads does not seem to improve their stability during storage or ripening (Gobbetti *et al.*, 1998), and may even be detrimental (Godward and Kailasapathy, 2003a; Kailasapathy and Masondole 2005). The same lack of beneficial effect of ME on viability during storage was noted in cream cheese and processed cheese, but with spray-coated cultures (Belvis *et al.*, 2006).

The ME benefits noted in yoghurt thus do not seem to extend to cheese. It may be hypothesized that cheese products could have lower free oxygen levels, and that this would not enable the expression of the ME benefit of lower redox level in the bead particle. Whatever the reasons for the discrepancies between data from cheese and yoghurt, these results show that the food matrix influences viability benefits of ME.

Ice cream and frozen desserts

Numerous studies show that probiotics entrapped in alginate or carrageenan beads have greater viability following freezing in various dairy desserts (Sheu and Marshall, 1993; Sheu *et al.*, 1993; Kebary *et al.*, 1998; Shah and Ravula, 2000; Godward and Kailasapathy, 2003b). This is particularly evident when cryo-protectants, such as glycerol, are added in the gel particles (Sheu *et al.*, 1993; Sultana *et al.*, 2000). ME enables the use of such protective compounds in the beads, without having to add them to the ice milk mix, where they would affect sensory properties. Therefore, the microenvironment which is provided by beads seems to be a critical aspect of the success of the application. This is in line with the data for yoghurt where, for that particular application, the oxygen micro-environment is involved. The milk fat in ice cream formulations may also act as encapsulating material for probiotic bacteria during the homogenization of the ice cream mix. The high total solids level in ice cream mix including the fat (emulsion) may provide protection for the bacteria (Kailasapathy and Sultana, 2003).

Unfermented milks

Losses in viability of probiotic bacteria during storage are much lower in unfermented milk than in yoghurt because of the higher pH. Nevertheless, for some strains sensitive to oxygen, significant drops in viability occur (Bolduc *et al.*, 2006). It has been shown that lower oxygen levels are encountered inside gel particles (Talwalkar and Kailasapathy, 2004), and this could potentially explain why bifidobacteria ME in alginate gels have proven to be more stable than free cells during incubation at 4 °C in 2% fat milk (Truelstrup-Hansen *et al.*, 2002).

A number of commercial unfermented milk products contain probiotics, and to our knowledge the cultures are added after pasteurization. It is always a concern to add ingredients after the heat treatment because of the potential contamination with pathogens. It was therefore examined whether micro-encapsulation in gel beads would enhance the survival of the cultures to heating. Although some protection was conferred by microentrapment when the cultures were pasteurized at 63 °C for 30 min, viability losses were high at 70 °C for 1 min (Kushal *et al.*, 2005). At this point, even if milk provides some protection to the cells during heating, it appears that the protection by ME in alginate beads is insufficient to enable a pre-pasteurization inoculation.

Powders

Adding probiotics to milk-based powders destined to be heated, such as for sauces, is arguably useless because of the viability losses encountered. However, a certain number of powdered dairy blends are consumed directly (Probio-Stick™) or freshly after rehydration (instant skim milk, infant formulas and instant pudding blends). Therefore, some studies have examined the stability of encapsulated probiotics in dried milk.

Dried cultures microentrapped in gel particles seem more stable during incubation at room temperature (Kailasapathy and Sureeta, 2004; Lee *et al.*, 2004; Capela *et al.*, 2006). However, spray-coating a freeze-dried culture is also effective in this instance (Siuta-Cruce and Goulet, 2001). With respect to the spray-coated cultures, it can be hypothesized that the lipid coating could constitute a barrier to water and oxygen entrance in the particle. Although ME significantly improved stability of bifidobacteria in powders during storage, the type of packaging as well as the addition of deoxidants and desiccants had still greater effects (Hsiao *et al.*, 2004).

In one study, the ME with starch was carried out using spray-drying technology (O'Riordan *et al.*, 2001), but was detrimental to the stability of bifidobacteria during storage at 19–24 °C. Therefore, not all ME technologies are successful in enhancing viability of cells during storage of dried products.

14.4.2 Meats

Contrary to the dairy sector, there is little research on encapsulation of probiotics in meat fermentation (Saucier and Champagne, 2005). A few studies conducted with starter and protective cultures are nevertheless useful in providing insight into the benefits of encapsulation in this environment.

Kearney *et al.* (1990) compared the fermentation of meat with free and immobilized cultures of *Lactobacillus plantarum* or *Pediococcus pentosaceus*. The fermentation rate was much faster with the encapsulated cells. The authors suggest that the greater fermentation performance of the immobilized cells was caused by the available nutrients (i.e. skim milk) and more protective rehydration conditions within the beads, although cell counts were equivalent in both groups. It could be argued that this could occur for probiotics as well.

Mandal *et al.* (2006) showed that *Lactobacillus casei* cells microencapsulated in alginate particles were more resistant to heat processing at 55–65 °C. It was also found that cultures microencapsulated in alginate beads were more resistant to a heat treatment of 55 °C for 15 min than were free cells, in a MRS broth acidified to pH 5.0 (Lemay *et al.*, 2002). These data suggested that cells microencapsulated in alginate gel particles could be used in meat processing which had moderate heat treatments. Unexpectedly, when microentrapped *Lactobacillus casei* were added in an acidified chicken meat model, the improved resistance conferred by ME in the MRS broth was not observed. These data demonstrate the potential of ME in gel particles for enhanced survival of lactic culture to a thermal processing step, but that the food matrix itself can affect the results.

Limited data suggest that ME in spray-coated particles could be even better than alginate beads. Goulet and Wozniak (2002) report enhanced survival of spray-coated lactobacilli to heating at 50 °C for up to 7 h in a simulated food process. However, as the data on *Lb. sakei* showed, the effect of matrix can be substantial, and assay in meats remain to be carried out.

14.4.3 Plant-based products

As with the dairy sector, most plant-based products which contain probiotics are the result of fermentation, and no ME is required to introduce the cultures into the food. The most frequent fermentations with probiotics are of soy products (Wang *et al.*, 2002), but the fermentation of cereals (Charalampopoulos *et al.*, 2003; Laine *et al.*, 2003) and vegetables is gaining interest (Savard *et al.*, 2003).

In a traditional African fermented maize product, ME in gellan/xanthan beads improved the stability of bifidobacteria during storage (McMaster *et al.*, 2005). This is in agreement with data on fermented milks.

Probiotics microencapsulated in whey particles were added to a pasteurized vegetable juice cocktail and the viability of the probiotic cultures was examined during storage. The microencapsulated *Lactobacillus rhamnosus* cultures were more stable during storage than free cells in the same whey-based medium. In the same study, it was found that ME improved the survival to freezing *Lb. rhamnosus* to freezing in a cranberry juice concentrate, as well as during storage of the frozen product (Reid *et al.*, 2007). In both the vegetable cocktail and frozen cranberry juice assays, data were in accordance with those of the dairy products. However, one important observation was also noted in these assays. Although the cells microencapsulated in a whey protein matrix were more stable than free cells in the comparable whey protein matrix, free cells which were freeze-dried in a milk-based medium showed the highest viability in both the vegetable cocktail and frozen cranberry juice environments, outperforming the whey-ME product. This shows that although ME can be of benefit to the stability of probiotics cultures, the way the bacteria are grown and dried for subsequent industrial use can have as much importance in promoting the viability of the cultures in food systems as ME itself can.

Probiotic lactobacilli show relatively good stability during storage of food products which have low water activity, such as peanut butter ($A_w = 0.24$). Although it was hoped that ME of *Lb. rhamnosus* by spray-coating would further enhance its viability during storage (in peanut butter incubated at 21 ºC) this did not occur (Belvis *et al.*, 2006). Theoretically, fats should constitute a barrier to moisture and acid, but the protection is unfortunately only transient (Ubbink and Kruger, 2006).

The heat treatment applied during biscuit or bread-making is very detrimental to viability of probiotics. ME by spray-coating did not improve the survival and stability of lactobacilli during the bread-making process (Belvis *et al.*, 2006). The use of a sporulated culture, for example *Bacillus subtilis* R0179, seems to be a more effective strategy than ME in enabling the delivery of probiotics in bread. However, ME in a whey-based protein particle was effective in enhancing the survival of probiotic lactobacilli to the heat treatment applied during biscuit manufacture (Reid *et al.*, 2007). Although ME cells had significantly higher viabilities in the biscuits than did free cells, viability losses were still too great to enable commercial application. Further improvements in ME, or by other means, are therefore needed for probiotic-containing biscuits or breads to be successfully marketed.

14.4.4 Other products

A few assays have been carried out for the delivery of probiotics into mayonnaise or cookie fillings. When bifidobacteria were microentrapped into alginate beads and added to mayonnaise, the stability of the cells during storage was greater than with free cells (Khalil and Mansour, 1998). However, when lactobacilli were encapsulated by spray-coating and added to cookie fillings, the cells did not show greater viability during storage. Rather, in the latter product, it was found that the A_w parameter had a greater effect on viability levels (Belvis *et al.*, 2006) than did coating.

14.5 Future trends

Prebiotics are increasingly mixed with probiotics in foods in order to enhance the functionality of the probiotic culture present. It can be expected that prebiotics could also be used as encapsulation materials. Adding fructooligosaccharides or isomaltooligosaccharides into alginate beads has helped improve the stability of bifidobacteria during storage in unfermented milk as well as in simulated gastric solutions (Chen *et al.*, 2005). It can be expected that prebiotics which enable the formation of glassy structures would be particularly useful because of their potential to stabilize biological structures through hydrogen bonding (Ubbink and Kruger, 2006). Prebiotics could also be used to microencapsulate probiotics in spray-drying processes (Ross *et al.*, 2005; Crittenden *et al.*, 2006). Using fats for both their coating and prebiotic properties has also been proposed (Naidu and Baksh, 2004).

The biological activity of probiotic bacteria is due in part to their ability to attach to enterocytes and prevent binding of pathogens by a process of competitive exclusion. Attachment of probiotic bacteria to receptors on the cell surface of intestinal epithelial cells can activate signalling processes leading to synthesis of cytokines that can activate the immune system (Kailasapathy and Chin, 2000). More research is needed to study the adhesion properties of the probiotic bacteria and screening of polymers that can trigger adhesion to targeted intestinal cells and utilize the polymers as capsular wall materials. This could achieve targeted delivery of probiotic bacteria within the gastrointestinal system.

Sensory aspects of foods are critical in acceptance of new products. Food scientists have generally tried to prevent sensory changes related to the addition of probiotics (Champagne *et al.*, 2005), but in many instances there are subtle modifications in texture or flavour (Kailasapathy, 2006). As the benefits of probiotics become increasingly known, it is to be expected that their presence in the product be better identified. There is already one commercial yoghurt product in Mexico which advertises the presence of capsules containing bifidobacteria, the presence of which can easily be recognized by mouthfeel. It can he expected that, in the future, encapsulation will not only be used for the controlled delivery of probiotics in FFN but also to differentiate the functional food from others on the shelves. Thus, ME could actually become a marketing tool for the FFN industry.

14.6 Sources of further information and advice

KAILASAPATHY K (2002) 'Microencapsulation of probiotic bacteria: technology and potential applications' *Curr Issues Intest Microbiol*, 3, 39–48.

SAARELA M (ed.) (2006) *Functional Dairy Products*, Vol. 2. Cambridge: Woodhead Publishing.

STRAND BL, MORCH YA AND SKJAK-BRAEK G (2000) 'Alginate as an immobilization matrix for cells'. *Minerva Biotechnologica*, Special Issue on Bioencapsulation, 12 (4), 223–233.

TALWALKAR A and KAILASAPATHY K (2004) 'A review of oxygen toxicity in probiotic yogurts: influence on the survival of probiotic bacteria and protective techniques' *Compr Rev Food Sci Food Safety*, 3, 117–124.

14.7 References

ADHIKARI K, MUSTAPHA A, GRUN IU and FERNANDO L (2000) 'Viability of microencapsulated bifidobacteria in set yogurt during refrigerated storage' *J Dairy Sci*, 83(9), 1946–1951.

ADHIKARI K, MUSTAPHA A and GRUN IU (2003) 'Survival and metabolic activity of microencapsulated *Bifidobacterium longum* in stirred yogurt' *J Food Sci*, 68, 275–280.

ANDERSON KI and SALYERS AA (1989) 'Genetic evidence that outer membrane binding of starch is required for starch utilization by Bacteroides thetaiotaomicron' *J Bacteriol*, 171, 3199–3204.

ANJANI K, IYER C and KAILASAPATHY K (2004) 'Survival of co-encapsulated complementary probiotics and prebiotics in yoghurt' *Milchwiss*, 59, 396–399.

ARAYA M, MORELLI L, REID G, SANDERS ME, STANTON C, PINEIRO M and BEN EMBAREK P (2002) 'Guidelines for the evaluation of probiotics in food'. Joint FAO/WHO Working Group Report on Drafting Guidelines for the Evaluation of Probiotics in Food, London (ON, Canada), 30 April and 1 May, pp. 1–11.

BELVIS J, TOMPKINS TA, WALLACE TA, CASAVANT L, FORTIN C and CARON C (2006) 'Stability of probiotic bacteria in foodstuffs' CIFST Meeting, Montreal, 30 May.

BOLDUC MP, RAYMOND Y, FUSTIER P, CHAMPAGNE CP and VUILLEMARD JC (2006) 'Sensitivity of bifidobacteria to oxygen and redox potential of non-fermented pasteurized milk' *Int Dairy J*, 16, 1038–1048.

CAPELA P, HAY TKC and SHAH NP (2006) 'Effect of cryoprotectants, prebiotics and microencapsulation on survival of probiotic organisms in yoghurt and freeze-dried yoghurt' *Food Res Internat*, 39, 203–211.

CHAMPAGNE CP (2006) 'Starter cultures biotechnology: the production of concentrated lactic cultures in alginate beads and their applications in the nutraceutical and food industries' *Chem Industry Chem Engin Quarterly*, 12, 11–17.

CHAMPAGNE CP, GARDNER N and DUGAL F (1994) 'Increasing the stability of immobilized *Lactococcus lactis* cultures stored at 4C' *J Indust Microbiol*, 13, 367–371.

CHAMPAGNE CP, GARDNER N and ROY D (2005) 'Challenges in the addition of probiotic cultures to foods' *Crit Revs Food Sci Nutr*, 45, 61–84.

CHAN ES and ZHANG Z (2002) 'Encapsulation of probiotic bacteria *Lactobacillus acidophilus* by direct compression' *Trans IChemE* 80, Part C, 78–82.

CHAN ES and ZHANG Z (2005) 'Bioencapsulation by compression coating of probiotic bacteria for their protection in an acidic medium' *Process Biochem*, 40, 3346–3351.

CHANDRAMOULI V, KAILASAPATHY K, PEIRIS P and JONES M (2003) 'Survival of Ca-alginate microencapsulated *Lactobacillus* spp in simulated gastric conditions' *Austral J Dairy Technol*, 58, 178.

CHANDRAMOULI V, KAILASAPATHY K, PEIRIS P and JONES M (2004) 'An improved method of microencapsulation and its evaluation to protect *Lactobacillus* sp. in simulated gastric environments' *J Microbiol Meth* 56, 27–35.

CHARALAMPOPOULOS D, PANDIELLA SS and WEBB C (2003) 'Evaluation of the effect of malt, wheat and barley extracts on the viability of potentially probiotic lactic acid bacteria under acidic conditions' *Int J Food Microbiol*, 82, 133–141.

CHEN KN, CHEN MJ, LIU CW and CHIU HY (2005) 'Optimization of incorporated prebiotics as coating materials for probiotic microencapsulation' *J Food Sci* 70, M260–M266.

CRITTENDEN R, LAITILA A, FORSSELL P, MATTO J, SAARELA M, MATTILA-SANDHOLM T and MYLLARINEN P (2001) 'Adhesion of bifidobacteria to granular starch and its implications in probiotic technologies' *Appl Environ Microbiol*, 67, 3469–3475.

CRITTENDEN R, SANGUANSRI L and AUGUSTIN MA (2005) 'Probiotic storage and delivery' International Patent WO 2005/030229 A1.

CRITTENDEN R, WEERAKKODY R, SANGUANSRI L and AUGUSTIN MA (2006) 'Synbiotic microcapsules that enhance microbial viability during nonrefrigereted storage and gastrointestinal transit' *Appl Environ Microbiol*, 72, 2280–2282.

DAVE RI and SHAH NP (1997) 'Viability of yoghurt and probiotic bacteria in yoghurts made from commercial starter cultures' *Int Dairy J*, 7, 31–41.

DEMBCZYNSKI R and JANKOWSKI T (2002) 'Growth characteristics and acidifying activity of *Lactobacillus rhamnosus* in alginate/starch liquid-core capsules' *Enz Microb Technol*, 31, 111–115.

DESMOND C, STANTON C, FITZGERALD GF, COLLINS K and ROSS RP (2002) 'Environmental adaptation of probiotic lactobacilli towards improvement of performance during spray drying' *Int Dairy J*, 12, 183–190.

DOLEYRES Y and LACROIX C. (2005) 'Technologies with free and immobilised cells for probiotic bifidobacteria production and protection' *Int Dairy J*, 15, 973–988.

DURAND H and PANES J (2001) 'Particules contenant des microorganismes vivants et procédé de production' Patent WO 01/68808 A1.

GARTI N and BENICHOU A (2004) 'Recent developments in double emulsions for food applications' in: *Food Emulsions*, Friberg SE, Larsson K and Sjöblom J (eds), New York: Marcel Dekker, 353–412.

GOBBETTI M, CORSETTI A, SMACCHI E, ZOCCHETTI A and DE ANGELIS M (1998) 'Production of Crescenza cheese by incorporation of Bifidobacteria' *J Dairy Sci*, 81, 37–47.

GODWARD G and KAILASAPATHY K (2003a) 'Viability and survival of free and encapsulated probiotic bacteria in Cheddar cheese' *Milchwiss*, 58, 624–627.

GODWARD G and KAILASAPATHY K (2003b) 'Viability and survival of free, encapsulated and co-encapsulated probiotic bacteria in ice cream' *Milchwiss*, 58, 161–164.

GOUIN S (2004) 'Microencapsulation: industrial appraisal of existing technologies and trends' *Trends Food Sci Technol*, 15, 330–347.

GOULET J and WOZNIAK J (2002) 'Probiotic stability: a multifaceted reality' *Innovations Food Technol*, February, pp 14–16.

GROBOILLOT AF, CHAMPAGNE CP, DARLING GD, PONCELET D and NEUFELD RJ (1993) 'Membrane formation by interfacial cross-linking of chitosan for the microencapsulation of *Lactococcus lactis*' *Biotechnol Bioengin*, 42, 1157–1165.

GUÉRIN D, VUILLEMARD JC and SUBIRADE M (2003) 'Protection of bifidobacteria encapsulated in polysaccharide-protein gel beads against gastric juice and bile' *J Food Protect*, 66, 2076–2084.

HEINZEN C (2002) 'Microcapsules solve time dependent problems for foodmakers' *European Food Drink Review*, Issue 3, 27–30.

HOU RC, LIN MY, WANG MM and TZEN JTC (2003) 'Increase in viability of entrapped cells of *Lactobacillus delbrueckii* ssp. *bulgaricus* in artificial sesame oil emulsions' *J Dairy Sci*, 86, 424–428.

HSIAO HC, LIAN WC and CHOU CC (2004) 'Effect of packaging conditions and temperature on viability of microencapsulated bifidobacteria during storage' *J Sci Food Agric*, 84, 134–139.

HUGHES DB and HOOVER DG (1995) 'Viability and enzymatic activity of bifidobacteria in milk', *J Dairy Sci*, 78, 268–276.

IWANA H, MASUDA H, FUJISAWA T, SUZUKI H and MITSUOKA T (1993) 'Isolation and identification of *Bifidobacterium* spp in commercial yoghurt sold in Europe. *Bifidobacteria Microflora*, 12, 39–45.

IYER C and KAILASAPATHY K (2005) 'Effect of co-encapsulation of probiotics with prebiotics in increasing the viability of encapsulated bacteria under *in vitro* acidic and bile salt conditions and in yogurt' *J Food Sci* ,70, M18–M23.

IYER C, KAILASAPATHY K and PEIRIS P (2004) 'Evaluation of survival and release of encapsulated bacteria in *ex vivo* porcine gastrointestinal contents using a green fluorescent protein gene-labelled *E. coli*' *Lebens Wissens Technol*, 37, 639–642.

IYER C, PHILLIPS M and KAILASAPATHY K (2005) 'Release studies of *Lactobacillus casei* strain shirota from chitosan-coated alginate-starch microcapsules in *ex vivo* porcine gastrointestinal contents' *Letters in Applied Microbiol*, 41, 493–497.

KAILASAPATHY K (2006) 'Survival of free and encapsulated probiotic bacteria and their effect on the sensory properties of yoghurt' *Lebens Wissens Technol*, 39, 1221–1227.

KAILASAPATHY K and CHIN JC (2000) 'Survival and therapeutic potential of probiotic organisms with reference to *Lactobacillus acidophilus* and *Bifidobacterium* spp.' *Immunology and Cell Biology*, 78, 80–88.

KAILASAPATHY K and MASONDOLE L (2005) 'Survival of free and microencapsulated *Lactobacillus acidophilus* and *Bifidobacterium* and their effect on texture of feta cheese' *Austral J Dairy Technol*, 60, 252–258.

KAILASAPATHY K and RYBKA S (1997) '*Lactobacillus acidophilus* and *Bifidobacterium* spp: their therapeutic potential and survival in yoghurt' *Austral J Dairy Technol*, 52, 28–35.

KAILASAPATHY K and SULTANA K (2003) 'Survival and beta-D-galactosidase activity of encapsulated and free *Lactobacillus acidophilus* and *Bifidobacterium lactis* in ice cream' *Austral J. Dairy Technol*, 58, 223–227.

KAILASAPATHY K and SUREETA BS (2004) 'Effect of storage on shelf life and viability of freeze-dried and microencapsulated *Lactobacillus acidophilus* and *Bifidobacterium infantis* cultures' *Austral J Dairy Technol*, 59, 204–208.

KEARNEY L, UPTON M and McLOUGHLIN A (1990) 'Meat fermentation with immobilized lactic acid bacteria' *Appl Microbiol Biotechnol*, 33, 648–651.

KEBARY KMK, HUSSEIN SA and BADAWI RM (1998) 'Improving viability of bifidobacteria and their effect of frozen ice milk' *Egyptian J Dairy Sci*, 26, 319–337.

KHALIL AH and MANSOUR EH (1998) 'Alginate-encapsulated bifidobacteria survival in mayonnaise' *J Food Sci*, 63, 702–705.

KIM HS, KAMARA BJ, GOOD IC and ENDERS GL (1988) 'Method for the preparation of stable microencapsulated lactic acid bacteria' *J Industrial Microbiol*, 3, 253–257.

KLAVER FAM, KINGMA F and WEERKAMP AH (1993) 'Growth and survival of bifidobacteria in milk' *Neth Milk Dairy J*, 47, 151–164.

KLEIN J, STOCK J and VORLOP KD (1983) 'Pore size and properties of spherical Ca-alginate biocatalysts' *Eur J App Microbiol Biotechnol*, 18, 86–91.

KRASAEKOOPT W, BHANDARI B and DEETH H (2003) 'Evaluation of encapsulation techniques of probiotics for yoghurt' *Int Dairy J*, 13, 3–13.

KRASAEKOOPT W, BHANDARI B and DEETH H (2004) 'The influence of coating material on some properties of alginate beads and survivability of microencapsulated probiotic bacteria' *Int Dairy J*, 14, 737–743.

KRASAEKOOPT W, BHANDARI B and DEETH HC (2006) 'Survival of probiotics encapsulated in chitosan-coated alginate beads in yoghurt from UHT- and conventionally treated milk during storage' *Lebens Wissens Technol*, 39, 177–183.

KURMANN JA and RASIC JL (1991) 'The health potential of products containing bifidobacteria' in: *Therapeutic Properties of Fermented Milks*, Robinson RK (ed.). London: Elsevier Science, 117–158.

KUSHAL R, ANAND SK and CHANDER H (2005) 'Development of a direct delivery system for a co-culture of *L. acidophilus* and *B. bifidum* based on micro-entrapment' *Milchwiss*, 60, 130–134.

LAINE R, SALMINEN S, BENNO Y and OUWEHAND AC (2003) 'Performance of bifidobacteria in oat-based media' *Int J Food Microbiol*, 83, 105–109.

LARISH BC, PONCELET D, NEUFELD RJ and CHAMPAGNE CP (1994) 'Microencapsulation of *Lactococcus lactis* subsp. *cremoris*' *J Microencapsul*, 11, 189–195.

LE-TIEN C, MILLETTE M, MATEESCU MA and LACROIX M (2004) 'Modified alginate and chitosan for lactic acid bacteria immobilization' *Biotechnol Appl Biochem*, 39, 347–354.

LEE JS, CHA DS and PARK HJ (2004) 'Survival of freeze-dried *Lactobacillus bulgaricus* KFRI 673 in chitosan-coated calcium alginate microparticles' *J Agric Food Chem*, 52, 7300–7305.

LEE KY and HEO TR (2000) 'Survival of *Bifidobacterium longum* immobilized in calcium alginate beads in simulated gastric juices and bile salt solution' *Appl Environ Microbiol*, 66, 869–873.

LEMAY MJ, CHAMPAGNE CP, GARIÉPY C and SAUCIER L (2002) A comparison of the effect of meat formulation on the heat resistance of free or encapsulated cultures of *Lactobacillus sakei*. *J Food Sci*, 67, 3428–3434.

LOURENS-HATTINGH A and VILJOEN BC (2001) 'Yoghurt as probiotic carrier in food' *Int Dairy J*, 11, 1–17.

McCLEMENTS DJ (2005) 'Emulsion stability' in: *Food Emulsions. Principles, Practices, and Techniques*, 2nd edn, McClements DJ, Boca Raton, FL: CRC Press, 185–233.

McMASTER LD, KOKOTT SA, REID SJ and ABRATT VR (2005) 'Use of traditional African fermented beverages as delivery vehicles for *Bifidobacterium lactis* DSM 10140' *Int J Food Microbiol*, 102, 231–237.

MANDAL S, PUNIYA AK and SINGH K (2006) 'Effect of alginate concentration on survival of microencapsulated *Lactobacillus casei* NCDC-298' *Int Dairy J*, 16, 1190–1195.

MUTHUKUMARASAMY P, ALLAN-WOJTAS P and HOLLEY RA (2006) 'Stability of *Lactobacilus reuteri* in different types of microcapsules' *J Food Sci*, 71, M20–M24.

MYLLARINEN P (2002) 'Starches – from granules to novel applications' *VTT Publications* 473. http://virtual.vtt.fi/inf/pdf/publications/2002/P473.pdf

MYLLARINEN P, FORSSELL P, VON WRIGHT A, ALANDER M, MATTILA-SANDHOLM T and POUTANEN K (2000) 'Starch capsules containing microorganisms and/or polypeptides or proteins and a process for producing them' Patent WO9952511.FI104405.

NAIDU SA and BAKSH B (2004) 'Prebiotic and preservative uses of oil-emulsified probiotic encapsulations' International Patent WO 2004/028460 A2.

O'RIORDAN K, ANDREW D, BUCKLE K and CONWAY P (2001) 'Evaluation of microencapsulation of a bifidobacterium strain with starch as an approach to prolonging viability during storage' *J Appl Microbiol*, 91, 1059–1066.

OUWEHAND AC and SALMINEN, SJ (1998) 'The health effects of cultured milk products with viable and non-viable bacteria' *Int Dairy J*, 8, 749–758.

OUWEHAND AC, SALVADORI BB, FONDÉN R, MOGENSEN G, SALMINEN S and SELLARS R (2003) 'Health effects of probiotics and culture-containing dairy products in humans'. Brussels: International Dairy Federation (IDF), Document #380, pp. 4–19.

PICOT A and LACROIX C (2003a) 'Optimization of dynamic loop mixer operating conditions for production of o/w emulsion for cell microencapsulation' *Le Lait*, 83, 237–250.

PICOT A and LACROIX C (2003b) 'Production of multiphase water-insoluble microcapsules for cell microencapsulation using an emulsification/spray-drying technology' *J Food Sci*, 68, 2693–2700.

RAO AV, SHIWNARAIN N and MAHARAJ I (1989) 'Survival of microencapsulated *Bifidobacterium pseudolongum* in simulated gastric and intestinal juices' *Can Institute Food Sci Technol J*, 22, 345–349.

REID AA, CHAMPAGNE CP, GARDNER N, FUSTIER P and VUILLEMARD JC (2007) 'Survival in food systems of *Lactobacillus rhamnosus* R011 microentrapped in whey protein gel particles' *J Food Sci.* 72, M031–M037.

ROSS RP, DESMOND C, FITZGERALD GF and STANTON C (2005) 'Overcoming the technological hurdles in the development of probiotic foods' *J Applied Microbiol*, 98, 1410–1417.

RYBKA S and FLEET GH (1997) 'Populations of *Lactobacillus delbrueckii* subspp *bulgaricus, Streptococcus thermophilus, Lactobacillus acidophilus* and *Bifidobacterium* spp in Australian yoghurts' *Food Australia*, 49, 471–475.

SAUCIER L and CHAMPAGNE CP (2005) 'Cell immobilization technology and meat processing' in: *Applications of Cell Immobilisation Bioechnology*. Nedovic V and Willaert R (eds), Dordrecht, The Netherlands: Springer-Kluwer series 'Focus on Biotechnology' Vol 8B, pp. 337–350.

SAVARD T, GARDNER N and CHAMPAGNE CP (2003) 'Croissance de cultures de *Lactobacillus* et de *Bifidobacterium* dans un jus de légumes et viabilité au cours de l'entreposage dans le jus de légumes fermenté' *Sciences des Aliments*, 23, 273–283.

SCHROOYEN PMM, VAN DER MEER R and DE KRUIF CG (2001) 'Microencapsulation: its application in nutrition' *Proc Nutr Soc*, 60, 475–479.

SHAH NP (2000) 'Probiotic bacteria: selective enumeration and survival in dairy foods' *J Dairy Sci*, 83, 894–907.

SHAH NP and RAVULA RR (2000) 'Microencapsulation of probiotic bacteria and their survival in frozen fermented dairy desserts' *Austral J Dairy Technol*, 55, 139–144.

SHEU TY and MARSHALL RT (1993) 'Microentrapment of lactobacilli in calcium alginate gels' *J Food Sci*, 54, 557–561.

SHEU TY, MARSHALL RT and HEYMANN A (1993) 'Improving survival of culture bacteria in frozen desserts by microentrapment' *J Dairy Sci*, 76, 1902–1907.

SHIN YS, BAIK BH and LEE JK (2002) 'Liquid yogurt with encapsulated lactic acid bacteria and method for producing the same' US Patent US 6,447,823 B1.

SIMMONS DL, MOSLEMY P, PAQUETTE GO, GUÉRIN D and JOLY MH (2004) 'Stable probiotic microsphere compositions and their methods of preparation' International Patent WO 2004/022031 A2.

SIUTA-CRUCE P and GOULET J (2001) 'Improving probiotic survival rates: microencapsulation preserves the potency of probiotic microorganisms in food systems' *Food Technol*, 55, 36–42.

SRIAMORNSAK P (1999) 'Effect of calcium concentration, hardening agent and drying condition on release characteristics of oral proteins from calcium pectinate gels' *Eur J Pharmaceut Sci*, 8, 221–227.

STANTON C, ROSS RP, FITZGERALD GF and SINDEREN DV (2005) 'Fermented functional foods based on probiotics and their biogenic metabolites' *Curr Opin Biotechnol* 16, 198–203.

SULTANA K, GODWARD G, REYNOLDS N, ARUMUGASWAMY R, PEIRIS P and KAILASAPATHY K (2000) 'Encapsulation of probiotic bacteria with alginate-starch and evaluation of survival in simulated gastrointestinal conditions and in yoghurt' *Int J Food Microbiol*, 62, 47–55.

TABRIZI CA, WALCHER P, MAYR UB, STIEDL T, BINDER M, McGRATH J and LUBITZ W (2004) 'Bacterial ghosts – biological particles as delivery systems for antigens, nucleic acids and drugs' *Curr Opin Biotechnol*, 15, 530–537.

TALWALKAR A and KAILASAPATHY K (2003) 'Effect of microencapsulation on oxygen toxicity in probiotic bacteria' *Austral J Dairy Technol*, 58, 36–39.

TALWALKAR A and KAILASAPATHY K (2004) 'A review of oxygen toxicity in probiotic yogurts: influence on the survival of probiotic bacteria and protective techniques' *Comp Rev Food Sci Food Safety*, 3, 117–124.

TRUELSTRUP-HANSEN L, ALLAN-WOJTAS PM, JIN YL and PAULSON AT (2002) 'Survival of Ca-alginate microencapsulated *Bifidofacterium* ssp. in milk and simulated gastrointestinal conditions' *Food Microbiol*, 19, 35–45.

UBBINK JB, SCHAER-ZMMARETTI P and CAVADINI C (2003) 'Probiotic delivery system' European Patent EP 1 244 458 A1.

UBBINK J and KRUGER J (2006) 'Physical approaches for the delivery of active ingredients in foods' *Trends Food Sci Technol*, 17, 244–254.

WANG YC, YU RC and CHOU CC (2002) 'Growth and survival of bifidobacteria and lactic acid bacteria during the fermentation and storage of cultured soymilk drinks' *Food Microbiol*, 19, 501–508.

15

Encapsulation of fish oils

S.-J. Lee, Massey University, New Zealand and D. Y. Ying, HortResearch, New Zealand

15.1 Introduction

Fish oil is the richest source of omega-3 long chain (LC) polyunsaturated fatty acids (PUFA), such as eicosapentaenoic acid (EPA, C20:5 n-3) and docosahexaenoic acid (DHA, C22:6 n-3). These two fatty acids have been extensively researched and are now well known for their beneficial roles for health promotion and disease prevention. Inadequate intake of omega-3 LC PUFA could possibly have been contributing to some health problems. As a consequence, an adequate consumption of fish oils or fresh fish has been emphasized to be essential for normal healthy life. However, there is a growing concern that current diets are deficient in omega-3 LC PUFA and intakes of fish or fish oil are very low in some populations. This could be partly due to culture, eating habits, unavailability of fresh fish, and unpleasant fishy smell and taste.

Incorporation of fish oils into commonly consumed foods has been considered as a potential strategy to increase intake of omega-3 LC PUFA. However, a high susceptibility of fish oil to oxidative deterioration can cause off-flavour development and shelf-stability reduction. Also, fish oil has objectionable fishy odour and taste, thereby altering sensory properties of foods. These problems have been a major hurdle for a wide use of fish oils in the food industry. Microencapsulation technology has been approached as a solution by protecting fish oil from oxidation and masking fishy flavour and taste, thus enabling it to be delivered to food products without affecting sensory properties. In this chapter, the range of microencapsulation technologies and encapsulant materials used to deliver fish oils and the application of encapsulated fish oils in food products are described. The benefits of fish oil and challenges of adding fish oils to foods are also discussed.

15.1.1 Benefits of fish oil and recommended levels of intake

Omega-3 LC PUFA, such as EPA and DHA, primarily found in marine fish oils, have been known to offer numerous health benefits relating to coronary heart disease, brain and eye function in infants, joint health, psychological mood, mental behaviour, cancer, diabetes, skin disorders, pregnancy, and lactation.[1–9] In particular, DHA is ascribed to be the essential nutrient for normal brain growth and development[7,10] and visual development in foetus, new born and infants.[7] Most of these health benefits are associated with omega-3 LC PUFA themselves, serving as important structural components of phospholipids in cell membranes, thus influencing membrane fluidity and function, and modulating the optimal synthesis of potent cellular mediators, eicosanoids (prostaglandins and leukotrienes),[4,6–8,11,12] which mediate a wide range of physiological functions such as blood pressure control, muscle contraction, and inflammation mediation.[9]

In contrast to increasing evidence of the health benefits of fish oil containing omega-3 LC PUFA, intake of omega-3 LC PUFA has been recognized to be very low in some populations and countries, being far below the optimal level required for health benefits. This has led scientific, nutritional and health organizations to advocate the increased consumption of omega-3 LC PUFA and to make a range of nutritional recommendations for optimal daily intake levels of omega-3 LC PUFA. The Scientific Committee for Food of the European Union recommends an intake of 1000 mg of omega-3 LC PUFA daily, representing about 0.5% of the total energy consumption.[13] A workshop on the 'Essentiality of and Recommended Dietary Intake for Omega-6 and Omega-3 Fatty Acids' held in 1999 at the National Institute of Heath (NIH), Maryland, USA, recommended the intake levels of EPA and DHA as at least 220 mg per day each, corresponding to 0.1% of total diet energy each.[14]

In 2002, the US Food and Drug Administration (FDA) approved a health claim for omega-3 fatty acids but limited its use only to dietary fish oil supplements.[15] The health claim states 'consumption of omega-3 fatty acids may reduce the risk of coronary heart disease'. To make this claim, a product must contain at least 600 mg but not more than 2 g per day of DHA plus EPA. The maximum allowed upper level limit is to prevent unnecessary over-consumption of omega-3 fatty acids which might have adverse effects on glycemic control, increased bleeding times and elevated low density lipoprotein (LDL) cholesterol level.[6,16] Fish oil dietary supplements in the form of soft gelatine capsules are currently sold widely in food and health stores and pharmacies all over the world. Based on the recommendations from different agencies and organizations, the optimum daily intake of omega-3 LC PUFA can be estimated between 600 mg and 1 g. It is not easy for consumers to actually know how much omega-3 LC PUFA they intake via consumption of fish meals. At least two or three fish meals per week (200–300 g fish per week) are generally recommended, which is estimated to be approximately equivalent to a minimum of 200–400 mg of EPA and DHA per day.[4]

In 2004, the FDA also announced that conventional foods containing EPA and DHA omega-3 fatty acids may also make a qualified health claim for reduced risk of coronary heart disease (CHD) on their label. The qualified claim for labelling should read: 'Supportive but not conclusive research shows that consumption of EPA and DHA omega-3 fatty acids may reduce the risk of coronary heart disease. One serving of [name of food] provides [x] grams of EPA and DHA omega-3 fatty acids. [See nutrition information for total fat, saturated fat and cholesterol content.]'.[17] This health claim allows health conscious consumers to be able to identify and differentiate foods containing these important nutrients.[17] Although official recommended dietary allowances (RDA) have not been set for omega-3 LC PUFA (EPA and DHA), it could be possible to establish them in the next 10 years because of increasingly strong evidence of health benefits of EPA and DHA from clinical, nutritional, medical and physiological studies.

As a result of increased consumer awareness of health promoting functions of omega-3 fatty acids, EPA and DHA, and therefore an increased demand for intakes of these nutrients, large gelatine-based soft gels (capsules) containing highly concentrated omega-3 LC PUFA from fish oil have been manufactured and sold as dietary supplements in health shops. The dietary supplements are recommended to be taken 1–3 capsules daily with a meal, depending on content of EPA and DHA. The fish oil supplements may be the best option to ensure an intake of the recommended daily dose of omega-3 PUFA. However, a frequent, regular administration of dietary supplements in the form of large capsules every day is not very appealing and can be inconvenient and troublesome to many people, especially children and adolescents.

In recent years, the food industry has also been attempting to incorporate fish oils into their food products to meet consumer demand for healthy foods, develop novel functional foods, and facilitate the general public's intake of omega-3 LC PUFA. This approach could potentially offer significant health benefits to more people. However, rapid oxidative deterioration and fishy flavours and odours have been a major technological hurdle in developing fish oil-enriched processed foods.[4,18] The incorporation of fish oils in foods requires special food formulations and delivery systems to overcome these issues. Microencapsulation technology has been explored as a vehicle for delivering fish oil into foods, thus protecting the oil from oxidation during food processing and storage and masking fishy taste and smell in food products. Some foods (bread, margarine, yoghurt, infant formula) fortified with fish oils have been introduced and are currently available in the market in some countries. More varieties of fish oil-enriched functional products are expected to be introduced with the development of more advanced encapsulation technology and more suitable and cost effective microencapsulating materials.

15.1.2 Sources of omega-3 LC PUFA

As described above, marine fish oils are the only major dietary source of omega-3 LC PUFA. Fish oils available on the current market are the oils extracted from anchovy, capelin, cod and cod liver, herring, mackerel, menhaden, salmon, sardine, tuna, and rainbow trout.[12] The fatty acid composition of fish oils varies considerably among species of fish. Table 15.1 shows the fatty acid composition of selected fish oils. The relative amount of EPA and DHA ranges within 2–28% and 4–33%, respectively.[19]

In recent years, concerns have been raised about consuming fish or their oils that are rich in mercury or might be contaminated with heavy metals, toxins or environmental pollutants. In March 2004, the FDA and the Environmental Protection Agency jointly announced a consumer advisory on fish intake.[20] Pregnant women, nursing mothers, and young children are recommended to avoid fish high in mercury including shark, swordfish, king mackerel or tilefish. The importance of fish consumption was also emphasized on the consumer advisory which recommends eating two fish meals (up to 12 oz (340 g) of a variety of fish and shellfish with a low level in mercury) weekly. This is in line with the daily minimum intake of omega-3 LC PUFA.[7] When producing dietary supplements or fortifying foods with fish oils, food grade fish oils extracted and refined from crude fish oils recognized as safe should be used to eliminate any risk of mercury, pollutant or toxin contamination.

Another known possible source of omega-3 LC PUFA is from alpha-linolenic acid (ALA, C18:3 n-3). ALA is also an essential omega-3 fatty acid, which cannot be synthesized in the human body. The plant seeds and oils of canola, flaxseed, perilla, walnut, soybean, and rapeseed are primary sources of ALA.[12] As a precursor of EPA and DHA, ALA can be converted to both EPA and DHA in the human body through elongation and desaturation of ALA.[12,16,21,22] However, the extent of its conversion is not clearly known[16] and remains controversial.[23] Diets rich in ALA are not a reliable means of obtaining high levels of EPA and DHA.

A more promising alternative source is from microbial oil which is derived from microalgae grown in controlled culture systems. Microbial oil is an emerging source of omega-3 LC PUFA, especially DHA. Recently, Martek Biosciences (Columbia, MD) have produced omega-3 oils rich in DHA (40%) from microalgae that have been granted GRAS (generally recognized as safe) status by the US government. Production and applications of algal oils rich in DHA are expected to grow markedly for both nutritional supplements and fortified foods as a vegetarian source of omega-3 LC PUFA. The possibility of production of omega-3 LC PUFA in transgenic plants using genetic engineering has been also explored as a sustainable and clean alternative to fish oil.[24]

Fish oils have been added into animal feeds to produce animal products (meats, eggs and milk) naturally containing a high amount of omega-3 LC PUFA by enriching their tissues with omega-3 LC PUFA.[7,25–36] This is another potential way of increasing natural sources of food materials rich in omega-3 LC PUFA.

15.1.3 Challenges in fish oil formulation

Fish oil is highly susceptible to oxidation because of the high degree of unsaturation of omega-3 LC PUFA that contains five or six double bonds.

Table 15.1 Typical composition of fatty acids of lipids from selected fish

Fatty acid	Capelin [134]	Cod flesh [134]	Cod liver [134]	Herring [134]	Menhaden [134]	Mackerel [12]	Salmon [12]	Tuna [12]	Anchovy [135]	Rainbow trout [135]	Sardine [135]
C14:0	5.9	1.4	3.6	6.1	7.3	8.0	6.0	3.0	7.4	2.0	6.7
C16:0	8.7	18.0	11.0	10.7	19.5	14.0	15.0	20.0	17.8	17.8	18.3
C16:1 *n*-7	10.3	3.3	11.7	7.3	9.1	5.0	7.0	4.0	9.0	6.1	6.0
C18:0	0.6	3.4	0.3	3.4	4.5	2.0	3.0	5.0	4.7	4.8	4.2
C18:1[a]	6.0	10.7	22.0	10.3	10.4	12.0	16.0	16.0	11.6	32.5	13.0
C18:2 *n-6*	0.5	1.0	1.5	1.0	1.3	2.0	3.0	2.0	1.2	14.3	1.2
C18:3 *n-3 (ALA)*	0.2	0.1	0.1	2.0	0.7	2.0	1.0	1.0	0.7	0.3	1.0
C18:4 *n-3*	1.2	0.2	1.0	3.2	2.7	7.0	2.0	1.0	3.1	0.4	3.1
C20:1[b]	17.6	2.6	9.1	13.4	1.5	12.0	15.0	2.0	1.5	2.6	4.3
C20:5 *n-3* (EPA)	9.3	19.1	12.2	7.5	18.3	10.0	8.0	7.0	17.0	2.0	11.0
C22:1[c]	27.8	0.7	3.2	21.3	1.6	1.0	2.0	–	1.1	1.3	3.8
C22:6 *n-3* (DHA)	4.1	32.6	12.7	6.8	9.6	14.0	10.0	26.0	8.7	10.3	13.0
Other fatty acids	7.8	6.9	11.6	7.3	13.9	11.0	12.0	13.0	16.2	5.6	14.4
Total omega-3 LC PUFA[d]	14.8	52.0	26.0	19.4	31.2	33.0	21.0	35.0	29.5	13.0	28.1

[a] C18:1 n-5, n-7 & n-9
[b] C20:1 n-9 & n-7
[c] C22:1 n-9 & n-11
[a,b,c] Monoenes with varying mixtures of n-5, n-9 and n-_1 isomers.
[d] EPA and DHA
References in square brackets

Hence, oxidation can occur very readily at any stage if precautions are not taken during fish oil manufacture, storage and transport and also during food processing and storage when fish oil is fortified into foods. Exposure to certain conditions (oxygen, light, heat, pH, moisture, metal ions and reactive compounds) that promote an initiation of oxidation cannot be avoided completely. Oxidation causes the development of rancid off-flavour and degradation of essential nutrients, thus affecting sensory and nutritional qualities and shelf-stability of fish oil-enriched foods. Also, oxidation induces the formation of potent toxic compounds and makes oils detrimental to human health.[37–39] Although hydrogenation can be employed in manufacturing fish oils to increase their oxidative stability by decreasing the degree of unsaturation, it is not desired because of the formation of nutritionally unfavourable trans isomers[40] and a loss of beneficial effects of omega-3 LC PUFA.[8,41]

Another issue associated with fish oils is their objectionable fishy aroma and taste. Simply adding fish oil to food products can produce unacceptable sensory profiles, even if the added fish oil has not been oxidized. The oxidative products derived from omega-3 fatty acids have been positively identified with fishy and metallic flavours. A report on milk and mayonnaise fortified with fish oil identified more than 60 different volatiles with a strong fishy odour, including alkenals, alkadienals, alkatrienals and vinyl ketones.[42] The specific compound identified as fishy off-flavour after oxidation of fish oils are two isomers, (trans, cis, cis) and (trans, trans, cis) of 2,4,7-decatrienal.[42,43] The most potent odours were identified as 1-penten-3-one (pungent, green odour), 4 (cis)-heptenal (fishy odour), 1-octen-3-one, 1,5-octadien-3-one, 2,4 (trans, trans)-heptadienal and 2,6 (trans, cis)-nonadienal (cucumber odour), with 1-penten-3-one (plastic, leather odour) reported as the major contributor to the unpleasant sharp-fishy off-flavour in fish oil.[42,44–46] These off-flavours, generally identified as fishy, metallic and rancid, are considered one of the major deterrents for the increased consumption of fish and fish oils.

The biggest challenge to the food formulators and manufacturers in developing food products fortified with fish oils is therefore to overcome rapid oxidative deterioration of fish oil and to mask fishy odour and taste in the final products. To remove the fishy flavour, a proprietary process called 'molecular distillation' has been developed. Addition of some flavours such as orange, mint, apple or lemon to fish oils can make fish oil taste more appetizing.[47] An enzymatic process for deodorization of fishy odour and taste in the production of fish oils has been developed by Pronova Biocare (Lysaker, Norway).[48]

One important consideration to minimize oxidation is to use food grade fish oils of the highest quality. Fish oils must be extracted from raw materials of good quality and manufactured under controlled conditions (i.e. high vacuum and flushing with nitrogen) of refining, deodorization, packaging and storage. If any of processing steps is not done carefully, the oils deteriorate faster than expected and still retain strong fishy odour and taste. Another possible factor to consider in incorporating fish oil in foods is to choose foods that would be stored at a cold storage temperature and consumed soon after the time of food manufacture.[8] Ideal types of foods for enrichment with fish oils are those consumed frequently, thus increasing the constant daily intake of EPA and DHA.[14] The level of fish oils to be added is another factor. It should not be too excessive or small. For foods that already contain a large amount of plant or vegetable oil or milk fat, some portions of fat can be replaced by fish oil without influencing the fat content of foods. Non-excessive fishy flavour in foods can be masked with other intense odour and taste (sweet).[4]

The absorption and efficacy of omega-3 LC PUFA from foods fortified with microencapsulated fish oil have been questioned. Studies on bioavailability have been conducted by measuring the composition of fatty acids, including omega-3 LC PUFA, in blood plasma after administration of microencapsulated fish oil enriched foods and fish oil dietary supplements to human subjects.[49–52] No significant differences in bioavailability of EPA and DHA have been shown between the two different delivery forms. Microencapsulated fish oil added to foods is thus as effective as fish oil delivered in a dietary supplement form.

15.1.4 Stabilization of fish oils with antioxidants

To stabilize fish oils against oxidative deterioration, synthetic or natural antioxidants have been added to fish oils during the oil manufacturing process. Synthetic antioxidants include butylated hydroxylanisole (BHA), butylated hydroxytoluene (BHT), and tertiary butylhydroquinone (TBHQ) used either alone or in combination. Natural antioxidants, such as tocopherols, ascorbyl palmitate, lecithin, carotenoids and rosemary extracts, have become preferred because they are naturally occurring and have fewer restrictions in food standards and regulations. Caffeic acid known as a natural phenol has also been demonstrated for its effectiveness as an antioxidant for cod liver oil.[53]

Effectiveness of an antioxidant can differ with the source of the fish oils,[54,55] being affected by chemical composition and natural antioxidants already present in fish oils.[7] Alpha tocopherol is the major naturally occurring antioxidant found in fish oils, but the content becomes very low during oil processing (refining, bleaching and deodorization) since these processing conditions remove non-triacylglycerol components from fish oils and cause a vaporization of heat labile tocopherol.[8] This implies that the addition of antioxidants should be made at the last stage of the oil manufacturing process prior to packaging which might not involve high temperatures. Several studies have shown that, depending on the levels of added alpha-tocopherol, it can behave as an antioxidant or a prooxidant.[39,56–58] This is believed to be partly affected by the amount of naturally occurring alpha-tocopherol that was already present in fish oils.

A mixture of ascorbyl palmitate, lecithin and tocopherol has been found to be most effective in protecting fish oil against oxidation because of synergistic effects.[1,7,59] This antioxidant system was patented by Löliger and Saucy[60] and is often referred to as ALT. The ALT antioxidant system was, however, shown to be more effective in oils that exist in an emulsion state than in a dried form because of the greater partitioning and orientation of the added antioxidants between water and oil phases of the emulsion.[61] Antioxidant effect for microencapsulated fish oil powder was shown to be greater with alpha-tocopherol, which is a lipophilic antioxidant, than with an amphiphilic antioxidant of ascorbyl palmitate, suggesting the result of partitioning.[62] It should be mentioned, however, that ascorbic acid can also act as a prooxidant in a food system where both oils/fats and iron are present by reducing Fe^{3+} (ferric iron) to Fe^{2+} (ferrous iron). Ferrous iron catalyzes the breakdown of lipid hydroperoxides at a much faster rate than ferric iron, thus accelerating lipid oxidation.[63,64]

It appears that antioxidants must be selected carefully according to chemical composition of fish oils and physical forms of oils (free oil, emulsion and microencapsulated oil). It should also be noted that even with antioxidants, oils might still be oxidized and deteriorated if they are exposed to air, light or heat for a certain period of time and if antioxidants are consumed for antioxidant activity, thus becoming no longer available for activity. Therefore, it is essential to eliminate oxygen, light and heat by storing fish oils under dark and cold conditions and in an oxygen-free nitrogen atmosphere.[8] Vacuum packaging and packaging material impermeable to oxygen need also to be considered in order to prolong the shelf-life of encapsulated fish oil during storage.[39] Nevertheless, the most important factor in the production of a high quality fish oil is the condition of the raw fish material at the beginning of processing. It should not have been contaminated or damaged and should have been held under chilled conditions so as to minimize the possible microbial, enzymatic and oxidative degradation of fish fatty tissue.

15.2 Encapsulation technologies used to deliver fish oils

Microencapsulation is often described as a process of packaging or coating of small droplets of liquid or solid particles (referred to as core, active or filler) with a protective shell (referred to as wall, coating, carrier or encapsulant) into small capsules,[65] which range in size from 1 to 1000 μm diameter. Over the past three decades, a variety of food ingredients including essential oils, fats, pigments, vitamins, minerals, enzymes, antioxidants, sweeteners, flavours, volatiles and probiotics have been encapsulated as core materials by using different encapsulating wall materials and encapsulation methods for various purposes: protection from deterioration, preservation of functional characteristics, extended shelf storage stability, controlled release, targeted delivery, and alleviation of formulation and processing problems. Despite such efforts, the successful utilization and practical implementation of encapsulation technology in the food industry has been relatively limited with the use of the technique of spray-drying as a main encapsulation method. This is because of production cost, economic infeasibility, scale-up difficulty, limited availability of suitable natural coating materials, and incompatibility with final products. Nevertheless, encapsulation technology still has significant potential as a novel solution to the problems associated with formulations, fortification, stabilization and delivery of labile food ingredients to foods which cannot be achieved simply by other methods.

The materials and methods that have been used for microencapsulation of fish oils are listed in Table 15.2. It is not possible to compare the results of oxidative stabilities of microencapsulated fish oils that have been reported by different research groups because of differences in materials and methods used, for example, types and composition of wall materials, sources of fish oils, homogenization and encapsulation conditions, sample storage environments, addition and types of antioxidants used, and different methods for measuring oxidation parameters.

A number of review articles on encapsulation relating to food ingredients and applications and methods have been published.[66–78] This chapter covers mainly techniques and materials that have been used for encapsulation of fish oils.

15.2.1 Emulsion preparation prior to microencapsulation of fish oils

Irrespective of encapsulation methods, the initial step in the process of fish oil encapsulation is to prepare an emulsion. Briefly, oil is emulsified as submicron-sized particles in an aqueous solution containing dissolved encapsulating wall material by a homogenizer or microfluidizer ranging in pressure from 2500 to 20 000 psi and with single or multiple passes. The matrix-forming wall material can be either a single component or a combination of several different polymers. If a wall material does not have a surface active property, an emulsifier must be added to a wall solution prior to homogenization. The prepared emulsion is then processed for its conversion into powder, particles, beads or pellets, depending on a subsequent encapsulation process used. For example, emulsion is dried into a powder form by spray drying or into pellets by freeze drying followed by grinding to convert it to granules (particles) or extruded into another aqueous or lipid phase for formation of beads followed by crosslinking wall matrix of the beads. The choice of an encapsulation method depends on types and properties of microcapsules to be achieved and their final targeted food applications.

Formation, structure and properties of microcapsules containing oils and oxidative stability of encapsulated oils are significantly influenced by physico-chemical properties of emulsions prior to their use for microencapsulation. Factors affecting emulsion properties include particle size and distribution of emulsified oil droplets, viscosity and total solid content of emulsion, type and concentration of carrier, ratio of core to carrier, type of emulsifier, and homogenization temperature and pressures. In other words, emulsion stability is one of the critical factors and a prerequisite for preparation and stability of microcapsules with desired properties and performance, especially for oxidative stability of encapsulated oil.

Generally, a fine stable emulsion ideal for oil encapsulation should have the following properties: small size and narrow distribution of oil droplets stable to agglomeration and coalescence, a high solid content for enabling them to form a

Table 15.2 Materials and methods used to microencapsulate fish oil

Method	Wall material	Oil	Sources
Spray drying	n-octenylsuccinate-derivatised (n-OSA) starch, glucose syrup, trehalose	Unknown	85, 104
	Sodium caseinate, maltodextrin	Menhaden or herring oil	39
	Sodium or calcium caseinate, skim milk powder, lactose	Sand eel oil	79, 81
	Sodium caseinate, whey protein isolate (WPI), maltodextrin, branched cyclic dextrin	Unknown	89
	Modified cellulose		110
	Methylocellulose (MC), hydroxypropyl methylocellulose (HPMC), maltodextrin	Unknown	109
	Cyclodextrin, corn syrup solid or maltodextrin	Seal blubber oil	40
	Gum arabic, maltodextrin, gelatine	Shark liver oil	103
	Chitosan, lecithin, corn syrup solid	Tuna oil	108
	Whey protein isolate or soy protein isolate, glucose syrup	Tuna oil	105
	Modified starch (Capsul from National Starch and Chemical company)	ROPUFA (fish oil unknown)	136
Spray drying, freeze drying	Egg white protein	Sardine oil	130
Spray drying (electrostatic layer-by-layer deposition process)	Corn syrup solids, chitosan, lecithin	Tuna oil	82
Freeze drying (transglutaminase crosslinking)	Soluble wheat protein, whey protein isolate, sodium caseinate, isolated soy protein	Marinol D-40 (fish oil unknown)	121
Freeze drying	Sodium caseinate, lactose, maltodextrin	Sand eel oil	87, 118
	Sodium caseinate, lactose		61, 137
Coacervation and spray drying	Hydroxyproply methylcellulose (HPMC), maltodextrin	Unknown	116
Coacervation	Gelatine, acacia	EPA ethyl ester	115
Coacervation	Pectin, chitosan	Shark liver oil	117
Ionotropic gelation	Chitosan	Krill oil	126

continuous matrix as a protective barrier where oil droplets are uniformly distributed and embedded or entrapped, and a low viscosity for easy flow, pump and spray. The higher the homogenization pressure, the finer the oil droplet size. Multiple passes also allow the oil droplet and particle size distribution to be small and in a narrow range. Homogenization should also be carried out at low temperatures below 10 ºC to minimize any possible thermal damage to fish oil during emulsification. Prepared emulsion should also be maintained under low temperatures until it is processed for microencapsulation.

The large size and broad distribution of emulsion droplets are adversely related to the high content of surface oil (free oil) in spray-dried powders. As a result, encapsulation efficiency and physical properties (flowability, bulk density, dispersibility) of spray-dried powder are adversely influenced. The presence of surface fat is correlated with the rate of lipid oxidation.[79,80] The size and size distribution of emulsion droplets are therefore important factors determining the properties and stability of spray-dried oil powder.

Also, free fat content is related to vacuole volume which is a common feature typically observed from spray-dried powder where air can be occluded, thus increasing the rate of lipid oxidation. Keogh *et al.*[79] and Kelly and Keogh[81] showed that vacuole volume could be positively related to oxidation of the spray-dried fish oil powders made from sodium caseinate and lactose or skim milk powder. They found that emulsions prepared using skim milk powder gave rise to powders with lower levels of vacuole volume compared to those in sodium caseinate/lactose-based fish oil powder. The result was described to be attributable to the low tendency of skim milk protein to foam during emulsification compared with sodium caseinate. This was in turn related to a higher degree of aggregation of micellar casein present in skim milk powder compared with sodium caseinate. This indicates that foaming commonly created during emulsification of oils can have a great effect on vacuole volume of spray-dried powder, thus affecting shelf-life of microencapsulated fish oil during storage.

In recent years, oxidative stability of emulsions could be improved by tailoring the interfacial membrane properties of emulsion droplets through formation of a multilayered interfacial membrane using an electrostatic deposition method (for example, binding of chitosan with lecithin emulsified oil droplets).[82–84] Use of this type of emulsion system for preparing microcapsules can be expected to enhance oxidative stability of the resulting encapsulated oil.

15.2.2 Criteria required for microencapsulating wall materials

A variety of natural polymers have been used as coating or encapsulating wall materials for encapsulation of oils or fats. These include proteins (gelatine, whey proteins, sodium caseinate, and soy proteins), carbohydrates (emulsifying starches and modified starches), gums (gum arabic), and small molecules of carbohydrates (sucrose, glucose, lactose, maltodextrin).[39,85–87]

The selection of a suitable wall material is critical for successful microencapsulation. In general, wall materials to be used for microencapsulation of oils must have emulsifying properties, high water solubility, low viscosity even at highly concentrated solution, and film forming and drying properties.[66,88–90] Moreover, wall materials should not react with core materials to be encapsulated and must be safe for use in foods, natural polymers being preferred in this aspect, and compatible with formulations in the final food product applications. Wall materials especially suitable for oil encapsulation must possess the ability to form and stabilize emulsions by providing a stable, thick, dense layer (film) around the surface of emulsion oil droplets during homogenization, and to form a continuous compact wall matrix acting as an oxygen barrier between oil droplets in microcapsules when the emulsion is subsequently processed for drying or solidification. In this aspect, whey proteins have been reported to be an excellent encapsulating agent for volatiles and oils/fats.[65,88–92] Sodium caseinate has also been demonstrated to serve as effective wall materials for encapsulation of orange peel oil.[92] Whey proteins are also known to have significant antioxidant properties and provide oxidative stability to oil emulsion stabilized by whey proteins.[93–95]

Gum arabic has been also used extensively as an encapsulating wall material for oils or volatiles by spray drying since this hydrocolloid exhibits good emulsifying, film forming and dehydration properties, high solubility and low solution viscosity.[96] Its expensive price is a main disadvantage, limiting its wider use in wall materials in the food industry. Starches have virtually no emulsification properties and have high viscosity. A chemical modification involving the addition of lipophilic groups (octenyl succinate) into their molecules, which has been approved by the FDA, renders modified starches exhibiting totally different characteristics (high solubility, low viscosity and good emulsifying properties) that are suitable for microencapsulation.[97] These emulsifying starches have some drawbacks with regard to cost, poor protection against oxidation and synthetic non-natural label.[98]

Unlike proteins and some carbohydrates (gum arabic and modified emulsifying starches) that have excellent emulsifying properties and provide relatively greater oxidative stability, most carbohydrates and hydrocolloids are unsuitable for use alone as a primary encapsulating wall material as they do not have emulsifying properties and, except for small molecules of monosaccharides and oligosaccharides, most carbohydrates and hydrocolloids are also highly viscous even at very low solid concentrations. Therefore they should be used in conjunction with proteins or emulsifying agents in order to encapsulate oil materials. Additionally, they have poor film-forming abilities around the droplets of oils.[74] Small molecules of carbohydrates, such as maltodextrins, corn syrup solids, glucose, and lactose, which are often referred to as secondary carriers (fillers), have been used to improve oxygen barrier properties of the wall matrix, to enhance drying property of the wall matrix,[90] and to improve physical properties of spray-dried oil powders. The main property of these components as an excellent oxygen barrier is related to their formation of amorphous state rather than crystalline state in the dried wall matrix of microcapsules during drying procedures such as spray drying.[97] Maltodextrins and corn syrup solids have been shown to exhibit a varying degree of protection of encapsulated ingredients from oxidation. A higher dextrose equivalent (DE) value of maltodextrins has a tendency to offer better protection to oxidation sensitive oils since the wall systems with the higher DE are less permeable to oxygen.[74] A greater number of reducing groups present in maltodextrins with a higher DE value has been also proposed to play a role in stabilizing oxidative sensitive oils by exerting an antioxidant environment in encapsulating the wall matrix system.[98]

Various studies have shown that a wall system consisting of protein in conjunction with carbohydrates is more effective in stabilizing encapsulated oil during long-term storage of spray-dried powder. The ratio of protein and carbohydrate and types (molecular weight, conformational structure and reducing groups) of carbohydrates have been shown to influence the physical properties and oxidative stability of encapsulated fish oil powders.[85] Carbohydrates in the state of amorphous structures in the wall matrix rather than as crystalline structures are more effective for oxidative stability of encapsulated fish oil.[85] Drusch *et al.* indicated that low molecular weight carbohydrates added in spray-dried microcapsules could cause caking and disrupt the structural integrity of spray-dried powders by crystallization of carbohydrates during storage. This can result in the release of some encapsulated oil from the powder matrix and induction of lipid oxidation.[85] Occurrence of carbohydrate crystallization is dependent on molecular weight and conformational structure of carbohydrates, their glass transition temperatures, and relative humidity. Crystallization of carbohydrates takes places readily at higher relative humidity, therefore, when carbohydrates are to be included in the wall matrix system containing oxidation sensitive substances, carbohydrates must be prevented from development of crystalline form and storage conditions must be chosen carefully. Prevention of crystallization can be achieved by use and addition of substances such as plasticizers into wall systems. These include sorbitol, glycerol, mannitol, propylene glycol, polyethylene glycol, sucrose, sucrose fatty acid esters, and lactones.[99]

Aside from those chemical properties required, other factors to consider in selecting wall materials are their compatibility with final food applications and their nutritional value and sensory properties.[59] It is desirable if wall materials that are already available in the final foods can be used so that food formulation and labelling constraints resulting from the addition of foreign substance can be minimized or avoided.[7] Cost is another factor[59] and wall materials should be readily available and cheap. Proteins are generally considered relatively expensive compared with carbohydrates. Any single wall material meeting all the requirements described is rarely available. To achieve desired properties and performance of microcapsules, more complicated and sophisticated combinations of multiple materials are needed.

15.2.3 Spray drying

Spray drying was originally developed as a dehydration process to remove water and make the aqueous products into a dry powder form.[100] Spray drying is a process involving an extremely rapid spontaneous evaporation of water from a mist of fine particles produced by atomizing a liquid feed (e.g. emulsions) into a spray dryer through a nozzle or rotary spinning wheel at very high pressure. Hot air flows either in a co-current or counter-current direction inside the spray dryer and contacts the atomized particles, evaporating the water and forming dried particles. The dried particles falling to the bottom of the dryer are collected. Temperatures used for spray drying are generally in a range of 180–220 °C and 60–80 °C for inlet and outlet air temperatures, respectively. When using spray drying for microencapsulation, core materials dissolved or suspended in a liquid feed become embedded or enclosed uniformly in the entire matrices of dried coating materials of particles. Although the high temperatures are used in the process, water evaporates rapidly from the coating material during its solidification, thus keeping the core materials temperature below 100 °C.[73] Residence time of dried powder in the drying chamber is very short, thus heat damage to spray-dried encapsulated core material is much lower than generally considered.

In the past several decades, spray drying has been also used as an encapsulation technique to microencapsulate a wide range of food ingredients such as flavours, vitamins, minerals, colours, fats and oils in order to protect them from their surrounding environment (heat, moisture, air) and extend shelf stability during storage. It is currently the most commonly used encapsulation method in the food industry. More than 90% of microencapsulated omega-3 oils are produced by spray drying.[101] The advantages over other encapsulation methods for use in the food industry are: easy availability of equipment, large scale production, simple continuous processing operation, low processing cost, easy manipulation of processing conditions, and good stability of finished dried product. In addition, the conversion of a liquid material into a dry powder form presents several other advantages to food manufacturers and formulators, including easy handling and storage of sensitive liquids, better uniform dry blending with other food ingredients in products (for examples, dry milk powder and infant formula), and protection from undesirable interaction with other incompatible components and from nutritional loss during storage. A few disadvantages are that it uses heat for evaporation of water, which it is not ideal for heat-sensitive materials. Depending on requirements, a very fine powder produced by spray drying needs further processing such as agglomeration or coating via fluidized bed coating to modify its physical properties, such as bulk density, flowability, wettability and dispersibility in water or lipid phase, and also to make encapsulated core material more stable.

Various wall materials have been tested using the spray-drying technique for their efficiency and suitability as protective wall materials for encapsulation of fish oil. Stability of encapsulated fish oil against oxidation has been shown to be determined by oxygen permeability of encapsulation wall system.[99] A mixture of sodium caseinate and maltodextrin were used as a wall matrix of spray-dried encapsulated fish oil powder.[39,89] The oil–wall ratio, protein–carbohydrate ratio and the dextrose equivalence (DE) value of maltodextrins both influence the oxidative stability of encapsulated fish oil and morphological structures of spray-dried powders. Differences in drying rate rising from different types and amounts of carbohydrates added and levels of oil loaded contribute noticeable structural differences between powders. Kagami *et al.* showed that addition of maltodextrin with DE 18 or highly branched cyclic dextrin (HBCD) to a wall matrix of sodium caseinate improved the oxidative stability of encapsulated fish oil powders.[89] The Maillard browning reaction and agglomeration of spray-dried powder made from a mixture of maltodextrin DE 18 and sodium caseinate were the only drawbacks.[89] Jonsdottir *et al.* also reported the oxidative stability of spray-dried fish oil could be achieved using different combinations of wall materials (gelatine or caseinate with sucrose, lactose or maltodextrin).[102] Hogan *et al.*, however, reported that oxidation had already occurred significantly during both emulsion preparation of fish oil and subsequent spray drying of the prepared emulsion containing fish oil.[39] In contrast, a combination of gum arabic, maltodextrin and gelatine as encapsulating agents were shown to significantly improve oxidative stability of spray-dried shark liver oil powder during storage compared with the corresponding unprotected crude oil.[103]

Recent studies have demonstrated the possibility of using modified starch, such as n-octenylsuccinate-derivatized (n-OSA) starch, as an encapsulating wall material in combination with glucose syrup or trehalose as filler agent for oxidative stability of spray-dried fish oil powder.[85,104] n-OSA starch has been approved for its use in foods by the FDA and permitted as an additive for infant formula and weaning foods in the EU. Drusch *et al.* reported that spray drying temperature and content of unencapsulated free oil (surface oil) of spray-dried powders were the main causes of lipid oxidation of spray-dried oil powders.[104] This implies that oxidative deterioration of spray-dried fish oil powders can be reduced to a certain degree by minimizing the formation of surface oils and maintaining a drying temperature as low as possible.

Shahidi and Wanasundara have shown the ability of three different encapsulating wall materials of beta-cyclodextrin, corn syrup solids and maltodextrin for encapsulation of seal blubber oil by spray drying using Tween-80 as emulsifier.[40] Beta-cyclodextrin and corn syrup solids were found to be suitable encapsulating materials while maltodextrin was not effective, suggesting that the latter alone is not recommended for use in protecting oils against oxidation.

Several studies have proposed the potential for improved oxidative stability of spray-dried encapsulated fish oil by means of the Maillard reaction products formed during heating a mixture of protein and carbohydrates prior to emulsion preparation.[105,106] The Maillard reaction products are known to exhibit good emulsifying and antioxidant properties and form a stable and robust shell around the oil droplets, thus making them suitable as encapsulating materials for stabilization of oil.[105,107] Augustin *et al.* reported the increased oxidative protection of spray-dried fish oil encapsulated into a wall matrix consisting of protein (sodium caseinate, whey protein isolate, soy protein isolate, or skim milk powder) and carbohydrate (glucose, dried glucose syrup, oligosaccharide) by the use of Maillard reaction products formed via heating (60–100 °C for 30–90 min) of aqueous wall solid mixtures of protein and carbohydrate prior to emulsification.[106] It was reported that there was no heat-induced adverse influence on encapsulation efficiency (in all cases more than 98% oil retention) while protection of microencapsulated fish oil from oxidation was improved. On the other hand, Rusli *et al.* showed the increased viscosity of emulsions and coalescence of emulsion oil droplets resulting from pre-heated protein and carbohydrate mixtures, which can be problems as this can reduce emulsion stability, thereby contributing to a decreased microencapsulation efficiency of spray-dried oil powders.[105]

Klinkesorn *et al.* showed that spray-dried carbohydrate-based powders embedding tuna oil droplets could be prepared from emulsions in which oil droplets were surrounded and stabilized by a multilayer of lecithin–chitosan membranes formed via an electrostatic layer-by-layer deposition process (i.e. through interaction between oppositely charged polymers).[82,108] The oil powders produced in this way were reported to have good physicochemical properties and dispersibility[82] and be more stable to oxidation than bulk oils,[108] suggesting the potential of using this technique for improved oxidative stability.

Kolanowski *et al.* evaluated the adaptability of modified starches, such as methylcellulose (MC) and hydroxypropyl methylcellulose (HPMC), to be used for microencapsulation and stabilization of fish oil by spray drying. Both materials have good emulsifying properties to form a stable emulsion.[109] This study showed significant improvement of oxidative stability of fish oil by microencapsulation with modified starches when compared with the bulk oil. However, MC was shown to be a better coating material than HPMC due to less air entrapped resulting from less tendency of foaming during emulsification.

Recent studies reported by Kolanowski *et al.* indicated that the oxidation of spray-dried microencapsulated fish oil powder prepared from fish oil emulsions using modified cellulose, skim milk powder or a mixture of gelatine and cornstarch occurred much faster compared with that of bulk fish oil, suggesting that a decrease of oxidative stability of fish oil powders was probably related to an increase in the surface area of powders in comparison to bulk fish oil.[110,111] This result was in contrast to many of those results that have been reported by others. It seems this was partly because of wall materials which were not able to form a continuous wall matrix to act as an oxygen permeable barrier. Although gelatine can be used to encapsulate oils as it has emulsifying and stabilizing properties, gelatine is not considered to be used for microencapsulation of oils by spray drying.[99] This is because gelatine is highly viscous even at low concentrations compared with other proteins, such as whey proteins and caseins, and forms gelation at low temperatures, thus requiring to be used at low concentration and maintained at an elevated temperature during and after emulsion preparation prior to spray drying. Partially hydrolyzed gelatine, however, does not exhibit those problems, thus making them suitable for spray-dried microcapsules.[99]

In addition to the importance of microencapsulating wall systems as barriers for oxygen, moisture, heat and shear for oxidative stability of fish oil, physical properties of spray-dried powders are also becoming more important attributes for food applications. Important physicochemical properties are particle size, flowability, dispersibility, bulk density and water activity. It is important to match the physical properties of spray dried powder to that of the powder with which it is to be dry blended in order to minimize or avoid physical separation or segregation of the ingredients during formulation, transport and storage. For liquid food applications, a liquid form (e.g. emulsion) is sometimes preferred, but a powder form that can be rehydrated and dispersed into a stable emulsion is also suitable.[59]

Spray drying for protection of fish oils against oxidation and for delivery of fish oils into food product is anticipated to be utilized continuously as a main encapsulation method according to anticipated introduction and development of new wall materials from plant, animal, marine, and microbial sources and also generation of enhanced wall materials from existing proteins or carbohydrates through chemical or enzymatic modification.

15.2.4 Coacervation

Generally, coacervation has been used for encapsulation of drugs in the area of pharmaceutical sciences. Coacervation is often considered to be the original and true method of microencapsulation in which a core material is completely entrapped or enclosed by a continuous coating of wall material.[112] Coacervation is divided into two methods, simple and complex, depending on the number of biopolymers used (e.g. one or more).[113] The process involves the separation of an aqueous phase of coating material from a polymeric solution by the addition of another chemical or oppositely charge colloid. This competes for solubility with or associates with coating material, forming or precipitating one liquid phase rich in coating material that surrounds the suspended core material. This is then solidified by using heat, crosslinking or solvents. This method is rather expensive. Its practical use in food applications is very limited because of a scarcity of food grade hardening agents.[114]

Lamprecht *et al.* have shown the preparation of microparticles containing eicosapentaenoic acid ethyl ester (EPA-EE) using gelatine and gum acacia by complex coaceravtion.[115] Briefly, EPA-EE was emulsified into a gelatine solution and coacervated by pouring it into an acacia aqueous solution followed by adjusting pH and temperature. Then, the coacervates were hardened using different processes, such as ethanol extraction, crosslinking with dehydroascorbic acid or glutaraldehyde and spray drying. They reported that hardening of the coacervates with ethanol resulted in the microparticles most protective against oxidation while microparticles crosslinked by dehydroascobic acid or produced by spray drying were less protective.

Wu *et al.* studied microencapsulation of fish oil by simple coacervation using HPMC followed by spray drying of the coacervate suspension in which fish oil was emulsified into a HPMC solution and then mixed with maltodextrin for coacervation. HPMC is a polysaccharide derived from cellulose and has good water solubility and emulsification and filming properties,[116] hence it can be used as a microencapsulating agent. Its solubility decreases when a more water soluble polymer is added, thereby allowing its use in microencapsulation via the coacervation process.[116] Fish oil was emulsified into a HPMC solution and then mixed with malt dextrin for coacervation. The coacervated suspension was then spray dried. Wu *et al.* reported that formation, retention efficiency and characteristics of coacervate were affected by various factors, such as concentration, proportion, molecular size and relative water solubility of colloids and polymers dissolved in aqueous solution and also the amount of oils to be encapsulated, so that all these parameters need to be optimized when microcapsules are prepared by coacervation.[116] They also reported that the microencapsulation efficiency and oxidation stability of encapsulated fish oil were improved significantly when replacing part of maltodextrin by gum arabic (acacia). This was due to its excellent emulsifying and film-forming abilities and very low viscosity at high concentration, thus facilitating the hardening and drying process.

Diaz-Rojas *et al.* demonstrated the encapsulation of shark liver oil by coacervation using various ratios of pectin and sodium alginate polymers as wall materials. Shark liver oil emulsions were dropwise added into a stirred calcium chloride solution to form gel beads. The beads were transferred to a solution of chitosan to form a membrane layer and submerged into a calcium chloride bath. The beads were then washed with ethanol and dried at room temperature.[117] Oxidative stability of encapsulated shark liver oil was not reported in this study.

15.2.5 Freeze drying

Freeze drying uses very low temperature and pressure conditions unlike spray drying that involves high temperatures. Thus, freeze drying is regarded as an ideal process for dehydration of heat sensitive materials. It is simple to use and a mild process suitable for encapsulation of water-soluble essences and natural aromatic compounds.[74] The main disadvantage is long dehydration time which often takes longer than 20 h.

A few studies have been conducted to apply a freeze-drying technique to microencapsulate fish oil for improved oxidative stability.[87,118] For the production of freeze dried microencapsulated fish oil, fish oil was emulsified in a solution of sodium caseinate with lactose or maltodextrin. The prepared emulsions were frozen at various rates using different freezing techniques and the frozen emulsion was then freeze dried under high pressure. After freeze drying, the freeze-dried emulsion pellet was ground in a grinding machine to form a free flowing powder for its better application in enrichment of foods.[118] Heinzelmann *et al.* reported that freeze-dried fish oil encapsulated particles made from sodium caseinate with lactose or maltodextrin had very porous internal and external structures with irregular pore sizes.[118] The presence of irregular shapes and sizes of pores can be problematic because the empty pores can be filled with oxygen, thus facilitating oxidation deterioration of the encapsulated fish oils. Another problem associated with the presence of pores is the creation of irregular shapes and sizes of freeze-dried powders when grinding the freeze-dried pallets. This will make it difficult to obtain uniform surface areas of particles for controlled dry blending with other ingredients during formulations. Interestingly, the ground powders were reported to show a better oxidation stability compared with the pellets despite the fact that the powders had undergone an additional exposure to air and mechanical stress.[87] The oil stability of freeze-dried fish oil against oxidation was affected by the mechanism of the freezing technique (e.g. freezing rate) prior to drying of the frozen emulsion.[87,118]

Although the use of an optimized freeze-drying process and the addition of antioxidants can improve oxidation stability, some technical issues relating to the effects of temperature and rate of freezing of emulsion need to be further investigated.[118]

15.2.6 Other techniques

Spray-dried powder or particles can be further treated for coating and/or agglomeration by a fluidized bed coating to improve oxidative stability of encapsulated oil and tailor physical properties of powders such as instantization, bulk density, flowability, wettability and dispersibility.[119] Coating materials used for fluidized coating can be either the same as or different from those used in the production of spray-dried powder, depending on desired types and characteristics of the resulting coated or agglomerated powders. To avoid possible oxidation of oils, nitrogen gas can be purged during the fluidized bed process. Also, a modified spray-drying process, which is a combination of spray drying and fluidized bed coating and often cited as spray congealing, is used for microencapsulation of oxidation sensitive and unpleasant tasting ingredients such as fat-soluble vitamins, carotenoids and fish oils.[120] In modified spray drying and congealing, an emulsion is sprayed into a spray-drying chamber followed by feeding dried products into anther chamber containing coating materials or an emulsion is sprayed directly into a fluidized layer loaded with a drying or coating medium, such as hydrophobic starch, dehydrating liquid and oil.

Microencapsulation of fish oil in protein wall matrices by techniques consisting of double emulsification (o/w/o) and subsequent crosslinking of proteins with several different methods (heat gelation, enzyme-induced crosslinking or a combination of both) has been demonstrated by Cho *et al.*[121] The encapsulated fish oil was shown to be significantly more stable against oxidation compared with the corresponding unencapsulated fish oil during storage. The size of microcapsules produced by this method is relatively small, ranging from 10 to 100 μm and is similar to the typical size of spray-dried powder. This type of microcapsule also has a very limited water solubility because of crosslinking of wall matrices, thus also enabling used for controlled release and targeted delivery applications.[121–123] One disadvantage of this encapsulation technique is that microcapsules formed must be washed with organic solvent such as ethanol after gelation in order to remove the oil used to form a double emulsification, resulting in a loss of some encapsulated oil and causing the formation of surface pores.[122,124]

A method described in the patent application by Kantor *et al.* showed the successful development of fish oil microcapsules with no fishy taste and odour by encapsulating fish oil within an enteric coating such as ethyl cellulose.[125] In this method, fish oil is emulsified into a basic solution of ethyl cellulose suspended in ammonium hydroxide and then the emulsion is atomized using a nitrogen stream into a stirred acidic solution. The resulting microcapsules range in size from 0.1 to 500 μm, are stable at room temperature and at 4 °C in the presence of light, and thus can be incorporated into a variety of aqueous, oil-based or solid dry food products such as gelatine, orange juice and yoghurt that have a pH of less than 7. The enteric coating only disintegrates under neutral and basic conditions and this enables encapsulated fish oil to bypass the stomach, to be delivered and released in the small intestines where omega-3 LC PUFA are absorbed.[125]

Bustos *et al.* has demonstrated the feasibility of using an emulsion phase separation technique involving an ionotropic gelation method to be used to prepare chitosan microcapsules stabilizing fish oil such as krill oil rich in both EPA and DHA. Emulsions were prepared from a mixture of chitosan, Tween 20 as an emulsifier and krill oil and then slowly added into a stirred solution of sodium hydroxide in ethanol for hardening of the wall matrix of dispersed emulsion particles.[126] Chitosan microcapsules prepared were reported to be not significantly effective in stabilizing the encapsulated krill oil against oxidation compared to the corresponding emulsion form, suggesting some modifications of the properties of wall materials to improve oxidative stability.

15.3 Using encapsulated fish oils in food products

Functional foods are defined as any foods that are fortified with an ingredient which has already been demonstrated through clinical studies to play significant health promoting functions in one or more target organs of the human body in addition to its nutritional value.[127] The functional ingredients include anti-oxidants, vitamins, minerals, fibre, omega-3 fatty acids and probiotics.[5] For the past decade, as a result of the growing awareness of dietary importance for healthy life and demand for more nutritious healthy foods by consumers, functional foods have been one of the hottest food categories on the food market and a rapidly expanding area of the food industry.

Omega-3 enriched foods have been introduced in the market in some countries. However, it should be realized that some of the products labelled as omega-3 enriched products are made by adding flaxseed oil or ground flaxseed rich in ALA rather than by adding fish oil. With fewer double bonds in ALA compared with EPA and DHA, the products are less sensitive to oxidative degradation. Although ALA has its own health benefits including reduced risk of cardiovascular disease,[128] its conversion to EPA and DHA in the human body is very limited, and so the same degree of health effects that could be obtained from EPA and DHA cannot be expected through intake of diets fortified with

ALA-rich sources. Also, it should be mentioned that some products have been incorporated with hydrogenated fish oils so as to be less challenged by oxidative stability issues during storage. Some omega-3 products are also derived from eggs of hens or milk from cows fed an animal feed supplemented with fish oil to increase the omega-3 PUFA content.

Foods fortified with microencapsulated fish oil have been attempted in recent years. A range of foods are reported to be able to be formulated with micro-encapsulated fish oils, including beverages (nutritional, dairy, smoothies, high viscosity juices), dressings and sauces (spreads, margarine, salad dressing, mayonnaise), baked goods (bread, muffins, cake, nutritional and sports bars, cereals), dairy products (flavoured milk, yoghurt, cheese, ice cream), and powdered products (soup powder, instant dry mix, dairy powders, infant formula). Some of these fish oil enriched products such as orange juice, milk drink, breads and margarines that are currently sold on the market are also manufactured via incorporation of fish oil in the form of emulsions by using special emulsification methods and technologies for production and stabilization of fish oil emulsions.

Although microencapsulated fish oils in the form of powders or particles can exist stably without oxidation, their stability can be lost dramatically when they are added to foods because of phase transition and exposure to harsh food processing conditions (heat, pressure, and shear) which can disrupt the physical integrity of microcapsules. A good encapsulation system therefore requires strong mechanical and physical strength that can withstand harsh food processing conditions and stresses. Mechanical strength of microcapsules can be affected by wall composition and thickness and microcapsule structure (e.g. elasticity and flexibility).[99] It is thus important to produce fish oil encapsulated powders or particles which can remain intact and/or insoluble during complex food processing and subsequent storage conditions. The special processing techniques used to add encapsulated fish oil need to be considered carefully to minimize and avoid any possible physical and chemical damage to the microcapsules.

There is scant literature available regarding foods enriched with micro-encapsulated fish oil except for reports from research laboratories of fish oil suppliers and manufacturers. Kolanowski *et al.* have studied the impact of the addition of fish oil (fish oil liquid or powder-microencapsulated form) on sensory properties and palatability of a variety of food products.[4] They reported that particular groups of food products, such as salad oil, fat spreads, flavoured yoghurt and instant powder-milk-based formula which contain high fat and low water, were useful for enrichment with fish oil as high flavour and sweetness intensity can mask the unpleasant fishy taste. However, orange juices and fruit beverages which are water-based liquid products at low pH were reported as unsuitable foods for enrichment with fish oil due to accelerated hydrolysis and oxidation of added fish oil unless additional technological treatment is used to decrease oxidation by removal of oxygen from the packaging and addition of appropriate levels of antioxidants. Even though microencapsulated fish oil can exhibit oxidative stability, it is recommended to store fish oil-enriched foods under vacuum at lower temperatures. By controlling storage conditions, a longer shelf-life can be expected. Kelly and Keogh have demonstrated the sensory quality of a number of foods fortified with microencapsulated fish oil powder and have identified several consumer food products, such as bread, biscuits and soup powder, as the most suitable delivery foods for the incorporation of microencapsulated fish oil powder.[81] Bread was ascribed to be suitable because of the presence of a carbon dioxide atmosphere. Soup powder was easy to dry blend with microencapsulated fish oil powder and does not require heating until just before consumption.

Some of the world leading manufacturers and suppliers of fish oils have introduced a range of premium products, such as concentrated omega-3 fish oil and microencapsulated omega-3 fish oil powder, by using proprietary special microencapsulation processes that can be used as ingredients to produce dietary supplements and functional foods. Dry powder containing highly concentrated microencapsulated fish oil offers the advantage of allowing food manufacturers to create functional foods with greater omega-3 levels per serving using less omega-3 fish oil powder.

15.3.1 Infant formula

In 2002, the FDA approved supplementation of both omega-3 and omega-6 fatty acids in infant formula since both fatty acids are potentially important in foetal and infant neural development. Therefore, there is a strong demand and trend for adding DHA and ARA to infant formula. One of the difficulties with infant formula applications is a limited selection of available encapsulating wall materials as any materials added must comply with infant formula standards and regulations. Ingredients already present and allowed for use in infant formula manufacture are considered to be the best option to use as encapsulating materials, avoiding conflicts with requirements of food standards and laws set for infant formula. Also, the powder containing encapsulated fish oil must be shelf stable for a long period of storage time (generally 3 years from manufacture), must not react with other ingredients, and be readily dispersible and soluble in aqueous medium upon its reconstitution without affecting flavour.

A number of infant formula products enriched with microencapsulated fish oil are now available in the commercial market of some countries. Infant formula containing microencapsulated tuna oil is manufactured by Heinz Farley Infant Formula, UK.[129] Powdered infant formula milk (Karicare) fortified with microencapsulated tuna oil powder has been developed by Food Science Australia for Clover Corporation Ltd. The product is currently available on the market in Australia and New Zealand. According to the company, microencapsulated tuna oil powder is stable to oxidation and has a water-soluble free flowing property without fishy smell.[78]

15.3.2 Bakery and cereal products

Baked products (e.g. bread) are one of the products most widely applying microencapsulated fish oil. The relatively short shelf-life (less than one week) requirement of bread makes the fortification of fish oil less challenging in terms of oxidative deterioration, but it requires encapsulated fish oil to be sustainable to mixing, heating and baking without being released from the microcapsules during the breadmaking process.[59]

Sliced bread enriched with microencapsulated tuna oil powder from Nu-Mega was released on the Australian market in 2002.[129] The encapsulated tuna oil powder with 25% oil load is described as stable under baking, toasting and freezing conditions and the smell and flavour of tuna oil is masked.[129] The bread is currently on the market in Australia and New Zealand. In 2003, bread enriched with microencapsulated tuna oil powder without affecting the bread taste was introduced by Clover Corporation Ltd and George Weston Foods in Australia. This bread contains 10 mg of omega-3 long chain PUFA per slice.[49]

In 2005, breads enriched with fish oils containing 40–45 mg of DHA and EPA per slice were launched by US supermarket chain Wegmans using MEG-3 encapsulated fish oils manufactured by Ocean Nutrition Canada.

Taguchi *et al.* reported that microencapsulated sardine oil in egg white protein by spray drying was stable against oxidation during prolonged storage, and cookies fortified with this powder had the same sensory quality as plain cookies with no added fish oil. This also implies no effect of baking on stability of spray-dried powders encapsulating fish oil.[130]

15.3.3 Dairy products

Yoghurt is considered as a suitable product for incorporation of micro-encapsulated fish oil because of its requirements for cold storage temperatures, relatively short shelf storage life and also its strong acidic and sour flavour that can help mask the fishy odours and taste. Drinking yoghurt enriched with fish oil powder (50% oil powder) was launched in Saudi Arabia under the Live Well brand.[129]

15.3.4 Margarine and spreads

Two different types of spreadable fats (soft margarine and mix of butter and vegetable oil) have demonstrated for their oxidative stability and sensory acceptability during storage after enriching them with varying amounts of two different types of microencapsulated fish oils (ROPUFA 30 and MARITEX 43-02).[14] A mix of butter and vegetable oil and margarine were reported to be stable during storage for up to 3 and 6 weeks, respectively, without significant oxidative deterioration. The level of enrichment possible without affecting sensory acceptability was reported up to 1% EPA and DHA. This means it would be possible to intake about 0.2–0.3 g EPA and DHA if 25–30 g of fish oil enriched spreadable fat is consumed daily.

15.4 Suppliers of microencapsulated fish oil powder

Ocean Nutrition Canada Ltd (Dartmouth, NS, Canada) produces a free-flowing omega-3 fish oil powder under the MEG-3 brand that is shelf-stable for 12 months.[48] This powder has been described as a tiny gelatine sack where hundreds of other smaller sacks containing fish oils are packed, thus masking the taste of fish oil and protecting the oil from oxidation.[131] A number of food manufacturers have been using MEG-3 as ingredients for creating functional products containing omega-3 fish oil, including tortillas by Cali-Wraps (St. Marys, ON, Canada), OMU milk by Trappist Dairy (Hong Kong), Tropicana orange juice product in the US and Canada by Pepsi Co Inc., Cardivia yoghurt in Canada by Danone, and omega-3-enriched breads by US supermarket chain Wegmans. Ocean Nutrition Canada also introduced a new version of its MEG-3 ingredients with a DHA:EPA ratio of 5:1 which simulates human breast milk containing a DHA to EPA ratio of 4:1.

Martek Biosciences (Columbia, MD) and Nutrinova (Frankfurt, Germany) produce DHA extracted from microalgae grown under controlled fermentation conditions. Although this is a rather expensive source of omega-3 LC PUFA, any risk of contamination resulting from oceanic pollutants that might be present in DHA derived from fish or fish oils is removed from this vegetarian source of DHA omega-3.

Roche (Basel, Switzerland) produces a range of omega-3 fatty acid products in refined fish oil, emulsion and powder formats. Its omega-3 product under the name of ROPUFA 75 contains high amounts of EPA and DHA and has gained GRAS status in the US.[132] According to the manufacturer, ROPUFA 75 can be used to fortify a variety of food products such as soups, spreads, sauces, frozen prepared meals, pasta, milk, drinks, yoghurts and nutrition bars without affecting the taste.

Nu-Mega Ingredients Pty Ltd manufactures omega-3 DHA as food ingredients, including refined tuna oil under the HiDHA® brand and micro-encapsulated dry powder of tuna oil under the Driphorm® brand. According to the producer, these functional ingredients can be dry blended with other dry ingredients, reconstituted as a stable emulsion in aqueous solution or added directly to wet food products. Some products fortified with Driphorm encapsulated tuna oil powder have been launched, including breads and orange juice in Europe, fruit-flavoured cereal bars (under the IQ3 Brainstorm brand) containing 150 mg of total DHA and EPA in UK by Biomedical Laboratories and pizza in the US by Nardone Bros Baking (Wilkes-Barre, PA).[131]

Lipid Nutrition (Channahon, IL, USA), a division of Loders Croklaan, has introduced two products containing fish oil omega-3 fatty acids in the US market under the brand names, Marino Omega-3 HS (High Stability) Powder and Marino DHA HS Powder. These powders contain a minimum of 24% omega-3 fatty acids with 10% EPA and 7% DHA. The DHA HS Powder offers a minimum of 17% DHA and a low concentration of EPA. Both products also contain a sugar alcohol mannitol which is used to improve the sensory characteristics and prevent the smell of fish oil, thus enabling them to be incorporated into food products without compromising taste and stability. According to the company, the powders can also be dry blended with other ingredients, making it possible to create baked goods enriched with omega-3 fatty acids as well as beverages.

15.5 Future trends

As a result of increasing consumer health consciousness and increasing evidence of the health benefits of omega-3 LC PUFA, the demand for healthy food products enriched with fish oil rich in omega-3 LC PUFA is expected to grow continuously. More varieties of novel products need to be developed to satisfy the growing marketplace of health conscious consumers. Microencapsulation has been proved to hold great promise for the incorporation of fish oils into foods and the stabilization of fish oil enriched foods. It is becoming more feasible for food manufacturers to incorporate fish oil into their products with advancement of processing technologies and formulation techniques and development of microencapsulation technologies that improve the palatability and stability of fish oils.

As knowledge in encapsulating materials and technological advances expand and new natural wall materials with improved functionality are found and developed, microencapsulated fish oil powders with enhanced stability to heat and abrasion are expected to be developed. This will certainly expand the scope of a range of liquid or solid foods that can be incorporated with fish oil rich in omega-3 LC PUFA. Foods designed for different age groups (infants, babies, the youth and elderly) are attracting more attention, so more foods targeted at specific age groups with specific health conditions and requirements are anticipated to be developed by incorporating fish oils with ideal ratios of EPA and DHA. More food products incorporated with DHA are expected to be developed for infants and growing children.

Achieving a meaningful dose of omega-3 LC PUFA per serving via consumption of fish oil-enriched foods without affecting original food formulations is becoming an important issue. Incorporation of a large amount of encapsulated fish oil can change sensory properties as well. It has been shown that the level of fish oil added has a significant effect on changes in sensory properties of enriched foods.[4,18,133] This means it is necessary to develop powders containing a high amount of fish oil (e.g. more than 60% oil on dry basis) with highly concentrated omega-3 LC PUFA without affecting oxidative stability. This will enable food manufacturers to add less of the microencapsulated fish oil ingredient into food products while achieving greater levels of omega-3 LC PUFA per severing, thus minimizing impact on production cost per serving without a significant alteration in its food formulations. Also, there is a growing trend to avoid or reduce the use of food additives including natural antioxidants in the food industry. It will be necessary to develop encapsulation techniques that can provide oxidative stability of encapsulated fish oil without using antioxidants and that is stable for a longer period of storage (longer than 12 months).

15.6 Sources of further information and advice

The quality of fish oil should be characterized and tested before its use in food applications. Although official standards and limits specific for fish oils have not been established, quality guidelines designed for edible plant-derived oils can be used as rough guidelines for a measure of quality parameters of fish oils to be used for food applications. Michelsen *et al.* summarized quality guidelines for characteristics of oils and powders containing omega-3 LC PUFA used for food enrichment, recommended the delivery format of fish oils (e.g. oily form or powder form) to be used in different types of food and drink and intake levels of different types of omega-3 and 6 PUFA recommended by various organizations.[7] For details of this information, see the relevant chapter in the book *Guide to Functional Food Ingredients* contributed by Michelsen *et al.* (2001).[7]

Since the FDA announcement of qualified health claims for omega-3 LC PUFA, more varieties of novel food products enriched with omega-3 fatty acids are becoming available on the market worldwide, particularly in Europe, Australia, Canada and South America. The US is one of the countries becoming more interested in introducing foods fortified with omega-3 LC PUFA. A comprehensive review of commercial products currently available and their omega-3 PUFA content can be found in a review article by Whelan and Rust.[16]

15.7 References

1. O'SHEA M, 'Health benefits of marine-based omega-3 fatty acids', *Cereal Food World*, 2003 **48**(5) 236–237.
2. OLSEN S F and SECHER N J, 'Low consumption of seafood in early pregnancy as a risk factor for preterm delivery: prospective cohort study', *Brit Med J*, 2002 **324**(7335) 447–450.
3. CONNOR W E, LOWENSOHN R and HATCHER L, 'Increased docosahexaenoic acid levels in human newborn infants by administration of sardines and fish oil during pregnancy', *Lipids*, 1996 **31** S183–S187.
4. KOLANOWSKI W, SWIDERSKI F and BERGER S, 'Possibilities of fish oil application for food products enrichment with omega-3 PUFA', *Int J Food Sci Nutr*, 1999 **50**(1) 39–49.
5. BLUM M, 'Designing foods for better health', *Food Ingredients*. 1996 **3** 25–29.
6. OH R, 'Clinical review: practical applications of fish oil (omega-3 fatty acids) in primary care', *J Am Board Family Pract*, 2005 **18**(1) 28–36.
7. MICHELSEN B, MADSEN S L and GOTFREDSEN P, 'Polyunsaturated fatty acids', in *Guide to Functional Food Ingredients*, J Young (ed.) 2001, Leatherhead, Surrey: Leatherhead Publishing, 141–195.

8. SHAHIDI F and KIM S K, 'Marine lipids as affected by processing and their quality preservation by natural antioxidants', *ACS Symposium Series*, 2002 **816** 1–13.
9. KOLANOWSKI W and LAUFENBERG G, 'Enrichment of food products with polyunsaturated fatty acids by fish oil addition', *Eur Food Res Technol*, 2006 **222**(3–4) 472–477.
10. VALENZUELA A, SANHUEZA B J and NIETO S, 'Docosahexaenoic acid (DHA), essentiality and requirements: why and how to provide supplementation', *Grasas Aceites*, 2006 **57**(2) 229–237.
11. KUROWSKA E M, DRESSER G K, DEUTSCH L, VACHON D and KHALIL W, 'Bioavailability of omega-3 essential fatty acids from perilla seed oil', *Prostaglandins Leukot Essent Fatty Acids*, 2003 **68**(3) 207–212.
12. KAMAL-ELDIN A and YANISHLIEVA N V, 'N-3 fatty acids for human nutrition: stability considerations', *Eur J Lipid Sci Tech*, 2002 **104**(12) 825–836.
13. SIMOPOULOS A P, 'Nutrition tid-bites – Essential fatty acids in health and chronic disease', *Food Rev Int*, 1997 **13**(4) 623–631.
14. KOLANOWSKI W, SWIDERSKI F, LIS E and BERGER S, 'Enrichment of spreadable fats with polyunsaturated fatty acids omega-3 using fish oil', *Int J Food Sci Nutr*, 2001 **52**(6) 469–476.
15. FDA, *Letter responding to a request to reconsider the qualified claim for a dietary supplement health claim for omega-3 fatty acids and coronary heart disease,* Office of Nutritional Products, Labeling, and Dietary Supplements, Center for Food Safety and Applied Nutrition, U.S. Food and Drug Administration, Docket no 91N-0103 2002 (accessed 10 January 2007); available from: http://www.cfsan.fda.gov/~dms/ds-ltr28.html.
16. WHELAN J and RUST C, 'Innovative dietary sources of n-3 fatty acids', *Annu Rev Nutr*, 2006 **26** 75–103.
17. FDA, *FDA Announces Qualified Health Claims for Omega-3 Fatty Acids*, PO4-89 2004 (accessed 10 January 2007); available from: http://www.fda.gov/bbs/topics/news/2004/NEW01115.html.
18. GIBNEY M J, 'Incorporation of n-3 polyunsaturated fatty acids into processed foods', *Brit J Nutr*, 1997 **78**(2) 193–195.
19. SAGLIK S and IMRE S, 'Omega 3-fatty acids in some fish species from Turkey', *J Food Sci*, 2001 **66**(2) 210–212.
20. FDA, *FDA and EPA Announce the Revised Consumer Advisory on Methylmercury in Fish,* US Department of Health and Human Services and US Environmental Protection, For Immediate Release, P04-33 2004 (accessed 10 January 2007); available from: http://www.fda.gov/bbs/topics/news/2004/NEW01038.html.
21. SINCLAIR A J, ATTAR-BASHI N M and LI D, 'What is the role of alpha-linolenic acid for mammals?' *Lipids*, 2002 **37**(12) 1113–1123.
22. LI D, SINCLAIR A, WILSON A, NAKKOTE S, KELLY F, ABEDIN L, MANN N and TURNER A, 'Effect of dietary alpha-linolenic acid on thrombotic risk factors in vegetarian men', *Am J Clin Nutr*, 1999 **69**(5) 872–882.
23. GERSTER H, 'Can adults adequately convert alpha-linolenic acid (18: 3n-3) to eicosapentaenoic acid (20: 5n-3) and docosahexaenoic acid (22: 6n-3)?' *Int J Vitam Nutr Res*, 1998 **68**(3) 159–173.
24. NAPIER J A, 'The production of n-3 long-chain polyunsaturated fatty acids in transgenic plants', *Eur J Lipid Sci Tech*, 2006 **108**(11) 965–972.
25. RYMER C and GIVENS D I, 'n-3 fatty acid enrichment of edible tissue of poultry: a review', *Lipids*, 2005 **40**(2) 121–130.

26. KORELESKI J and SWIATKIEWICZ S, 'The influence of dietary fish oil and vitamin E on the fatty acid profile and oxidative stability of frozen stored chicken breast meat', *J Anim Feed Sci*, 2006 **15**(4) 631–640.
27. BOU R, GUARDIOLA F, TRES A, BARROETA A C and CODONY R, 'Effect of dietary fish oil, alpha-tocopheryl acetate, and zinc supplementation on the composition and consumer acceptability of chicken meat', *Poultry Sci*, 2004 **83**(2) 282–292.
28. LOPEZ-FERRER S, BAUCELLS M D, BARROETA A C and GRASHORN M A, 'n-3 enrichment of chicken meat. 1. Use of very long-chain fatty acids in chicken diets and their influence on meat quality: fish oil', *Poultry Sci*, 2001 **80**(6) 741–752.
29. CASTILLO-BADILLO C, VAZQUEZ-VALLADOLID J L, GONZALEZ-ALCORTA M, MORALES-BARRERA E, CASTILLO-DOMINGUEZ R M and CARRILLO-DOMINGUEZ S, 'The tuna oil as omega-3 fatty acids source for egg of laying hens', *Grasas Aceites*, 2005 **56**(2) 153–159.
30. PACETTI D, HULAN H W, SCHREINER M, BOSELLI E and FREGA N G, 'Positional analysis of egg triacylglycerols and phospholipids from hens fed diets enriched with refined seal blubber oil', *J Sci Food Agr*, 2005 **85**(10) 1703–1714.
31. WISTUBA T J, KEGLEY E B and APPLE J K, 'Influence of fish oil in finishing diets on growth performance, carcass characteristics, and sensory evaluation of cattle', *J Anim Sci*, 2006 **84**(4) 902–909.
32. PETIT H V, DEWHURST R J, SCOLLAN N D, PROULX J G, KHALID M, HARESIGN W, TWAGIRAMUNGU H and MANN G E, 'Milk production and composition, ovarian function, and prostaglandin secretion of dairy cows fed omega-3 fats', *J Dairy Sci*, 2002 **85**(4) 889–899.
33. LACASSE P, KENNELLY J J, DELBECCHI L and AHNADI C E, 'Addition of protected and unprotected fish oil to diets for dairy cows. I. Effects on the yield, composition and taste of milk', *J Dairy Res*, 2002 **69**(4) 511–520.
34. COOPER S L, SINCLAIR L A, WILKINSON R G, HALLETT K G, ENSER M and WOOD J D, 'Manipulation of the n-3 polyunsaturated fatty acid content of muscle and adipose tissue in lambs', *J Anim Sci*, 2004 **82**(5) 1461–1470.
35. AHNADI C E, BESWICK N, DELBECCHI L, KENNELLY J J and LACASSE P, 'Addition of fish oil to diets for dairy cows. II. Effects on milk fat and gene expression of mammary lipogenic enzymes', *J Dairy Res*, 2002 **69**(4) 521–531.
36. KEADY T W J, MAYNE C S and FITZPATRICK D A, 'Effects of supplementation of dairy cattle with fish oil on silage intake, milk yield and milk composition', *J Dairy Res*, 2000 **67**(2) 137–153.
37. COUPLAND J N and McCLEMENTS D J, 'Lipid oxidation in food emulsions', *Trends Food Sci Tech*, 1996 **7**(3) 83–91.
38. PARK Y, KELLEHER S D, McCLEMENTS D J and DECKER E A, 'Incorporation and stabilization of omega-3 fatty acids in surimi made from cod, *Gadus morhua*', *J Agr Food Chem*, 2004 **52**(3) 597–601.
39. HOGAN S A, O'RIORDAN E D and O'SULLIVAN M, 'Microencapsulation and oxidative stability of spray-dried fish oil emulsions', *J Microencapsul*, 2003 **20**(5) 675–688.
40. SHAHIDI F and WANASUNDARA U N, 'Oxidative stability of encapsulated seal blubber oil', *ACS Symposium Series*, 1995 **610** 139-151
41. MULLER H, JORDAL O, SELJEFLOT I, KIERULF P, KIRKHUS E, LEDSAAK O and PEDERSEN J I, 'Effect on plasma lipids and lipoproteins of replacing partially hydrogenated fish oil with vegetable fat in margarine', *Brit J Nutr*, 1998 **80**(3) 243–251.
42. VENKATESHWARLU G, LET M B, MEYER A S and JACOBSEN C, 'Chemical and olfactometric characterization of volatile flavor compounds in a fish oil enriched milk

emulsion', *J Agr Food Chem*, 2004 **52**(2) 311–317.
43. LINDSAY R C, 'Fish flavors', *Food Rev Int*, 1990 **6**(4) 437–455.
44. KARAHADIAN C and LINDSAY R C, 'Low-temperature deodorizations of fish oils with volatile acidic and basic steam sources', *J Am Oil Chem Soc*, 1990 **67**(2) 85–91.
45. JACOBSEN C, HARTVIGSEN K, LUND P, MEYER A S, ADLER-NISSEN J, HOLSTBORG J and HOLMER G, 'Oxidation in fish-oil-enriched mayonnaise 1. Assessment of propyl gallate as an antioxidant by discriminant partial least squares regression analysis', *Eur Food Res Tech*, 1999 **210**(1) 13–30.
46. JACOBSEN C, HARTVIGSEN K, LUND P, ADLER-NISSEN J, HOLMER G and MEYER A S, 'Oxidation in fish-oil-enriched mayonnaise 2. Assessment of the efficacy of different tocopherol antioxidant systems by discriminant partial least squares regression analysis', *Eur Food Res Tech*, 2000 **210**(4) 242–257.
47. QUILICI-TIMMCKE J, *Improving omega-3 oils*, 2004 (accessed 15 January 2007); available from: http://www.ffnmag.com/NH/ASP/strArticleID/343/strSite/FFNSite/articleDisplay.asp.
48. SCHATZMAN D, *Food manufacturers are using new food technology to make better-tasting omega-3 products,* 2006 (accessed 15 January 2007); available from: http://www.nutritionaloutlook.com/article.php?ArticleID=2095.
49. YEP Y L, LI D, MANN N J, BODE O and SINCLAIR A J, 'Bread enriched with micro-encapsulated tuna oil increases plasma docosahexaenoic acid and total omega-3 fatty acids in humans', *Asia Pac J Clin Nutr*, 2002 **11**(4) 285–291.
50. HIGGINS S, CARROLL Y L, O'BRIEN N M and MORRISSEY P A, 'Use of microencapsulated fish oil as a means of increasing n-3 polyunsaturated fatty acid intake', *J Hum Nutr Diet*, 1999 **12**(4) 265–271.
51. MAKI K C, DAVIDSON M H, DICKLIN M R, INGRAM K A, CYROWSKI M, UMPOROWICZ D M, BELL M and ELLIOTT J G, 'Bioavailability of eicosapentaenoic and docosahexaenoic n-3 polyunsaturated fatty acids in salmon patties compared with capsules', *J Food Sci*, 2003 **68**(3) 761–764.
52. WALLACE J M W, McCABE A J, ROBSON P J, KEOGH M K, MURRAY C A, KELLY P M, MARQUEZ-RUIZ G, McGLYNN H, GILMORE W S and STRAIN J J, 'Bioavailability of n-3 polyunsaturated fatty acids (PUFA) in foods enriched with microencapsulated fish oil', *Ann Nutr Metab*, 2000 **44**(4) 157–162.
53. DE LEONARDIS A and MACCIOLA V, 'Effectiveness of caffeic acid as an anti-oxidant for cod liver oil', *Int J Food Sci Tech*, 2003 **38**(4) 475–480.
54. KULAS E and ACKMAN R G, 'Protection of alpha-tocopherol in nonpurified and purified fish oil', *J Am Oil Chem Soc*, 2001 **78**(2) 197–203.
55. KULAS E and ACKMAN R G, 'Properties of alpha-, gamma-, and delta-tocopherol in purified fish oil triacylglycerols', *J Am Oil Chem Soc*, 2001 **78**(4) 361–367.
56. HAMILTON R J, KALU C, McNEILL G P, PADLEY F B and PIERCE J H, 'Effects of tocopherols, ascorbyl palmitate, and lecithin on autoxidation of fish oil', *J Am Oil Chem Soc*, 1998 **75**(7) 813–822.
57. HUANG S W, HOPIA A, SCHWARZ K, FRANKEL E N and GERMAN J B, 'Antioxidant activity of alpha-tocopherol and Trolox in different lipid substrates: Bulk oils vs oil-in-water emulsions', *J Agr Food Chem*, 1996 **44**(2) 444–452.
58. ZUTA P C, SIMPSON B K, ZHAO X and LECLERC L, 'The effect of alpha-tocopherol on the oxidation of mackerel oil', *Food Chem*, 2007 **100**(2) 800–807.
59. SANGUANSRI L A and AUGUSTIN M A, Microencapsulation and delivery of omega-3 fatty acids, in *Functional Food Ingredients and Nutraceuticals Processing Technologies*, Shi J. (ed.), 2006, Boca Raton, FL: CRC Press, 297–327.

60. LÖLIGER J and SAUCY F, *Mélange antioxydant synergique*, European Patent EP 0326829 B1. 1989.
61. VELASCO J, DOBARGANES M C and MARQUEZ-RUIZ G, 'Oxidation of free and encapsulated oil fractions in dried microencapsulated fish oils', *Grasas Aceites*, 2000 **51**(6) 439–446.
62. BAIK M Y, SUHENDRO E L, NAWAR W W, MCCLEMENTS D J, DECKER E A and CHINACHOTI P, 'Effects of antioxidants and humidity on the oxidative stability of microencapsulated fish oil', *J Am Oil Chem Soc*, 2004 **81**(4) 355–360.
63. JACOBSEN C, TIMM M and MEYER A S, 'Oxidation in fish oil enriched mayonnaise: ascorbic acid and low pH increase oxidative deterioration', *J Agr Food Chem*, 2001 **49**(8) 3947–3956.
64. JACOBSEN C, ADLER-NISSEN J and MEYER A S, 'Effect of ascorbic acid on iron release from the emulsifier interface and on the oxidative flavor deterioration in fish oil enriched mayonnaise', *J Agr Food Chem*, 1999 **47**(12) 4917–4926.
65. ROSENBERG M, YOUNG S L, BROOKER B E and COLOMBO V E, 'Whey proteins as microencapsulating agents – microencapsulation of anhydrous milkfat – structure evaluation', *Food Struct*, 1993 **12**(1) 31–41.
66. RE M I, 'Microencapsulation by spray drying', *Dry Technol*, 1998 **16**(6) 1195–1236.
67. SCHROOYEN P M M, VAN DER MEER R and DE KRUIF C G, 'Microencapsulation: its application in nutrition', *P Nutr Soc*, 2001 **60**(4) 475–479.
68. UBBINK J and KRUGER J, 'Physical approaches for the delivery of active ingredients in foods', *Trends Food Sci Tech*, 2006 **17**(5) 244–254.
69. CHEN L Y, REMONDETTO G E and SUBIRADE M, 'Food protein-based materials as nutraceutical delivery systems', *Trends Food Sci Tech*, 2006 **17**(5) 272–283.
70. POTHAKAMURY U R and BARBOSA-CANOVAS G V, 'Fundamental aspects of controlled release in foods', *Trends Food Sci Tech*, 1995 **6**(12) 397–406.
71. GIBBS B F, KERMASHA S, ALLI I and MULLIGAN C N, 'Encapsulation in the food industry: a review', *Int J Food Sci Nutr*, 1999 **50**(3) 213–224.
72. GOUIN S, 'Microencapsulation: industrial appraisal of existing technologies and trends', *Trends Food Sci Tech*, 2004 **15**(7–8) 330–347.
73. DZIEZAK J D, 'Microencapsulation and encapsulated ingredients', *Food Technol,* 1988 **42**(4) 136–151.
74. HAN X Q and SHAHIDI F, 'Encapsulation of food ingredients', *Crit Rev Food Sci Nutr*, 1993 **33**(6) 501–547.
75. MADENE A, JACQUOT M, SCHER J and DESOBRY S, 'Flavour encapsulation and controlled release – a review', *Int J Food Sci Tech*, 2006 **41**(1) 1–21.
76. DESAI K G H and PARK H J, 'Recent developments in microencapsulation of food ingredients', *Dry Technol*, 2005 **23**(7) 1361–1394.
77. AUGUSTIN M A, 'The role of microencapsulation in the development of functional dairy foods', *Aust J Dairy Technol*, 2003 **58**(2) 156–160.
78. AUGUSTIN M A, SANGUANSRI L, MARGETTS C and YOUNG B, 'Microencapsulation of food ingredients', *Food Aust*, 2001 **53**(6) 220–223.
79. KEOGH M K, O'KENNEDY B T, KELLY J, AUTY M A, KELLY P M, FUREBY A and HAAHR A M, 'Stability to oxidation of spray-dried fish oil powder microencapsulated using milk ingredients', *J Food Sci*, 2001 **66**(2) 217–224.
80. GRANELLI K, FALDT P, APPELQVIST L A and BERGENSTAHL B, 'Influence of surface structure on cholesterol oxidation in model food powders', *J Sci Food Agr*, 1996 **71**(1) 75–82.
81. KELLY P M and KEOGH M K, *Fish oil powder ingredient,* 2000 (accessed 15 January

2007); available from: http://www.teagasc.ie/research/reports/dairyproduction/4340/eopr-4340.htm.

82. KLINKESORN U, SOPHANODORA P, CHINACHOTI P, DECKER E A and McCLEMENTS D J, 'Characterization of spray-dried tuna oil emulsified in two-layered interfacial membranes prepared using electrostatic layer-by-layer deposition', *Food Res Int*, 2006 **39**(4) 449–457.
83. KLINKESORNA U, SOPHANODORA P, CHINACHOTI P, DECKER E A and McCLEMENTS D J, 'Encapsulation of emulsified tuna oil in two-layered interfacial membranes prepared using electrostatic layer-by-layer deposition', *Food Hydrocolloid*, 2005 **19**(6) 1044–1053.
84. KLINKESORN U, SOPHANODORA P, CHINACHOTI P, DECKER E A and McCLEMENTS D J, 'Encapsulation of emulsified tuna oil in two-layered interfacial membranes prepared using electrostatic layer-by-layer deposition', *Food Hydrocolloid*, 2005 **19**(6) 1044–1053.
85. DRUSCH S, SERFERT Y, VAN DEN HEUVEL A and SCHWARZ K, 'Physicochemical characterization and oxidative stability of fish oil encapsulated in an amorphous matrix containing trehalose', *Food Res Int*, 2006 **39**(7) 807–815.
86. LIN C C, LIN S Y and HWANG L S, 'Microencapsulation of squid oil with hydrophilic macromolecules for oxidative and thermal stabilization', *J Food Sci*, 1995 **60**(1) 36–39.
87. HEINZELMANN K and FRANKE K, 'Using freezing and drying techniques of emulsions for the microencapsulation of fish oil to improve oxidation stability', *Colloid Surface B*, 1999 **12**(3-6) 223–229.
88. REINECCIUS G A, 'Spray drying in food flavors', in *Flavor Encapsulation*, Reineccius G A and Risch S J (eds), 1988, Washington, DC: American Chemical Society, 55–66.
89. KAGAMI Y, SUGIMURA S, FUJISHIMA N, MATSUDA K, KOMETANI T and MATSUMURA Y, 'Oxidative stability, structure, and physical characteristics of microcapsules formed by spray drying of fish oil with protein and dextrin wall materials', *J Food Sci*, 2003 **68**(7) 2248–2255.
90. SHEU T Y and ROSENBERG M, 'Microencapsulation by spray-drying ethyl caprylate in whey-protein and carbohydrate wall systems', *J Food Sci*, 1995 **60**(1) 98–103.
91. MOREAU D L and ROSENBERG M, 'Porosity of microcapsules with wall systems consisting of whey proteins and lactose measured by gas displacement pycnometry', *J Food Sci*, 1999 **64**(3) 405–409.
92. KIM Y D and MORR C V, 'Microencapsulation properties of gum arabic and several food proteins: spray-dried orange oil emulsion particles', *J Agr Food Chem*, 1996 **44**(5) 1314–1320.
93. ELIAS R J, McCLEMENTS D J and DECKER E A, 'Antioxidant activity of cysteine, tryptophan, and methionine residues in continuous phase beta-lactoglobulin in oil-in-water emulsions', *J Agr Food Chem*, 2005 **53**(26) 10248–10253.
94. KELLERBY S S, McCLEMENTS D J and DECKER E A, 'Role of proteins in oil-in-water emulsions on the stability of lipid hydroperoxides', *J Agr Food Chem*, 2006 **54**(20) 7879–7884.
95. OSBORN H T and AKOH C C, 'Effect of emulsifier type, droplet size, and oil concentration on lipid oxidation in structured lipid-based oil-in-water emulsions', *Food Chem*, 2004 **84**(3) 451–456.
96. PORZIO M, 'Flavor encapsulation: a convergence of science and art', *Food Technol,* 2004 **58**(7) 40–47.

97. RUNGE F E, 'Multiple-core encapsulation: encapsulation materials', in *Microencapsulation of Food Ingredients*, Vilstrup P (ed.), 2001, Leatherhead, Surrey: Leatherhead Publishing, 133–145.
98. REINECCIUS G A, 'Multiple-core encapsulation: the spray drying of food ingredients', in *Microencapsulation of Food Ingredients*, Vilstrup P (ed.), 2001, Leatherhead, Surrey: Leatherhead Publishing, 151–185.
99. MARRS W M and DOWNES M J, 'Microencapsulation of fish oil – a review', Research Report No. 785. 2002, Leatherhead, Surrey: Leatherhead Publishing.
100. REINECCIUS G A, Spray-drying of food flavors, in *Flavour Encapsulation*, Risch S J and Reineccius G A (eds), 1988, Washington, DC: American Chemical Society, 55–66.
101. SUBRAMANIAN S, *Harnessing the powder of omega-3s,* 2005 (accessed15 January 2007); available from: http://www.ffnmag.com/NH/ASP/strArticleID/659/strSite/FFNSite/articleDisplay.asp.
102. JONSDOTTIR R, BRAGADOTTIR M and ARNARSON G O, 'Oxidatively derived volatile compounds in microencapsulated fish oil monitored by solid-phase microextraction (SPME)', *J Food Sci*, 2005 **70**(7) C433–C440.
103. GARCIA E, GUTIERREZ S, NOLASCO H, CARREON L and ARJONA O, 'Lipid composition of shark liver oil: effects of emulsifying and microencapsulation processes', *Eur Food Res Technol*, 2006 **222**(5-6) 697–701.
104. DRUSCH S, SERFERT Y and SCHWARZ K, 'Microencapsulation of fish oil with n-octenylsuccinate-derivatised starch: flow properties and oxidative stability', *Eur J Lipid Sci Tech*, 2006 **108**(6) 501–512.
105. RUSLI J K, SANGUANSRI L and AUGUSTIN M A, 'Stabilization of oils by micro-encapsulation with heated protein-glucose syrup mixtures', *J Am Oil Chem Soc*, 2006 **83**(11) 965–972.
106. AUGUSTIN M A, SANGUANSRI L and BODE O, 'Maillard reaction products as encapsulants for fish oil powders', *J Food Sci*, 2006 **71**(2) E25–E32.
107. GOUIN S, 'Microencapsulation: industrial appraisal of existing technologies and trends', *Trends in Food Sci Tech*, 2004 **15**(7–8) 330–347.
108. KLINKESORN U, SOPHANODORA P, CHINACHOTI P, McCLEMENTS D J and DECKER E A, 'Stability of spray-dried tuna oil emulsions encapsulated with two-layered interfacial membranes', *J Agr Food Chem*, 2005 **53**(21) 8365–8371.
109. KOLANOWSKI W, LAUFENBERG G and KUNZ B, 'Fish oil stabilisation by microencapsulation with modified cellulose', *Int J Food Sci Nutr*, 2004 **55**(4) 333–343.
110. KOLANOWSKI W, ZIOLKOWSKI M, WEISSBRODT J, KUNZ B and LAUFENBERG G, 'Micro-encapsulation of fish oil by spray drying-impact on oxidative stability. Part 1', *Eur Food Res Technol*, 2006 **222**(3–4) 336–342.
111. KOLANOWSKI W, JAWORSKA D and WEISSBRODT J, 'Importance of instrumental and sensory analysis in the assessment of oxidative deterioration of omega-3 long-chain polyunsaturated fatty acid-rich foods', *J Sci Food Agr*, 2007 **87**(2) 181–191.
112. SOPER J C, 'Utilization of coacervated flavors', in *Encapsulation and Controlled Release of Food Ingredients*, Risch S J and Reineccius G A (eds), 1995, Washington, DC: American Chemical Society, 104–112.
113. SCHMITT C, SANCHEZ C, DESOBRY-BANON S and HARDY J, 'Structure and techno-functional properties of protein-polysaccharide complexes; a review', *Crit Rev Food Sci Nutr*, 1998 **38**(8) 689–753.
114. RISCH S J, 'Encapsulation: overview of uses and techniques', in *Encapsulation and*

Controlled Release of Food Ingredients, Risch S J and Reineccius G A (eds), 1995, Washington, DC: American Chemical Society, 2–7.

115. LAMPRECHT A, SCHAFER U and LEHR C M, 'Influences of process parameters on preparation of microparticle used as a carrier system for omega-3 unsaturated fatty acid ethyl esters used in supplementary nutrition', *J Microencapsul*, 2001 **18**(3) 347–357.
116. WU K G, CHAI X H and CHEN Y, 'Microencapsulation of fish oil by simple coacervation of hydroxypropyl methylcellulose', *Chinese J Chem*, 2005 **23**(11) 1569–1572.
117. DIAZ-ROJAS E I, PACHECO-AGUILAR R, LIZARDI J, ARGUELLES-MONAL W, VALDEZ M A, RINAUDO M and GOYCOOLEA F M, 'Linseed pectin: gelling properties and performance as an encapsulation matrix for shark liver oil', *Food Hydrocolloid*, 2004 **18**(2) 293–304.
118. HEINZELMANN K, FRANKE K, JENSEN P and HAAHR A M, 'Protection of fish oil from oxidation by microencapsulation using freeze-drying techniques', *Eur J Lipid Sci Tech*, 2000 **102**(2) 114–121.
119. FUCHS M, TURCHIULI C, BOHIN M, CUVELIER M E, ORDONNAUD C, PEYRAT-MAILLARD M N and DUMOULIN E, 'Encapsulation of oil in powder using spray drying and fluidised bed agglomeration', *J Food Eng*, 2006 **75**(1) 27–35.
120. VILSTRUP P, 'Multiple-core encapsulation: modified spray congealing/spray drying of aqueous dispersions', in *Microencapsulation of Food Ingredients*, Vistrup P (ed.), 2001, Leatherhead, Surrey: Leatherhead Publishing, 185–197.
121. CHO Y H, SHIM H K and PARK J, 'Encapsulation of fish oil by an enzymatic gelation process using transglutaminase cross-linked proteins', *J Food Sci*, 2003 **68**(9) 2717–2723.
122. LEE S J and ROSENBERG M, 'Whey protein-based microcapsules prepared by double emulsification and heat gelation', *Lebensm Wiss Technol*, 2000 **33**(2) 80–88.
123. LEE S J and ROSENBERG M, 'Preparation and some properties of water-insoluble, whey protein-based microcapsules', *J Microencapsul*, 2000 **17**(1) 29–44.
124. LEE S J and ROSENBERG M, 'Microencapsulation of theophylline in whey proteins: effects of core-to-wall ratio', *Int J Pharm*, 2000 **205**(1–2) 147–158.
125. KANTOR M L, STEINER S S and PACK H M, 'Microencapsulation of fish oil', United States Patent 4895725. 1990.
126. BUSTOS R, ROMO L, YANEZ K, DIAZ G and ROMO C, 'Oxidative stability of carotenoid pigments and polyunsaturated fatty acids in microparticulate diets containing krill oil for nutrition of marine fish larvae', *J Food Eng*, 2003 **56**(2–3) 289–293.
127. DIPLOCK A T, AGGETT P J, ASHWELL M, BORNET F, FERN E B and ROBERFROID M B, 'Scientific concepts of functional foods in Europe consensus document', *Brit J Nutr*, 1999 **81**(4) S1–S27.
128. METCALF R G, JAMES M J, MANTZIORIS E and CLELAND L G, 'A practical approach to increasing intakes of n-3 polyunsaturated fatty acids: use of novel foods enriched with n-3 fats', *Eur J Clin Nutr*, 2003 **57**(12) 1605–1612.
129. AUSTRALIA F S, *Microencapsulated fish oil makes a splash Australia,* Food Facts no. 11 2003 (accessed 10 January 2007); available from: http://www.foodscience.afisc.csiro.au/foodfacts/foodfacts11.pdf
130. TAGUCHI K, IWAMI K, IBUKI F and KAWABATA M, 'Oxidative stability of sardine oil embedded in spray-dried egg-white powder and its use for N-3 unsaturated fatty-acid fortification of cookies', *Biosci Biotech Bioch*, 1992 **56**(4) 560–563.
131. McGUIGAN P, *Taking the 'fishy' out of fish oil,* 2005 (accessed 15 January 2007); available from: http://www.ffnmag.com/NH/ASP/strArticleID/778/strSite/FFNSite/

articleDisplay.asp.

132. NEWS HEADLINES CONSUMER TRENDS, *Roche gains GRAS for omega-3 ingredient,* 2003 (accessed 15 January 2007); available from: http://www.nutraingredients-usa.com/news/ng.asp?id=24957-roche-gains-gras.
133. ROCHE H and GIBNEY M J, 'The effect of consumption of fish oil-enriched spreadable fats on platelet phospholipid fatty-acid composition in human volunteers', *Int J Vitam Nutr Res*, 1994 **64**(3) 237–242.
134. SHAHIDI F and MIRALIAKBARI H, 'Marine oils: compositional characteristics and health effects', in *Nutraceutical and Specialty Lipids and Their Co-Products*, Shahidi F (ed.), 2006, New York: Taylor and Francis Group, 227–250.
135. ZAMOR R and HIDALGO F J, 'Fatty acids', in *Handbook of Food Analysis*, Nollet L M L (ed.), 2004, New York: Marcel Dekker, 221–274.
136. TAN L H, CHAN L W and HENG P W S, 'Effect of oil loading on microspheres produced by spray drying', *J Microencapsul*, 2005 **22**(3) 253–259.
137. MARQUEZ-RUIZ G, VELASCO J and DOBARGANES C, 'Evaluation of oxidation in dried microencapsulated fish oils by a combination of adsorption and size exclusion chromatography', *Eur Food Res Technol*, 2000 **211**(1) 13–18.

16

Encapsulation approaches for proteins

A. Millqvist-Fureby, YKI Institute for Surface Chemistry, Sweden

16.1 Introduction

Proteins are abundant in nature, and each protein generally has a given purpose such as catalytic action (enzymes), interactions (e.g. receptors), antibodies, structure or nutrition. Owing to the different characteristics of proteins, they have found widespread use in a variety of applications including foods, biosensors, pharmaceuticals and nutrition. In foods, the main functions of proteins are nutritional, action as emulsifiers and to provide structure and body to a food. Antibodies and antibody fragments are finding increasing applications in therapy and bioanalysis ('lab-on-a-chip'). In the pharmaceutical industry, therapeutic proteins are becoming more common, among the best known we find insulin and human growth hormone. Particular difficulties encountered in protein therapy are the metabolism of proteins in the body and the efficiency of the uptake. The oral route is the preferred route of administration from a patient perspective, but this is essentially precluded due to the efficient degradation of proteins in the gastrointestinal (GI) tract and the poor uptake of proteins into the blood stream. Hence, the most suitable route of administration of protein drugs is injections, but alternatives such as inhalation and skin therapy are actively sought. Moreover, the formulation of the bioactive proteins is essential, e.g. liquid or solid formulation, both for storage stability and delivery of the product. Encapsulation of proteins can aid in targeting in the body, enable different delivery routes, and improve storage stability.

The development of modern biotechnology, which allows genetic engineering and efficient fermentation processes, has opened up wide fields of large-scale use of proteins, including in food processing, organic synthesis, detergents, treatments of paper pulp, etc. These proteins are often engineered to increase the stability (e.g. thermal or pH), change the activity profile and substrate specificity, and the enzymes are expressed in a microorganism that is well adapted for cultivation on a large scale. In some processes, these industrial enzymes are used as catalysts, and reuse of the enzymes is of interest. This can be accomplished by, for example, immobilization or encapsulation in 'micro-reactors'. A controlled release of the active proteins can also be desired, e.g. in food products or in cleaning applications, and there is potential to accomplish this through application of encapsulation technologies.

Encapsulation of bioactive proteins can expand their fields of applicability, through, for example, improved stability, sustained or triggered release, targeting and immobilization. Several different approaches to achieve the desired properties of the encapsulated protein have been developed based on, for example, coating or formation of microcapsules. This chapter will discuss solid formulations for proteins, and in particular how spray drying can be used to encapsulate bioactive materials in solid formulations, and the application of polyelectrolyte multilayer microcapsules to protein encapsulation in liquid formulations.

16.2 Solid formulations for proteins

Through the advent of genetic engineering and the prospects of using proteins as pharmaceuticals, the interest in dry formulation of proteins has increased. Dry formulations are often preferred for proteins since denaturing reactions are slower (aggregation, deamidation, oxidation, etc.) and thus the stability is improved (Wang, 2000). In addition, the requirements for storage conditions can often be relaxed, since dry formulations often can be stored and shipped at room temperature, in contrast to the need for frozen or chilled conditions for liquid formulations. Genetically engineered human insulin is generally marketed in liquid formulations, but the insulin is typically present as crystals in suspension, which improves the stability. However, a dry protein formulation has to be carefully optimized to provide high retention of activity and sufficient shelf-life. To ensure preservation of the protein structure, an amorphous or at least partially amorphous formulation is desired (Chang *et al.*, 2005), which implicates the storage stability since amorphous materials are inherently unstable. Also for food applications, dry proteins are of interest (e.g. milk proteins, whey proteins and egg white proteins), and in all these instances, it is essential that the functionality of the protein is maintained, and generally a performance close to that of the fresh product is desired.

There are several drying techniques available for proteins, and we will here consider those that do not involve crystallization of the protein as an integral part. The most common techniques are spray drying (particularly in the food industry) and freeze drying (particularly in the pharmaceutical industry). An additional technique is vacuum drying, possibly following protein precipitation by ethanol or acetone, for example. Some more experimental techniques include supercritical fluid technologies such as SEDS (solution enhanced dispersion by supercritical fluid) and spray freeze drying. Each method has its advantages and disadvantages, and is appropriate for different applications. The production method extensively influences the physical powder properties, such as flowability, hygroscopicity and dissolution behaviour, due to variation of the physical state (amorphous vs crystalline) and composition at the particle surface. In addition, the biological activity of proteins included in the particles is influenced by the drying process. In the following sections, the more common drying techniques of spray drying and freeze drying are presented, together with a combination of the two, i.e. spray freeze drying.

16.2.1 Commonly used drying techniques

Spray drying

In the spray-drying process, small droplets are created by the nozzle, which can be of different types, suited for different feed and particle characteristics. These droplets are contacted with warm air, and water starts to evaporate from the droplet surface. The dryer may operate co-currently or counter-currently, and the former is more common for heat sensitive products, since the thermal load on the drying particles is less. During drying, the droplets are maintained at the wet bulb temperature as long as water evaporates freely from the drops (see Fig. 16.1 which illustrates the droplet/particle temperature and diameter during drying). As soon as a surface layer with restricted water transport is created, the evaporation

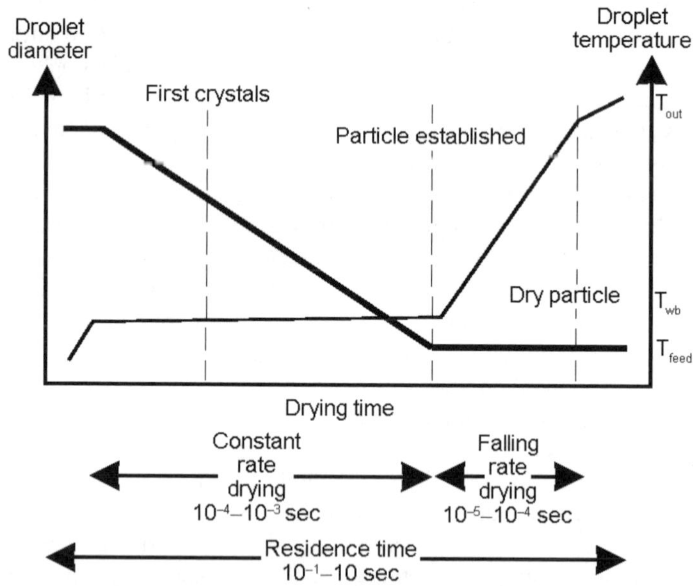

Fig. 16.1 Diagram showing the change in droplet size and droplet temperature during spray drying.

rate is reduced and the drop/particle temperature may start to increase. However, the particle temperature will not rise above the outlet temperature of the dryer. These events explain why spray drying is suited to thermosensitive materials, such as proteins and whole cells. The solids in the system are not heated as high as might be expected from the (typically) high inlet temperature for the air. In addition, the stability of proteins is generally improved by a reduced water content (or rather water activity), and hence the drying material becomes more heat stable as the drying process proceeds and the particle temperature increases. Spray-drying of bioactive protein has been studied by many authors (e.g. Broadhead *et al.*, 1994; Maa *et al.*, 1998; Maury *et al.*, 2005, Adler *et al.*, 2000; Millqvist-Fureby *et al.*, 1999b), and the consensus is that bioactivity can be well preserved in formulations with suitable excipients that prevent degradation processes. A detailed description of the spray-drying process and its applications can be found in Masters (1991). The most detrimental effects on proteins in spray drying include denaturation due to interactions with the gas/liquid interface in the spray, heating, shearing and dehydration.

Freeze drying

In the freeze-drying process, several steps occur: first the solution or suspension to be freeze dried is frozen, the temperature during the freezing process may be controlled to achieve so-called annealing, and finally the frozen water is sublimated under vacuum. The freezing process in itself is an important part of the process, since it influences the structure of the final product. During freezing, water freezes as pure ice and the remaining constituents are concentrated to a point where no further freezing of ice occurs, i.e. a maximally freeze concentrated material is obtained. For process control purposes it is important to know the glass transition temperature (T_g') of this material. It is essential that the temperature of the solid is kept below this temperature during the freeze-drying process in order to prevent collapse when the drying material cannot support its own weight. The glass transition temperature (T_g') is usually determined by differential scanning calorimetry. An alternative method for obtaining this critical temperature is to observe the collapse temperature by freeze-drying microscopy. Generally, the collapse temperature is a few degrees centigrade above T_g', thus imposing less strict process control. If the frozen material is maintained at a temperature above T_g' but below the melting point, growth of the ice crystals occurs, and the molecules in the freeze-concentrated material can more easily diffuse. This treatment, so-called annealing, changes the structure of the material, and the final freeze-dried product will contain larger pores and have smaller specific surface area (Sarciaux *et al.*, 1999). The effects of freezing and drying can impair the protein stability through cold denaturation, surface interactions, and dehydration.

Spray freeze drying

Spray freeze drying is a fairly new technique that can be used to create highly porous particles without involving milling or any other particle size reduction

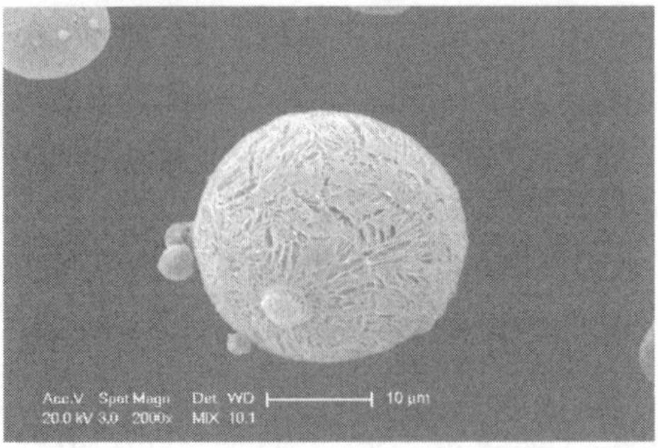

Fig. 16.2 Spray-freeze-dried particles composed of trehalose (95% w/w) and lysozyme (5% w/w).

techniques. The process has been used to create porous particles containing proteins for different therapy purposes (e.g. Webb *et al.*, 2003; Costantino *et al.*, 2000, 2002, Liao *et al.*, 2004, Maa and Costantino, 2004). These particles can be further processed by encapsulation in e.g. PLGA (e.g. Carrasquillo *et al.*, 2001; Lam *et al.*, 2001). In this process a liquid feed is sprayed into a cold medium, typically liquid nitrogen, and the droplets are instantly frozen to form beads. These beads are separated from the liquid nitrogen and freeze dried by the conventional procedure described in Section 16.2.1. Care must be taken when transferring particles from the freezing medium to the freeze-dryer so that the beads are not melted and agglomerates created. As in spray drying, the bead size depends on the settings of the atomizer and the feed composition. The droplets can be created above the surface of the freezing medium or with the spray orifice below the surface of the freezing medium (spray freezing into liquid). This may in the case of proteins have some influence on the protein stability (Yu *et al.*, 2006). Although the process appears complicated, it is possible to operate spray freeze drying on an industrial scale. Figure 16.2 shows a typical spray-freeze-dried particle. The effects of this drying method on the protein stability include cold denaturation, surface induced denaturation, shear induced denaturation and dehydration.

16.3 Surface formation in drying

The surface layer of a powder determines many of its functional properties such as flowability and dissolution rate, and in microencapsulation this surface is of great interest to control and design to achieve the desired powder properties. This requires that the process of forming the surface layer in relation to formulation and processing is well understood, and a natural step is then to analyse the surface composition. In order to analyse the surface composition of a powder only few techniques are available that provide quantitative analysis of a thin surface layer. X-ray photoelectron spectroscopy (XPS) and time-of-flight secondary ion mass spectrometry (ToF-SIMS) are two methods available, with different surface sensitivity (5–10 nm and 1–2 nm, respectively). The principle of XPS is that the sample, placed under high vacuum, is irradiated by monochromatic X-rays, which excite electrons in the different orbitals, and these emit photoelectrons with different kinetic energy. Photoelectrons will be emitted from the entire sample, but due to energy dissipation during the passage through the solid material, only the photoelectrons originating from the surface layer can escape and be detected, hence the surface sensitivity of the technique. The kinetic energy of the photoelectrons is analysed, and by knowing the energy of the X-ray radiation, the binding energy of the emitted electrons can be calculated, thus providing the atomic surface composition. When this is measured for samples and the pure molecular components in the sample, the surface composition of the sample can be estimated in terms of molecular components, rather than atomic surface composition, by use of different calculation methods (Fäldt *et al.*, 1993; Ernstsson *et al.*, 1999; Millqvist-Fureby and Smith, 2007). ToF-SIMS is also a technique that operates under high vacuum, and the analytical principle is based on the use of a pulsed proton beam to desorb and ionize species form the sample surface. Only the outermost (1–2 layers of molecules) are analyzed. The resulting secondary ions are accelerated into a mass spectrometer, where the mass is analysed by time-of-flight from the sample surface to the detector. The secondary ions are then identified as different molecular species.

16.3.1 Surface formation in spray drying

As described in Section 16.2.1, the time from droplet formation in spray drying until water evaporation is limited due to a crust or film forming at the surface and is very short – typically 5–20 ms, depending on droplet size. Thus it could be imagined that the components with low solubility will precipitate first and dominate the particle surface. However, there are other processes occurring in the droplet during this first period: any surface active components will adsorb to the air–liquid interface, and the extent of this can be anticipated to be related to the transport rate and the adsorption kinetics. There is an intricate interplay between drying rates, adsorption rates, degree of supersaturation and precipitation rates, which together determine the particle structure and properties. Thorough understanding of these processes enables the preparation of spray-dried particles with controlled surfaces.

The application of XPS for surface composition analysis, on spray-dried powders of different composition has shown that the surface is dominated by the surface active species in the feed (e.g. Fäldt and Bergenståhl, 1994, Adler *et al.*, 2000; Millqvist-Fureby *et al.*, 1999b; Elversson and Millqvist-Fureby, 2006) for different powders (emulsions, protein solutions). This is illustrated in Fig. 16.3, where the enrichment of protein (casein micelles) at the powder surface, in particular at low protein content, is evident. When several surface active species are present, a competitive situation arises (Landström *et al.*, 2000; Elversson and Millqvist-Fureby, 2006), and the outcome of the competition for the surface is determined rather by the adsorption kinetics, than by the ability of the different species to reduce the surface tension under equilibrium conditions (Elversson

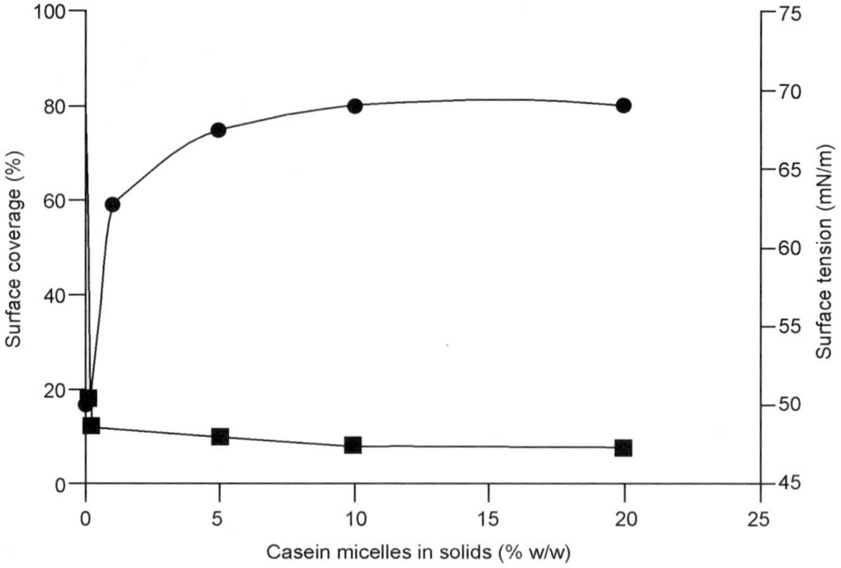

Fig. 16.3 Surface composition estimated by XPS analysis of spray-dried powders composed of casein micelles and lactose, and surface tension of the corresponding solutions.

and Millqvist-Fureby, 2006). Considering the droplet life-time and the diffusion process for a typical protein ($D \approx 6 \times 10^{-11}$ m^2/s), it is unlikely that diffusion is the only transport mechanism at play, since the diffusion time for a protein molecule from the centre of a 10 μm droplet to the surface is 20 ms, while the estimated drying time is about 4.5 ms.

16.3.2 Surface formation in freeze drying and spray freeze drying

The same mechanisms are involved with freeze drying as in spray drying (i.e. adsorption of surface active species to interfaces), although the effect is usually much less pronounced due to the slow transport rates in the freeze-concentrated material. Different investigations have shown that protein and surfactants indeed are enriched at the powder surface (Millqvist-Fureby *et al.*, 1999a; Webb *et al.*, 2002, 2003). Further, it is known that annealing during freezing of samples reduces the specific surface area and reduces the degree of aggregation of bovine IgG (Sarciaux *et al.*, 1999). It has also been shown that annealing increases the surface coverage of protein on freeze-dried powders (Millqvist-Fureby *et al.*, 1999a).

Spray freeze drying presents a mixture of the events occurring during spray drying (i.e. adsorption of surface active species to the droplet interface), and freeze drying (i.e. slow transport in the frozen material). It has been shown that surface active materials indeed accumulate at particle surface during spray freeze drying, albeit not to the same extent as in spray drying (Webb *et al.*, 2002), but to a higher extent than observed for freeze-dried powders (Millqvist-

Fureby *et al.*, 1999a). The surface adsorption of protein is also affected by annealing prior to freeze drying. However, it remains unresolved whether the surface accumulation at the internal surface (ice–water interface) is the same as at the outer surfaces (water–air interface, being replaced by water–liquid nitrogen interface).

16.4 Coating in spray drying

Coatings are applied to powders for a variety of reasons, such as encapsulation of ingredients, controlled release functionality, taste masking, prevention of oxidation, powder handling properties, etc. The most common way to do this is by spray-coating in a fluidized bed system (Jono *et al.*, 2000).The coating becomes relatively thick, may not coat completely, and agglomeration may occur simultaneously. A new approach to this problem can be developed from the behaviour of surface-active materials in spray drying, and considering the non-equilibrium state in the spray droplets, as described in the previous section.

16.4.1 *In-situ* coating of spray-dried powders

Surface-active materials in the liquid will adsorb to the air–liquid interface, and in a competitive situation the component that adsorbs fastest and most efficiently will tend to dominate the surface. This approach has been studied for spray drying of proteins in carbohydrate matrices using low molecular weight surfactants (e.g., Millqvist-Fureby *et al.*, 1999b; Millqvist-Fureby and Smith, 2007; Adler *et al.*, 2000) and surface-active polymers (Elversson and Millqvist-Fureby, 2006). The addition of polysorbate to trypsin co-dried with lactose improved the retained activity and decreased the surface coverage of protein, which can be explained by the large difference in diffusion coefficient between polysorbate and trypsin (Millqvist-Fureby *et al.*, 1999b). Other investigations have shown that addition of polysorbate 20 or 80 to recombinant human growth hormone decreased the formation of aggregates (Maa *et al.*, 1998; Mumenthaler *et al.*, 1994), which was attributed to less interaction of protein molecules with the air–liquid interface. The addition of the surfactant also influenced the particle structure but not the particle size (Millqvist-Fureby *et al.*, 1999a).

When polymers (of similar size as proteins) are used as coating substances, the transport of the coating polymer and a protein to the interface can be expected to occur at similar rates. The outcome of this competition for the interface will be determined mainly by the relative adsorption rates for these polymers, since an equilibrium situation cannot be expected due to the short life-time of the spray droplets. This can be studied by dynamic surface tension analysis, see Fig. 16.4(a), where poly(vinyl alcohol) (PVA) and insulin are used as an example. Although the protein may give a lower equilibrium surface tension (comparing insulin and PVA, for example), the polymer reduces the

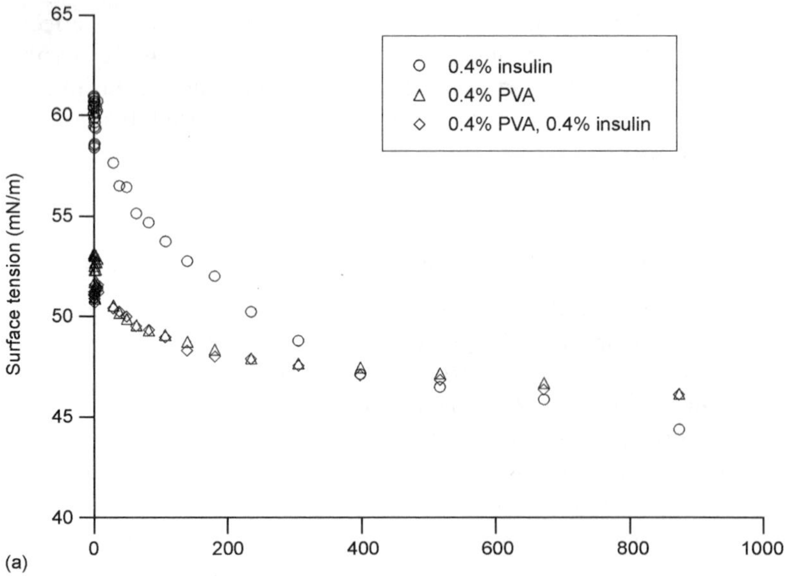

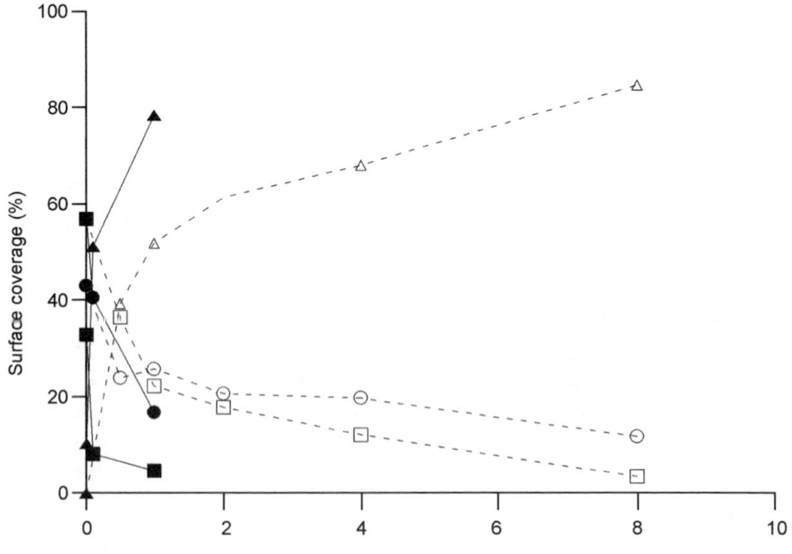

Fig. 16.4 (a) Dynamic surface tension analyses, using the pendant drop method, of PVA, insulin and mixed PVA and insulin. (b) The surface composition of spray-dried powders composed of PVA/insulin/trehalose (open, dashed line), and poloxamer/insulin/trehalose (filled, solid line). The insulin content was 4% w/w in the solids. (c) Comparison of coating efficiency for different surface active additives to protein/trehalose formulations, (◇) Polysorbate 80/insulin, (□) Poloxamer 188/insulin, (■) Poloxamer 188/BSA, (Δ) PVA/insulin, (●) HPMC/BSA. The insulin content was 4% w/w and the BSA content was 5% w/w of the total solids.

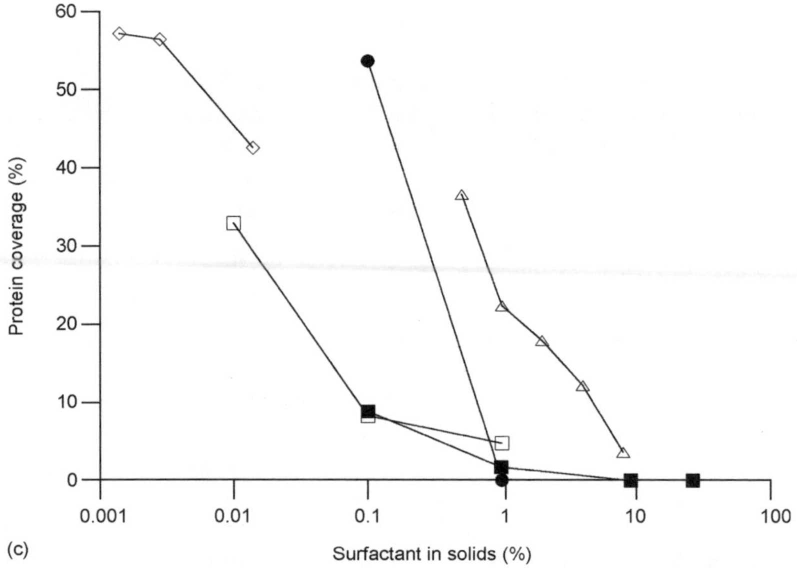

(c)

Fig. 16.4 *Continued*

surface tension more rapidly, and in a mixture of protein and polymer, the final surface tension is determined by the polymer rather than the protein, observed as identical curves for the surface tension of PVA and PVA/insulin solutions. This indicates that the replacement of an adsorbed polymer by the more surface active protein is a very slow process. The behaviour during dynamic surface tension measurements can be correlated with the surface composition of corresponding spray-dried samples (see Fig. 16.4(b)), although the time scale in spray drying is much shorter than in the surface tension measurements when using the pendant drop method. A comparison of PVA and Poloxamer (triblock copolymer poly(propylene glycol)-poly(ethylene glycol)-poly(propylene glycol)), showed that Poloxamer was much more efficient in coating the powders than was PVA, which correlates with the considerably higher surface activity of Poloxamer compared to PVA. In a different study, two surface-active polymers (Poloxamer 188 and hydroxypropyl methyl cellulose, HPMC) were used to coat particles containing BSA and trehalose (Elversson and Millqvist-Fureby, 2006), and the coating efficiency of these polymers is compared to that of PVA and Polysorbate 80 in Fig. 16.4(c), where it can be noted that the polymer with the highest surface activity (i.e. Poloxamer 188) is most efficient in coating the particles. The relatively small effect of Polysorbate 80 is in part related to the smaller size of this molecule and that it adsorbs as a monolayer with an approximate thickness of 2.5 nm, which is less than the depth of analysis in XPS, and thus the surface composition signal is a combination of the (presumed) monolayer at the surface and underlying material.

It was found that the coating did not influence the protein structure in the case of HPMC and Poloxamer 188 (Elversson and Millqvist-Fureby, 2006), while

BSA was negatively affected when dried with PVA. However, the powder properties were affected: the dissolution rate was decreased by the polymer coating, and flow properties of coated powders were also affected.

It can be anticipated that this technique can be applied for several coating purposes, when a thin coating is required, e.g. altering particle–particle interactions, dissolution properties, and preventing protein adsorption to the particle surface. The efficiency of this coating technology in preventing oxygen penetration remains to be studied.

16.4.2 Encapsulation in aqueous two-phase systems

The coating achieved by *in-situ* coating is very thin, most likely only a few nm thick. A thicker coating may be desired and this can also be accomplished in spray drying by using a different formulation approach. It is well known that when two neutral polymers are mixed in a common solvent, phase separation will occur spontaneously due to the unfavourable entropy of mixing for large molecules combined with an enthalpic contribution that favours phase separation (Flory, 1953). Such systems, called aqueous two-phase systems (ATPS), can also be formed by, for example, PEG/phosphate (Albertsson, 1986) or protein/polysaccharide under appropriate conditions (Grinberg and Tolstoguzov, 1997). The interfacial tension in ATPS is very low, and thus sensitive biological structures are not generally negatively influenced by the interface. The most described uses of ATPS include separation systems for biological entities (whole cells, organelles, proteins), mainly using PEG/dextran, and as extractive systems for synthesis or fermentation, where products are removed from one phase into the other. Here we describe the use of ATPS as an encapsulation system for biologically active materials, such as proteins and probiotic bacteria. This is schematically illustrated for a spray-dried system in Fig. 16.5. The ATPS behaves as a water-in-water emulsion, where one of the phases is dispersed in the other. The phase separation persists after drying, and thus results in double encapsulation of the materials that partition to the dispersed phase. Since the interfacial tension is very low, only a small input of energy is required to obtain a fine dispersion. For an optimal ATPS for encapsulation several requirements need to be fulfilled:

- The selected polymers form a two-phase system.
- The bioactive material partitions strongly to one of the phases.
- The viscosity of the ATPS must be suitable for spray drying, i.e. not too high and no substantial extensional viscosity.
- The polymers must not interact negatively with the bioactive material.

A number of different ATPS have been used for encapsulation of proteins. The system PVA/dextran has been studied in some detail by Elversson and Millqvist-Fureby (2005) for encapsulation of BSA. BSA partitions strongly enough to the dextran-rich phase to be a useful model protein. It was found that the two-phase system persists in the dried state, manifested as two separate glass

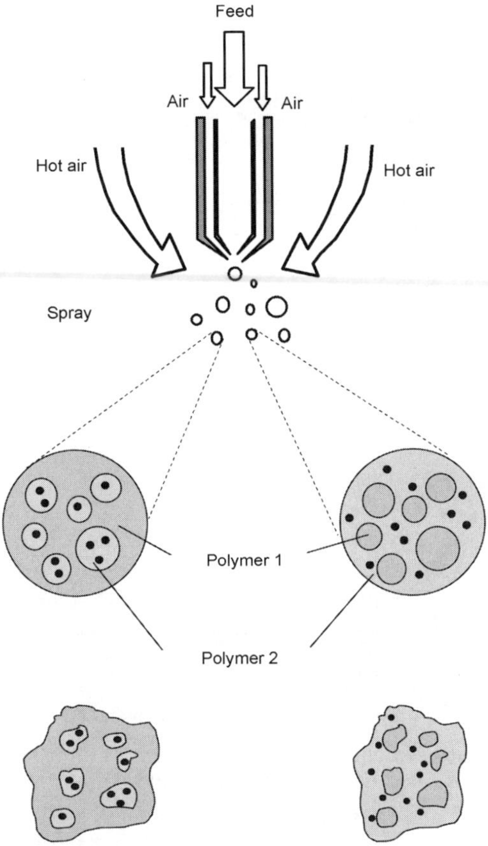

Fig. 16.5 Schematic description of ATPS in spray drying.

transitions observed by differential scanning calorimetry, corresponding to PVA and dextran, respectively. It might be expected that the surface composition of the powder would be related to the structure of the two-phase system, with a low surface coverage of PVA in the dextran-continuous region. However, it turns out that, as in the case of *in-situ* coating, PVA dominates the surface already for a dextran continuous system (Fig. 16.6). This is due to the surface activity of PVA, and the polymer's ability to compete favourably with BSA for the air–liquid interface. The dextran-rich phase contains sufficient PVA to coat the powder surface. This is also manifested in the dissolution properties, the dissolution time ranges from 10 min for dextran as the sole polymer to 60 minutes for PVA as the only polymer. In this particular case, the structure of BSA was disturbed after drying, which was attributed to specific interaction of BSA and PVA. However, the concept as such is regarded as useful for encapsulation and preservation of biological structures, as shown for live probiotic bacteria (Millqvist-Fureby *et al.*, 2000; Elofsson and Millqvist-Fureby, 2001).

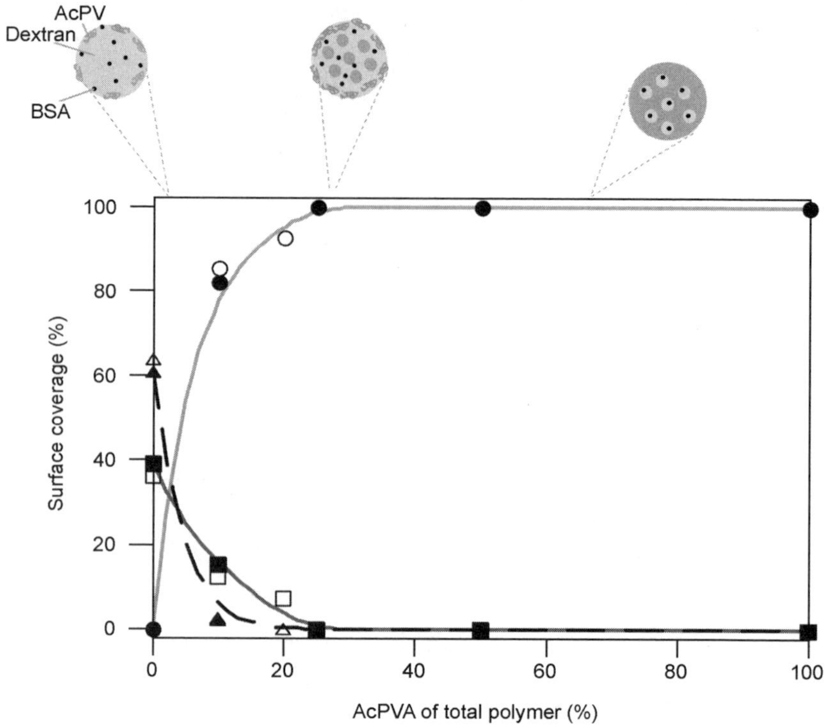

Fig. 16.6 The surface composition as estimated from XPS data for spray-dried PVA/dextran with encapsulated BSA, and schematic illustrations of the internal particle structure.

16.5 Polyelectrolyte microcapsules for protein encapsulation

The idea to use polyelectrolytes to create surface coating was first advanced by Iler (1966), and was further developed by Decher *et al.* during the 1990s (1991, 1992). The creation of the coatings relies on overcompensation of charges when a polyelectrolyte is adsorbed to a surface of opposite charge. Thus a new surface, with charge reversal, is formed and forms the template for the adsorption of a polyelectrolyte of the same charge as the original surface. On a small scale and for macroscopic surfaces, the technique is simple to carry out by dipping the substrate surface in alternating baths containing the coating polymers. It has been established that multilayers can be built with over 1000 polyelectrolyte layers. So far, no restrictions have been found regarding the choice of polyelectrolyte, as long as a pair of opposite charge is chosen, and these will adsorb to a substrate with charged surface. Also particles, cells, lipids, biomolecules, and virus can be used in the layers, and hence complex structures can be built. The thickness of the layers can be finely tuned by adjusting pH, ionic strength, temperature and rinsing conditions, and the thickness of a typical polyelectrolyte layer is in the order of 2 nm.

The multilayer of the microcapsules can be made porous or permeable to macromolecules without destroying the multilayer structure by adjusting the solution conditions. In order to render the multilayer permeable, the pH can be adjusted to decrease the charge of one of the polyelectrolytes, thus influencing the inter-polyelectrolyte interactions (Kabanov *et al.*, 1984; Philipp *et al.*, 1989). In this way, the multilayer can be reversibly opened up to allow passage of macromolecules over the membrane. The multilayer structure is also influenced by the ionic strength, since this influences the strength of the interactions between adjacent layers. Increasing the ion strength implies more shielding of the charges on the polyelectrolytes, thus weakening the interactions between polyelectrolyte layers. The temperature also has an effect on the multilayer stability.

16.5.1 Building multilayer microcapsules

The technique of coating surfaces with polyelectrolytes has been extended to coating solid microparticles (e.g. Caruso *et al.*, 1998) as well as liquid droplets (e.g. Gu *et al.*, 2005), and the principle is shown in Fig. 16.7. The solid colloidal core can be dissolved and thus empty capsules can be formed, which allow the study of polyelectrolyte multilayers at the liquid–liquid interface, but also provide a loadable vehicle for various substances (e.g., dyes, proteins, or drug substances). The dipping technique used for macroscopic surface has to be abandoned when working with microparticles. Instead, the particles must be separated from the polymer solution after each adsorption and washing step, by

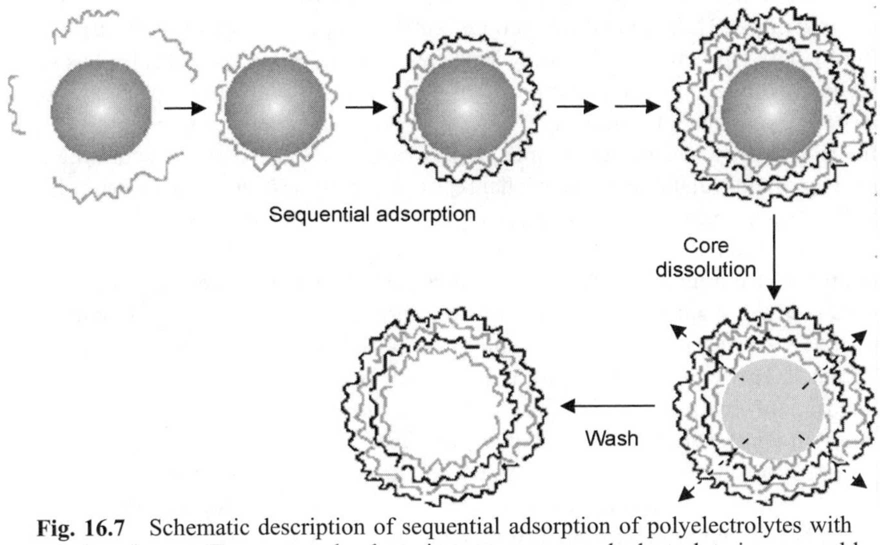

Fig. 16.7 Schematic description of sequential adsorption of polyelectrolytes with alternating charges. Between each adsorption step excess polyelectrolyte is removed by washing. After completion of the multilayer the core is removed by changing the solvent conditions. Finally, the remains of the core material is removed by washing, and empty microcapsules are obtained, ready for loading.

centrifugation or membrane filtration. Alternatively, the amount of polyelectrolyte to cover the particle surface can be calculated, and only a sufficient amount is added. These techniques have all been used for the preparation of microparticles on laboratory scale, and have different advantages. They were compared in a paper by Voigt *et al.* (1999), who found that they all share the problem of creating aggregates during the adsorption process. In the case of the centrifugation method, the particles must be completely resuspended after each centrifugation step, and the efficiency of this is determined by the aggregation strength of the particles. Since aggregates are formed, material is lost and, as an example, 80% of the material was lost on the process of preparing 20-layer microcapsules (Voigt *et al.*, 1999). Both the sequential adsorption with calculated amounts of polymer and the membrane filtration method resulted in higher yields of microcapsules: 50% and >97%, respectively. The main difficulty in the membrane filtration method is the accumulation of microparticles and polymer on the membrane itself, and it is required that the particles are maintained in suspension at all times. The process is also improved by using a minimal pressure for the filtration step – free membrane filtration is preferred, however, not always practical. This method has the obvious benefit of being possible to operate continuously and appears to be most suitable for scaling up.

16.5.2 Protein encapsulation by loading into empty microcapsules

Polyelectrolyte microcapsules have been built with a variety of core particles, including melamine formaldehyde (MF) particles (e.g. Radtchenko *et al.*, 2000), inorganic and organic crystals (Dai *et al.*, 2001), protein aggregates (Balabushevitch *et al.*, 2001, Trau and Renneberg, 2003) and protein crystals (Caruso *et al.*, 2000). The fabrication of multilayer particles on MF template particles has been studied extensively and the structure of both particles with core particles as well as the microcapsules after the core particles have been removed has been investigated by several techniques. It has been clearly shown by zeta potential measurements that the surface charge of the particles changes in accordance with the adsorption of the polyelectrolytes. Fluorescent labelling of the polymers has been used to illustrate the location of the polyelectrolytes by means of confocal scanning laser microscopy (CSLM). The inert core particles can be dissolved by adjusting the solvent conditions. In the case of MF particles, 0.1 M HCl is used to hydrolyse the core particles, and thus empty shells are obtained. However, fragments of the MF polymers are retained in the core due to insufficient hydrolysis, and this influences the loading properties and the environment for molecules transferred into the core. It has been shown that these 'empty' capsules frequently are spherical capsules when filled with liquid, and that they are impenetrable to macromolecules, unless the solution conditions are such that the membrane structure is opened up.

The multilayer capsules have been investigated as 'containers' for macromolecules such as proteins and enzyme, either as drug delivery vehicles, as microreactors, or as components in biosensors. The benefit of the microcapsules is that the enzyme is immobilized in an aqueous solution where it can display its activity. When the microcapsules are intended as drug delivery vehicles, the functionality relies on the controlled release properties of the microcapsules, the encapsulation efficiency, and the possibilities of targeting.

One of the most extensively studied systems for multilayer capsules is that composed of poly(styrene sulfonate) (PSS)/poly(allylamine hydrochloride) (PAH). These polymers are adsorbed in a sequential manner on core particles, the core is dissolved and removed by diffusion. The microcapsules obtained can then be prepared for loading with enzyme by e.g. mixing with 2 parts 96% ethanol and 1 part enzyme solution (Lvov *et al.*, 2001) and incubating for 10 minutes, or by adjusting the pH. It has been shown that the membrane composed of PSS/PAH is 'open' at pH < 6 and 'closed' at pH > 8 with respect to macromolecules such as dextran (MW 75 000) and BSA (open/closed transition at pH 5–7) (Antipov *et al.*, 2002). However, it has been reported (Dähne and Peyratout, 2004) that it has not been possible to repeat the successful loading of MF templated PAH/PSS microcapsules with protein as described by Antipov *et al.* (2002). The encapsulation efficiency in a system such as PSS/PAH microcapsules is often rather low, values of 7% have been reported for small molecules that easily diffuse through the 'opened' membrane, although this can be improved by adding sequesterants in the templates that remain in the interior of the capsule (Khopade and Caruso, 2002; Dähne *et al.*, 2001). Hence, the loading procedure is a limiting step for the application of multilayer microcapsules. The loading possibilities also depend on the quality of the MF template particles, their age and the dissolution conditions, since some polymer fragments will remain in the microcapsule after dissolution of the template particles, as these are too large to escape from the microcapsule. Further, loading of empty microcapsules with protein often results in adsorption or capture of the protein in the walls of the microcapsules, rather than in the core volume. It has been shown that as much as 90% of the captured protein resides in the capsule walls (Tiourina *et al.*, 2001).

16.5.3 Protein core particles for multilayer microparticles

When protein crystals (Caruso *et al.*, 2000) or aggregates/microparticles (Trau and Renneberg, 2003; Balabushevitch *et al.*, 2001) are used as templates, the difficulties of the often adverse loading conditions can be circumvented. The drawback of this approach is that the protein must be precipitated or crystallized, which is not suitable for all proteins. Further, the protein concentration in the final microcapsules cannot be controlled. Typically, the recovery of the bioactivity is less than 85% (Dähne and Peyratout, 2004). To reduce the dissolution of the protein during the coating procedure high salt concentrations can be applied (Balabushevitch *et al.*, 2001). When using crystals, the solution conditions are limited to a generally narrow range of pH where the crystals are stable, and this must be compatible with the pH conditions for formation of multilayers for the polyelectrolyte pairs chosen for encapsulation. The use of aggregates or protein microparticles obtained by lyophilization, precipitation and grinding allows a broader range of solution conditions for the assembly procedure, and thus the polyelectrolyte selection is broader.

16.5.4 Pre-loaded core particles for multilayer microparticles

An advancement of the protein microparticle approach to creating core particles containing a bioactive protein was developed by Sukhorukov *et al.* (2004), who used compound template particles composed of $CaCO_3$ and protein. These particles are subjected to multilayer coating by PAH/PSS, and finally the $CaCO_3$ is removed by addition of EDTA. These conditions are mild compared to the pH or solvent variations used to load empty microcapsules with proteins. These core particles are porous, and it is possible that the first polymer layers will extend into the interior of the particles, and create more complex microcapsule structures after dissolution of the core particles (Volodkin *et al.*, 2004).

A further opportunity is to use mesoporous silica particles loaded with protein (Wang and Caruso, 2005; Yu *et al.*, 2005), in which case the protein loading of the template particles needs to be carried out in an efficient way. This requires that the mesopores are large enough to allow efficient transport of protein into the particles, and furthermore that the protein is not denatured by the adsorption to the silica surface. In this paper the mesopores in the range 10–40 nm were used for catalase adsorption. The pre-loaded microparticles were coated with a number of polyelectrolyte layers, whereafter the silica core was dissolved by exposure to hydrofluoric acid/ammonium fluoride buffer at pH 5. The final activity of catalase in the microparticles was 66% of that in solution, and the protein loss was very small during the multilayer assembly and core dissolution steps. This appears to be an efficient method for encapsulation of proteins, in particular large proteins such as catalase, that do not easily diffuse through multilayers even when these are made permeable to macromolecules.

A different approach to pre-loaded core particles was used by Bergholtz (2004) and in this case, the bioactive materials are spray dried with other materials to create core particles with a charged external surface. This opens the opportunities for combining enzymes as required for different applications as well as in desired concentrations. In order to preserve the bioactivity of the macromolecules during drying, excipients are added, e.g. trehalose or mannitol. These materials are water soluble, which makes it difficult to retain the particles when the first layers at deposited on the core particles. The solubility issue can to some extent be controlled by adding a charged polymer to the spray dryer feed. In accordance with Section 16.3.1, this will lead to surface coating of the particles by this polymer. This polymer may be a polymer with weak charge or a protein. Further, the solutions used for dissolving the polyelectrolytes also contain dissolved carbohydrates to reduce the rate of dissolution of the amorphous core particles. PSS/PAH layers were built on core particles composed of lactose, FITC-dextran and PVA, using the membrane filtration method and short adsorption cycles (2 min) followed by short rinsing cycles (2 min). The short

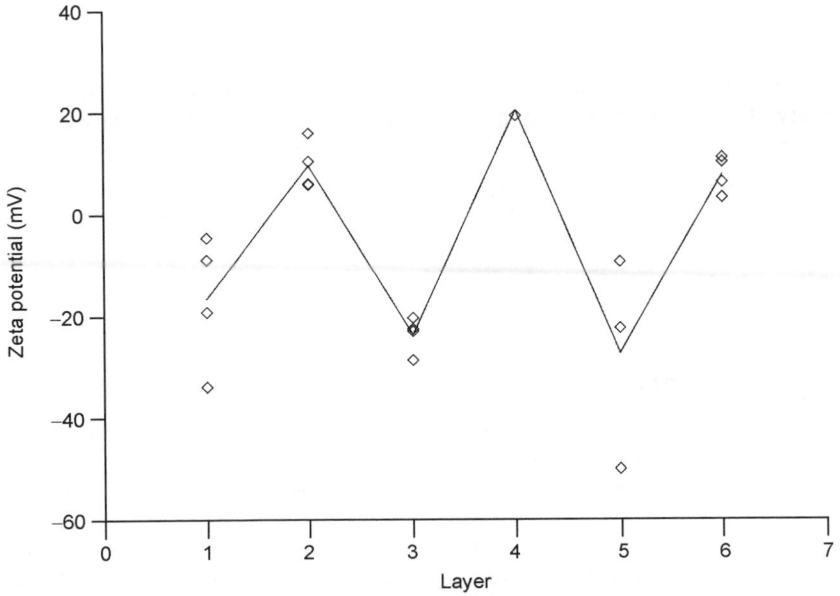

Fig. 16.8 Zeta potential during adsorption of PSS and PAH on spray-dried core particles composed of trehalose and PVA. The polymers were adsorbed from a solution containing 20% lactose, 10% glucose, and 0.2% polymer in a membrane reactor. After adsorption for 5 minutes, rinsing was carried out twice for 5 min using a solution composed of 20% lactose, 10% glucose.

adsorption time is sufficient to complete the adsorption process (Halthur and Elofsson, 2004). Even with all these precautions, the yield was very low. However, the concept was proven by the change in zeta-potential during the coating process (see Fig. 16.8), and the coated particles could be found by microscopy. This approach to loading macromolecules into microcapsules provides a means to finely tune the composition and protein concentration in the microcapsule core, and the saccharides used as carriers in the drying process are easily removed from the core by diffusion, which can be expected to occur quantitatively.

16.6 Summary and future trends

In this chapter a number of approaches to encapsulation of proteins in particles and microcapsules have been described. As a background, different drying technologies are presented, together with the mechanisms behind the surface formation. Through an understanding of the physical process occurring in the drying material, ways of coating proteins and microorganisms through spray drying have been developed. Both *in-situ* coating and ATPS are still investigated for their effects in encapsulation of proteins and probiotics, and the resulting powder properties. Microcapsules prepared by layer-by-layer coating of polyelectrolytes are discussed as carriers for proteins. The major difficulties here lie in the loading of empty microcapsules by adjusting the solution conditions, which often contradicts the protein's requirements for preserved structure and activity. Furthermore, the loading efficiency is rather poor in many studies. Attractive alternatives are provided by different ways of creating pre-loaded core particles, where the carrier materials can be removed, once the polyelectrolyte multilayers have been created. These systems also allow the use of combinations of proteins and control of the protein concentration.

16.7 References

ADLER, M, UNGER M, LEE, G, 'Surface composition of spray-dried particles of bovine serum albumin/trehalose/surfactant', *Pharm. Res.* 2000, **17**, 863–870.

ALBERTSSON P-Å, *Partition of Cell Particles and Macromolecules*, 3rd edn. New York, Wiley, 1986.

ANTIPOV AA; SUKHORUKOV GB, LEPORATTI S, RADTCHENKO IL, DONATH E, MÖHWALD H, 'Polyelectrolyte multilayer capsule permeability control', *Coll. Surf. A: Physicochem. Engng. Aspects*, 2002, **198–200**, 535–541.

BALABUSHEVITCH NG, SUKHORUKOV GB, MOROZ NA, VOLODKIN DV, LARIONOVA NI, DONATH E, MÖHWALD H, 'Encapsulation of proteins by layer-by-layer adsorption of polyelectrolytes onto protein aggregates: factors regulating the protein release', *Biotechnol. Bioengng*, 2001, **76**, 207–213.

BERGHOLTZ C, *Multilayer particles for encapsulation of bioactive proteins*, Master's thesis, Royal Institute of Technology, Stockholm, Sweden, 2004.

BROADHEAD J, EDMOND ROUAN SK, HAU I, RHODES CT, 'The effect of process and formulation variables on the properties of spray-dried β-galactosidase', *J. Pharm. Pharmacol.*, 1994, **46**, 458–467.

CARRASQUILLO KG, STANLEY AM, APONTE-CARRO JC. DEJESUS P, COSTANTINO HR, BOSQUES CJ, GRIEBENOW K, 'Non-aqueous encapsulation of excipient-stabilized spray-freeze dried BSA in poly(lactide-*co*-glycolide) microspheres results in release of native protein', *J. Control. Release*, 2001, **76**, 199–208

CARUSO F, CARUSO RA, MÖHWALD H, 'Nanoengineering of inorganic and hybrid hollow spheres by colloidal templating', *Science*, 1998, **282**, 1111–1114.

CARUSO F, TRAU D, MÖHWALD H, RENNEBERG R, 'Enzyme encapsulation in layer-by-layer engineered polymer mulitlayer capsules', *Langmuir*, 2000, **16**, 1485–1488.

CHANG L, SHEPHERD D, SUN J, OUELLETTE D, GRANT KL, TANG X, PIKAL MJ, 'Mechanism of protein stabilization by sugars during freeze-drying and storage: native structure preservation, specific interaction and/or immobilization in a glassy matrix?', *J. Pharm. Sci.*, 2005, **94**, 1431–1444.

COSTANTINO HR, FIROUZABADIAN L, HOGELAND K, WU C, BEGANSKI C, CARRASQUILLO KG, CÓRDOVA M, GRIEBNOW K, ZALE SE, TRACY MA, 'Protein spray-freeze drying. Effect of atomization conditions on particle size and stability', *Pharm. Res.*, 2000, **17**, 1374–1383.

COSTANTINO HR, FIROUZABADIAN L, CARRASQUILLO KG, GRIEBNOW K, ZALE SE, TRACY MA, 'Protein spray freeze drying. 2. Effect of formulation variables on particle size and stability', *J. Pharm. Sci.*, 2002, **91**, 388–395.

DÄHNE L, PEYRATOUT CJ, 'Tailor-made polyelectrolyte microcapsules: from multilayers to smart containers', *Angew. Chem. Int. Ed.*, 2004, **43**, 3762–3783.

DÄHNE L, LEPORATTI S, DONATH E, MÖHWALD H, 'Fabrication of micro reaction cages with tailored properties', *J. Am. Chem. Soc.*, 2001, **123**, 5431–5436.

DAI Z, VOIGT A, DONATH E, MÖHWALD H, 'Novel encapsulated functional dye particles based on alternately adsorbed multilayer of active oppositely charged macromolecular species', *Macromol. Rapid Commun.*, 2001, **22**, 756–762.

DECHER G, HONG JD, 'Buildup of ultrathin multilayer films by a self-assembly process. 2. Consecutive adsorption of anionic and cationic bipolar amphiphiles and polyelectrolytes on charged surfaces', *Phys. Chem. Chem. Phys.*, 1991, **95**, 1430–1434.

DECHER G, HONG JD, SCHMITT J, 'Buildup of ultrathin multilayer films by a self-assembly process. 3. Consecutively alternating adsorption of anionic and cationic polyelectrolytes on charged surfaces', *Thin Solid Films*, 1992, **210**, 831–835.

ELOFSSON U, MILLQVIST-FUREBY A, 'Drying of probiotic micro-organisms in aqueous two-phase systems', *Minerva Biotechnol.*, 2001, **12**, 279–286.

ELVERSSON J, MILLQVIST-FUREBY A, 'Aqueous two-phase systems as a formulation concept for spray-dried protein', *Int. J. Pharm.*, 2005, **294**, 73–87

ELVERSSON J, MILLQVIST-FUREBY A, '*In situ* coating – an approach for particle modification and encapsulation of proteins during spray-drying', *Int. J. Pharm.*, 2006, **323**, 52–63.

ERNSTSSON, M., CLAESSON, PM, SHAO, SY, 'Characterization of adsorptions sites on a quartz powder from ESCA analysis of adsorbed fatty diamine', *Surf. Interface Analysis*, 1999, **27**, 915–929.

FÄLDT P, BERGENSTÅHL B, 'The surface composition of spray-dried protein-lactose powders', *Colloids Surf. A: Physicochem. Eng. Aspects*, 1994, **90**, 183–190.

FÄLDT P, BERGENSTÅHL B, CARLSSON B, 'The surface coverage of fat on food powders analyzed by ESCA (Electron spectroscopy for chemical analysis)', *Food Structure*,1993, **12**, 225–234.

FLORY P, *Principles of Polymer Chemistry*, Ithaca, NY, Cornell University Press, 1953.

GRINBERG VY, TOLSTOGUZOV VB, 'Thermodynamic incompatibility of proteins and polysaccharides in solutions', *Food Hydrocolloids*, 1997, **11**, 145–158.

GU YS, DECKER AE, McCLEMENTS DJ, 'Production and characterization of oil-in-water emulsions containing droplets stabilized by multilayer membranes consisting of beta-lactoglobulin, iota-carrageenan and gelatin', *Langmuir*, 2005, **21**, 5752–5760.

HALTHUR T, ELOFSSON U, 'Multilayers of charged polypeptides as studied by *in situ* ellipsometry and quartz crystal microbalance with dissipation', *Langmuir*, 2004, **20**, 1739–1745.

ILER RK, 'Multilayers of colloidal particles', *J. Colloid Interface Sci.*, 1966, **21**, 569–594.

JONO K, ICHIKAWA H, MIYAMOTO M, FUKOMORI Y, 'A review of particulate design for pharmaceutical powders and their production by spouted bed coating', *Powder Technol.*, 2000, **113**, 269–277.

KABANOV VA, ZEZIN AB, 'Soluble interpolymeric complexes and the a new class of synthetic polyelectrolytes', *Pure Appl. Chem*, 1984, **56**, 343–354.

KHOPADE AJ, CARUSO F, 'Stepwise self-assembled poly(amidoamine) dendrimer and poly(styrenesulfonate) microcapsules as sustained delivery vehicles', *Biomacromolecules*, 2002, **3**, 1154–1162.

LAM XM, DUENAS ET, CLELAND JL, 'Encapsulation and stabilization of nerve growth factor into poly(lactide-*co*-glycolide) acid microspheres', *J. Pharm. Res.*, 2001, **90**, 1356–1365.

LANDSTRÖM K, ALSINS J, BERGENSTÅHL B, 'Competitive protein adsorption between bovine serum albumin and α-lactoglobulin during spray-drying', *Food Hydrocolloids*, 2000, **14**, 75–82.

LIAO Y, BROWN MB, MARTIN GP, 'Investigation of the stabilization of freeze-dried lysozyme and the physical properties of the formulations', *Eur. J. Pharm. Biopharm.*, 2004, **58**, 15–24.

LVOV Y, ANTIPOV AA, MAMEDOV A, MÖHWALD H, SUKHORUKOV GB, 'Urease encapsulation in nanoorganised microshells', *Nanoletters*, 2001, **1**, 125–128.

MAA Y, COSTANTINO HR, 'Spray freeze-drying of biopharmaceuticals', in *Lyophilization of Biopharmaceuticals*, Costantino HR and Pikal MJ (eds), Washington, DC, American Association of Pharmaceutical Scientists Press, 2004.

MAA Y-F, NGUYEN P-A, HSU SW, 'Spray-drying of air-liquid interface sensitive recombinant human growth hormone', *J. Pharm. Sci.*, 1998, **87**, 152–159.

MASTERS K, *Spray Drying Handbook*, Longman, Harlow, Essex, 1991.

MAURY M, MURPHY K, KUMAR S, MAUERER A, LEE G, 'Spray-drying of proteins: effects of sorbitol and trehalose on aggregation and FT-IR amide I spectrum of an immunoglobulin G', *Eur. J. Pharm. Biopharm.*, 2005, **59**, 251–261.

MILLQVIST-FUREBY A, SMITH P, '*In-situ* lecithination of dairy powders in spray-drying for confectionery applications', *Food Hydrocolloids*, 2007, **21**, 920–927.

MILLQVIST-FUREBY A, MALMSTEN M, BERGENSTÅHL B, 'Surface characterisation of freeze-dried protein:carbohydrate mixtures', *Int. J. Pharm.*, 1999a, **191**, 103–114.

MILLQVIST-FUREBY A, MALMSTEN M, BERGENSTÅHL B, 'Spray-drying of trypsin – surface characterisation and activity preservation', *Int. J. Pharm.*, 1999b, **188**, 243–253.

MILLQVIST-FUREBY A, MALMSTEN M, BERGENSTÅHL B, 'An aqueous polymer two-phase system as carrier in the spray-drying of biological material', *J. Colloid Interface Sci.*, 2000, **225**, 54–61.

MUMENTHALER M, HSU CC, PEARLMAN R, 'Feasibility study on spray drying protein pharmaceuticals: recombinant human growth hormone and tissue-type plasminogen activator', *Pharm. Res.*, 1994, **11**, 12–20.

PHILIPP B, DAUTZENBERG H, LINOW KJ, KÖTZ J, DAWYDOF W, 'Poly-electrolyte complexes – Recent developments and open problems', *Prog. Poly. Sci.*, 1989, **14**, 91–172.

RADTCHENKO IL, SUKHORUKOV GB, LEPORATTI S, KHOMUTOV GB, DONATH E, MÖHWALD H, 'Assembly of alternated multivalent ion/polyelectrolyte layers on colloidal particles. Stability of the mutilayers and encapsulation of macromolecules into polyelectrolyte capsules', *JCIS*, 2000, **230**, 272–280.

SARCIAUX JM, MANSOUR S, HAGEMAN MJ, NAIL SL, 'Effects of buffer composition and processing condtions on aggregation of bovine IgG during freeze-drying', *J. Pharm. Sci.*, 1999, **88**, 1354–1361.

SUKHORUKOV GB, VOLODKIN DV, GUNTHER AM, PETROV AI, SHENOY DB, MÖHWALD H, 'Porous calcium carbonate microparticles as templates for encapsulation of bioactive compounds', *J. Mater. Chem.*, 2004, **14**, 2073–2081.

TIOURINA OP, ANTIPOV AA, SUKHORUKOV GB, LARIONOVA NI, LVOV Y, MÖHWALD H, 'Entrapment of α-chymotrypsin into hollow polyelectolyte microcapsules', *Macromol. Biosci.* 2001, **1**, 209–214.

TRAU D, RENNEBERG R, 'Encapsulation of glucose oxidase microparticles within a nanoscale layer-by-layer film: immobilization and biosensor applications', *Biosensors & Bioelectronics*, 2003, **18**, 1491–1499.

VOIGT A, LICHTENFELD H, SUKHOROKOV GB, ZASTROW H, DONATH E, BäUMLER H, MÖHWALD H, 'Membrane filtration for microencapsulation and microcapsules fabrication by layer-by-layer polyelectrolyte adsorption', *Ind. Eng. Chem. Res.*, 1999, **38**, 4037–4043.

VOLODKIN DV, PETROV AI, PREVOT M, SUKHORUKOV GB, 'Matrix polyelectrolyte microcapsules: new system for macromolecule encapsulation', *Langmuir*, 2004, **20**, 3398–3406.

WANG W, 'Lyophilization and development of solid protein pharmaceuticals', *Int. J. Pharm.*, 2000, **203**, 1–60.

WANG Y, CARUSO F, 'Mesoporous silica spheres as supports for enzyme immobilization and encapsulation', *Chem. Mater.*, 2005, **17**, 953–961.

WEBB SD, GOLLEDGE SL., CLELAND JL, CARPENTER JF, RANDOLPH TW, 'Surface adsorption of recombinant human interferon-γ in lyophilized and spray-lyophilized formulations', *J. Pharm. Sci.*, 2002, **91**, 1474–1487.

WEBB SD, CLELAND JL, CARPENTER JF, RANDOLPH TW, 'Effects of annealing lyophilized and spray-lyophilized formulations of recombinant human interferon-γ', *J. Pharm. Sci.*, 2003, **92**, 715–729.

YU A, WANG Y, BARLOW E, CARUSO F, 'Mesoporous silica particles as templates for preparing enzyme-loaded biocompatible microcapsules', *Adv. Mater.*, 2005, **17**, 1737–1741.

YU Z, JOHNSTON KP, WILLIAMS RO, 'Spray freezing into liquid versus spray-freeze drying: influence of atomization on protein aggregation and biological activity', *European J. Pharm. Sci.*, 2006, **27**, 9–18.

Part IV

Regulatory issues and future trends

17

Regulatory aspects of nutrient delivery systems

E. Acosta, University of Toronto, Canada

17.1 Introduction

The laws and regulations pertaining to food product and nutraceutical supplements are, in principle, specific to each country, and it is difficult to make generalizations on the details of such regulations. However, in this chapter, the recommendations of international organizations such as the Food and Agriculture Organization (FAO) of the United Nations, the World Health Organization (WHO), and the International Conference on Harmonization (ICH) will be briefly reviewed and discussed. Additionally, some of the pertinent regulations established by the Food and Drug Administration (FDA) of the United States, and the recommendations of the United States Pharmacopoeia (USP) will also be discussed. The reader should bear in mind that the purpose of this chapter is to give an introduction to these regulatory issues and provide the appropriate references for additional details.

This chapter is divided into three main parts. Section 17.2 discusses the selection of food grade ingredients and their acceptable daily intake (ADI). Additionally, the procedures and documentation necessary to register new food additives are introduced in view of the issue that additives typically used in controlled released applications are not approved for food products, and therefore need to undergo this registration process. Section 17.3 addresses the regulatory issues regarding the stability of the product, especially in terms of biological, chemical and physical stability. Section 17.4 discusses the regulatory aspects concerning the effectiveness (release and absorption) of the formulation. It should be noted (and is later discussed in more detail) that at this moment the regulations described in Section 17.4 do not apply to food products or even nutraceutical supplements; however, there is growing pressure by consumer groups and regulatory agencies to adapt these regulations to nutraceutical supplements.

17.2 Selection of ingredients

One of the challenges of formulating controlled release technologies for food applications is the relatively small pool of ingredients (when compared to pharmaceutical applications) that are acceptable in food products. Additionally, even if a particular ingredient has food additive status, the concentration of this ingredient in the food product (or the acceptable daily intake) might also be regulated. In this section, these two aspects will be discussed in light of the regulations established by the Codex Alimentarius (a join FAO and WHO organization). In addition, the FDA definitions of GRAS (generally recognized as safe) and food additive status will be introduced.

17.2.1 Codex Alimentarius (FAO/WHO)

The Codex Alimentarius Commission was inaugurated in 1963 with the mandate of establishing and maintaining international standards on food safety (mainly concerned with chemical and microbiological aspects). Even before the establishment of the Codex Alimentarius, there was a Joint (FAO/WHO) Expert Commission on Food Additives (JECFA) that assessed the safety of food additives and their ADI. JECFA was initially established in 1955 and continues to meet on a regular basis. Although JECFA is a separate entity from the Codex Alimentarius Commission, it provides expert scientific advice to Codex that ultimately establishes the regulation through the General Standard for Food Additives (GSFA). To date, JECFA has evaluated more than 1500 food additives; however, only 273 food additives and 24 food additive groups have been approved and listed in GSFA.[1]

Approved ingredients and average daily intake (ADI)

The updated GSFA list of approved ingredients can be found online on the Codex Alimentarius website. It is beyond the scope of this chapter to present this complete list, but it is appropriate to summarize the regulatory status of common food additives that are used or could potentially be used in controlled release formulations. Tables 17.1–17.4 present the name, description and uses, and the ADI values for lipid-based emulsifiers, polymers and solvents, carbohydrates and gums, and for electrolytes and polyelectrolytes, respectively.

The ADI qualifications of 'not limited' or 'not specified' in Tables 17.1–17.4 mean that the additive can be used in foods in general with no limitation, except that the use of this additive should follow the Good Manufacturing Practices (GMP) guidelines. The 'not limited' or 'not specified' qualifications mean that 'on the basis of the available data (chemical, biochemical, toxicological, and other) the total daily intake of the substance arising from its use at the levels

Table 17.1 GSFA-listed lipid-based emulsifiers

Additives	Description/uses[a]	ADI[a,b] (mg/kg b.w.)
Beeswax, white and yellow, Candelilla wax	Mixture of esters of fatty acids and fatty alcohols – glazing agent; release agent; stabilizer; texturizer, clouding agent	No toxicological concern[c]
Calcium, sodium stearoyl lactylate	Emulsifier, stabilizer	0–20
Carnauba wax	Glazing agent, carrier	0–7
Diacetyltartaric and fatty acid esters of glycerol	Emulsifier	0–50
Dioctyl sodium sulfosuccinate	Emulsifying, wetting agent	0–0.1
Glycerol ester of wood rosin	Di- and triglycerols of abietic acids and other fatty acids – oil-soluble emulsifier, stabilizer	0–25
Lactic, acetic, citric and fatty acid esters of glycerol	Oil-soluble emulsifier	Not limited[d]
Lecithins	Phospholipids, emulsifier	Not limited
Mono- and diglycerides, e.g. glyceryl monostearate, glyceryl monooleate, etc.	Emulsifier	Not limited
Phosphatidic acid, ammonium salt	Emulsifier	0–30
Polyglycerol esters of fatty acids	Emulsifiers	0–25
Polyglycerol esters of interesterified ricinoleic acid	Emulsifiers	0–7.5
Polyoxyethylene sorbitan esters of fatty acids, e.g. Polyoxyethylene[20] sorbitan monooleate, etc.	Emulsifiers, dispersing agents	0–25
Propylene glycol esters of fatty acids	Emulsifier	0–25
Salts of myristic, palmitic, stearic, and oleic acids (NH4, Ca, K, Na)	Emulsifier	Not specified[d]
Sorbitan monolaurate, monooleate, monopalmitate, monostearate, tristearate	Emulsifier, stabilizer	0–25
Stearyl citrate	Emulsifier, sequestrant	0–50
Stearyl tartrate	Flour treatment agent	0–500
Sucroglycerides, sucrose esters of fatty acids	Mono and di-esters of sucrose and fatty acids – emulsifier	0–30
Thermally oxidized soya bean oil with mono- and diglycerides of fatty acids (TOSOM)	Emulsifier, antispattering agent	0–30

Notes
[a] Description, uses and ADI obtained from JECFA monographs online.[1]
[b] ADI expressed in mg of the additive per kilogram of body weight.
[c] JECFA has not found evidence of toxicological effect for the uses described.
[d] 'Not limited' is an outdated term that is still valid in the standard but that has been substituted in newer standards by 'Not specified'. These terms are equivalent. In both cases, the formulator needs to use the additive according to GMP principles for food additives described later in this section.

Table 17.2 GSFA-listed polymers and solvents

Additive	Description/uses	ADI (mg/kg b.w.)
Edible gelatin	Partially hydrolyzed collagen – stabilizer, gelling agent, emulsifying agent, crystallization inhibitor	Not limited
Glycerol	Humectant, solvent, bodying agent, plasticizer	Not specified
Insoluble polyvinylpyrrolidone	Colour stabilizer, colloidal stabilizer, clarifying agent	Not specified
Polyethylene glycol	Carrier solvent, excipients	0–10
Propylene glycol	Solvent, glazing agent, humectant	0–25
Polyvinyl alcohol	Coating, binder, sealing agent and surface-finishing agent.	0–50
Polyvinylpyrrolidone	Clarifying agent, stabilizer, tableting adjunct, dispersing agent	0–50

Note: See Table 17.1

Table 17.3 GSFA-listed polysaccharides, gums and other carbohydrates

Additive	Description/uses	ADI (mg/kg b.w.)
Agar	Polysaccharide – emulsifier, thickener, stabilizer	Not limited
Alginic acid	Acidic/anionic polysaccharide – emulsifier, thickener, stabilizer, gelling agent	Not specified
Carob (locust) bean gum, guar gum	Polysaccharide – thickener, stabilizer, emulsifier	Not specified
Carrageenan	Polysaccharide/polyelectrolyte – thickener, gelling agent, stabilizer	Not specified
Cross-linked carboxymethyl cellulose	Tabletting agent	Not specified
Curdlan	Polysaccharide, firming agent, gelling agent, stabilizer, thickener	Not specified
Cyclodextrin, beta-	Encapsulation agent for food additives, flavouring and vitamins	0–5
Ethyl hydroxyethyl cellulose	Emulsifier, stabilizer, thickener	Not specified
Gellan gum	Polysaccharide – polyelectrolyte, thickener, gelling agent, stabilizer	Not specified
Gum arabic	Polysaccharide/polyelectrolyte – thickener, gelling agent, stabilizer	Not specified

Table 17.3 *Continued*

Additive	Description/uses	ADI (mg/kg b.w.)
Isomalt	Mixture of mono- and disaccharides, sweetener, bulking agent, anticaking agent	Not specified
Karaya gum	Acetylated polysaccharides, emulsifier, stabilizer, thickener	Not specified
Maltitol and maltitol syrup	Sweetener, humectant, stabilizer	Not specified
Mannitol	Sweetener, humectant, stabilizer	Not specified
Microcrystalline cellulose	Coating agent, binder	Not limited
Modified starches: Dextrin roasted-, acid treated-, alkaline treated-, bleached-, oxidized-, enzyme-treated-, monostarch phosphate, distarch phosphate, phosphated distarch phosphate, acetylated distarch phosphate, starch acetate, acetylated distarch adipate, hydroxypropyl-, hydroxypropyl distarch phosphate, starch sodium octenylsuccinate, acetylated oxidized starch, dextrins	Modified starch – thickener, stabilizer, binder, emulsifier	Not specified
Modified celluloses: hydroxypropyl cellulose, hydroxypropyl methyl cellulose, methyl cellulose, methyl ethyl cellulose	Emulsifier, thickener, stabilizer, binder, suspension agent, film coating	Not specified
Pectins	Methyl esters of polygalacturonic acid and its salt – gelling agent, thickener, stabilizer, emulsifier	Not specified
Powdered cellulose	Anticaking agent, dispersing agent, texturizing agent	Not specified
Polydextroses A and N	Bulking agent, humectant, stabilizer, thickener	Not specified
Processed eucheuma seaweed	Semi-refined carrageenan	Not specified
Propylene glycol alginate	Stabilizer, thickener, emulsifier	0–70
Sodium carboxymethyl cellulose	Thickening agent, stabilizer, suspending agent	Not specified
Sucrose acetate isobutyrate	Density adjusting, cloud-producing agent in non-alcoholic beverages	0–20
Tara gum	Thickening agent, stabilizer	Not specified
Tragacanth gum	Emulsifier, stabilizer, thickener	Not specified
Xantham gum	Thickener, stabilizer, emulsifier, foaming agent	Not specified
Xylitol	Sweetener, humectant	Not specified

Notes: See Table 17.1

Table 17.4 GSFA-listed electrolytes and polyelectrolytes

Additive	Description/uses	ADI (mg/kg b.w.)
Adipic acid	Acidity regulator	0–5
Ammonium, potassium, sodium and calcium alginate	Ammonium and calcium salts of alginic acid with same uses	Not specified
Ammonium, sodium, potassium, and calcium polyphosphates	Polyelectrolyte, sequestrant, moisture retainer	70 as P – phosphorous[a]
Bone phosphate	Mixture of calcium phosphates – sequestrant, moisture retainer	70 as P
Calcium carbonate	Anticaking agent	Not limited
Calcium citrates	Sequestrant, buffer, firming agent	Not limited
Calcium dihydrogen diphosphate, dicalcium, disodium diphosphate	Stabilizer, leavening agent, neutralizing agent	70 as P
Calcium lactate	Buffer, yeast food	Not limited
Calcium gluconate	Acidity regulator, firming agent, sequestrant, nutrient supplement	Not specified
Calcium hydroxide	Neutralizing agent, buffer, firming agent	Not limited
Calcium sulphate	Yeast food, dough conditioner, firming agent, sequestrant	Not limited
Dipotassium, disodium orthophosphate	Stabilizer, leavening agent, neutralizing agent	70 as P
Disodium EDTA	Sequestrant, antioxidant synergist, preservative	0–2.5
Monopotassium, monosodium orthophosphate	Buffer, neutralizing agent, sequestrant, yeast food	70 as P
Potassium, sodium dihydrogen citrate	Buffering agent, sequestrant	Not limited
Sodium lactate	Antioxidant synergist, humectant	Not limited
Tetrapotassium, tetrasodium diphosphate	Buffering agent, sequestrant	70 as P
Tripotassium, trisodium citrate	Sequestrant, stabilizer, buffer	Not limited
Tripotassium, trisodium orthophosphate	Buffer, emulsion stabilizer, sequestrant	70 as P

Notes: See Table 17.1
[a] Expressed as mg of elemental phosphorous in the additive per kilogram of body weight.

necessary to achieve the desired effect and from its acceptable background in food does not, in the opinion of the Committee, represent a hazard to health'.[2] If a particular additive is meant to be used for a purpose other than that specified in the JECFA monograph, it may be necessary to consult JECFA to verify whether the new application is covered within the scope of the approved monograph.

The Good Manufacturing Practices (GMP) for food additives are described in Section 3.3 of the Preamble of the Codex General Standard for Food Additives.[3]

According to this regulation all food additives listed in the GSFA list shall be used under conditions of good manufacturing practice (unless otherwise specified), which include: (a) the quantity of the additive added to food shall be limited to the lowest possible level necessary to accomplish its desired effect; (b) the quantity of the additive that becomes a component of food as a result of its use in the manufacturing, processing or packaging of a food and which is not intended to accomplish any physical, or other technical effect in the food itself, is reduced to the extent reasonably possible; and (c) the additive is prepared and handled in the same way as a food ingredient.

The reader should be aware that the ADI values in Tables 17.1–17.4 should be used as a guideline to estimate the maximum concentration of the ingredient in the product. Consider, for example, the case of β-cyclodextrin. According to Table 17.3, the maximum daily intake of this ingredient is 5 mg/kg b.w. In order to determine how this ADI value translates into maximum β-cyclodextrin content in the product, it is necessary to consider the average weight of the consumer, the number of servings per day, and the presence of other sources of β-cyclodextrin in the average diet. After considering all these factors, the maximum content of the additive in the product can be calculated. Table 1 of the Codex General Standard for Food Additives (Additives Permitted for Use Under Specified Conditions in Certain Food Categories or Individual Food Items)[3] establishes this maximum value for some of the GSFA-listed additives in specific products. In the case of β-cyclodextrin, these maximum contents are given in Table 17.5.

As expected, the maximum content of β-cyclodextrin in a product with a small serving size, such as chewing gum is larger than the maximum content in a product with larger serving size. To illustrate the calculations behind these maximum additive contents, consider the following scenario. The ADI of β-cyclodextrin is 5 mg/kg, assuming that a 70 kg (body weight, BW = 70 kg) person consumes a 200 ml serving a day of a water-based flavoured drink (daily serving DS = 0.2 kg), and that there are no other sources of β-cyclodextrin in the average intake (AI = 0 mg/kg). In this case:

$$\text{Estimated maximum content} = (\text{ADI} - \text{AI}) * \text{BM/DS} \qquad (17.1)$$

Table 17.5 Maximum β-cyclodextrin content permitted as binder and stabilizer

Food Cat. No	Food category	Max level	Comments	Year adopted
05.3	Chewing gum	20 000 mg/kg		2001
14.1.4	Water-based flavored drinks, including 'sport', 'energy', or 'electrolyte' drinks and particulate drinks	500 mg/kg		2001
15.1	Snacks – potato, cereal, flour or starch based (from roots and tubers, pulses and legumes)	500 mg/kg		2004

According to equation (17.1), the estimated maximum β-cyclodextrin content in flavoured drinks would be 1750 mg/kg. Certainly, the assumptions regarding BW and AI are questionable. If one consider a young child with BW $= 20$ kg, then the maximum content matches the value in Table 17.5.

After reviewing Tables 17.1–17.4 and the restriction on additive content in foods (exemplified in Table 17.5), it becomes obvious that the formulator has to be creative in order to guarantee that the controlled release formulation uses only GSFA-listed ingredients. However, if there are sufficient reasons to support the introduction of a specific food additive, it is possible to submit to the consideration of JECFA new food additives. The process of submitting a monograph is certainly onerous, and there is no guarantee of success (a 1/6 chance according to the statistics of submitted vs approved monographs). The next section introduces the basic aspects of the process of preparing the monographs for toxicology studies and intake studies for new ingredients.

Toxicology and intake studies for new additives

In order to seek a food additive status from Codex, it is necessary to prepare two monographs: one that provides data on toxicological studies (also called toxicological working papers) and a second one that provides data on intake of the additive. An example of the table of contents of a toxicological working paper is presented in Table 17.6.[4] According to Table 17.6, there are three main areas covered in toxicological working papers: biochemical aspects, toxicological studies, and observation in humans.

The information compiled in the section on biochemical aspects is used by the expert committee to understand the fate of the additive and its byproducts in the body. Based on these assessments, the expert committee forms an opinion about the potential mechanism of toxicity (if any). This information also serves as a background to interpret the toxicological data.

The section of toxicological studies represents the core of the working paper. There are five types of mandatory studies (items 2.2.1–2.2.5 in Table 17.6): acute toxicity, short-term toxicity, long-term toxicity/carcinogenicity, genotoxicity, and reproductive toxicity. Typically, if there are any toxicity issues, these studies tend to indicate target organs that are then considered in special studies. The information collected for each of the five routine studies is reported using toxicological summaries. Such summaries contain the following information:

- purpose or objective of the experiment
- identification and purity of the additive
- animal species and strain employed in the study (typically these studies include a larger number of small rodents – studied first – and decreasing numbers of mammals)
- number of animals in test and control groups
- sex
- dose levels (in mg/kg b.w. per day)

Table 17.6 Example table of contents for toxicological working papers. Adapted from Ref. 4

1. Explanation
2. Biological data
    - 2.1 Biochemical aspects
        - 2.1.1 Absorption, distribution, and excretion
        - 2.1.2 Biotransformation
        - 2.1.3 Effects on enzymes and other biochemical parameters
    - 2.2 Toxicological studies
        - 2.2.1 Acute toxicity
        - 2.2.2 Short-term toxicity
        - *Mice, Rats, Hamsters, Rabbits, Dogs, Pigs, Monkeys*
        - 2.2.3 Long-term toxicity/carcinogenicity
        - 2.2.4 Genotoxicity [in tabular form; see Appendix D]
        - 2.2.5 Reproductive toxicity
            - 2.2.5.1 Multigeneration reproductive toxicity
            - 2.2.5.2 Developmental toxicity
        - 2.2.6 Special studies on cardiovascular effects
        - 2.2.7 Special studies on immune responses
        - 2.2.8 Special studies on macromolecular binding
        - 2.2.9 Special studies on metabolites
        - 2.2.10 Special studies on no-hormonal effect levels
        - 2.2.11 Special studies on ocular toxicity
        - 2.2.12 Special studies on photoisomerization products
        - 2.2.13 Special studies on thyroid function
    - 2.3 Observations in humans
3. Comments
4. Evaluation
5. References

- route(s) of administration
- duration of the experiment
- biological parameters examined
- effects observed (percentage survival should be included in long-term toxicity/carcinogenicity studies if affected by treatment)
- author's conclusions
- reference(s).

The main conclusion from these studies is a proposed dose (intake) where a no-observed-effect-level (NOEL) is obtained. In other words, this is the dose where no significant difference in biological response is obtained between the test group and the control group. Such conclusions from the authors may be rebutted or modified by a temporary advisor appointed by JECFA. Furthermore, the credibility of the laboratory conducting the study is paramount. It is best to contract these services to a laboratory that has Good Laboratory Practice (GLP) certification, or alternative, the laboratory could be subject to a quality assurance audit by JECFA representatives.

Initially, JECFA used the toxicological monographs to determine the ADI (typically the NOEL divided by a safety factor). However, the problem in doing so is that the maximum content of the additive could be calculated using an equation similar to equation (17.1), using the assumption that the background or average intake (AI) is zero. If the same assumption is made for other products, it is then possible that the average consumer who uses these products may intake more than what is regulated according to the ADI established in the GSFA listing. The purpose of the intake monograph is to produce an estimation of this background intake and propose a correction factor to the ADI obtained from the toxicological working paper (this does not apply to substances with 'not specified' or 'not limited' status).

The challenge of producing these monographs is that it is necessary to establish some average intake for an average diet. As expected, this is a value that varies from country to country because it depends on the socioeconomic and cultural characteristics of each country. One method to 'screen' the potential effect of background intake is to estimate the consumption of the additive based on average intake of calories – this is called the budget method.[5,6] If the maximum additive content in solid and liquid foodstuff calculated by the budget method is lower than that calculated according to the proposed ADI values, then there is no reason for concern about the effect of background intake, otherwise other intake estimations need to be carried out. Various methods for additive intake can be undertaken, including total mass balance of the additive {(annual production + import – export – non-food uses)/(total population × 365)} also known as 'poundage' evaluation; assessments based on data from household economic surveys and sales data; assessments based on model diets; and assessments based on individual dietary records. The conclusion of this report should indicate whether the ADI established based on the toxicity evaluation (NOEL) stands or if this value should be further reduced accordingly.

In addition to the safety of chemical additives discussed above, the evaluation of the safety and effectiveness of new biological agents such as probiotic cultures (discussed in other chapters in this book) is being considered for forthcoming regulations.[7]

17.2.2 Food and Drug Administration (FDA)

The Codex Alimentarius is used to regulate international trade of food products, but it does not have purview on individual countries. In that case, the laws/regulations of the country apply. In the United States, the Food and Drug Administration (FDA) is in charge of setting the standards for food additives. The principles and spirit of the FDA regulations are the same as those of Codex and JECFA; however, the terminology and procedures are different. The FDA regulation has two different categories for food additives, one is direct food additives generally recognized as safe (GRAS) and the second category is for food additives permitted for direct addition to food.

The intake of GRAS additives or their maximum content in food products is not specified (similar meaning to ADI 'not specified' for GSFA-listed additives). However, their use in food products should follow GMP guidelines.

The current list of GRAS additives is published annually in the Code of Federal Regulations of the United States under title 21, part 184 (21 CFR 184 – see ref. 8). Currently, the list contains 214 compounds and it is similar to the list of GSFA compounds with 'not limited' or 'not specified' ADI.

The second category of food additives corresponds to food additives whose use in certain food products is allowed, but they are regulated to a certain maximum content in such products. These food additives permitted for direct addition to food are listed (including details of maximum content in foods) under the regulation 21 CFR 172.[9] This list contains 147 compounds or family of compounds.

Traditionally, the FDA conducted their own evaluations for specific additives, following a priority list of food additives. However, in 1997 the FDA introduced the possibility of accepting GRAS notices from interested parties. Such notices have to be accompanied by supporting information (similar to the working papers described for JECFA) to substantiate the GRAS claim. The FDA then reviews the supporting information and produces an answer which typically falls into one of three categories:

1. the FDA has no questions
2. notice does not provide a basis for a GRAS determination
3. the FDA has no questions, but includes additional correspondence (specific modifications to the original submission).

To date, there have been 219 notifications, and most of them have been approved.[10]

In general, the regulations established in the Codex Alimentarius regarding ADI are more restrictive than the regulations of individual countries. However, translating ADI values into maximum additive content in food products varies from country to country. In addition to these international regulations, there are additional considerations regarding the source of the additive if a food product is to be used in special diets such as kosher, halal or vegetarian diets.

17.3 Stability of the product

17.3.1 The International Conference on Harmonization (ICH)

The current international standards on stability testing are set by the International Conference on Harmonization (ICH). These guidelines have been adopted by numerous countries as the basis to produce regulations for good manufacturing practices (GMPs). Section 'Q1' of these guidelines pertains to stability testing of active ingredients. Although the guidelines are, in principle, designed for pharmaceutical products, countries such as Canada use the same guidelines for active ingredients formulated into food products.

Section Q1 of the ICH guidelines has five parts (ICH Q1A–ICH Q1E).[11] The first part, now in its second revision – ICH Q1A(R2), also referred to as the parent guideline – establishes the protocol for simulated storage conditions and frequency of testing for new drugs and drug products. The second part, ICH

Q1B, describes the exposure protocol and available light sources to conduct photostability testing of new drugs or drug products. The third part, ICH Q1C, establishes a simplified stability protocol for new dosage (including controlled release) for existing and approved products. Part ICH Q1D provides guidelines for bracketing and matrixing design for stability tests. Part ICH Q1E provides guidelines for the interpretation of the stability tests in terms of shelf-life and labeling (shelf-life under specific storage conditions). In addition to these guidelines, there are guidelines pertinent to the stability of biotechnological products (such as proteins) which are summarized in ICH Q5C.

The simulated storage conditions prescribed by ICH Q1A(R2) for drug products (which include controlled release formulas) are summarized in Table 17.7. The spirit of the guideline is to provide a standard protocol that applies to Type I and II climate zones (the climatic zones of the United States, the European Union and Japan – the three founding members of the ICH). One note of caution about Table 17.7 is that there are intermediate stability tests (typically of 6 months duration) that can be used in certain cases.[11]

Long-term tests are meant for products with a shelf-life of more than one year, and the conditions are meant to simulate storage in the plant, and the point of distribution. Accelerated tests are meant to evaluate extreme conditions which may arise during shipping and handling of the product. The guideline also suggests that open container storage tests should be performed to simulate the storage of the product at the point of use (after being opened) and that the manufacturer should use that information to warn the customer against mishandling of the product.

The case of the semi-permeable packing (Table 17.7) applies to liquid products (mostly aqueous suspensions) to evaluate the possibility of water

Table 17.7 Testing protocol for product stability evaluation according to ICH Q1A(R2)

Case	Study	Storage condition	Frequency, months
Powders, general case	Long term	30 ± 2 °C , 65 ± 5% RH	0, 3, 6, 9, 12, 18, 24, annually
	Accelerated	40 ± 2 °C , 75 ± 5% RH	0, 3, 6
Liquids, semi-permeable packing	Long term	30 ± 2 °C , 35 ± 5% RH	0, 3, 6, 9, 12, 18, 24, annually
	Accelerated	40 ± 2 °C , 25 ± 5%RH	0, 3, 6
Impermeable packing	Long term	30 ± 2 °C	0, 3, 6, 9, 12, 18, 24, annually
	Accelerated	40 ± 2 °C	0, 3, 6
Refrigerated products	Long term	5 ± 3 °C	0, 3, 6, 9, 12, 18, 24, annually
	Accelerated	25 ± 2 °C , 60 ± 5%RH	0, 3, 6
Frozen products	Long term	−20 ± 5 °C	0, 3, 6, 9, 12, 18, 24, annually

evaporation when stored in low humidity conditions. Moisture is not considered as a variable in the case of impermeable packing because there is negligible solvent evaporation or moisture absorption.

The ICH Q1A(R2) guideline also considers that other stability protocols may be applied if there is enough scientific support, and appropriate label information to justify the protocol. In general, the guideline defines as a 'significant change' in the drug product quality (during the course of the stability test) as:

- a change of 5% in the initial active ingredient concentration, or failure to meet the acceptance criteria for potency when using biological tests (except dissolution tests or bioavailability tests – discussed in the next section). Guideline ICH Q6A summarizes the test procedure and acceptance criteria for drug products
- the presence of degradation products above their acceptable level (procedure described in the guideline ICH Q3B)
- failure to meet the acceptance criteria for appearance, physical attributes, and functionality test (such as color, phase separation, wettability, etc.). The guideline considers that reversible changes due to accelerated test conditions are still acceptable
- failure to meet the acceptance criteria for pH stability
- failure to meet the dissolution criteria.

Guideline ICH Q1E provides a detailed account of the statistical methods employed to determine the shelf-life of the product. The spirit of the ICH Q6A guideline is to prompt the manufacturer to produce a document proposing a set of specifications for the product and the adequate test protocols to determine whether the product adheres to these specifications. The document should contain a section where the manufacturer justifies the suitability of the proposed test protocols. In addition to the tests proposed by the manufacturer, the guideline has a set of tests of its own. These tests are separated into universal and specific tests.

Universal tests

- *Description*: it entails a description of the dosage form such as color, form (powder, liquid, etc.), texture, odor or taste and other pertinent information that could be perceived without specialized equipment. If some of these characteristics, in particular color (typically linked to oxidation of active ingredient or carrier), change during the course of the stability test, then a quantitative procedure to track this change should be adopted.
- *Identification*: the guideline requires that the manufacturer should identify a method to determine the presence of the active ingredient in the formula, such as infrared or near infrared spectroscopy. The guideline discourages the use of retention time for a single HPLC separation as a method of identification, unless more than one chromatography technique is used to confirm the presence of the active ingredient.
- *Assay*: this item refers to the quantification of the active ingredient. Liquid chromatography methods (e.g. HPLC) can be used as quantification methods

as long as they have been properly validated (guideline ICH Q2R1). Titration methods are discouraged unless they are combined with other test methods that could verify the absence of potential interfering species in the sample.
- *Impurities*: detection and quantification of degradation products and impurities following the ICH Q3B and ICH Q3C guidelines.

Specific tests
- *Dissolution*: for fast release formulas (formulas where more than 80% of the active ingredient dissolves in 15 minutes at pH 1.2, 4.0 and 6.8), the concentration of the active ingredient in the dissolution media can be determined at a predetermined time (single point measurement). However, for controlled release formulas, the dissolution profile should be obtained (mass of active ingredient dissolved as a function of time). For delayed release formulas, a two-stage dissolution test should be used to evaluate the effectiveness of the formula. The guideline does not prescribe a particular composition of the diluent solution, but the manufacturer should select an appropriate diluent solution representative of the release conditions.

  The spirit of the guideline is to use the dissolution test as a measure of bioavailability. This concept will be revisited in Section 17.4. The dissolution test is best suited for an active ingredient whose bioavailability has been shown to be highly dependent on the dissolution rate of the active ingredient. The relation between the dissolution rate and bioavailability should be used to define a minimum acceptable dissolution rate. For controlled release products, the *in vivo/in vitro* correlation data may be used to establish the acceptance criteria. If such data are not available, the permitted variability in the degree of release at half-life should not exceed a total numerical difference of ±10% (i.e. 40–60% release at half-life), unless the manufacturer can support the validity of a larger variability using bioequivalence studies.
- *Disintegration*: this test measures the rate of disappearance of the carrier as opposed to the dissolution of the active ingredient. In general the guideline prioritizes the use of dissolution tests, but also accepts disintegration tests if there is previous data showing a relation between these two, or if the disintegration test is a suitable method to detect variations in the formulation or production method that may impact the bioavailability of the active ingredient.
- *Hardness/friability*: the guideline suggests using this parameter as an in-process control which does not need to be included in the specification unless it has an impact on the quality of the product and its ability to release the prescribed amount of active ingredient per dose. This might be especially important for powdered formulas containing powdered premixes where particle attrition and segregation may affect the uniformity of dose per serving.
- *Uniformity of dosage unit*: it measures the amount of active ingredient per dose or serving, or the total mass of the dose or servings. The in-process

blending uniformity analysis (BUA) discussed above serves to address this point.

- *Water content*: this test should be included in the set of specifications of the product when the presence of water in the product may have an effect on the stability or effectiveness of the active ingredient. Gravimetric methods could be used when properly validated, but detection methods specific for water (such as Karl Fischer titration) are preferred.
- *Microbial limits*: the criteria should be set for the total count of aerobic microorganisms, the total count of yeasts and molds, and the absence of objectionable bacteria. The protocol of exposure should be consistent with current pharmacopoeial procedure. There are certain testing exemptions for solid products, but the manufacturer must make the case that these exemptions apply to the product.

For liquid products there is a similar set of specific tests; however, because degradation reactions are usually faster in liquids, there are additional tests required to guarantee the stability of the product. These specific tests include pH, microbial limits, antimicrobial preservative content, antioxidant preservative content, extractables (presence of impurities that have leached out of the container and into the liquid product), alcohol or cosolvent content (when applicable), dissolution (maintain integrity upon dilution), particle size distribution (for suspensions and emulsions), redispersability, rheological properties, reconstitution time and water content.

The manufacturer should also consult the GMP guidelines of the country where the product is going to be sold. Most countries adapt the ICH guidelines to design their own GMP guidelines.

17.4 Bioavailability of the active ingredient

Currently, there are no international regulations that require the testing of the effectiveness of nutraceutical supplements and related products. Generally, these products are only required to comply with GMP guidelines that apply to food products, to ensure that the product contains the ingredients claimed on the label, and that the product is safe. Recently, various scientific groups and regulators have started to look at the appropriate methods to understand the mechanisms of action of nutraceuticals and related products, and more importantly, to determine the appropriate method to regulate these products.[12] One of the developments that has resulted from these intiatives is the introduction, by the United States Pharmacopoeia (USP) of the 'USP verified' seal of approval for dietary products.[13] To obtain this seal, the products and ingredients have to pass a series of USP verification requirements that include a GMP audit, product and ingredient testing, manufacturing documentation review, and a simulated dissolution test. Participation in the program is voluntary.

Although the nutraceutical market is still growing at a healthy pace, there are indications that for certain nutraceuticals, the growth in installed production capacity has surpassed the growth of the market, which only means that the market is nearing its maturity, and that competition for prices, but especially quality, will be the key for success.

The rapidly changing environment for nutraceuticals and related products is likely to evolve into a new set of rules designed to test the effectiveness of these products. In the spirit of offering a point of reference, the current FDA testing guidelines to evaluate the bioavailability (uptake of the active ingredient) of pharmaceutical products are briefly reviewed and discussed.

17.4.1 FDA regulation on bioavailability and bioequivalence

Bioavailability and bioequivalence tests are regulated (in the United States) by the code 21 CFR 320 'Bioavailability and bioequivalence requirements'. According to the code, bioavailability is defined as the rate and extent to which the active ingredient or active moiety is absorbed from a drug product and becomes available at the site of action. Bioequivalence, on the other hand, is defined as the absence of a significant difference in the rate and extent to which the active ingredient or the active moiety in alternative but equivalent formulations becomes available at the site of drug action when administered at the same molar concentration of the active ingredient. For controlled release products, the condition of rate of absorption is waived (this will be defined by the specific claim of controlled release of the formula), but the extent of absorption should match the extent of release of an equivalent (standard) uncontrolled release formulation. According to the regulation, bioavailability is determined and reported, but bioequivalence has to be demonstrated.

The regulation does consider that, under certain circumstances, the *in vivo* bioavailability tests could be waived or estimated using alternative *in vitro* studies (covered in Section 17.4.2). However, in most cases, new or reformulated products have to undergo the *in vivo* bioavailability tests.

In section 320.24 of the regulation, it is specified that the preferred method of evaluating *in vivo* bioavailability is the use of *in vivo* tests in humans, where the concentration of the active ingredient or its metabolite in whole blood, plasma, serum or other appropriate biological fluid is measured as a function of time. Alternatively, a properly validated *in vitro* test method (using *in vitro–in vivo* correlations or IVIVC) could be used instead. Next on the scale (of decreasing preference), the regulation accepts *in vivo* tests where the concentration of the active ingredient or its metabolite is measured as a function of time, using finely divided time scales. The regulation also accepts *in vivo* tests that measure the pharmacological effect of the active ingredient or its metabolite as a function of time. Clinical trials could be acceptable if they are carefully designed to minimize the uncertainties, and if dose-effects are documented and properly measured.

For extended release formulations, section 320.25 of the regulation requires (a) that the *in vivo* test sustains the extended release claims made by the applicant, (b) that the test rules out the occurrence of 'dose dumping' (excessive fast release due to a malfunctioning controlled release), (c) that the drug produces the same steady state performance as other non-extended release formulas, and (d) that the product produce consistent pharmacokinetic performance between dosages.

Section 320.26 of the regulation provides the applicant with guidelines as to how the bioavailability test should be designed for single dose, and section 320.27 provides the guidelines for multiple dose studies. Common to both studies is the fact that normal (healthy) adults should be used as the subject of the study. In the case of single dose, the product should be taken in fasting state, but for multiple doses fasted and fed conditions should be evaluated. The comparison between the tested product and the reference one should be designed using a crossover strategy allowing for an elimination period in between. The length of the duration of the experiment should cover at least three times the half-life of the active drug ingredient, and the frequency of sampling must be designed in a way that allows an accurate determination of the maximum concentration of the active ingredient and the area under the curve (AUC). For extended release products, the regulation recommends carrying out multi-dose experiments to determine the levels of the active drug at steady state condition and determine the equivalence between the extended release product and the non-extended one.

Typically, the blood or plasma concentrations data is interpreted using the Wagner–Nelson (one compartment) model where there is a single expression for the rate of elimination such that:[14,15]

$$Fa = \frac{C_T + K_E \int_0^T C dt}{K_E \int_0^T C dt} \qquad (17.2)$$

where Fa is the fraction absorbed, C_T is the concentration of the active ingredient in the blood or plasma at time T, K_E is the first-order elimination constant and C is the concentration at any time t. Most authors warn about the misinterpretation of the data on the potential onset of 'flip-flop' towards the latter part of the curve, which occurs if the rate of elimination is equal or higher than the rate of absorption.[15]

One of the important outcomes of applying a pharmacokinetic model such as the Wagner–Nelson model or the modified multi-compartment model of Loo–Riegelman[16] is to obtain the fraction of drug absorbed as a function of time. This is the measure of bioavailability, and the parameter that is used to decide whether the formulation (being controlled release or not) is effective. As will be discussed later, this is also the parameter that is used in the validation of *in vitro* tests.

17.4.2 FDA regulation on *in vitro* and *in vivo* correlation

An *in vitro-in vivo* correlation (IVIVC) is a mathematical expression that can be used to describe an *in vivo* response (in particular bioavailability) using parameters obtained with an *in vitro* test method (e.g. fraction of active ingredient released). The FDA guidelines for developing IVIVC for extended release oral dosage formulas describe four correlation levels (A–D, see ref. 17).

Level A correlation (the highest level) applies to *in vitro* dissolution tests where the fraction of active ingredient dissolved (as a function of time) is directly proportional (if not the same) to the fraction of the active ingredient absorbed (as a function of time) obtained after deconvoluting a plasma concentration study (using an appropriate model such as the Wagner–Nelson model (equation 17.2). Level A correlation is an acceptable level to obtain a 'biowaiver' (a waiver for an *in vivo* bioequivalence study required in compliance of FDA regulations).

Level B correlation applies to *in vitro* dissolution tests were the mean *in vitro* dissolution time is comparable to the mean *in vivo* residence time or the mean *in vivo* dissolution time. Although this method uses all the experimental points to calculate the mean dissolution time, it is not a point-to-point correlation with *in vivo* absorption data. Level B correlation is not acceptable as a biowaiver alternative.

According to Level C correlation, one dissolution time point (e.g. $t_{50\%}$) or the fraction dissolved at a particular time is correlated to a pharmacokinetic parameter (AUC, t_{max}, C_{max}). Because level C does not represent the complete dissolution curve, it is not an acceptable alternative to obtain a biowaiver. However, a multiple-point Level C correlation may be justified as an alternative to Level A if the selected points cover the complete dissolution range (e.g. $t_{5\%}$, $t_{50\%}$, and $t_{95\%}$) and compare to more than one pharmacokinetic parameter.

A Level D correlation implies that there is a qualitative but not a quantitative relation between the dissolution time and the *in vivo* bioavailability. It has no value from the regulatory point of view, but it may be used as a screening tool during the early formulation stages.

In order to evaluate these dissolution times, the United States Pharmacopoeia (USP) has specified four types of apparatus: apparatus 1 (rotating basket), apparatus 2 (paddle-stirred apparatus), apparatus 3 (reciprocating cylinder), and apparatus 4 (flow through cell).[18] Although apparatus 1 and 2 are commonly used to evaluate conventional formulas, apparatus 3 and 4 are recommended for controlled-release formulations.[15]

The composition of the aqueous media may vary, but must ultimately be justified for the particular application. In general the FDA guidelines recommend three aqueous media: degassed water only, 0.1 N HCl (to simulate stomach conditions), and USP buffer (pH of 6.8). For poorly soluble drugs, the addition of 1% sodium dodecyl sulfate may be acceptable.[17] Neither the USP nor the FDA recommends the use of enzymes in the aqueous solutions, or the use of bile salts. However, for emulsions and self-microemulsifying drug delivery systems (SMEDDS), it is imperative to use the Fasted/Fed Simulated State Intestinal Fluids (FaSSIF/FeSSIF) which include the necessary bile salts and lecithin necessary to stabilize the delivery system.[19]

It must be highlighted that all the correlations considered so far do not consider absorption and permeation through the intestinal wall as a variable. To understand the apparent discrepancy between the FDA guidelines for IVIVC and the factors that affect the *in vivo* absorption, it is necessary to review the Biopharmaceutical Classification System (BCS) introduced by Amidon *et al.*[20] According to the BCS, there are three basic factors that affect the bioavailability of a formulation: the solubility of the active ingredient, the dissolution conditions (mean residence time) and the permeability of the active ingredient (which influences the absorption time). These three factors are defined by three dimensionless numbers, the absorption number (An = mean residence time/mean absorption time), the dose number (Do = dose concentration/solubility), and the dissolution number (Dn = mean residence time/mean dissolution time). For a set mean residence time, there are two properties of the active ingredient that influence its *in vivo* absorption: solubility and permeability. There are four possible combinations of these two factors which correspond to a class of biopharmaceutical. A BCS class I active ingredient has high solubility and high permeability. A BCS class II active ingredient has a high permeability but low solubility. A BCS class III active ingredient has low permeability and high solubility. Finally, a BCS class IV active ingredient has low permeability and low solubility. A controlled release formulation of a BCS I active ingredient typically complies with an IVIVC Type A (or multiple C) correlation because its release is controlled by the rate of dissolution. A BCS II active ingredient typically complies with an IVIVC type A or multiple C correlation. BCS III and IV would not comply with the FDA guidelines for IVIVC. BCS IV active ingredients are the one that cause the most concern and should be reformulated (using various approaches described at the beginning of this chapter) to shift the formulation to a BCS I or BCS II.

The IVIVC guidelines neglect the effect of the active ingredient permeability on bioavailability. Fortunately, newer FDA biowaiver guidelines based on the BCS classification system take into account the permeability of the active ingredient; however, these guidelines only recognize BCS class I materials as eligible for biowaivers.[21] Unfortunately, only a small fraction of all active ingredients belong in the BCS I class.[22]

17.5 Future trends

Nutraceuticals, nutritional supplements, and related products are considered, in principle, as food products that may improve the health of the consumer. The fact that they are considered as food products and not as pharmaceuticals exempts them from having to undergo bioavailability assessment. However, this situation may change in the near future, not only due to the growing interest of regulatory agencies, but also due to the increasing competitiveness of the market. At the moment, demonstrating the effectiveness of the formulation, even if through simple dissolution tests included in the USP-verified program, is a marketing tool that manufacturers can use to set themselves apart from the competition.

While the regulations concerning the effectiveness of nutraceutical products may be changing in the near future, it is unlikely that the list of approved direct food additives would be significantly expanded to include additives used in sophisticated controlled release formulations. Perhaps the best alternative to cope with these scenarios is to concentrate on innovative delivery and controlled release formulations produced with existing food-grade additives.

17.6 Sources of further information and advice

The GSFA online database (http://www.codexalimentarius.net/gsfaonline/index.html) has the most up-to-date information regarding the status of food additives and links to JECFA summary monographs. Additionally, the website of the international program for chemical safety (IPCS) has a more complete version of JECFA monographs and studies (http://www.inchem.org/pages/jecfa.html). The book *Food Chemical Safety*, edited by David H. Watson (Woodhead Publishing, 2002), is also a good source of information on the subject of risk assessment, and ingredient selection for food products.

The reader is also encouraged to visit the website of the International Conference on Harmonization (http://www.ich.org/cache/compo/276-254-1.html) to learn more about the stability and safety testing protocols which are, in principle, designed for pharmaceutical products but that some countries also apply to food products under good manufacturing practices (GMP) standards. For more information regarding the stability of food products, the book *The Stability and Shelf Life of Products*, edited by David Kilcast and Persis Subramanian (Woodhead Publishing, 2000), presents detailed information on stability evaluations. To explore the ICH guidelines in more detail, the book *International Stability Testing*, edited by David J. Mazzo (Interpharm Press Inc., 1999), has detailed accounts of various stability tests.

The FDA website (www.fda.gov) should also be consulted for more information on the codes 21 CFR 211 for in-process sampling and testing, and 21 CFR 320 for bioavailability and bioequivalence testing.

17.7 References

1. GSFA online at http://www.codexalimentarius.net/gsfaonline/additives/index.html#F (accessed April 2007).
2. World Health Organization (1987). *Principles for the Safety Assessment of Food Additives and Contaminants in Food* (Environmental Health Criteria, No. 70), p. 83.
3. Codex General Standard For Food Additives. Codex Stan 1992–1995, Rev. 7-06. http://www.codexalimentarius.net/download/standards/4/CXS_192e.pdf (accessed April 2007).
4. JECFA Guidelines for the preparation of toxicological working papers. www.who.int/ipcs/food/jecfa/en/tox_guidelines.pdf (accessed April 2007).
5. HANSEN S.C. (1979) Conditions for use of food additives based on a budget method

for an acceptable daily intake. *J. Food Protection*, 42, 429–434.
6. DOUGLASS J.S., BARRAJ L.M., TENNANT D.R., LONG W.R., CHAISSON C.F. (1997) Evaluation of the budget method for screening food additive intake. *Food Additives Contaminants*, 14 (8), 791–802.
7. Joint FAO/WHO Working Group Report on Drafting Guidelines for the Evaluation of Probiotics in Food (London, Ontario, Canada), April 30 and May 1, 2002. Available online at http://www.who.int/foodsafety/fs_management/en/probiotic_guidelines.pdf (accessed April 2007).
8. 21 CFR 184 – Direct Food Substances Affirmed As Generally Recognized As Safe. Available online at http://www.access.gpo.gov/nara/cfr/waisidx_06/21cfr184_06.html (accessed April 2007).
9. 21 CFR 172 – Food Additives Permitted For Direct Addition To Food For Human Consumption. Available online at: http://www.access.gpo.gov/nara/cfr/waisidx_06/21cfr172_06.html (accessed April 2007).
10. FDA – Office of Food Additive Safety. Summary of all GRAS notices. March 2007. Available online at: http://www.cfsan.fda.gov/~rdb/opa-gras.html (accessed April 2007).
11. International Conference on Harmonization website (http://www.ich.org/cache/compo/276-254-1.html) (accessed January 2007).
12. RAPAKA R.S., COATES P.M. (2006) Dietary supplements and related products: A brief summary. *Life Sciences* 78, 2026–2032.
13. U.S. Pharmacopoeia – USP Verified Dietary Supplements. Available online at http://www.usp.org/USPVerified/dietarySupplements/ (accessed April 2007).
14. WAGNER J.G., NELSON E. (1964) Kinetic analysis of blood levels and urinary excretion in the absorptive phase after single doses of drug. *J Pharm Sci*, 53, 1392–1403.
15. EMAMI J. (2006) *In vitro–in vivo* correlation: from theory to applications. *J. Pharm. Sci.* 9 (2), 169–189.
16. WAGNER J.G. (1983) Pharmacokinetic absorption plots from oral data alone or oral/intravenous data and an exact Loo-Riegelman equation. *J Pharm Sci*, 72 (7), 838–842.
17. Guidance for Industry, extended release oral dosage forms: development, evaluation and application of *in vitro*/*in vivo* correlations. FDA, CDER, 1997.
18. U.S. Pharmacopoeia. *In vitro* and *in vivo* evaluations of dosage forms, 27th edition. Mack Publishing Co., Easton, PA, 2004.
19. PORTER C.J.H., CHARMAN W.N. (2001) *In vitro* assessment of oral lipid based formulations. *Adv. Drug Deliv. Rev.*, 50, S127–S147.
20. AMIDON G.L., LENERNAS H., SHAH V.P., CRISON J.R. (1995) A theoretical basis for a biopharmaceutical drug classification: the correlation of *in vitro* drug product dissolution and *in vivo* bioavailability. *Pharm. Res.* 12(3), 413–419.
21. Waiver of *In Vivo* Bioavailability and Bioequivalence Studies for Immediate-Release Solid Oral Dosage Forms Based on a Biopharmaceutics Classification System. FDA, CDER, 2000.
22. ABRAHAMSSON B., LENNERNÄS H. (2003) 'Application of the biopharmaceutical classification system now and in the future', in van de Waterbeemd H., Lennernäs H., and Artursson P. *Drug Bioavailability*. Weinheim (Germany), Wiley-VCH, 495–531.

18

The future of controlled release and delivery technologies

N. Garti, The Hebrew University of Jerusalem, Israel

18.1 Introduction

In the last few decades the food industry has undergone major concept revolutions leading to products that might not resemble the foods we consumed some 50 years ago. The food that is produced today is much more delightful, sophisticated, attractive, and nutritional. Some might consider today's foods to be 'synthetic', 'artificial', less tasty, not delightful, and less nutritional and healthy.

The food composition, texture, and taste have been changed so much that there is a tendency not to recognize the products loved in childhood. A very noticeable example is that of the today's industrial bread. It has lost so much of the aroma, freshness, and flavour in comparison to the old-fashioned country bread that we must frequently toast it or cover it with a thick layer of spread. Yet, close examination reveals that while the industrial bread of the 1950s became stale very fast, today's product is edible even after being stored for a long time in the open air. Similarly, the olden days' products based on baked dough were non-edible after a few days because of fast growth of Rope and fungi, while modern products have a long shelf-life.

Was the progress in the technology beneficial to the consumer or has it caused some losses in flavour and quality? Are the modern products stable for prolonged periods of time because they are composed of extra chemicals that are beneficial to the food producer rather than to the consumer? Is the variety of food additives jeopardizing humans' long-term health?

Close examination reveals that the industry is following some general trends – dictated to some extent by consumers' needs and requirements, but in others by their own need for mass production.

The industrialization of food was aimed at being very beneficial to the consumer, mainly since that very long list of food additives was approved by the Food and Health Administrations in various countries. These additives allowed manufacturers to produce food with longer shelf-life, food that will keep its organoleptic properties longer, and will protect the consumer from health hazards derived from spoilage and environmental chemical reactions (e.g. oxidation, hydrolysis, hydrogenation, amination).

In previous centuries food was based on naturally occurring ingredients that were processed together, in most cases by cooking, grilling, baking, or frying. The major concern of the food scientists and technologists was how to keep ingredients from deteriorating, decomposing, undergoing enzymatic transformations, and breakage, and how to make them more appealing to the consumer. The results in many cases were products that were treated heavily with preservatives, acidulants, complexing agents, colorants, antioxidants, etc., that in some products altered the taste of the original food. Similarly, new processing methods caused formation of products with different taste and consistency.

The producers had and still have the dilemma of making a product that will deal with hunger and malnutrition or making a product that will appeal to the hectic modern world with no free time to spend on the preparation of food or producing luxurious, delightful food products for the wealthy population that is searching for just such a delight.

Once the world was hungry and the population suffered from malnutrition and lack of sufficient energy sources to maintain the basic nutritional needs of the body; there was an urgent need for industrial products geared to provide basic nutritional food based on proteins, carbohydrates, and lipids. These basic products were supplemented with essential vitamins and minerals. In addition, the fast growth of the world's population created long transportation lines over which there was a need to maintain food quality.

Scientists made significant efforts to better understand the characteristics and interactions of the major food ingredients, to learn how to stabilize them and prevent their spoilage. In food research, special effort was made to learn the macro properties of proteins, carbohydrates, and lipids, and their chemical and physical interactions during the processing conditions.

Today's world is less hungry and its basic needs have changed. The modern food industry is trying not just to supply basic food but also to *bridge the gap between the consumer's demands and needs and the industry's capabilities and interests*. This communication gap with the consumer is based on a claim that an average modern consumer would like to see food as a delight.

18.1.1 Food is delight

Food science and food technology have to meet the challenges in providing the basis for the production of delightful products. Professor Watzke from the Nestlé Research Centre, in one of his public presentations, summarized consumer needs in a very interesting term – abbreviated as CHIEF:

- Convenience
- Health
- Indulgence
- Excitement
- Freshness

The five slogans or goals very clearly define the future challenges of the modern food industry and of the food scientists and technologies.

In order to achieve high standards of delight or CHIEF requirements, the industry must deal with diverse challenges in the various stages of food production.

Raw materials
Scientists should be searching for less expensive or better cost-performance raw materials. This means that special effort will have to be made on agricultural aspects related to the sources of the raw materials (seeds, growing conditions, harvesting, etc.); the quality of the raw materials (fewer and lower levels of contaminants and impurities, less oxidized or hydrolysed raw materials), more availability of the raw materials, etc. Selection of the raw materials should reduce cost and improve performance.

In order to better meet these requirements there is, in many cases, a need to use processing additives (emulsifiers, acidulants, complexing agents, etc.). Naturally occurring substances do not always perform satisfactorily and the gap was filled by 'synthetic food additives'. The synthetic food additives, after prolonged health studies with experimental methods that were available at that time, were approved by the health authorities. Such additives were considered to be safe and in most cases they were allowed for use in a limited number of products and at restricted levels.

Today we know that many of these synthetic additives that were approved in the past might cause some health concerns and there is strong pressure to ban them. On the other hand, advanced studies have shown that natural ingredients can be good alternatives to synthetic additives, with similar performance. The natural ingredients are often extracted from plant roots, leaves, flowers, etc., using special vehicles necessary to retain their performance. Some such examples are antioxidants (carotenoids and polyphenols) or colorants (carotenoids and xanthins), sweeteners (Stevia), aromas and flavouring agents (essential oils), processing aids (enzymes).

We are now facing a new challenge, to develop modern physical vehicles (carriers) to improve the performance of natural and safer food additives.[1–4] The delivery systems became the bottle-neck for the utilization of methods that will improve the performance of the raw materials.

Ingredient preparation
In this stage of food production it is essential to find engineering techniques and equipment to compose the ingredients such as mixing, blending, homogenization, extruding cooking, frying, to obtain the required texture, organoleptics, sensory properties, etc. Chemical engineers are responsible for improving the

machinery's performance but in many cases these improvements are heavily dependent on the use of 'processing aids' that are synthetic chemicals (filter aids, emulsification and/or de-emulsification agents, thickening and/or thinning agents, foamers and/or defoamers, bleaching agents, pH controllers, aroma stability retainers, temperature controllers, antioxidants, etc.).

The processing aids require special vehicles that will facilitate diffusion, contact, separation, mixing, etc. There is a strong need for special physical *delivery vehicles* to enhance environmental protection of the active ingredients and to ensure controlled release and delivery. Encapsulation techniques will help to utilize the processing aids in the most efficient way.

Processing steps
Any industrial food product undergoes complex processing before it ends up as a final product. It is common practice that during the processing stage the producers rely heavily on various additives to control heat transfer, mass transfer, homogeneity, etc. The producers must guarantee optimal incorporation of the food additives and best functionality. The food additives must be soluble or dispersible in the mass. Synthetic additives were designed to cope with these requirements. However, once natural or less hazardous food additives have to be used because of health constraints, it is essential to entrap or encapsulate these ingredients into protecting capsules or controlled release vehicles. The vehicles will better control the processes of cooling, heating, applying pressure, evaporation, cooking, extruding, frying, coating, enrobing, etc.

Packing and packages
Packing materials and the packing process are always guided by methods to protect the product from strong environmental attack as well as be appealing to the consumer. The trend in recent years is to utilize special and advanced packaging materials. It is, therefore, important to include ingredients that will prevent contact of the food product with the environment (prevent oxygen and moisture penetration) as well as release of flavour from the packaging materials to the product during storage or use. Similarly, it is important to be able to release compounds that will absorb undesired flavours, oxidized species, degradation or byproducts, etc. The utilization of delivery systems that will enhance the efficacy of the 'packaging aid' compounds is very beneficial. Efforts are made to identify methods to embed, entrap, and encapsulate aromas and other compounds within the packaging polymers or plastics or to add them at the head-space of the containers.

Storage and shelf-life
The long transportation distances and the need to reach remote sites and to maintain large stocks dictate utilization of storage materials and storage spaces that will protect the final product. Some of these transportation vehicles (ships, trucks, etc.) are equipped with cooling devices, humidity regulation units, etc. Producers spend much effort to find efficient ways to protect the delicate processed foods from spoilage and environmental chemical reactions during

transportation and storage. Companies use 'shelf-life enhancers' and 'special storage aid carriers' to maintain sterility, to remove or minimize oxygen intake, or to avoid humidity and hydrolysis conditions. The active ingredients are formulated in encapsulation devices, carriers, coated systems. These require new technologies and advanced solid or liquid vehicles that lately are utilized more and more in foods during storage and transportation.

All these increasingly important food processing aids are leading to a new concept of production of industrial food – the so-called '*integral high performance products*'. The integral high performance products must fulfil all the 'Delight' requirements of optimal sensation and functionality, quality, safety, and value-for-money characteristics.

18.2 The food additives

Food additives are very efficient active compounds to help maintain the 'integral high performance products' requirements with strong emphasis on texture, organoleptics, physical and chemical stability, taste, biological reactivity, etc. Food additives such as preservatives, emulsifiers, antioxidants, acidulants, complexing agents, thickeners, gelling agents, colorants, aromas, etc., are constantly used by food producers. The well-designed synthetic food additives answered most of the producers' needs. Therefore, producers were less interested in new encapsulation and delivery methods for food additives. Synthetic additives for aqueous-based products as well as for oil-based products were always available with minor restrictions on their levels of use. Some of the most common examples are water-soluble and oil-soluble synthetic antioxidants, water-soluble and oil-soluble colorants, aroma products, etc. In cases where the food additives proved to perform poorly in any of the processing stages, their consumption levels were increased significantly. We could, therefore, find products with very high levels of synthetic preservatives, high levels of antioxidants, and high levels of foaming or antifoaming agents. With the lack of sufficient 'delivery performance improving vehicles', the health authorities were almost obliged to accept and authorize high levels of use. With the lack of evidence of hazardous effects, the food additives were permitted with certain limitations. It was claimed that the food additives, even though they were synthetic compounds, had no health risks.

Some of these additives such as antifoams (siliconic compounds in dairy, confectionery, and beverages), weighting agents (brominated vegetable oils in beverages), antioxidants (hydroquinones such as BHA, BHT, TBHQ), emulsifiers (polysorbates), and oxidizing agents (bromates in baked goods) were found after years of investigation to have some very negative effects on our health and might risk our quality of life.

The progress in food science and engineering research and development in the modern food industry of sophisticated machinery and improved processing techniques is today managing to make food with excellent 'delight' without some of these food additives and with full replacement of these additives by natural ingredients delivered to the food products by modern delivery systems.

Yet, in many cases there is still a strong need for the food additives and once the old synthetic compounds are slowly replaced by natural ingredients, there is a stronger need for delivery vehicles.[5–8]

Convenience is part of the delight scheme and is a key factor in our selection of the food that we buy. The result is the development of a fast food industry which in many cases tends to be convenient, tasty, and appealing, but non-nutritional or not healthy food. The development of ready-made food, fast food, and junk food saved the consumer time and appealed to his taste but did not serve his long-term health interest. These types of foods were always easy to consume, relatively inexpensive, and provided an easy and convenient solution to any family or individual. However, it helped develop the modern world's chronic disease known as obesity.

The health authorities are making very significant efforts to attract the attention of the population to their future health. We restrict our intake of fast and non-balanced nutrition not just to look better, but also to be healthier.

The first step was to educate the consumer and the producers to restrict the quantities of salts (blood pressure), fats and oils (triglycerides and cholesterol), and sugars (excess calories and diabetes) in the food products. This fast solution led, in the last decade, to production of food products that had lost the 'delight' concept. Many of the newly introduced products suffered from a lack of taste because the levels of salt were dramatically reduced. The products were tasteless because the quantities of fats and oil were also reduced. Low or reduced calorie products were introduced into the marketplace without finding a good solution for the missing fat in the product. Similarly, the sugar was replaced by a variety of artificial flavours and hydrocolloids. The products became gummy and sticky.

Once these major ingredients were decreased in our food, the products lost their appeal and the food with reduced calories or low in calories, the food with reduced fats and reduced cholesterol, and the foods with less sugars or less starch were considered to be less tasty and suffered from a low quality image. To make a low calorie bread, low calorie baked goods, or low calorie spreads or dressing was a difficult task. The health aspects were less stressed and in many cases even hidden.

However, it is now very clear to all of us that the food we consume must not necessarily be of reduced calories but it is essential that it will not damage our long-term life quality and will provide decent protection against health hazards. The producers' slogan has slowly changed to: food must provide the nutritional value but also must bring some healthy benefits.

18.2.1 Food is delightful, nutritional, and healthy

The health benefits can be derived from going 'back to nature' and consuming more fresh and unprocessed food, or food derived from raw materials unexposed to chemicals or pesticides (organic food). Such food will not have all the food additives or food processing aids. However, such an approach is good for those who can afford it and are ready to give up some of the 'delight' characteristics of food. The producers are therefore starting to apply a different approach, adding back into the food the ingredients that were lost during processing, or fortifying the food with additives that are generally less common in modern processed food. This approach very quickly extended into making food rich with ingredients that are recognized by the public as healthy and nutritional.

A new term was coined: 'food supplements' or 'food enriched with nutritional and healthy components'. The term food supplements originally meant vitamins and minerals, but it is now extended to a variety of other natural ingredients with health benefits. Another term – nutraceuticals – was coined by those who wanted to stress the benefits of these additives. Nutraceuticals are food additives extracted mainly from natural sources that can enhance human health and reduce the health risk factors. Most of these additives were studied by scientists around the world but were not systematically examined by the Food and Drug Administration in the USA or other major health authorities in the world. Therefore, these additives do not have official health benefits and the products enhanced with them can not have a clear supportive claim of their benefits. Many foods companies, therefore, declare their presence in various foods but give a general and vague statement on their activity.

Nutraceuticals are in some cases well-documented chemicals naturally existing in our body, but in other cases they are extracts from plants leaves, flowers, fruits, seeds, roots, or from tree barks or fungi, that are not known to the body, but it is believed (or shown in some limited number of studies) that they act similarly to other components that are present in the human body.

In recent years hundreds of new compounds or mixtures of chemicals or unidentified 'water or alcohol extracts' have emerged in the marketplace. Very fast and massive incorporation of these nutraceuticals into the human diet has been seen. Many are taken in the form of tablets, capsules, soft gels, etc., but others are incorporated directly into our food. The food supplements are sold in drug stores, grocery stores, supermarkets, health stores, and pharmacies. These food supplements are becoming so popular that the market is estimated to grow by more than 10% annually.

The food supplements or nutraceuticals are being slowly introduced into many foods that are mass produced. In many cases consumers prefer food enriched with these supplements in the belief that they will help to improve their health and quality of life.

18.3 Bioavailability and performance – the need for new vehicles

Bioactives can be incorporated into our food systems in solid form (powdered, granulated, micronized, lyophilized, etc.) and/or in liquid form (liquids, solutions, emulsions), or as dispersions in liquids (suspensions and dispersions).

Most of the past work concentrated on finding the best ways to incorporate the active ingredients as solid (crystalline or amorphous) powders that are directly dispersed into aqueous phases or directly added into the food during processing. It was well recognized that it is essential to reduce the particles of water-insoluble powders to micron or submicron range in order to achieve good homogeneous dispersions and to avoid effects such as after-taste and graininess. Since many of the additives are very sensitive to environmental reactions (oxidation, hydrolysis, degradation) it was essential to find encapsulation methods for protection against external reactivity or internal self-destruction of the active matter. Many such technologies were developed and are well studied. It is beyond the scope of this book to discuss them.

However, once coatings, encapsulation, and protection were achieved, it was essential to study how to release the encapsulated active matter from the covering shells. The scientists studied methods to control the release upon contacting the capsules or spheres with aqueous phases and also learned to trigger and control the rate of release. These technologies are quite mature and will not be discussed here.

Vitamins are either water- or oil-soluble. The water-soluble vitamins (C, B) are well recognized and known to consumers and often are supplemented in many of our foods. The oil-soluble vitamins (A, K, E, D, and F) are less known to consumers and are added only into specific foods. The less common use is due, in some cases, to possible health damage that excess vitamin intake can cause and in other cases because of lack of a convenient way to incorporate them into food products. The present technology does not provide good and efficient methods or 'vehicles' (carriers) to incorporate them into water-based liquid foods.

There is always a struggle between the competence and capabilities of the industry to incorporate nutraceuticals based on known technologies and the strong public demand to fortify foods with healthy supplements. Similarly, any extra cost that such supplements incur must be fully justified to the customer in terms of performance, otherwise consumers will not pay for the enriched product. But the most critical limitation is the lack of knowledge of how to physically add the nutraceuticals into the food product without damaging its properties and the difficulty in demonstrating that the fortified products indeed have health benefits.

Recent studies repeatedly show that apart from vitamins there are many ingredients that exist in plants, flowers, roots, etc., that have shown significant beneficial effects to our nutrition and health and in many cases do exist as part of our diet. Yet, the modern world's diet and nutrition is not well balanced and in many cases the food that we consume is lacking these ingredients. Are we consuming sufficient quantities of fresh vegetables and fruits? It seems that the food pyramid, as recommended by the health authorities, is followed by only a small portion of the population. Most of us are not getting sufficient levels of these healthy food supplements.

Several families of natural occurring carotenoids such as, beta-carotenes, luteins, astaxanthins, and lycopene are very efficient lipophilic antioxidants that are lacking in our daily diet. Simple- and polyphenols are very efficient water-soluble antioxidants but since we do not consume sufficient levels of fruits we are not getting enough of them in our diet. Similarly, several phytosterols are effective cholesterol management bioactives existing in soybean and related legumes, but we do not consume sufficient quantities of them.

The list of food supplements is very long, and almost daily another naturally occurring compound that has some health benefits is announced. The consumer is very often confused since he lacks the means to have a balanced diet mainly because he lacks the knowledge of what is good for him. An easy way is, therefore, to take some pills or capsules of these food supplements. It is a costly and non-regulated way to intake these products and in many cases excess intake can be hazardous. Is it easier to control the intake of these food additives by simply incorporating them in a regulated way into our food?

However, apart from the conceptual dilemma and the debatable health concern issues that always exist and will be discussed by the health authorities, there remains a key question of how to incorporate these ingredients into our food products in an efficient way without damaging the 'delight' issue and in a manner that they will indeed be beneficial to our health (effective bioavailability). Many of the nutraceuticals are not soluble in water and as a result are not absorbed from our gut into the bloodstream.

The challenges of future food producers is therefore to maintain the 'delight' concept and not to give up any of the characteristics that the consumer wants and is willing to pay for, but to load his food with some extra nutraceuticals. The utilization of nutraceuticals in raw materials, preparation of ingredients, processing, packing, and storage is essential and beneficial and we must find ways to incorporate these ingredients into the food without damaging its delight qualities. The nanovehicles and nanocapsules or nanospheres are excellent vehicles to carry those ingredients that are not soluble or not easily incorporated into the industrial food product.[1,3,5–8]

18.4 Bioavailability

Another very important challenge is to maximize the solubilization capacity, bioavailability, and bioefficacy of the enriched foods. Many of the food additives or food nutraceuticals must be incorporated into our food in large quantities. One such example is phytosterols that are known as cholesterol reducing agents. The FDA has approved their use at levels of 700 mg/day for the free phytosterols or 1300 mg/day for the corresponding fatty acid esters.

Phytosterols are not water-soluble and not oil-soluble, and therefore, their incorporation into aqua-based products is impossible. We need to develop vehicles that can solubilize the phytosterols with high solubilization capacity in order to introduce it into beverages or other liquid foods. Similarly, we do not have clear beverages that will contain tocopherols (vitamin E) since the d-α-tocopherol or its acetate derivative are not water soluble and will separate once incorporated into the beverage. Similar concerns exist with lycopene, lutein, and CoQ_{10}. The new vehicles must solubilize these nutraceuticals at high capacities. This is indeed a very significant challenge that has been tackled by the author of this chapter.

The question of bioavailability is the most critical question. We incorporate the nutraceuticals in paste form, in powdered form and in solution, but in many cases the bioavailability of the nutraceuticals is very limited and in some cases almost zero or negligible. We must find ways to improve the bioavailability of these ingredients. This is a big challenge and requires development of liquid nanovehicles that will solubilize the active matter, bring it to the gut membrane, and efficiently cross the gut–blood barrier. The challenge of bioavailability is the key to controlling health and related performance properties. Bioavailability has to be controlled in structured food matrices with reactive interaction/binding sites.

18.5 Nanoscience, nanovehicles, and controlled release

Scientists are encouraged to search for novel carriers or vehicles to optimize the incorporation of the nutraceuticals into our food and to optimize transport methods of the bioactives across the gut–blood barrier. Nanoscience has sufficient knowledge to be applied to food science in order to optimize the solubilization of the bioactives together with the transport phenomena. The food industry must adopt nanotechnology methods, vehicles to entrap the nutraceuticals without damaging the delight of the food product.

The last challenge that we can easily foresee is how to carry out organic and enzymatic reactions at interfaces to generate novel aroma compounds. Such efforts have been discussed in the past by several scientists and it seems increasingly essential to put more effort in this direction. The Maillard reaction was studied by several authors and it was demonstrated that reacting furfural with cysteine in aqueous solution will lead to the formation of mainly one product. However, if the same two components were treated in microemulsions of water-in-oil or oil-in-water or in cubic phases, different products would be formed with different flavours (Fig. 18.1).

18.6 Specific targets

In order to improve the solubilization of nutraceuticals using microemulsions, one must construct the microemulsions from food-grade ingredients. Unfortunately, the list of available food emulsifiers is very restricted and similarly the number of organic solvents to be used as oil phase is also very limited.

It is well known that formation of dilutable microemulsions based on vegetable oils, which are triacylglycerides (TAG), is a difficult task. Many preparations of W/O microemulsions made from TAG oils are available and well documented. These include systems into which ca. 40 wt% water can be solubilized in the oil/surfactant mixtures. Similarly there are studies showing solubilization of ca. 20 wt% O/W microemulsions but there are only a few oil-based concentrates that can be fully diluted with aqueous phase containing water and other cosolvents (polyols, hydrotropes); there are no food-grade microemulsions based on vegetable oils that are fully dilutable with simple water phase. This remains a challenge to scientists in this area. Some recent

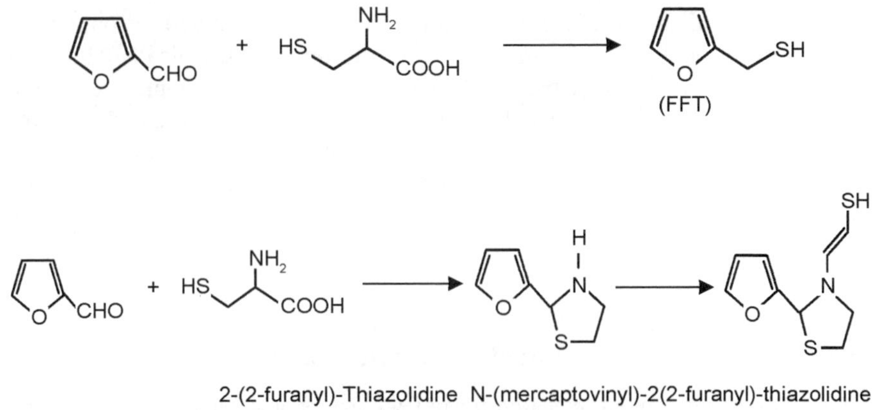

Fig. 18.1 Two different sets of products obtained from a Maillard reaction of furfural and cysteine: the upper reaction is conducted in aqueous phase while the lower process is carried out in a W/O microemulsion.

work by Sabatini and Acosta *et al.*[9–11] shows that if oil and/or water extenders are incorporated as part of the surfactant system, the triglyceride concentrates of vegetable oils can be water-diluted. Unfortunately, these extenders are not food grade and the problem remains a challenge to future work.

Another important open issue is how to solubilize, within or at the interface of the microemulsions, insoluble molecules such as lycopene, phytosterols, and CoQ_{10} in large loading capacities. Yet another challenge is the solubilization of high molecular weight guest molecules such as proteins, hydrocolloids, and other biopolymers.

18.7 The future of controlled release systems

The delivery systems in food systems can be divided into four categories of technologies:

1. solid and powdered delivery technologies, including microspheres, microcapsules, nanospheres, and micro- or nanoparticles
2. liquid micro- and nanovehicles that include emulsions, miniemulsions, nanoemulsions, and microemulsions
3. organo and aqueous gels
4. lyotropic liquid crystals.

The solid micro- and nano-dispersed particles and crystals are well documented and studied. The liquid dispersions are more challenging and more difficult to stabilize and keep stable.

The numerous techniques to encapsulate bioactives mainly for protective purposes was recently diverted into engineering efforts to make such coatings and encapsulation as efficient in cost and performance as possible. Such efforts are in the area of spray drying, spray cooling, freeze drying, and many other variations. There is very significant progress to layer-by-layer coat the solid powdered bioactives and to maximize the protection and at the same time to trigger and control the release patterns of the encapsulated matter. Similarly, efforts are being made to select new coating and encapsulation chemicals to make the release more controlled. Since the most effective coating and microsphere particle formations are biopolymers derived from proteins and/or hydrocolloids, it is very clear that efforts are in the area of incorporating and integrating the two types of biopolymers into each other so that their performance will be enhanced. Complexes, hybrids, adducts, and integrated compounds are only some of the options. The readers can find an excellent review of these methods in other chapters in this book.

Recently, other combinations of biopolymers with small amphiphilic molecules have been suggested and are showing some promise. Electrostatic and lipophilic interactions between two charged amphiphilic molecules can provide unique properties to the coated ingredients. Yet, some of these structures form soft or hard gels of micro- or nanosize that can alternatively be used for encapsulation of insoluble bioactives.

There is much more engineering and research work to do to integrate the knowledge that we have on the behaviour of amphiphilic molecules and their capabilities to adsorb on solid surfaces and to encapsulate the bioactives. Each of the methods has its advantages and disadvantages. We have reached the stage where technologists have a great selection of options and they have to tailor the best technology for the individual product. We have reached the stage where the details are making the performance difference.

It is expected that the future will bring many new products that are based on loading various mixtures of additives, flavours, nutrients, and nutraceuticals that are very sensitive to harsh environmental conditions or are interacting with each other, or are easily decomposed by the product processing stages. These ingredients will be segregated and separated from each other or completely compartmentalized into nanocapsules so that they can be released in their own time and will provide some synergistic effects that were not achievable by past procedures and protection techniques.

It should be clear that health authorities are very reluctant to allow incorporation of new derivative biopolymers and therefore the activity and performance will have to be achieved by '*in-situ* formation of adduct products with enhanced activity'.

However, the recent trend in liquid food systems is to make transparent, clear beverages (tea, beer, wines, soft drinks, energy drinks, and dairy low calorie beverages). It is a very big challenge to fortify these systems with nutraceuticals and food supplements that are not soluble in water. This is therefore a big future challenge.

Once the solubilization and the bioavailability technology become available, it will be time to move into the next stage of controlled, triggered, and slow release of nutraceuticals and nutritional additives from the produced food product into the final food that we drink or eat. In a later stage the scientist must concentrate on building within the vehicles certain characteristics or to embed within the vehicles functional molecules that will control the release of the bioactive from the food product into our gut and thereafter across the gut–blood barrier.

It seems that we have long way to go. The pharmaceutical industry has made much progress in controlling the release of bioactives and we can learn from them. Yet, the food systems are by far more complex, have many more restrictions, and are difficult to control because of ingredient restrictions, compatibility issues, and mainly cost problems.

We hope that many of the controlled release aspects that are as yet hardly touched will be the challenge of future studies. Once these technologies become available, the food systems will become healthier and can in some cases also replace some drugs.

18.8 Encapsulation with organogels and lyotropic liquid crystals

Aqueous gels based on hydrocolloids and proteins, and the related glycolipids and glycoproteins and lipoproteins, are well known in the food industry as thickening and gelling agents, and provide many other texture and bulk properties of aqueous-based products. However, these systems are not very popular as encapsulation systems.

This category of lyotropic liquid crystalline delivery systems is not well known to most food scientists, mainly because of the characteristic physical properties of these structures.[12,13] Lamellar, hexagonal, cubic, sponge, and the corresponding dispersed particulate products (hexosomes, cubosomes) were first made some 20 years ago but were not utilized in the food industry except for some lamellar phases that are used as emulsion stabilizing interfaces (Fig. 18.2).

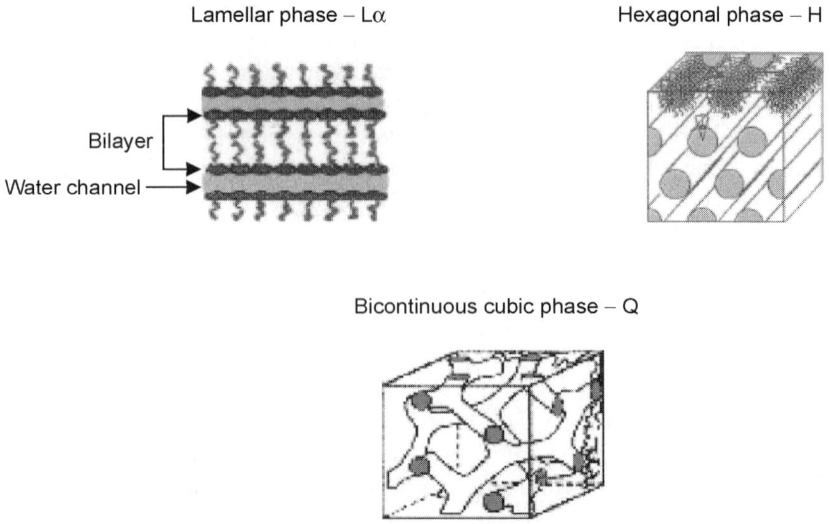

Fig. 18.2 Three different lyotropic liquid crystalline mesophases.

However, recently some major modifications were made on these systems by using a combination of emulsifiers and by introducing a third component into the mixture, resulting in less-ordered structures with fluid properties. This technology opens new options of utilizing these structures that are based on generally recognized as safe (GRAS) compounds with no toxicity or other ingredient–additive interactions. We will see many more of these structures in the literature and eventually also in food products.

18.9 Reactivity at interfaces

This book deals mainly with solubilization issues related to nutraceuticals. This is still a significantly large challenge for most scientists but one should also take advantage of the very large interfacial area that the microemulsions provide and it is expected that the food-grade microemulsions will serve as microreactors for various regioselective reactions.[14–17]

18.10 Conclusions

The present food conferences are devoting much more attention and many more sessions to nanotechnology and to delivery systems. The term nanotechnology is not always very popular when dealing with health authorities and big companies that are very sensitive to the 'delight' and health flags.

In some countries health authorities are advising minimising the use of insoluble solid nanoparticles that have been shown to cause some significant damage to the healthy growth of various cells and to decrease the viability of others. However, companies and those dealing with future foods prefer to concentrate on delivery methods and technologies and to refrain from using the term nanosized particles. The more commonly used slogan is therefore: 'Structured food systems and advanced delivery technologies'.

Yet, all the technologies will involve utilization of new encapsulation and entrapment of bioactives and other food additives by advanced encapsulation micro- and nanovehicles.

It seems advisable to put more effort into liquid rather than solid vehicles once the entrapped bioactives are molecularly and interfacially incorporated into the food and have a cleaner label. Their performance and bioavailability properties are more pronounced and they offer less health threat and more nutritional and health benefits together with controlled release and structural advantages.

18.11 References

1. GARTI N, ASERIN A, SPERNATH A and AMAR I, 'Nano-sized self-assembled structured fluids', US Patent Application No. 10/173,508, 2006.

2. GARTI N, ASERIN A and EFRAT R, 'Stable ill-defined cubic nanosized particles in ternary aqueous phase', Patent Application No. 60/533230, 2003.
3. SPERNATH A, ASERIN A, ZIMERMAN L, DANINO D and GARTI N, 'Phosphatidylcholine embedded microemulsions: physical properties and improved Caco-2 cell permeability', *J Controlled Release*, 2007 **119**(3) 279–290.
4. EFRAT R, ASERIN A and GARTI N, 'Synergistic solubilization of mixed nutraceuticals', in Dickinson E and Leser M E (eds), *Modified Discontinuous Micellar Cubic Structures, Food Colloids: Self-Assembly and Material Science*, Cambridge, Royal Society of Chemistry, 2007, 87–102.
5. GARTI N, ZAKHARIA I, SPERNATH A, YAGHMUR A, ASERIN A, HOFFMAN R E and JACOBS L, 'Solubilization of water-insoluble nutraceuticals in nonionic microemulsions for water-based use', *Progress in Colloid Polymer Science,* 2004 **126** 184–189.
6. AMAR I, ASERIN A and GARTI N, 'Solubilization patterns of lutein and lutein esters in food-grade nonionic microemulsions', *J Agr Food Chem,* 2003 **51** 4775–4781.
7. SPERNATH A, YAGHMUR A, ASERIN A, HOFFMAN R and GARTI N, 'Food grade microemulsions based on nonionic emulsifiers: media to enhance lycopene solubilization', *J Agric Food Chem*, 2002 **50** 6917–6922.
8. GARTI N, YAGHMUR A, ASERIN A, SPERNATH A, ELFAKESS R and EZRAHI S, 'Solubilization of active molecules in microemulsions for improved solubilization, and environmental protection', *Colloids and Surfaces, A: Physicochemical and Engineering Aspects*, 2003 **230** 183–190.
9. SALAGER J L, ANTON R E, SABATINI D A, HARWELL J H, ACOSTA E J and TOLOSA L I, 'Enhancing solubilization in microemulsions – state of the art and current trends', *J Surfactants Detergents*, 2005 **8** 3–21.
10. WITTHAYAPANYANON A, ACOSTA E J, HARWELL J H and SABATINI D A, 'Formulation of ultralow interfacial tension systems using extended surfactants', *J Surfactants Detergents*, 2006 **9** 331–339.
11. ACOSTA E J, SZEKERES E, SABATINI D A and HARWELL J H, 'Net-average curvature model for solubilization and supersolubilization in surfactant microemulsions', *Langmuir*, 2003 **19** 186–195.
12. EFRAT R, ASERIN A, KESSELMAN E, DANINO D, WACHTEL E J and GARTI N, 'Novel discrete micellar cubic phase form mixture of GMO/ethanol/water', *Australian J Chem*, 2005 **58** 762–766.
13. AMAR-YULI I, WACHTEL E, BEN-SHOSHAN E, DANINO D, ASERIN A and GARTI M, 'Hexosome and hexagonal phases mediated by hydration and polymeric stabilizer', *Langmuir*, 2006 **23** 3637–3645.
14. YAGHMUR A, ASERIN A and GARTI N, 'Furfural-cysteine model reaction in food-grade nonionic O/W microemulsions for selective flavor formation', *J Agric Food Chem*, 2002 **50** 2878–2883.
15. YAGHMUR A, FANUN M, ASERIN A and GARTI N, 'Food grade microemulsions based on nonionic emulsifiers as microreactors for selective flavor formation by Maillard reaction', in Robinson B H (ed.), *Self-Assembly*, Amsterdam, IOS Press, 2003, 144–151.
16. YAGHMUR A, ASERIN A, ABBAS A and GARTI N, 'Reactivity of furfural-cysteine model reaction in food-grade five-component nonionic O/W microemulsions', *Colloids and Surfaces, A: Physicochemical and Engineering Aspects*, 2005 **253** 223–234.
17. FANUN M, LESER M, ASERIN A and GARTI N, 'Sucrose ester microemulsions as microreactors for model Maillard reaction', *Colloids and Surfaces, A: Physicochemical and Engineering Aspects*, 2001 **194**(1–3) 175–187.

Index

absolute absorption 15, 17–18
absorption 4, 22
  factors affecting *in vivo* absorption 75–7
  in vitro tests 91–5
  mathematical models for *in vitro* absorption 80–6
  measuring 5–13
  of nutraceuticals 173–4
  structured lipids 140–1
  studies of 13–19
absorption, distribution, metabolism and excretion (ADME) 4
  ADME study 17
absorption number 81, 82–3, 447
acacia gum *see* gum arabic
accelerated tests 440
acetate phthalate 354
acidolysis 137
active transport 76
acute studies 14–15
adhesion
  in vitro tests 91–5
  mucoadhesion *see* mucoadhesion
  probiotics to intestinal cells 363
adsorption theory of mucoadhesion 37
AG-protein complex (AGP) 243–4
age groups 394
albumin microparticles 262
alcohol 157
alcoholysis 137
algal oil-in-water emulsion delivery system 200–2
alginate 41, 242
  beads 266
    microencapsulation of folic acid 335–7
    microentrapment of probiotics 346–7
alginate-whey protein microspheres 261, 262, 265
alpha gliadins 245–6
alpha-linolenic acid (ALA) 184, 185–6, 188, 373, 389–90
  see also omega-3 fatty acids
alpha-tocopherol 300, 301, 376, 377
ALT antioxidant system (ascorbyl palmitate, lecithin, tocopherol) 376–7
American Heart Association (AHA) 188
amylases 288, 289
amylopectin 279–80
amylose 279–80, 288–9
  microencapsulation 282, 283–5
  nanoencapsulation 285–8, 289
    amylose-CLA complexes 286–8, 289
anchovy 372, 374
anionic polysaccharide-cationic polysaccharide complexes 242–3
annealing 407, 410
anti-ICAM nanoparticle system 306
antioxidants 297–330, 457–8
  distribution of antioxidant molecules in a cell 298–9
  encapsulation by liposomes 216–17
  in vivo tests of bioavailability 74–5, 77

nanoencapsulation 304–19
advantages 304–9
characterisation methods 313
controlled release 313–17
future trends 318–19
top-down techniques 309–12
natural antioxidants 300–4
protecting human health 298–304
stabilisation of fish oils 376–7
approved ingredients 430–6
aqueous gels 460, 462
aqueous two-phase systems (ATPS) 414–16, 421
arabinogalactan (AG) 243
AG-protein complex 243–4
arachidonic acid 136, 186
archaeosomes 216–17
area under the curve (AUC) 19
problem of partial AUC 11
aromas
fish oils 375
solubilisation and microemulsions 163–4
artificial membranes 86–7
ascorbic acid (vitamin C) 20, 301, 303
assay 441–2
astaxanthins 457–8
average daily intake (ADI) 430–6, 437–8, 439

bakery products 391, 392
see also biscuits; cookies; bread
beta-carotene 23, 301, 302, 457–8
beta-cyclodextrin 61–2, 384
ADI and estimated maximum content 435–6
beta-lactoglobulin (BLG or β-lg) 237, 254–6
chitosan nanoparticles coated in 269
emulsion gels 257–8
beverages 307–8
bicontinuous phases
cubic liquid crystalline phase 107, 108, 110, 111, 117–19, 217, 463
reversed 110, 113, 117–19
microemulsions 153, 154, 155
bifidobacteria 266
starch and encapsulation of 283–5
bilayer discs 125
bile salts 140, 173
binding properties 254–6
bioavailability 3–25, 172–4, 175
concept 4–5
controlled absorption and product development 22–3
future trends 24
health response 21–2
hydrophobic compounds 19–20
measuring absorption, metabolism and tissue targeting 5–13
nanoencapsulation of antioxidants and vitamins 304–6
need for new vehicles 456–9
regulation and bioavailability of the active ingredient 443–7
study design and interpretation 13–19
biodegradation properties 254, 255
bioequivalence 444–5
Biopharmaceutical Classification System (BCS) 83, 447
biotechnology 404–5
biscuits 362
blend uniformity analysis (BUA) 67–8
bottom-up nanoencapsulation techniques 309
bovine serum albumin (BSA) 267, 414–16
bread 362, 391, 392, 450
budget method for intake 438
bulk oils 189–90
burst effect 313, 314

C-terminal ends 32–3, 34
Caco-2 cell cultures 87–9
calcium 21
dissolution tests 78
in vivo tests of bioavailability 74
calcium carbonate 420
cancer 23, 172
capelin 372, 374
caplets 354–5
Caprenin 141
Captex 141
carbohydrates 22
encapsulation of fish oils 380, 381, 382
small molecules 381–2
GSFA-listed 430, 432–3
side chains of mucin molecules 31, 32
see also starch
Carbopol/carbomer 39, 41
cardiac arrhythmias 187
carnosine 301, 303
carotenoids 7–8, 301, 302, 457–8
see also under individual types
carrageenan 242
catalase 420
cationic structured lipids 146
cell cultures 87–9, 96
cellular uptake, of antioxidants and vitamins 306
centrifugal atomisers 67
centrifugal extruders 67

centrifugation 418
ceramide 136
chain flexibility 40
charge regulation 238
Cheddar cheese 338–40
cheese 215–16
  encapsulation of folic acid 334, 338–40
  encapsulation of probiotics 359
chelators 193
chemical bonds 38–9
chemical interestification 138
CHIEF requirements 451–2
chitosan 241, 242–3, 389
  beta-lactoglobulin-coated chitosan nanoparticles 269
  coating of alginate beads 347
  lecithin-chitosan membranes 197–8, 385
  mucoadhesion 41, 42–3
cholesterol 136, 168, 169, 208, 216
chronic studies 14
chylomicrons 19–20, 140, 174, 300
clearance rate 15, 18
clouding point 115
Clover Corporation 391, 392
coacervation 58–9, 236
  complex *see* complex coacervation
  encapsulation of fish oils 379, 386–7
  protein micro- and nanoparticles 259, 263–4
  simple 58, 236, 263
coating 65–7
  alginate beads with entrapped probiotics 347
  in spray drying of proteins 411–16, 421
cod 372, 374
cod liver oil 372, 374
Codex Alimentarius 430–8, 439
  approved ingredients and ADI 430–6
  toxicology and intake studies for new additives 436–8
Coenzyme Q_{10} (CoQ_{10} or ubiquinone) 171–2, 300–2, 304, 459
cold-set gels 256–7
colloidal mills 60
colloidosomes 240–1
colloids 111–12
colon 337, 338
colon cancer 23
column chromatography 215
comparative absorption studies 18–19
compartmental modelling 8–13, 445
complex coacervation 58, 59, 64, 236–43, 264
  factors affecting formation and properties of the complexes 238–40
  studies of mechanisms 236–8
complexes 234–43
  and coacervates 236–43
    encapsulation and delivery of bioactives 240–3
  encapsulation using amylose 285–9
  future trends 246
  of a protein with a polysaccharide-protein-conjugate 245–6
compression extrusion 353
concentration
  mucin molecules and mucus layer 35–6
  mucoadhesion and 40
cone-jet-mode electrosprays 263
conjugated linoleic acid (CLA) 146
  nanoencapsulation by starch 286–8, 289
conjugates 234–6, 243–6
  complexes of a protein with a polysaccharide-protein-conjugate 245–6
  future trends 246
  man-made 244–5
  nanoencapsulation of antioxidants and vitamins 317
  naturally occurring 243–4
constant release 313, 314
consumer needs 451–2
continuous flow dialysis (CFD) 79
controlled clinical studies 16–17
convenience 452, 455
cookies 392
  fillings 363
copolymer ratio 315–16
corn syrup solids 198–9, 381–2, 384
coronary heart disease (CHD) 185, 186–7, 371–2
crosslinking 39
crossover design 14, 15
crystallinity of starch 280
crystallisation of carbohydrates 382
cubic phases 107–8, 110, 111, 210, 211, 217–24
  bicontinuous 107, 108, 110, 111, 117–19, 217, 463
    reversed 110, 113, 117–19
  characterisation 221–2
  introducing guest molecules 222–4
  structure and properties 217–19
cubosomes 108, 117, 211, 219–24, 462
  characterisation 221–2
  introducing guest molecules 222–4
cysteine 175–7, 459, 460

cysteine-rich domains in mucin molecules 31–2, 33
cystic fibrosis 26

dairy products
encapsulation of fish oil 392
encapsulation of folates 338–40
encapsulation of probiotics 359–61
liposomes 215–16
see also under individual products
Debye-Huckel theory 236
definitive absorption studies 17–18
delight, food as 451–4, 455–6
delivery systems evaluation 53–106
drying/coating 65–7
future trends 95–6
in vitro tests 77–95, 444
in vivo tests 70–7, 444
post processing 68–70
solid processing 67–8
wet (liquid) processing 54–65
dendrimers 62
description 441
dialysis method for dissolution tests 78–9
dietary fibre 23
diffusion 313
transport through mucin gels and mucus layer 44–6
diffusion theory of mucoadhesion 38
digestion 22, 75–7
see also gastrointestinal system
diglycerol monooleate 223
dilution line 154
dioleoylphosphatidylcholine (DOPC) 121–3
direct addition food additives 439
direct compression extrusion 353
direct imaging cryo-TEM 222
disintegration test 442
dispersing agents 114
dissolution 75–6
mathematical models for *in vitro* dissolution 80–6
tests 65, 78–80, 442
dissolution number 81, 82, 83, 447
distribution 16–19
disulphide bridges 34–5
DNA 241–2
docosahexaenoic acid (DHA) 136, 146, 184, 186–9, 370, 389–90
and health 186–8, 371–2
recommended intake 371–2
sources 184, 372–3
see also omega-3 fatty acids
dose number 81, 82, 447
dose response 12–13
double emulsification 57, 311, 312, 388
downstream metabolites 9, 10
Driphorm encapsulated fish oil 393
drug content 314
drug delivery 4–5
release profiles 313–17
structured lipids in 143–6
drying 65–7
proteins 405–16
coating in spray drying 411–16, 421
surface formation in drying 408–11
techniques 405–8
spray drying *see* spray drying
duodenum 76
dynamic light scattering 215
dynamic surface tension analysis 411–13

early nutrition programming 24
egg-protein conjugates 244–5
eicosapentaenoic acid (EPA) 136, 184, 186–9, 370, 389–90
and health 186–8, 371–2
recommended intake 371–2
sources 184, 372–3
see also omega-3 fatty acids
eicosapentaenoic acid ethyl ester (EPA-EE) 386
electrolytes, GSFA-listed 430, 434
electronic theory of mucoadhesion 37
electrostatic interactions 45–6
electrostatic layering 235, 379, 385
in oil-in-water emulsions 196–8
see also polyelectrolyte microcapsules
elimination rate 84, 85
ellipsometry 120–3
empty microcapsules 418–19, 422
emulsification *see* emulsions/emulsification
emulsifiers (surfactants) 149, 152–3, 157
GSFA-listed 430, 431
emulsion diffusion method 309, 310
emulsion droplet interfacial engineering 184–206
dietary omega-3 fatty acids 185–6
and health 186–8
potential in functional foods 188–90
electrostatic layering 196–8
engineering to stabilise omega-3 fatty acids 200–2
interfacial engineering 194–5
lipid oxidation 190–4
in emulsions 193–4
mechanisms of 190–2

microencapsulation of oil-in-water emulsion systems 198–200
emulsion evaporation 309, 310, 311–12
emulsion gels 242, 257–8
emulsion stability studies 59
emulsions/emulsification 54–8
emulsification-stabilisation technique for protein microparticles 259, 260–1, 262, 264
preparation prior to microencapsulation of fish oils 378–80
probiotics 350, 353
microentrapment in gel particles 346–7
encapsulation 334
antioxidants *see* antioxidants
biopolymer complexation 240–3
fish oils *see* fish oils
folic acid *see* folic acid/folate
microencapsulation *see* microencapsulation
nanoencapsulation *see* nanoencapsulation
probiotics *see* probiotics
proteins *see* proteins
starch *see* starch
vitamins *see* vitamins
encapsulation load 56
enteral nutrition 142
enteric coatings 281–2, 389
entrapment efficiency 314–15
environmental conditions 316–17, 318–19
enzymatic digestion experiments 288, 289
enzymatic interesterification 138
epidemiological studies 23
essential fatty acids 141–2
estimated maximum content 435–6
ethanol 219
ethics committees 13
ethosomes 144
ethyl cellulose 389
ethylenediamine tetraacetic acid (EDTA) 193, 268–9
European Union Scientific Committee for Food 371
ex vivo intestinal models 89–91
excitement 452
excretion 16–19
extrinsic tagging method 72
extrusion 68
folic acid encapsulation 335–7
probiotics 350, 352–3
microentrapment in gel particles 346, 347
protein micro- and nanoparticles 259, 260, 264

F127 113–14, 115, 219, 221
falling film technique 335–7
fast food 455
Fasted/Fed Simulated State Intestinal Fluids (FaSSIF/FeSSIF) 446
fatty acids
composition of fish oils 373, 374
essential 141–2
omega-3 *see* omega-3 fatty acids
unsaturated and microparticles 267–8
fermented dairy products 338
fibre, dietary 23
filamentous gels 254, 255, 257
film forming (wall) materials 55–6, 380–2
firmly adherent mucus layer 28–9
fish 188–9
recommended intake 373
fish oils 184–5, 370–403
benefits 371–2
challenges in formulation 373–6
encapsulation technologies 377–89
future trends 394–5
recommended levels of intake 371–2
sources of omega-3 LC PUFA 372–3
stabilisation with antioxidants 376–7
suppliers of microencapsulated fish oil powder 393–4
using encapsulated fish oils in food products 389–92
flavonoids (polyphenols) 23, 301, 302–3, 458
flavours
fish oils 375
solubilisation and microemulsions 163–4
Flexosome 117
Flory-Huggins model 236, 237
flow-through diffusion cell 79–80, 93–5
fluidised bed coating 67, 388, 411
folic acid/folate 20, 331–43
applications of encapsulation in functional foods 338–40
controlled release in gastrointestinal system 337–8, 339
effect of thermal processing 332, 333
encapsulation technologies 334–7
fortification 332, 340
future trends 340–1
food additives 451, 454–6
and stages of food production 452, 453, 454
Food and Agriculture Organization (FAO) 429, 430

see also Codex Alimentarius
Food and Drug Administration (FDA) 68, 340, 429
health claims for PS 139
recommendations on fish intake 373
recommendations on fish oils 371–2
regulation on bioavailability and bioequivalence 444–5
regulation on *in vitro* and *in vivo* correlation 445–7
selection of ingredients 438–9
food industry 450–4
challenges in stages of food production 452–4
CHIEF requirements 451–2
food processing 451
microencapsulated fish oils and 390
steps 453
thermal processing 332, 333
and viability of probiotics 358
food-protein-derived materials *see* proteins
Food Science Australia 391
food supplements 150, 456, 458
fish oil 371–2, 376
fraction absorbed 81, 82–3, 84, 85, 445
fractional absorption 15
fracture theory of mucoadhesion 38
fragrance oil 163–4
free energy of microemulsion formation 155
'free-living' individuals studies 16–17
free radical scavengers (FRS) 193–4
'free' water 44
freeze drying
encapsulation of fish oils 379, 387–8
proteins 405, 407
surface formation 410–11
freeze-fracture cryo-TEM (FF-TEM) 222
freshness 452
frozen products 440
cranberry juice 362
desserts 360
fruit juices 362, 390
fucose 31, 32
functional foods 162, 363–4
encapsulated fish oils 389–92
encapsulated folic acid 338–40
encapsulation of probiotics 358–63
potential for omega-3 fatty acids in 188–90
functional groups 34, 40
furfural 175–7, 178, 459, 460
furfurylthiol 175, 176
galactose 31, 32
gastrointestinal mucus layer 26–52
mucin molecules *see* mucin molecules
mucoadhesion *see* mucoadhesion
physico-chemical properties 27–30
transport through mucin gels and mucus layer 44–6
gastrointestinal system 22, 26
controlled release of folate 337–8, 339
probiotics in gel beads
controlled release 357–8
survival 356–7
structured lipids 140–1
gel beads 266
microentrapment of probiotics 346–51
production of microencapsulated probiotic cultures in 355–8
gel to liquid-crystalline transition temperature 209
gelatin 385
complexes 234, 236–7, 241–2
hydrogels 256
microparticles 260–1
gels 460, 462
emulsion gels 242, 257–8
gel phase in lipid/water systems 110
hydrogels *see* hydrogels
mucin 35–6, 44
General Standard for Food Additives (GSFA) 430
approved ingredients and ADI 430–6
generally recognized as safe (GRAS) additives 438–9
genetic engineering 404–5
genetics 24
George Weston Foods 392
glass transition temperature 70, 407
glucose 22
glycerol 161–2
glycerol monooleate (GMO)
GMO/F127 cubic phase dispersions 113–14, 115
GMO-water-ethanol system 218
LCNPs 117–19
interaction with model membranes 120–4
glyceryl monooleyl ether (GME) 117–19
glycoprotein 244
good manufacturing practices (GMP) 68–70, 430, 434–5, 439, 443
growth 21
gum arabic (gum acacia) 58, 243–4, 381
complexes 234, 236, 237, 239, 245–6
gums, GSFA-listed 430, 432–3

HACS 282–3, 284
'hairy layer' 243
hardness/friability test 442
health 24, 452, 455–6
  antioxidants, vitamins and protecting 298–304
  infant health 188
  omega-3 fatty acids and 186–8, 371–2
  potential benefits of probiotics 344, 345
  response and bioavailability 21–2
health claims 139, 371–2
heat treatment 115, 332, 333
Heinz Farley Infant Formula 391
heparin 145
herring 372, 374
hexagonal phases 110–11, 210, 211, 217–24, 462, 463
  characterisation 221–2
  introducing guest molecules 222–4
  reversed 110, 113, 117, 119
  structure and properties 217–19
hexosomes 117, 211, 219–24, 462
  characterisation 221–2
  introducing guest molecules 222–4
HiDHA encapsulated fish oil 393
high-shear energy input techniques 115
homoeostatic set point 12, 13, 21
homocysteine 332, 339–40
hormetic dose response curve 12–13
host-controlled release 89
HT-29 cell line 88–9
humidity 70
hydration, degree of 40
hydrochloric acid 30
hydrocortisone 46
hydrogels 242, 252–8, 271
  environment-sensitive hydrogels 253–6
    binding properties 254–6
    biodegradation properties 254, 255
    swelling properties 253–4
  folic acid encapsulation 334–7
  formation 256–8
hydrogen abstraction 190, 191
hydrophilic ingredients
  antioxidants 193–4
  in vivo tests of bioavailability 71–4, 77
hydrophobic compounds 19–20
  coatings 70
  in vivo tests of bioavailability 74–5, 77
hydrophobic interactions 45–6, 209
hydroxypropyl methylcellulose (HPMC) 41, 385, 386–7, 412, 413–14
hyperhomocysteinemia 339–40

ice cream 360
identification 441
ileostomy volunteers 15, 16, 22
immobilisation 334
impermeable packing 440, 441
impurities 442
in-situ coating 411–14, 421
in vitro-in vivo correlation (IVIVC) 444, 445–7
in vitro tests 77–95, 444
  adhesion/absorption tests 91–5
  dissolution tests 65, 78–80, 442
  mathematical models for dissolution and absorption 80–6
  permeability tests 86–91, 96
in vivo tests 70–7, 444
  bioavailability of antioxidants 74–5, 77
  bioavailability of minerals 71–4, 77
  factors affecting *in vivo* absorption 75–7
inclusion complexation 61–2
  starch as encapsulation material 285–9
  see also complexes
indulgence 452
infant formula 141–2, 391
infant health 188
infinite periodic minimal surfaces (IPMS) 217
inflammation 187–8
infusions 6–7
ingredient preparation 452–3
ingredient selection, regulation of 430–9
injections 6–7
insulin 405, 411–13
intake diaries 17
intake studies 436–8
integral high performance products 454
integrated formulation 96
interesterification 137, 138, 139
interfaces, reactivity at 463
interfacial engineering *see* emulsion droplet interfacial engineering
International Conference on Harmonization (ICH) 68–70, 429, 439–43
intestinal models, everted 89–91
intrinsic labelling method 72
inverted lipid self-assembled structures 209, 210
Invirase 63
ionotropic gelation 379, 389
iron 257
  in vivo tests of bioavailability 73–4
  supplementation 73–4
isoflavones 63–4

jejunum 337–8, 339
Joint (FAO/WHO) Expert Commission on Food Additives (JECFA) 430, 434
junk food 455

Karicare 391
krill oil 389

lactating women 188
Lactobacillus acidophilus 357
Lactobacillus casei 357–8, 361
Lactobacillus rhamnosus 362
lactose 387
lamellar phase 110, 211, 462, 463
  colloids 111–12
large unilamellar vesicles (LUVs) 60, 212
lecithin 56, 197–8
  lecithin-chitosan membranes 197–8, 385
lectins 43–4, 77, 270–1
levels of correlation 446
linear polymers 39
linker-based microemulsions 65, 159
linoleic acid 185–6
lipases 123–4, 138
lipid-based spray-coating 351–2, 354–5
lipid disperse systems (LDSs) 143–4
lipid hydroperoxides 192, 193
Lipid Nutrition 393–4
lipid oxidation 189
  in emulsions 193–4
  fish oils 373–5
  interfacial engineering in emulsion droplets 194–5
  mechanisms 190–2
    initiation 190–1
    propagation 191–2
    termination 192
lipids 135–6
  and diffusion through mucus layer 46
  lyotropic liquid crystals *see* lyotropic liquid crystals
  malabsorption 142
  phospholipids *see* phospholipids
  solid lipid nanoparticles 62–3
  structure and properties of 208–9
  structured *see* structured lipids
lipolysis 125–6
lipophilic antioxidants 193–4
lipophilic linkers 65, 159
lipoproteins 7
liposomes 59–61, 107, 111–12, 144, 211–17, 224
  applications in food systems 215–17
  characterisation and evaluation of performance 214–15
  classification 211–13
  controlling self-assembly 214
  formation 211, 212
  preparation methods 213–14
liquid crystalline nanoparticles (LCNPs) 108–9, 112–24, 217–24
  formation mechanisms 113–14
  functionality 120–4
  nanostructure 117–20
  preparation methods 115–17
liquid micro- and nanovehicles 460–2, 464
liquid processing *see* wet processing
Live Well yoghurts 392
long-term stability tests 440
loosely adherent mucus layer 28–9
low density lipoprotein (LDL) 187
lutein 169–71, 301, 457–8, 459
lutein ester 169–71
lycopene 23, 166–8, 223, 457–8, 459
lyotropic liquid crystals 107–31, 207–33, 460, 462–3
  cubic and hexagonal mesophases and their dispersed nanoparticles 217–24
  future trends 125–6, 224–5
  lamellar phase colloids 111–12
  liposomes *see* lipsomes
  liquid crystalline phases 109–11, 209–11
  nonlamellar liquid crystalline lipid-based nanoparticles *see* liquid crystalline nanoparticles (LCNPs)
  self-assembly and microstructures of lipid systems 209–11
lysine-arginine-phenylalanine-lysine 267
lyoszyme 238

mackerel 372, 374
Madin-Derby canine kidney cells (MDCK) 88
Maillard reactions 175–7, 178
  products 384–5, 459, 460
maltodextrins 198–9, 381–2, 383–4, 387
man-made conjugates 244–5
mannitol 262
margarine 392
Marino DHA HS Powder 393–4
Marino Omega-3 HS Powder 393–4
Martek Biosciences 393
mass balance
  mathematical models for *in vitro* dissolution and absorption 80–6

studies 15–16
maximum content, estimated 435–6
maximum solubilisation dilution line 154
mayonnaise 363
meats 361
mechanical agitation 115
mechanical bonds 38–9
mechanical theory of mucoadhesion 38
medium chain fatty acids (MCFA) 142
MEG-3 encapsulated fish oil 392, 393
melamine formaldehyde (MF) 418
membrane filtration 418
menhaden 372, 374
mesoporous silica particles 420
metabolic balance 71–2, 74–5
metabolism 4
measuring 5–13
structured lipids 140–1
studies 13–19
methyl mercury 189
methylcellulose (MC) 385
2-methyl-3-furanthiol (MFT) 175
5-methyltetrahydrofolic acid (5-MTHF) 20
micellar cubic mesophases 110, 111, 217
reversed 110, 113, 117, 120
micelles 153, 154
microbial limits test 443
microbial oil 373
microelectromechanical systems (MEMS) 89
microemulsion catalysis 175–7, 178
microemulsions 63–5, 149–83, 211, 459–60
adhesion/absorption tests 93–5
bioavailability of nutraceuticals 172–4, 175
composition-structure dependence 155–62
loaded 162–72
solubilisation of flavours and aromas 163–4
solubilisation of nutraceuticals 164–72
as microreactors 174–9
preparation 152–5
terminology 152
thermodynamic properties 155
microencapsulation 331, 377
oil-in-water emulsion systems 198–200
use of starch to control delivery 280–5
microencapsulation efficiency 56–7
microentrapment
folic acid in hydrogel beads 334
probiotics in gel particles 346–51
microfluidics 95
microfluidisation 213
microorganisms *see* probiotics
microparticles 258–70, 271, 460
controlled release systems 264–70
preparation techniques 259, 260–4
microreactors 174–9
microvilli 76
milk 216, 360
milk-based powders 360–1
minerals
encapsulation by liposomes 216–17
in vivo tests of bioavailability 71–4, 77
model membranes 120–4
modified spray drying and congealing 388
molecular configuration (*cis*/*trans*) 173
molecular distillation 375
molecular weight 39, 46, 316
monoolein-aqueous system 107–8
MTT 88
mucin gels 35–6
transport through 44
mucin molecules 27, 30–6, 91–2
chemical composition 30–3
conformation of single mucin molecules in solution 33–4
hydrophobic and electrostatic interactions between diffusing molecules and 45–6
mucoadhesion 36–44
physico-chemical properties 34–5
structure and arrangement within the mucus layer 35–6
mucoadhesion 36–44, 76–7
definition 36–7
factors influencing strength of 39–40
first generation of mucoadhesives 40, 41–3
in vitro tests 91–5
molecules improving mucoadhesion of controlled-release formulations 40–4
role in controlled release 36–7
second generation of mucoadhesives 40, 43–4
theories and factors affecting 37–9
mucus layer *see* gastrointestinal mucus layer
multilamellar vesicles (MLVs) 60, 212–13
multilayer emulsions 196–8
multilayer microcapsules *see* polyelectrolyte microcapsules
multilayered interfacial membrane 380
multiple-layer spray-coating 352

multivesicular vesicles (MVVs) 212

N-acetylgalactosamine 31, 32
N-acetylglucosamine 31, 32
n-octenylsuccinate-derivatised (n-OSA) starch 384
N-terminal ends 32–3, 34
naked regions 31–2, 33
nano-emulsions 57–8, 149–83
nanoencapsulation 246
 antioxidants *see* antioxidants
 complexation 240–3
 starch as material for 285–8, 289
 vitamins *see* vitamins
nanoparticles 62–3, 234, 236, 460, 463
 characterisation methods 313
 protein 258–70, 271
  controlled release systems 269–70
  preparation techniques 259, 260–4
 size and release profile 316, 318
nanoprecipitation 309, 310, 311
nanoscience 459
nanotechnology 151, 225, 246, 459, 463
nanovehicles 459
natural antioxidants 300–4
natural food additives 452
naturally occurring conjugates 243–4
near infrared (NIR) imaging 67
near infrared (NIR) spectroscopy 67–8
negative charge 34
Neobee 141
Neoral 63
no-observed effect level (NOEL) 437–8
non-communicable diseases (NCD) 21
nonlamellar liquid crystalline nanoparticles *see* liquid crystalline nanoparticles (LCNPs)
nonlinear polymers 39–40
non-metabolisable organic compounds 9–10
nonpolar lipids 109
normal lipid self-assembled structures 209, 210
Norvir 63
Nu-Mega Ingredients 392, 393
nutraceuticals 150–1, 251, 443–4, 456, 458
 bioavailability *see* bioavailability
 encapsulation of probiotics 354–8
 solubilisation and microemulsions 162–3, 164–72
nutrigenomics 23
Nutrinova 393
nutrition 140–1, 455–6

Ocean Nutrition Canada 392, 393
oil-in-water (O/W) emulsions 110, 190
 electrostatic layering 196–8
 microemulsions 153, 154
 microencapsulation 198–200
 nanoencapsulation of antioxidants and vitamins 311, 312
 stabilisation of omega-3 fatty acids 200–2
oil-in-water-in-oil (O/W/O) emulsions 57
oleic acid 223
oligosaccharides 344, 363
omega-3 fatty acids 184–206, 268
 emulsion droplet engineering to stabilise 200–2
 foods enriched with 389–90
 and health 186–8, 371–2
 long chain (LC) PUFA 370
  benefits 371–2
  recommended levels of intake 371–2
  sources 372–3
 potential in functional foods 188–90
 see also fish oils
omega-6 fatty acids 185
one compartment (Wagner-Nelson) model 445
optical microscopy 92
oral drug delivery 144–6
organogels 460, 462
overloads 9–10
oxidation
 lipids *see* lipid oxidation
 reactive oxygen species and 298–300
 stability of amylose-CLA complexes 287–8
 stability of emulsions 380
 stability of microencapsulated fish oil 390–1
oxidative stress 298
 body's defence system 298, 299
oxygen 359

packaging 440, 453
packaging aid compounds 453
packing parameter 110, 209–10, 211
palmitic acid 125
parallel artificial membrane permeability (PAMPA) assay 86–7
parenteral nutrition 142
partial AUC problem 11
particle counting methods 92
particle size 173
 distributions 66, 116
 nanoencapsulation of vitamins and antioxidants 316, 318

particulate gels 254, 255, 257
passive transport 76
pea globulins 245–6
peanut butter 362
pectin 198, 240, 242–3, 282–3, 284
PEO-PLGA nanoparticles 315
peptides 267
permeability 447
permeability tests 86–91
  cell cultures 87–9, 96
  ex vivo intestinal models 89–91
  PAMPA 86–7
permeation-enhancing agents 143–4
peroxyl radicals 191–2
pH
  enteric coating materials and 281–2
  mucin molecules and mucus layer 35–6
  nanoencapsulation of antioxidants and vitamins 316
  swelling properties of hydrogels 253–4
  varying within the gastrointestinal mucus layer 29–30
pH-stat titration 123–4
phosphatidyl choline (PC) 143
phophatidyl serine (PS) 139
phospholipids 29, 46, 136, 208–9
  structured 138–40
photon correlation spectroscopy 214
physiological response to food 11
phytantriol (PtOH) 117, 218
phytosterols 168, 169, 223, 458
plant-based products 361–2, 457–8
plasma concentration 72
  profile 84–6
plasticisers 382
Pluronic F127 113–14, 115, 219, 221
polar antioxidants 200
  paradox 193
polar lipids 109, 208
  see also lyotropic liquid crystals
polarised light microscopy 221
Poloxamer 412, 413–14
polycarbophil 41
polyelectrolyte microcapsules 416–22
  building multilayer microcapsules 417–18
  loading into empty microcapsules 418–19, 422
  pre-loaded core particles 420–1, 422
  protein core particles 419–20
  see also electrostatic layering
polyelectrolytes 35
  GSFA-listed 430, 434
polyesters 308
polyglycerol esters 156
polylactide-co-glycolide (PLGA) 308, 315–16
polymers
  GSFA-listed 430, 432
  nanoencapsulation of vitamins and antioxidants 315–16
polymetacrylates 281
polyols 161–2
polyphenols 23, 301, 302–3, 458
polysaccharides
  complexes *see* complexes
  conjugates *see* conjugates
  GSFA-listed 430, 432–3
polysorbate 411
  Polysorbate 80 116–17, 413
polystyrene sulphonate/polyallylamine hydrochloride (PSS/PAH) microcapsules 419, 420–1
polyunsaturated fatty acids (PUFAs) 190
  nanoencapsulation by starch 286–8, 289
  see also fish oils; omega-3 fatty acids
polyvinyl alcohol (PVA) 411–13
  PVA/dextran ATPS 414–16
positively charged submicron emulsions 145
post processing 68–70
powders 440, 457
  encapsulation of probiotics 354–5
    milk-based powders 360–1
  mixing 67–8
  powdered delivery technologies 460–2
  spray-dried microencapsulated fish oils 385–6
prebiotics 344, 363
pregnancy 331–2
pre-loaded core particles 420–1, 422
probiotics 88–9, 95–6, 344–69
  applications of encapsulation 354–63
    in functional foods 358–63
    in nutraceuticals 354–8
  encapsulation technologies 346–53
  future trends 363–4
  potential health benefits 344, 345
  protein microparticles 266
  starch and encapsulation of 283–5
processing aids 453
proline 31, 33, 34
Pronova Biocare 375
propofol 109
propylene glycol (PG) 161–2
prostate cancer 23
protein-anionic polysaccharide complexes 240–1

protein-cationic polysaccharide complexes 241
protein core particles 419–20
protein-DNA complexes 241–2
protein-protein complexes 241–2
proteins
backbone of mucin molecules 31, 33, 34
complexes *see* complexes
conjugates *see* conjugates
encapsulation approaches 404–25
coating in spray drying 411–16, 421
future trends 421–2
polyelectrolyte microcapsules 416–22
solid formulations 405–8
surface formation in drying 408–11
encapsulation of fish oils 380, 381, 382
food protein-derived materials 251–78
future trends 271
hydrogels 252–8
micro- and nanoparticles 258–70
functions in foods 404
therapeutic 404
pullulan acetate 306

Q_L mesophase 218, 223
quality control tests 68–70
quartz crystal microbalance with dissipation monitoring (QCM-D) 121–3

radioisotopes 72
rainbow trout 373, 374
random phase approximation (RPA) 237
raw materials 452
reaction media 174–9
reactive oxygen species (ROS) 297, 298–300
recommended dietary allowances (RDA) 371–2
reduced calorie/salt/sugar foods 455
refrigerated products 440
regulation 429–49
bioavailability of the active ingredient 443–7
FDA regulation on bioavailability and bioequivalence 444–5
FDA regulation on *in vitro* and *in vivo* correlation 445–7
future trends 447–8
product stability 439–43
selection of ingredients 430–9
Codex Alimentarius 430–8
FDA 438–9
relative humidity 70
release medium 317
release profiles 313–14, 315
encapsulated probiotics 357–8
folate in gastrointestinal system 337–8, 339
nanoparticles and 313–17
active component 314–15
complexation 317
environmental conditions 316–17
nanoparticle size 316
polymer 315–16
quantification methods 313–14, 315
replacement studies 17
repletion, rate of 72
resistant starch (RS) 280, 282–3, 284, 348
retention (core retention) 56–7
retinol 265–6
retro-design 96
retrogradation 280
reversed phases
bicontinuous cubic 110, 113, 117–19
hexagonal 110, 113, 117, 119
micellar cubic 110, 113, 117, 120
riboflavin 262, 265
risk-benefit analysis 23
Ritonavir 63
Roche 393
ROPUFA 75 encapsulated fish oil 393

safety 307–9
salad dressings 201, 202
SALATRIM 141
salmon 372, 374
salting out 309, 310, 311
Saquinavir 63
sardine 373, 374
secretory mucin 34
self-assembled monolayers 95
self-(micro)emulsifying drug delivery systems (SMEDDS or SEDDS) 63–4, 144–5
semi-permeable packing 440–1
sensory properties 363
serine 31, 33
shelf-life 453–4
shelf-life enhancers 454
sialic acid 31, 32, 34, 44
silica 120–1
simple coacervation 58, 236, 263
simple dissolution test 78
single emulsion evaporation method 311, 312
small angle X-ray scattering (SAXS) 221–2

small intestine 22, 76, 173
controlled release of encapsulated probiotics 357–8
controlled release of folate 337–8, 339
small unilamellar vesicles (SUVs) 212
sodium carboxymethyl cellulose (CMC) 41
sodium caseinate 239, 381, 383–4, 387
sodium caseinate-dextran (SC-Dex) conjugates 245
sodium dodecyl sulphate (SDS) 198, 261
sodium oleate 223
sodium tripolyphosphate 241
solid delivery technologies 460–2, 463–4
solid lipid nanoparticles 62–3
solid processing 67–8
drying techniques for proteins 405–8
solubility 447
solution enhanced dispersion by supercritical fluid (SEDS) 406
solvent diffusion/displacement 309, 310, 311
solvents, GSFA-listed 430, 432
soup powder 391
soybean water-soluble polysaccharide (SSPS) 244
spatial configuration 39–40
specialty dietary fats 141
specific tests 442–3
sphingosine 136
sponge phase 111, 113, 116–17
LCNPs 119–20
spray chilling 66
gel beads with microentrapped probiotics 347–8
spray cooling 66
spray coating 349, 351–2
spray drying 54–5, 65–6, 199, 200, 377
encapsulation of fish oils 379, 382–7
encapsulation of probiotics 349, 352
combined with emulsification 353
protein micro- and nanoparticles 259, 261–3, 264
proteins 405, 406–7
coating 411–16
surface formation 409–10
spray freeze drying 406, 407–8
surface formation 410–11
spreads 392
stability testing 68–70, 439–43
stable isotopes 72–3
starch 279–93, 381, 385
microentrapment of probiotics in starch granules 348–51
as nanoencapsulation material 285–8, 289
uses to control delivery in microencapsulation 280–5
sterols 140, 208
stomach 22, 75–6
controlled release of folate 337, 338
storage 390–1, 453–4
ICH storage conditions 440
storage aid carriers 454
structured lipids 135–48
drug delivery 143–6
food applications 141–2
nutrition, absorption, transport and metabolism 140–1
synthesis 138–40
sucrose 157, 158
sudden cardiac death 187
supplementation studies 17
surface activity 29
surface formation 408–11
in freeze drying and spray freeze drying 410–11
in spray drying 409–10
surface plasmon resonance 93
surfactants *see* emulsifiers
surrogate approach 6
swelling properties, of hydrogels 253–4
synthetic food additives 452, 454
synthetic polymers 308

tablets 355
targeting
antioxidants and vitamins 306
tissue targeting *see* tissue targeting
temperature
effect on stability 70
gel to liquid-crystalline transition temperature 209
glass transition temperature 70, 407
and mucin gel structure 36
nanoencapsulation of antioxidants and vitamins 316
tensile tests 91–2
termination reactions 192
ternary phase diagram 153–4
testosterone 45, 136
tetradecane 223
thermal processing 115, 332, 333
thermodynamic properties 155
threonine 31, 33
time-of-flight secondary ion mass spectrometry (ToF-SIMS) 408, 409
tissue targeting 4–5
measuring 5–13
tocopherols 165–6, 300, 458–9

alpha-tocopherol 300, 301, 376, 377
tocopheryl acetate 165–6
tocotrienols 300
tomato lectin (TL) 269
top-down nanoencapsulation techniques 309–12
toxicological studies 436–8
tracers 7
transdermal drug delivery 143–4
transglutaminase crosslinking 379, 388
transition metals 192, 193
transmission electron microscopy (TEM) 215, 222
transportation 453–4
triacylglycerols (TAG) (triglycerides) 136–7, 146, 157, 223, 459–60
Tris 145
trypsin 411
tuna oil 199, 373, 374, 391, 392

U-type microemulsions 64–5, 159–62, 165–72
ubiquinone (Coenzyme Q_{10}) 171–2, 300–2, 304, 459
unfermented milks 360
uniformity of dosage unit test 442–3
unilamellar vesicles (ULVs) 212
United States
  Environmental Protection Agency 373
  FDA *see* Food and Drug Administration
United States Pharmacopoiea (USP) 429, 446
  'USP verified' seal of approval 71, 95, 443
universal tests 441–2
unsaturated fatty acids 267–8

V-form amylose 286, 287
vacuole volume 380
vacuum drying 405
vegetable juice cocktail 362
vesicles 111–12
viscoelasticity 28–9
viscosity 44
viscous fingers 30
vitamins 77, 297–330, 457
  encapsulation by liposomes 216–17
  folic acid *see* folic acid/folate
  microemulsions and 165–6
  nanoencapsulation 304–19
    advantages 304–9
    characterisation methods 313
    controlled release 313–17
    future trends 318–19
    top-down techniques 309–12
  protecting human health 298–304
  protein microparticles 265–6
  vitamin C (ascorbic acid) 20, 301, 303
  vitamin E 300, 301
    bioavailability and microemulsions 174, 175
    nanoencapsulation 307–8
  vitamin H 306

Wagner-Nelson model 445
wall materials 55–6, 380–2
water, 'free' 44
water activity (relative humidity) 70
water content test 443
water-in-oil (W/O) emulsions 57, 110
  microemulsions 153, 154
water-in-oil-in-water (W/O/W) emulsions 57
  nanoencapsulation of antioxidants and vitamins 311, 312
Wegmans 392
wet processing 54–65
  coacervation 58–9
  emulsification 54–8
  inclusion complexation 61–2
  liposome encapsulation 59–61
  microemulsions 63–5
  solid liquid nanoparticles 62–3
wetting theory of mucoadhesion 38
whey proteins 58, 239, 240, 245, 381
  conjugates 245
  encapsulation of probiotics 362
  microparticles 260, 262, 265
  nanoparticles 270
World Health Organization (WHO) 429, 430
  see also Codex Alimentarius

X-ray photoelectron spectroscopy (XPS) 408–9
xanthan gum 239, 240

yoghurt
  microencapsulated fish oil 392
  microencapsulated probiotics 359, 363

zein 240
zero order release 81, 83
zeta potential 92–3, 421